SOLAR ENERGY

FOURTH EDITION

About the Authors

S P Sukhatme is Professor Emeritus, Department of Mechanical Engineering at the Indian Institute of Technology in Mumbai. He has been a faculty member at the Institute since 1965 and has served as the Director from 1995 to 2000. He also served as the Chairman of the Atomic Energy Regulatory Board from 2000 to 2005. Professor Sukhatme is known for his contributions to teaching and research in the fields of heat transfer and energy, and has been the recipient of many honours. He was awarded the Padma Shri by the Government of India in 2001.

J K Nayak is Emeritus Fellow in the Department of Energy Science and Engineering at the Indian Institute of Technology in Mumbai. He has been a faculty member at the Institute since 1982. Prior to that he received his Ph.D from IIT Delhi and served as a project staff in the Centre for Energy Studies there. Professor Nayak is known for his research contributions in thermal applications of solar energy. He is the author of the *Handbook on Energy Conscious Buildings* dealing with passive solar principles in building design.

SOLAR ENERGY

FOURTH EDITION

Suhas P Sukhatme
Professor Emeritus
Department of Mechanical Engineering
Indian Institute of Technology Bombay

J K Nayak
Emeritus Fellow
Department of Energy Science and Engineering
Indian Institute of Technology Bombay

McGraw Hill Education (India) Private Limited

Published by McGraw Hill Education (India) Private Limited
Registered Office: Anjana Complex No: 5/90 A, Butt Road, St. Thomas Mount, Chennai – 600016

Solar Energy, 4/e

Seventh reprint 2023

Print Edition
ISBN-13: 978-93-5260-711-2
ISBN-10: 93-5260-711-2

e-Book Edition
ISBN-13: 978-93-5260-712-9
ISBN-10: 93-5260-712-0

08 09 10 11 12 RAJ 27 26 25 24 23

Managing Director: *Lalit Singh*
Head—Production: *Pankaj Manohar*
Sr. Production Manager: *Reji Kumar*

Typeset at Text-O-Graphics, B-1/56 Arawali Apartment, Sector 34, Noida 201301 and printed and bound in India at Rajkamal Electric Press, Kundli, Sonepat, Haryana.

Visit us at: www.mheducation.co.in

Write to us at: info.india@mheducation.com

CIN: U80302TN2010PTC111532
Toll free Number: 1800 103 5875

ध्येयः सदा सवितृमण्डलमध्यवर्ती।
नारायणः सरसिजासनसन्निविष्टः।।
केयूरवान्मकरकुण्डलवान् किरीटी।
हारी हिरण्यमयवपुर्धतशंखचक्रः।।

—श्री भविष्योत्तर पुराण

'Always focus your attention at the centre of the solar system where the Sun, the supreme power of the universe, resides'

Contents

Preface to the Fourth Edition

It is more than 40 years since the first oil shock in 1973. Since then the words 'energy crisis' and 'energy security' continue to dominate the news. Added to these worries now are the issues of climate change. In spite of efforts to promote and develop renewable sources of energy and other new sources, fossil fuels (coal, oil and natural gas) continue to dominate the energy scene throughout the world. Even today in India, fossil fuels supply 80 percent of our electricity, renewable energy sources supply 17 percent and nuclear energy supplies about 3 percent. With renewable sources, large hydroelectric schemes supply 12 percent and the category of 'all other renewable sources' (comprising solar, wind, small hydroelectric schemes and others) supply about 5 percent. Forty years ago, the category 'all other renewable sources' supplied a negligible amount of electricity. Thus, although renewable energy sources are now playing a small and useful role in meeting our electricity and energy needs, there is a long way to go.

The first edition of this book was published in 1984 and focused essentially on devices used for collection and storage of solar energy for thermal applications. In the second edition (published in 1996), two chapters were added. This was done on the basis of feedback received which suggested that the scope of the book be enlarged beyond thermal applications. Thus, one of the new chapters was devoted to describing other direct and indirect methods for solar energy utilization like photovoltaic conversion, wind energy, energy in biomass, etc. The other chapter described techniques for making economic assessments of renewable energy systems. This layout of chapters was retained in the third edition in 2008.

The fourth edition takes into account important developments which have occurred since 2008. These have been described by adding or modifying content in all the chapters. At the same time, it is now obvious that certain applications or devices which showed promise in the past have not been successful essentially for economic reasons. A clear example is the solar pond which was described and analysed in detail in a separate chapter in earlier editions. In this edition, only the principle of working of the solar pond has been described in Chapter 2 for historical reasons. Thus, Chapter 8 (3rd edition) on the solar pond has now been deleted because of lack of relevance. At the same time, the importance of photovoltaic conversion has grown so much that a new separate chapter has been devoted to this topic. This is now Chapter 8 in the fourth edition.

Some of the additions/modifications made in the fourth edition are as follows:

(i) Chapter 1 describes the overall energy scenario in the world and in India and discusses various alternative energy options. The figures and tables giving data on the consumption and production of various commercial sources of energy have been updated. A discussion on India's future needs of electricity as it develops economically and the need for an equitable distribution in the population is also added.

(ii) Chapter 2 is an overview of solar thermal applications. Changes in this chapter include an introduction to the Linear Fresnel Reflector (LFR) and an update of solar thermal power plants installed all over the world. These include power plants using parabolic trough concentrators, linear Fresnel reflectors as well as central receiver stations with heliostats.

(iii) In Chapter 3 on Solar Radiation, a new method called the ASHRAE-2013 method is described. This replaces the ASHRAE-1972 method for the predicting radiation under clear skies.

(iv) Chapter 4 is on Liquid Flat-Plate Collectors. The classical analysis of flat-plate collectors originated by Hottel and modified by others has been retained. However, developments in selective coatings and testing procedures have been described. In addition, some new collector designs are described.

(v) In Chapter 5 on Solar Air Heaters, the classical analysis of a conventional air heater is again retained. Recent work on two-pass solar air heaters, unglazed transpired solar air heaters and evacuated tube solar air heaters is presented.

(vi) Chapter 6 is on Concentrating Collectors. Apart from updating the existing sections on cylindrical parabolic collectors and central receiver collectors, a new section has been added on the design and development of linear Fresnel reflectors.

(vii) With the increasing use of solar thermal energy, the importance of thermal storage is being understood. This is particularly true for high temperatures associated with energy conversion heat engine cycles. Thus Chapter 7 on Thermal Energy Storage now includes a discussion on long term storage and new materials for use in concentrated solar power plants. Two tank liquid storage is also discussed.

(viii) Chapter 8 is on Photovoltaic Conversion. Apart from modifying and updating the write-up on photovoltaic conversion, which was in Chapter 9 of the third edition, a few topics of interest for energy engineers are now added. These include the description of a number of PV systems, stand alone and grid connected. Procedures for sizing such systems or calculating their performance are also presented through numerical examples. The subject of hybrid power systems is also broached by considering a system in which a PV system is run in tandem with a CSP system.

(ix) Chapter 9 is now renamed as 'Other Renewable Energy Sources'. Apart from modifying the information on wind energy and energy from biomass, a small section on small hydroelectric power units is added for completeness. Having described all the renewable energy sources which are capable of generating electricity, the chapter concludes with a separate section that asks the pertinent question "Can renewable energy sources meet India's future needs of electricity?"

(x) Chapter 10 of the book concludes with presenting ways of performing an economic analysis of renewable energy systems.

In all the chapters, some new problems have been added and some more examples solved. In addition, a few photographs and many useful references have been added at the end of each chapter. These will be particularly useful to students who wish to do projects and engineers who are doing research and development work.

The book is targetted for the use of final year undergraduate students studying for their B.E./B.Tech degrees and first year postgraduate students working for their M.E./M.Tech degrees. Suitable parts of the book have to be selected for study for each category of students. Thus, for undergraduate students on engineering who have to study the subject of Non-Conventional Energy Sources or Renewable Energy Sources, the syllabus would be covered if one taught Chapters 1, 2, 8, 9 and 10 fully and Chapters 3 to 7 partially. For undergraduate or postgraduate students studying only the subject of Solar Energy, one would cover Chapters 2, 3, 8 and 10 fully, Chapters 1 and 4 to 7 partially and delete Chapter 9. For a more specialized subject like Solar Energy and its Thermal Applications, one would study Chapters 2 to 7 fairly fully, Chapters 1, 8 and 10 partially and delete Chapter 9. Over the years, it has been a source of satisfaction for us to receive queries from students working on their M. Tech dissertations and from PhD research scholars. From the point of view of these students, it is worth noting that in many topics the latest references have been given in this edition.

It is highly gratifying that the first three editions of the book have been widely appreciated. It is our hope that readers will welcome the fourth edition and benefit from the changes made.

OLC Supplements

The text is supported by a website accessible at: *http://www.mhhe.com/sukhatme/se4e* with the following supplements for instructors:

- *Solutions Manual*
- *PowerPoint Lecture Slides*

SUHAS P SUKHATME

JAYANTA K NAYAK

August 2017

Publisher Note

McGraw Hill Education (India) invites suggestions and comments from you, all of which can be sent to *info.india@mheducation.com* (kindly mention the title and author name in the subject line). Piracy-related issues may also be reported.

Acknowledgements

It is a pleasure for us to acknowledge with humility the generous help, advice and encouragement which we have received from our faculty colleagues in the Departments of Energy Science and Engineering, Mechanical Engineering and Electrical Engineering. First and foremost, we are specially grateful to our colleagues, Professor U N Gaitonde for numerous discussions on many topics, Professor Rangan Banerjee for many general discussions and a detailed review of Chapter 10 on Economic Analysis, and Professors Vivek Agarwal, C S Solanki and D Suryanarayana for numerous discussions and a detailed review of Chapter 8 on Photovoltaic Conversion. Professors S L Bapat, Shireesh Kedare, Santanu Bandyopadhyay, Suneet Singh, Sagar Mitra, Shaibal Sarkar, Shyamprasad Karagadde and Vishal Sardeshpande (CTARA) have also helped in various ways, discussed specific topics and have given useful information. We are indebted to all of them.

Next to faculty, the support which we have received from students and former students has been extensive. We would particularly like to thank Dr Nishith B Desai, Shri N G R Kartheek, Ms. Surabhi Pendharkar, Shri Nikhil Salunkhe, Shri Pardeep Shahi, Shri Akash Srivastava, Shri Rohan Gaikar and Ms. Shweta Kamat for collecting relevant literature, plotting data, drawing figures and typing. We are also indebted to our former student, Dr Vashi Mant Sharma (now with IIT Kanpur) for his detailed review of the section on LFR in Chapter 6.

The assistance which we have received from our administrative staff in the Departments of Mechanical Engineering and Energy Science and Engineering has been impeccable. In particular, we would like to thank Shri Praveen Naik for his very prompt typing.

Officials of the Ministry of New and Renewable Energy, Government of India have been particularly generous. Specifically, we wish to thank Dr Ashvini Kumar (Managing Director, Solar Energy Corporation of India), Shri B K Bhatt (Adviser, MNRE), Dr O S Sastry (Ex-Director General, NISE) and Dr A K Singhal (National Project Manager, UNDP Project, MNRE). The websites of MNRE and its associated organizations have also yielded a great deal of useful information.

From other organizations, we acknowledge the information provided by

(i) Dr B C Jain (Ankur Scientific Energy Technologies Pvt Ltd, Vadodara) on biomass gasifiers,

(ii) Shri G Venugopal (DynaLab Weather Technologies Pvt Ltd, Pune) on solar radiation measuring instruments,

(iii) Dr H C Barshilia (National Aerospace Laboratories, Bengaluru) on NAL-developed selective coatings, and

(iv) Professor Arun Kumar (IIT Roorkee) on small hydropower.

Many friends have supplied us with photographs for use in our book. We would like to acknowledge the help of the following:

- Dr C L Gupta and Mrs. Shipra Gupta (Sri Aurobindo Ashram, Puducherry) and Mr. John Harper and Mr. Gilles Guigan for the photograph of the solar bowl at Auroville.
- Professor R L Sawhney (ex-professor, DAVV, Indore) for the photograph of the evacuated tube collector.
- Dr C L Palaniappan (Planters Energy Network) for the photograph of solar dryers.
- Shri Deepak Gadhia (Gadhia Solar Energy Systems, Valsad) for the photographs of solar cooking systems using Scheffler concentrators.
- Shri A D Paranjape (Clique Solar, Mumbai) for the photograph and information on their dish concentrator.
- Shri S Malik and Shri A Aggarwal (Megawatt Solutions, New Delhi) for the photograph and information on their dish concentrator.

Many individuals associated with manufacturing and installing solar equipment have also provided information. We thank them for their cooperation. In particular, we would like to acknowledge the help received from Shri Zubin Bhumgara and Shri C N Achari of Standard Products Manufacturing Co., Taloja and Shri Sanjay Jinturkar and Shri Amit Bharambe of Sudarshan Saur Shakti Pvt. Ltd, Aurangabad for providing information on the cost of flat-plate collectors, evacuated tube collectors and water heating systems.

Shri V Shandilya and Shri A Dubey of Godawari Green Energy Ltd., Raipur have provided useful information on their 50 MW PTC plant. Dr C S Sureshkumar and Dr S P Vishwanathan of KGDS Renewable Energy Pvt Ltd, Coimbatore and Shri K Deshpande and Shri Y Joshi of Thermax Pvt Ltd, Pune provided data on their LFR plants. Shri J Jacob (ACME Cleantech Solutions Pvt Ltd, Gurugram) has provided data on the solar tower plant, the only one of its kind installed in India so far. We are grateful to all of them.

The Pergamon Press (USA) has permitted the use of some figures and tables which have appeared in the Solar Energy journal. This is acknowledged at appropriate places in the text.

We would also like to thank the following faculty members from various educational institutions for reviewing the manuscript of this edition and providing us with their helpful feedback.

1. Shri Tushar Agarwal, Rajasthan Institute of Engineering and Technology, Jaipur, Rajasthan.
2. Dr Ramesh Chandra Borah, Assam Kaziranga University, Jorhat, Assam.

3. Shri Umesh Kumar Dewadi, Amity University, Noida, Uttar Pradesh.
4. Shri Kuldeep Singh, Radharaman Engineering College, Bhopal, Madhya Pradesh.
5. Dr L K Sreepathi, Jawaharlal Nehru National College of Engineering, Shivamogga, Karnataka.
6. Shri Sharad R Valvi, Sardar Patel College of Engineering, Mumbai, Maharashtra.
7. Dr Sudhansu S Sahu, Department of Mechanical Engineering, CET, Bhubaneswar, Odisha.

Writing a book is possible only if the right ambience exists both at the work place and at home. IIT Bombay has been our work place for many years. We have grown up with the Institute, worked for its cause and its good, and savoured the pleasure of having wonderful colleagues and students. We consider ourselves fortunate to have lived in the scenic surroundings of Powai and gratefully acknowledge the infrastructural support received from the Institute in the process of writing this book. In particular, we thank the group which efficiently manages the e-journals facility of the library. At home, our wives, Shobha and Purnima, have supported us, borne with our shortcomings, and tolerated our late hours. No words are adequate to describe our indebtedness to them.

Suhas P Sukhatme

August 2017 **Jayanta K Nayak**

Preface to the First Edition

Since 1973, the word "energy" has been continuously in the news. There have been shortages of oil in many parts of the world and the price of this commodity has increased steeply. It is by now clear that the fossil fuel era of non-renewable resources is gradually coming to an end. Oil and natural gas will be depleted first, followed eventually by coal.

In India the energy problem is very serious. In spite of discoveries of oil and gas off the west coast, the import of crude oil continues to increase and the price paid for it now dominates all other expenditure. This year the country will spend more than Rs 5000 crores for the import of oil. This amount forms a major part of India's imports bill. The need for developing energy alternatives is thus evident and considerable research and development work is already in progress in this direction.

One of the promising options is to make more extensive use of renewable sources of energy derived from the sun. Solar energy can be used both directly and indirectly. It can be used directly in a variety of thermal applications like heating water or air, drying, distillation, and cooking. The heated fluids can in turn be used for applications like power generation or refrigeration. A second way in which solar energy can be used directly is through the photovoltaic effect in which it is converted to electrical energy. Indirectly, the sun causes winds to blow, plants to grow, rain to fall, and temperature differences to occur from the surface to the bottom of the oceans. Useful energy can be obtained for commercial and noncommercial purposes through all these renewable sources.

In this book, we are primarily concerned with the collection and storage of solar energy for thermal applications. The purpose of writing this book is to provide a suitable text for teaching the subject to engineering and science students, as well as a reference book for practising professionals. The material is based on the author's research experience and his experience of teaching the subject for a number of years to postgraduate and final-year undergraduate engineering students. It is assumed that as background, the students have undergone basic elementary courses in engineering thermodynamics, fluid mechanics and heat transfer. In case this has not been done, a few supplementary lectures on these topics will be necessary.

The literature on the subject of solar energy is already large and is expanding at a fast pace. The purpose of the book therefore is not to provide a comprehensive compilation of the literature, but to give basic and fundamental information in the form of a text so that the reader will be in a position to understand and evaluate the literature on thermal collection and storage on his own. This information has been given keeping the Indian context in mind. Thus, many of the solved examples are based on Indian data, while the photographs are of Indian devices and installations. In addition, many problems are given for solution. Answers to these problems are also provided.

The fact that this book is concerned with the thermal aspects of solar energy does not imply that the other applications like photovoltaics or biomass are less important or less promising. It does not also imply that the other energy alternatives like the nuclear option, are less important. For this reason, the first two chapters are of a general introductory nature.* Chapter 1 is concerned with outlining the dimensions of the energy crisis and with describing the various energy alternatives. In Chapter 2, the many ways of using solar energy, both directly and indirectly, are described by considering a variety of applications. Particular emphasis is laid on familiarizing the reader with thermal applications. Chapter 3 is concerned with the fundamentals of solar radiation and its availability. The analysis of flat-plate collectors is taken up in Chapters 4 and 5. Chapter 4 is concerned with the liquid flat-plate collector and Chapter 5 with the solar air heater. Following this, concentrating collectors are discussed in Chapter 6. The different methods of thermal storage are described in Chapter 7, while the last chapter is on the solar pond, in which collection and storage is done simultaneously. I have found this order helpful in developing the subject.

Many persons have contributed towards the writing of this book by their encouragement and help, and in closing I would like to thank them. The Director of IIT Bombay, Prof. AK De has been a source of encouragement for me ever since I first came in touch with him in 1965. I am deeply grateful to him for all his help.

I am particularly indebted to the authorities of the Tata Energy Research Institute for the generous financial grant given for the preparation of the book.

Sincere thanks are due to my colleagues and students in the Mechanical Engineering Department. I am specially grateful to Dr UN Gaitonde and to Mr DG Motwani for helping me in the preparation of the text, Mr AA Poulose for typing the manuscript, and Mr DC Sakhalkar for drawing the illustrations.

This book is dedicated to my wife, Shobha, for sustaining and inspiring me at all times with her love and affection.

August 1983 **SUHAS P SUKHATME**

---

* Teachers wishing to use this book as a text for thermal collection and storage only could skip parts of Chapters 1 and 2. They need cover only Sub-section 1.4.2 of Chapter 1 and Sections 2.1 and 2.2 of Chapter 2 before passing on to Chapter 3.

The Energy Scenario

CHAPTER 1

INTRODUCTION 1.1

Man has needed and used energy at an increasing rate for his sustenance and well-being ever since he came on the earth a few million years ago. Primitive man required energy primarily in the form of food. He derived this by eating plants or animals which he hunted. Subsequently, he discovered fire and his energy needs increased as he began to make use of wood and other biomass to meet the energy needs for cooking as well as for keeping himself warm. With the passage of time, man started to cultivate land for agriculture. He added a new dimension to the use of energy by domesticating and training animals to work for him. With further demand for energy, man began to harness the wind for sailing ships and for driving windmills, and the force of falling water to turn water wheels. Till this time, it would not be wrong to say that the sun was supplying all the energy needs of man either directly or indirectly and that man was using only renewable sources of energy.

The Industrial Revolution, which began with the discovery of the steam engine (AD 1700) brought about a great many changes. For the first time, man began to use a new source of energy, that is coal, in large quantities. Later, the internal combustion engine (AD 1870) was invented and other fossil fuels, oil and natural gas began to be used extensively. The fossil fuel era of using nonrenewable sources had begun and energy was now available in a concentrated form. The invention of heat engines and the use of fossil fuels made energy portable and introduced much needed flexibility in man's movement. For the first time, man could get the power of a machine where he required it and was not restricted to a specific site like a fast-running stream for running a water wheel or a windy hill for operating a windmill. This flexibility was enhanced with the discovery of electricity and the development of central power generating stations using either fossil fuels or water-power.

A new source of energy–nuclear energy–came on the scene after the Second World War. The first large nuclear power station was commissioned more than 60 years ago, and already, nuclear energy is meeting a small but significant amount of the energy requirements of many countries.

Thus today, every country draws its energy needs from a variety of sources. We can broadly categorize these sources as commercial and noncommercial. The commercial sources include the fossil fuels (coal, oil and natural gas), hydroelectric power, nuclear power and wind power, while the noncommercial sources include wood, animal wastes and agricultural wastes. In an industrialised country like USA, most of the energy requirements are met from commercial sources, while in an industrially less developed country like India, the use of commercial and noncommercial sources is about equal.

In the past few years, it has become obvious that fossil fuel resources are fast depleting and that the fossil fuel era is gradually coming to an end. This is particularly true for oil and natural gas. It will be useful, therefore, to first examine the rates of consumption of the different sources of energy and to give some indications of the reserves available. First, the study will be done for the world as a whole (Section 1.2) and then for India in particular (Section 1.3). The figures so obtained will help in estimating the time periods for which the existing sources will be available. The need for alternative energy options will thus be established and these options will then be briefly described (Section 1.4).

While man's large scale use of commercial energy has led to a better quality of life, it has also created many problems. Perhaps the most serious of these is the harmful effect on the environment. The combustion of fossil fuels leading to localised release of large amounts of harmful gases into the atmosphere is one of the main causes of air pollution. It is also the main contributor to the phenomenon of global warming which is now a matter of great concern. According to the 2007 report of the Intergovernmental Panel on Climate Change (IPCC), the atmospheric concentration of carbon dioxide increased from a pre-industrial era value of about 280 ppm to a value of 379 ppm in 2005 primarily due to fossil fuel use [1]. This along with increases in the concentration of methane and nitrous oxide has resulted in an increase in the global average temperature. The consequences of these increases are already evident through the increase of the global average sea level and the decrease in the Northern Hemisphere snow cover. In their 2013 report [2], IPCC indicate that global warming over a 60-year period from 1951 to 2012 has averaged 0.11°C per decade. However, the report also states that over the last 15 years of the above period, i.e. from 1998 to 2012, the trend has only been 0.04°C per decade. This decrease in global warming has been termed by IPCC as a 'hiatus'. However, recently Karl *et al.* [3] have shown after a detailed study that there has been no hiatus and that the rate of warming even during the last 15 years (2000 to 2014) is essentially the same as the rate observed in the second half of the previous century, that is 0.11°C per decade.

Other environmental issues of concern are the release of large amounts of waste heat from power plants. This has caused thermal pollution in lakes and rivers leading to the destruction of flora and fauna. There is also concern over the possibility of radioactive emissions from the nuclear plants being released into the atmosphere in the event of an accident and over the long-term problems of disposal of radioactive wastes from these plants. The gravity of most of these environmental problems had not really been foreseen. Now, however, as man embarks on the search for alternative sources of energy, he would indeed have to keep the environment always in mind.

WORLD'S PRODUCTION AND RESERVES OF COMMERCIAL ENERGY SOURCES 1.2

1.2.1 Fossil Fuels

❑ *Coal* Coal is the end product of a natural process of decomposition of vegetation buried in swamps and out of contact with oxygen for thousands of years. The word 'coal' denotes a wide variety of solid fuels. The varieties in approximate order of their formation are peat, lignite, bituminous coal and anthracite.

The rate of coal production of the world from the time it has been used in significant amounts is shown in Fig. 1.1 [4-6]. Coal production showed a steady increase of 4.6 per cent per year during the period from about 1860 to the beginning of the First World War in 1914. Thereafter, for a period of 31 years till the end of the Second World War in 1945, the changes in the production rate were erratic. The production declined in some years and increased in others, the average increase over the whole time span being only 0.25 per cent per year.

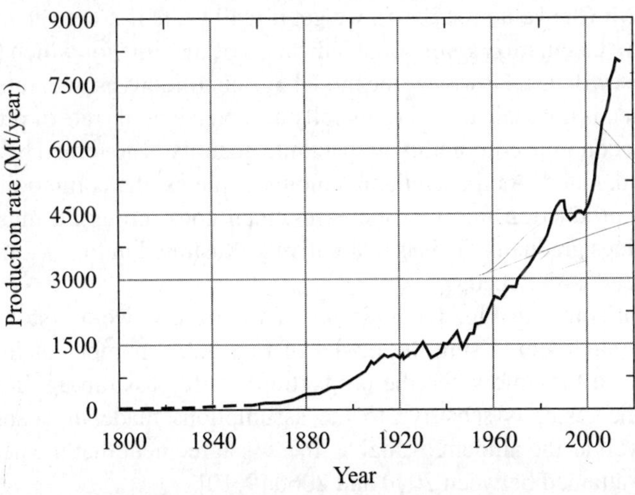

Fig. 1.1 *World production of coal [4-6]*

After 1945 till 1990, the annual production rate again increased (more or less steadily) at a rate ranging between 2 and 4 per cent per year. The production in 1980 was 3792 Mt, in 1985, 4438 Mt and in 1990, 4849 Mt. From 1991 to 2002, the annual production stagnated and fluctuated between a low of 4350 Mt in 1993 and a high of 4876 Mt in 2001, the average for the 11-year period being 4557 Mt. However, after 2001, the production rate has again been increasing. The total amount of coal produced and consumed up to the year 2014 (the area under the curve in Fig. 1.1) is approximately 350 000 Mt.

Many estimates have been made regarding the amount of coal still available in the earth. Before giving these, it will be useful to distinguish between the terms ‘resources’ and ‘proved reserves’. The term ‘resources’ implies an estimate of the total quantity available that may eventually be successfully exploited and used by man. The term ‘proved reserves’ generally refers to that portion of the resources which has been proved to exist and can be economically recovered with available techniques. It is important to note that the estimates for these quantities change with time. It is currently estimated that the world’s coal resources are 22.4×10^6 Mt [7], while the proved reserves have been estimated to be about 891 530 Mt [8]. 72 per cent of the reserves lie in five countries, USA, Russia, China, Australia and India.

It is of interest to estimate approximately the time for which coal would be available. Assuming a continuous growth rate in production of r per cent, it is easy to show that the number of years n, for which coal will last is given by the expression

$$R = P'\left[\left\{\left(1+\frac{r}{100}\right)^{n+1}-1\right\}\Big/\left\{\frac{r}{100}\right\}\right] \tag{1.1}$$

where R is the amount available and P', the present production rate. Assuming the value of R to be the current proved reserves of 891 530 Mt and substituting $P' = 8022$ Mt (the value for 2014), we get $n = 49$ years if $r = 3$ per cent.

The above calculation gives some indication of the time for which the world’s coal reserves will last. However, the model of a continuous growth rate which has been assumed is not realistic. It is generally accepted that the rate of production of an existing economic commodity with a finite quantity is governed by the laws of supply and demand. As the available amount depletes, the commodity becomes costlier; its production rate first increases, then goes through a maximum and finally declines gradually. This is qualitatively illustrated in Fig. 1.2 in which the expected trend is sketched.

A comparison with Fig. 1.1 indicates that currently, we are still on the left side of the maximum. Using the model of Fig. 1.2, a number of investigators have estimated the time when the production would pass through a maximum. Although the result is sensitive to the assumptions made in quantifying the model, as well as the amount available, there is agreement that the peak value is likely to be attained between 2040 and 2060 [9, 10].

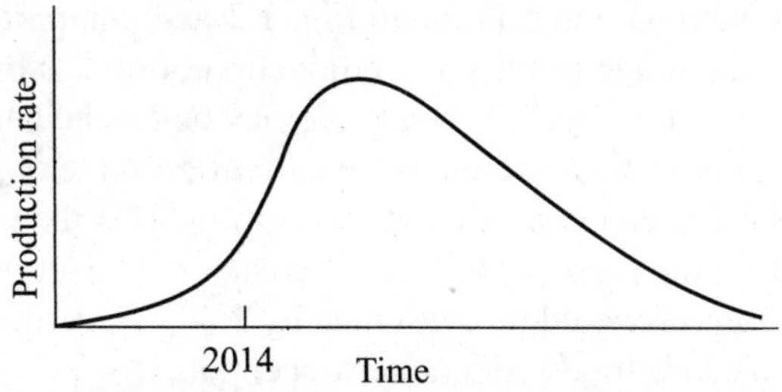

Fig. 1.2 *Expected trend in the world's production rate of coal*

❑ *Oil* The world's rate of production of crude oil (in billions of barrels* per year) is shown in Fig. 1.3 [4,11]. It is seen that oil began to be used in significant quantities around 1900 and that there was an almost steady increase in its production all through and even during the world wars. The production increased at the average rate of over 7 per cent per year from 1945 to 1973 and reached a value of 19.96 billion barrels in 1973. Thereafter with the beginning of the oil crisis, the annual production fluctuated for 12 years from 1973 to 1985 before starting to increase more or less steadily from 1985 onwards. The average rate of increase of production from 1985 to 2014 has been about 1.3 per cent. The total amount of oil produced and consumed up to 2014 comes out to be approximately 1200 billion barrels.

It is estimated that the world's resources of oil are about 3100 billion barrels [7]. Values of the proved reserves are of more interest. In the late 1970s and early 1980s, most estimates of the reserves were around 600 billion barrels. However, lately there has been an increase in the estimate in spite of the approximate consumption of 200 to 250 billion barrels every ten years. This increase is because of more discoveries. Proven oil reserves in 2014 are estimated to be about 1653 billion barrels [11], a substantial part of this quantity (about 49 per cent) being in the Middle East.

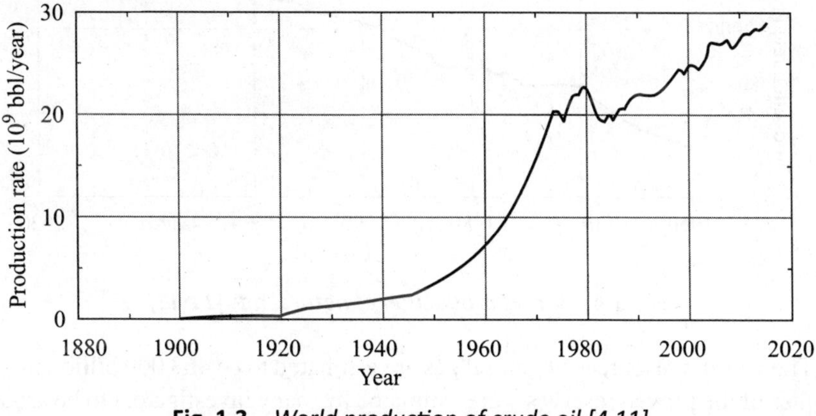

Fig. 1.3 *World production of crude oil [4,11]*

---

* 1 ton of crude oil is equivalent to 7.32 barrels.

Ignoring local maxima and minima in Fig. 1.3 and comparing the data with the model of Fig. 1.2, it appears that the production of oil is slowly approaching the maximum indicated in Fig. 1.2. This peak may be reached around 2021 if no significant discoveries (which add on to the current estimate of proved reserves) are made [10]. These predictions are not as alarming as some of the predictions made in the past. At the time of the first oil crisis in 1973, it had been forecast that most of the reserves would be consumed by 2020. Nevertheless, the present prediction is still a cause for considerable concern.

❑ *Natural Gas* Natural gas is a mixture of various compounds of carbon and hydrogen (hydrocarbons) and small quantities of non-hydrocarbons existing in the gaseous phase, or in solution with oil in natural underground reservoirs. It is classified into two categories:

1. Associated gas
2. Non-associated gas

Associated gas is natural gas originating from fields producing both liquid and gaseous hydrocarbons simultaneously. On the other hand, non-associated gas is natural gas which is obtained independently. It is generally found in the space above an oil reservoir or an aquifer. Typically the major constituents of associated natural gas are methane (about 50 per cent by volume), ethane (about 20 per cent) and propane (about 10 per cent). The major constituent of non-associated natural gas is methane (more than 90 per cent by volume).

The world's rate of production of natural gas from 1960 to 2013 is shown in Fig. 1.4 [11-13]. It is seen that the production has been increasing more or less continuously at the rate of about 3.5 per cent per year.

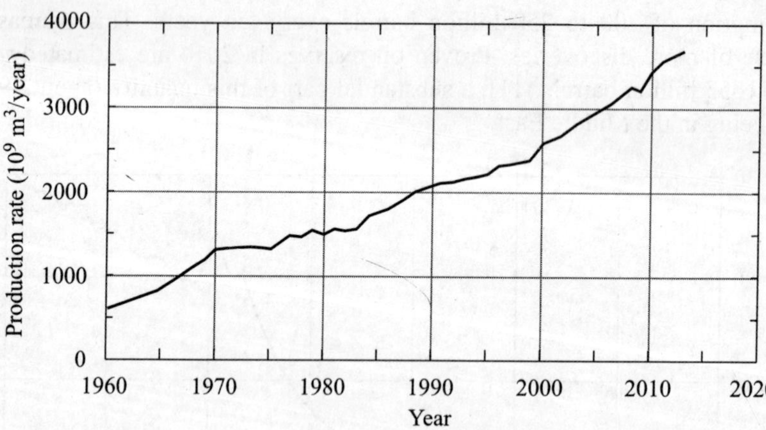

Fig. 1.4 *World production of natural gas [11-13]*

The world's resources of natural gas are estimated to be 463 000 billion m³ [7]. Values of the proved reserves were estimated by many investigators to be around 50 000 billion m³ in the seventies. However, with more discoveries in the 1980s and 1990s, the proved reserves are currently estimated to be 232 000 billion m^3[7]. How long will these reserves last? The answer depends upon the category of the gas. Associated gas will last for approximately the same time as crude oil. However, the availability of non-associated gas should help. Thus the peak in the production of natural gas may occur between 2024 and 2046 [10].

1.2.2 Water-Power

Water (or hydroelectric) power is developed by allowing water to fall under the force of gravity. It is used almost exclusively for electrical power generation and currently provides about 16 per cent of the world's production of electricity. In fact, the generation of water-power on a large scale became possible only around the beginning of the twentieth century with the development of electrical power transmission. Prior to that, water-power plants were of small capacities, usually less than 100 kW.

Data on the installed water-power capacity, the electricity production and the annual plant load factor for four representative years (1925, 1966, 1971 and 1980) is presented in Table 1.1 [11,14-16]. This data along with annual data from 1980 onwards is also plotted in Fig. 1.5. It is seen that from 1925 to 1980, the annual increase in capacity ranged between 4 and 5 per cent. Thereafter, from 1980 to 2014, the annual increase has generally ranged between 2 and 3 per cent.

Table 1.1 Installed water-power capacity, electricity produced, and annual plant load factor in the world

	1925 [14]	1966 [15]	1971 [16]	1980 [11]
Installed capacity (MW)	26 400	230 000	307 131	462 000
Electricity produced (GWh)	78 678	956 000	1 307 257	1 723 000
Annual plant load factor	0.340	0.474	0.486	0.426

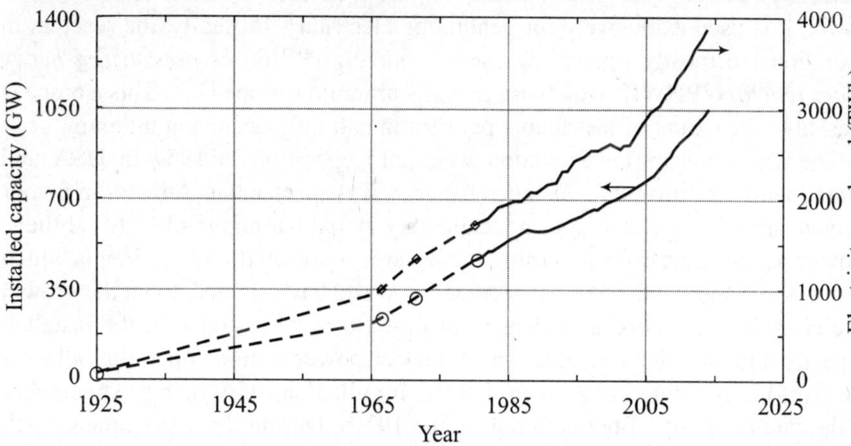

Fig. 1.5 *World installed capacity and electricity generation from water-power [11, 14-18]*

The *annual plant load factor* (APLF) is the ratio of the electricity produced in an year to the electricity which would have been produced if the plant had produced its installed capacity throughout the year. For the world as a whole, APLF for water-power had an average value of 0.42 from 2004 to 2014. The low value is attributed to shutdowns for maintenance, and the non-availability of the optimum level of water required in the dams for generating the installed capacity throughout the year.

Estimates of the total capacity which can be installed in the world vary considerably and depend upon the assumptions made. One method adopted is to state the maximum generation usable 95 per cent of the year. By this method, it is estimated that the total capacity which can be installed is 3×10^6 MW [19]. Recent estimates made by the World Energy Council show that the technically exploitable potential of water-power is 10 000 TWh/year [20]. Much of the capacity, which is yet to be installed exists in the developing countries. Thus, it is clear that there is considerable scope for augmenting the production of hydroelectric power.

Hydroelectric power is the major renewable energy source of the present commercial scene and does not contribute to global warming. However, many environmental and social concerns are being raised whenever large dams are to be constructed for generating hydroelectric power. One of the environmental issues is that certain biological species are threatened with destruction which would be harmful to the ecosystem. Another concern is that the large body of water stored behind a the dam may trigger more seismic activity in the region. Social issues are largely concerned with the displacement of people who live in areas which get submerged when a new dam is being built.

1.2.3 Nuclear Power

Nuclear power is generated by fission reactions in nuclear reactors. Like water-power, it is used exclusively for generating electricity. Presently, the reactors in operation are mostly *pressurized water reactors* (PWR) or *pressurized heavy water reactors* (PHWR) which use only the uranium isotope U^{235}. This isotope is present to the extent of less than 1 per cent in naturally occurring uranium.

The first nuclear power station went into operation in 1957 in USA and subsequently the growth in installed capacity was spectacular. A factor in favour of nuclear power is that it does not cause any global warming unlike fossil fuels. However, apprehensions regarding the hazards associated with accidents which might occur and the disposal of radioactive waste have slowed down the growth rate considerably. This is evident from Fig. 1.6 in which data on the installed capacity and electricity production of nuclear power stations operating all over the world is presented. It is seen that the installed capacity increased annually at the rate of about 24 per cent during the 1970s. During the 1980s, the growth rate decreased to about 8 per cent, while subsequently, the growth rate has been less than 1 per cent. The total installed capacity in 2014 was 376 200 MW [21]. Figure 1.6 also shows that the production of electricity dropped sharply in 2011 because many nuclear stations in Japan were shut down after the accident at Fukushima. Production has gradually began to pick up after a lapse of about 3 years. It may be noted that the annual plant load factor prior to the Fukushima accident was around 0.80.

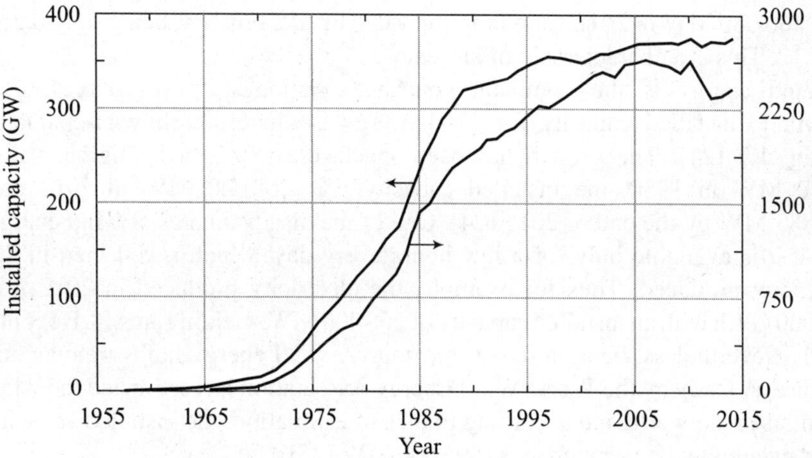

Fig. 1.6 *World installed capacity and electricity generation from nuclear power [11, 12, 21]*

The world's resources of uranium are reported based on the cost of production. Identified resources consisting of reasonably assured resources and inferred resources recoverable at a cost of less than US $ 260/kgU are estimated to bc 7.64 Mt [22]. Out of this, the reasonably assured resources are 4.59 Mt and the inferred resources are 3.05 Mt. The corresponding numbers at a cost of less than US $ 130/kgU are 5.90 Mt, 3.70 Mt and 2.20 Mt. Nearly 77 per cent of these resources lie in seven countries, Australia, Kazakhstan, Russia, Canada, Niger, Namibia and South Africa. The present installed capacity of about 376 200 MW requires 66 000 t of uranium every year [23]. Thus the reasonably assured resources of uranium costing less than US $ 260/kgU and US $ 130/kgU respectively would be adequate for running the present installed capacity for a period of about 70 years and 56 years. In fact, the resources would last for a lesser period if there is an increase in the installed capacity. It is evident that a difficult situation may arise between 2070 and 2090 if the present fission technology continues to be used. It may be mentioned that recovering uranium as a minor by-product from various unconventional sources such as phosphate/phosphorite deposits, lignites, seawater, etc., are being explored [23].

1.2.4 Other Sources

Apart from fossil fuels, water-power and nuclear power, there are a number of other sources of energy which have begun to contribute, though in a small way, to the world's commercial energy scene for generating electrical power. These are essentially renewable sources like wind energy, photovoltaic conversion, solar thermal power generation, bio-mass, etc, and one non-renewable source, viz. geothermal energy. Figure 1.7 shows the growth in installed capacity and the electricity produced from these sources from 1980 onwards [11]. It is seen that the installed capacity has grown from less than 10 000 MW in the 1980s to more than 460 000 MW in 2012. However, the value of APLF is low, being only 0.268 in 2012. This is a characteristic of all renewable sources.

Wind energy is the dominant contributor amongst these sources. The growth in installed capacity for wind energy is, therefore shown separately in Fig. 1.7 [24]. The growth has been spectacular; from a negligible value of 10 MW in 1980, the installed capacity was 268 000 MW in 2012 and 432 883 MW by the end of 2015 [24]. One of the disadvantages of wind energy is that it is available only for a few hours every day, which is reflected in the electricity produced. Thus for example, the electricity produced in 2012 was 519 000 GWh with an installed capacity of 268 000 MW, yielding an APLF of only 0.221. Nevertheless, being a renewable source, wind energy holds tremendous promise. A study by the World Wind Energy Association has estimated the wind potential in the world and according to a 2014 estimation, the installed capacity could eventually be as much as 94 895 000 MW [25].

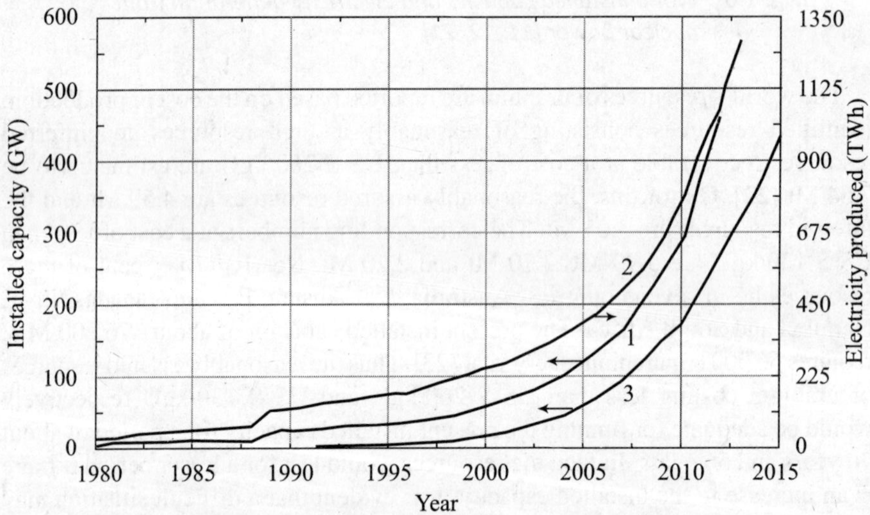

Fig. 1.7 *World installed capacity (curve 1) and electricity generation (curve 2) from other sources and installed capacity of wind power (curve 3) [11, 24]*

1.2.5 Electricity Production in the World

A significant amount of the energy available from commercial energy sources is used for the production of electricity. Data on the installed capacity and the electricity produced in the world from these sources is presented in Fig. 1.8 [11, 12, 26]. The curve shows that the electricity produced has increased more or less linearly from 6117 TWh in 1973 to 23 322 TWh in 2013, the average annual increase being about 430 TWh every year. The percentage contribution of various energy sources to the total for two years, 1973 and 2013 is shown in Table 1.2 [26]. It is seen that some interesting changes have taken place over a forty-year time span. These trends are indicative of changes which may occur in the future.

Table 1.2 Percentage contribution of various sources to the total electricity production of the world [26]

	1973	2013
Coal (per cent)	38.3	41.3
Crude oil (per cent)	24.8	4.4
Natural gas (per cent)	12.1	21.7
Water-power (per cent)	20.9	16.3
Nuclear power (per cent)	3.3	10.6
Other sources (per cent)	0.6	5.7
Total (TWh)	6131	23 322

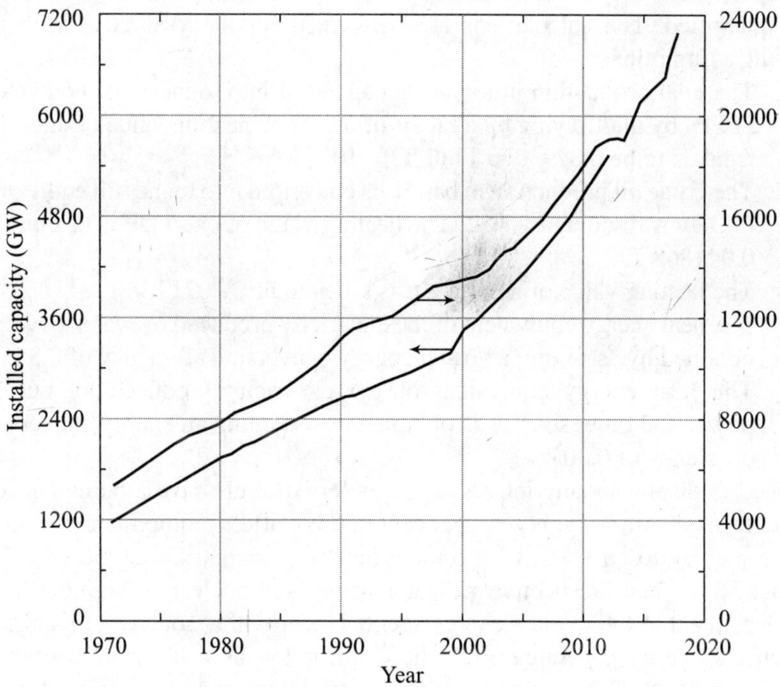

Fig. 1.8 *World installed capacity and electricity generation from all commercial sources [11, 12, 26]*

At the time of the first energy crisis in 1973, oil contributed 24.8 per cent to the electricity produced. In 2013, the percentage contribution has declined significantly to 4.4 per cent. Coal continues to be a dominant contributor, the percentage has increased from 38.3 per cent in 1973 to 41.3 per cent in 2013. The percentage contribution from gas has increased from 12.1 per cent to 21.7 per cent. Adding up the three percentages, we see that fossil fuels still play a dominant role in supplying the energy needs for production of electricity. However, their contribution has declined from 75.2 per cent in 1973 to 67.4 per cent in 2013.

In 1973, the other major contribution was from water-power (20.9 per cent). In 2013, this declined a little to 16.3 per cent. On the other hand, the contribution from nuclear power increased significantly from 3.3 to 10.6 per cent. Also, the contribution from other sources (essentially wind and solar power) has increased a little from being less than 1 per cent to 5.7 per cent.

1.2.6 Annual Production of Energy for the World

In previous sections, the world's production of various energy sources has been stated in the units appropriate to that energy source. Thus, for example, coal production has been given in tonnes, while crude oil production has been given in barrels. It is of interest to calculate the heat energy content associated with these production figures, so that the different sources of commercial energy can be compared on a uniform scale and the extent of their contribution to the world scene quantified. The calculations are performed for the year 2013 with the following assumptions:

1. The coal production in tonnes is converted into tonnes coal equivalent (TCE) by multiplying by a factor of 0.9†. The heating value of one TCE is taken to be 0.0293076 TJ (1 TJ = 10^{12}J).
2. The crude oil production in barrels is converted into tonnes oil equivalent (TOE) by dividing by 7.32. The heating value of one TOE is taken to be 0.041868 TJ.
3. The heating value of natural gas is taken to be 39 021 kJ/m^3.
4. The heat energy equivalent of the electricity produced by water-power is obtained by assuming a virtual energy conversion efficiency of 0.35.
5. The heat energy equivalent of the electricity produced by nuclear power and other sources is obtained by assuming an energy conversion efficiency of 0.30.

Table 1.3 displays many interesting facts. Fossil fuels have a dominant role in the current situation. Eighty-six per cent of the world's requirement of energy is being met by fossil fuels, with oil and natural gas contributing 50 per cent, and coal 36 per cent, respectively. Water-power and nuclear power contribute only 11 per cent to the world's energy scene. Under 'Other sources', in addition to electrical energy generated, only the contribution of geothermal energy as heat is considered. The direct use of geothermal energy as heat amounting to 91 000 GWh [27] has been added to the electricity generation of 1 273 000 GWh. This amounts to a contribution of about 3 per cent.

---

†The factor of 0.9 is the ratio of the heating value of the coal produced $(\overline{HV})$ and the heating value of one TCE. $(\overline{HV})$ is the weighted average of the heating values of the three main types of coal, viz. anthracite, bituminous coal and lignite.

Table 1.3 World energy production from commercial energy sources-Year 2013

Energy source	Production	Energy equivalent (in 10^{15} J)	Per cent contribution
Coal	8076 Mt	213 019	35.61
Crude oil	27.81×10^9 barrels	159 064	26.59
Natural gas	3675×10^9 m^3	143 402	23.98
Water-power	3761 TWh	38 685	6.47
Nuclear power	2364 TWh	28 368	4.74
Other sources	{1273 TWh (electricity) 91 TWh (heat)	15 604	2.61
Total		598 142	100.00

INDIA'S PRODUCTION AND RESERVES 1.3

1.3.1 Fossil Fuels

❑ *Coal* The rate of production of coal in India over the last 65 years is shown in Fig. 1.9 (Curve 1). It can be seen that there has been an eighteen-fold increase in production from 1951 to 2014 and that the average annual growth rate has been about 4.7 per cent. In 2014, India's production was 612 Mt, which was about 7.6 per cent of the world's production. In recent years, a significant amount of non-coking coal is being imported to meet the growing domestic demand. This is shown in Curve 2. The total consumption of non-coking coal from 1999 onward is shown in Curve 3 [28-30].

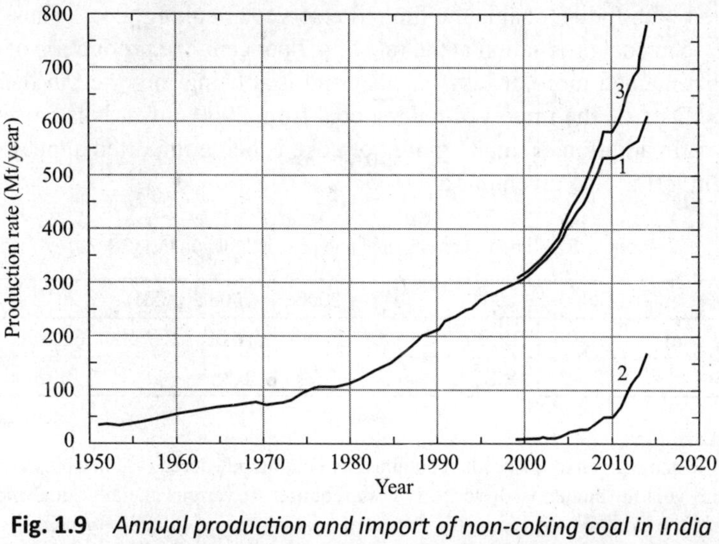

Fig. 1.9 *Annual production and import of non-coking coal in India*
1. Domestic production, 2. Import, 3. Total [28–30]

India has fairly large reserves of coal. Estimates made by the Geological Survey of India over the years are given in Table 1.4. It can be seen that the reserves have been increasing because of the location of more coal fields. The 2015 estimates give the total reserves to be 306 595 Mt and the proved reserves to be 131 614 Mt. This value is about 14.8 per cent of the world's proved reserves [28].

Table 1.4 Coal reserves in India (in Mt) [28]

Year	Proved reserves	Indicated and inferred reserves	Total reserves (Resources)
1972	21 360	59 590	80 950
1981	27 912	87 490	115 402
1985	35 030	120 870	155 900
1992	64 800	129 000	193 800
2006	95 866	157 435	253 301
2015	131 614	174 981	306 595

Using the variation shown in Fig. 1.2, rough estimates have been made of the time when the production could pass through a maximum. These show that coal production in India may be expected to peak between 2040 AD and 2080 AD. These predictions are not very different from those made for the world as a whole.

Although India has large reserves of coal, there are number of associated problems which are note worthy. The first is the high ash content in the coal, which may be very often as much as 30 to 40 per cent. In the absence of large scale beneficiation, the high ash content adds to the transportation cost and places an undue burden on the equipment in which the coal is being used. Another problem associated with Indian coal is the limited reserves of coking‡ coal. Thus, inspite of increase in coal production at the rate of 4.7 per cent, the production of coking coal has remained more or less stagnant and it is being imported in increasing amounts. Data on the import of coking coal from 2000 onwards is presented in Table 1.5. In addition, a small amount of coke is being imported annually. Data on this import is also given in Table 1.5.

Table 1.5 Import of coking coal and coke in India (in Mt) [28-30]

	2000	2002	2004	2006	2008	2010	2012	2014
Coking coal	11.06	12.95	16.93	17.88	21.08	19.48	35.56	43.71
Coke	2.42	2.25	2.84	4.69	1.88	1.49	3.08	3.29

‡Coke is the name given to the product obtained by heating powdered coal in a closed chamber so that the volatile matter is driven off. It is required in various metallurgical industries, particularly the iron and steel industry. Forms of coal which are suitable for being converted into coke are referred to as coking coal.

❑ *Oil* Figure 1.10 presents data on the annual consumption of petroleum products in India (Curve 3) from 1951 onwards. It also shows the variation in the domestic production of crude oil (Curve 1) and the import of crude oil (Curve 2) over the years [31].

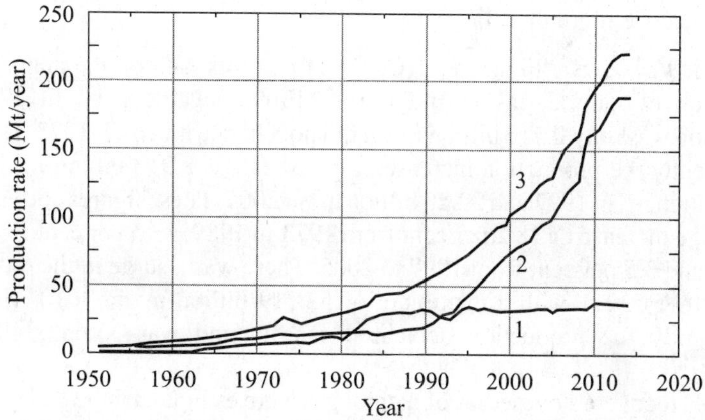

Fig. 1.10 *Annual production, import and consumption of oil in India*
1. Domestic production of crude,
2. Import of crude,
3. Consumption of products [31]

In 1951, the consumption of petroleum products was only 3.89 Mt, most of which were imported. In contrast to this, the consumption was 17.59 Mt in 1970, 30.90 Mt in 1980, 55.04 Mt in 1990, 103.44 Mt in 2000, 194.82 Mt in 2010 and 221.14 Mt in 2014 with most of the products being refined in India. From 1951 to 1970, the annual increase in the consumption was 8.3 per cent, from 1970 to 1990, it was 5.8 per cent, from 1990 to 2000, it was 6.5 per cent, while from 2000 to 2010, it was 6.5 per cent. After 1973, these percentages are much higher than the values for the world as a whole (Section 1.2). As a result, in 2014, India consumed 5.7 per cent of world's production. About fifty years earlier, in 1961, it consumed only 0.7 per cent of the world's production. The continuous increase in the consumption of petroleum products is related to the good economic growth of the country in the last 25 years. However, it is a source of concern, because a substantial part of the crude oil required has to be imported. This import costs the country thousands of crores of rupees every year and raises the issue of energy security.

The import of crude oil is proving to be necessary in spite of the increase in the domestic production of crude as seen in Fig. 1.10. In 1951, domestic production was only 0.27 Mt. It rose sharply in the 1960s with discoveries of oil onshore in Assam and Gujarat, and was a little over 7 Mt from 1971 to 1974. Thereafter, it again increased considerably with discoveries offshore (in regions like Bombay High). It was 33.3 Mt in 1990 and has fluctuated between 27 and 38 Mt since then. Present indications are that domestic production of crude may not increase much. Thus the country will have to resort to increasing imports of crude oil as the demand for petroleum products keeps on increasing.

The position regarding reserves is rather uncertain. Proved recoverable reserves were estimated to be 469 Mt in 1980, 806 Mt in 1990, 786 Mt in 2005 and 635 Mt in 2014 [31]. Extensive prospecting is in progress and more oil discovery is expected offshore. In the event that no substantial discoveries are made, the position would become rather serious. The present domestic production cannot be sustained for more than 17 years.

❑ *Natural Gas* Figure 1.11(Curve 1) presents data on the annual useful production of natural gas in India from 1969 onwards [31]. In 1969, the production was only 0.516 billion m^3. It did not change much till 1973. However, subsequently the production increased rapidly. It was 8.913 billion m^3 in 1989, 24.52 billion m^3 in 1997 and 32.20 billion m^3 in 2005. These figures correspond to an average increase of 18.2 per cent from 1973 to 1989, 13.5 per cent from 1989 to 1997 and 3.5 per cent from 1997 to 2005. There was a surge in the production from 2009 to 2011 with a maximum of 52.219 billion m^3 in 2010. However, subsequently the production decreased sharply and was 33.656 billion m^3 in 2014.

Proved recoverable reserves of natural gas were estimated to be 352 billion m^3 in 1980 and 686 billion m^3 in 1990 [33]. Major gas fields were discovered subsequently off the Andhra Pradesh coast in the Krishna-Godavari basin. These along with other finds have raised the proved reserves estimate for the country as a whole to 1331.94 billion m^3 in 2014 [31]. However, the prospects of finding more gas reserves are not so promising and indicate that the domestic reserves of gas will last only a few years longer than oil.

One of the major problems concerning natural gas in India is that a significant amount is in the form of associated gas. This gas is often flared because facilities for using it have not been developed. The data presented in Fig. 1.11 does not show the flared quantities.

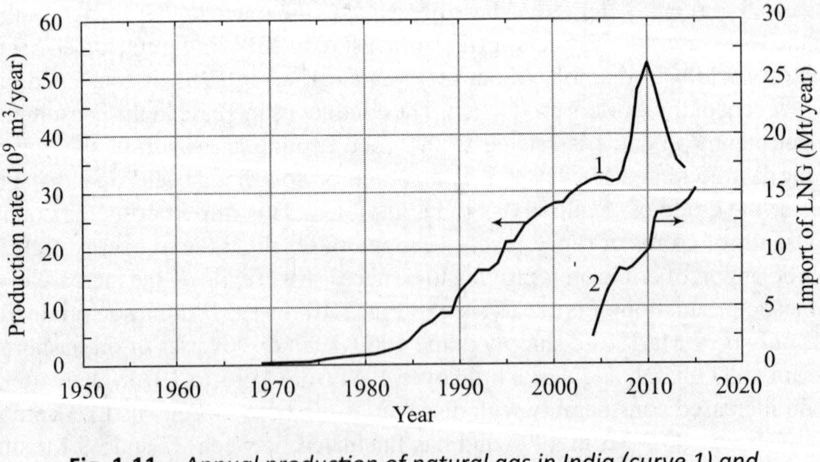

Fig. 1.11 *Annual production of natural gas in India (curve 1) and import of liquefied natural gas (curve 2) [31,32]*

Finally, it should be mentioned that since the current demand for natural gas in India far exceeds the domestic supply, the shortage is made up by importing *liquefied natural gas* (LNG). Data on the imports from 2004 onwards is presented in Fig. 1.11 (Curve 2) [32]. The LNG is re-gasified and supplied mainly to industries. Currently, there are only four terminals on the west coast at Dahej, Hazira, Kochi and Dabhol for importing LNG [34]. Efforts are being made to eastablish more terminals. It is interesting to compare the heat energy content of LNG imported and the natural gas produced in the country. In 2014, 14.092 Mt of LNG was imported. Taking 49 404 kJ/kg as the lower heating value of LNG, the energy content of the LNG becomes 6.962×10^{17} J. On the other hand, the energy content of natural gas produced in 2014 is 1.313×10^{18} J. It is seen that the energy content of imported LNG is about 53 per cent of the energy content of domestic natural gas production. Clearly, the imported LNG contributes significantly to natural gas needs of the country.

1.3.2 Water-power

Data on the installed capacity of hydroelectric power in India and the electricity produced from it from 1947 onwards is presented in Fig. 1.12. It will be seen that the installed capacity was only 508 MW at the time of independence and 560 MW at the beginning of the First Five-Year Plan in 1951. It increased rapidly in the 1950s at an average rate of about 14 per cent. Thereafter, the growth rate has been slower, but nevertheless impressive. Over a fifty-four year time span, starting from 1960, the installed capacity of large hydro-power (units with capacities greater than 25 MW) increased at an average rate of about 5.8 per cent from 1917 MW to 41 262 MW in 2014. This excludes 4785.6 MW of pumped storage units. Corresponding to the increase in installed capacity, the electricity produced increased from 2195 GWh in 1947 to 129 244 GWh in 2014 [35]. The APLF in 2014 was 0.36.

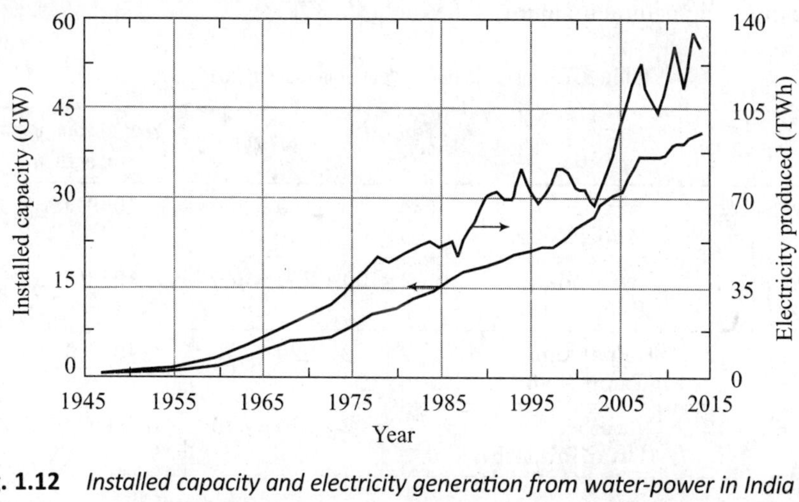

Fig. 1.12 *Installed capacity and electricity generation from water-power in India [35]*

It may be noted that units with capacities less than 25 MW are referred to as small hydro-power units. The installed capacity of these units is included later under Other Sources.

The water-power reserves of India have been estimated by the Central Electricity Authority. It is estimated that there is potential for installing large units with a total capacity of about 148 700 MW. The region-wise break-up is shown in Table 1.6. It is evident that only about 20 per cent of the potential has been used so far and there is considerable scope for generating more power by hydroelectricity. This is particularly true in the northeastern region of India where much of the reserve lies untapped.

Table 1.6 Water-power potential in India [35]

Region	Potential (MW)
Northeast	58 971
East	10 949
North	53 395
West	8 928
South	16 458
Total	148 701

1.3.3 Nuclear Power

The installed capacity of nuclear power plants in India in 2015 is shown in Table 1.7. It is seen that 21 units having a total capacity of 5780 MW are in operation. TAPS 1 and 2 at Tarapur are boiling water reactors, KKNPP1 at Kudankulam is a pressurized water reactor and all the other units are pressurized heavy water reactors using natural uranium as the fuel [36, 37].

Table 1.7 Installed nuclear power in India in 2015 [37]

Unit name	Location	Capacity (MW)	Year of commercial operation
TAPS–1&2	Tarapur (Maharashtra)	$2 \times 160 = 320$	1969, 1969
RAPS–1&2	Rawatbhata (Rajasthan)	$100 + 200 = 300$	1973, 1981
MAPS-1&2	Kalpakkam (Tamil Nadu)	$2 \times 220 = 440$	1984, 1986
NAPS–1&2	Narora (Uttar Pradesh)	$2 \times 220 = 440$	1991, 1992

(Contd.)

KAPS–1&2	Kakrapara (Gujarat)	$2 \times 220 = 440$	1993, 1995
KAIGA–1&2	Kaiga (Karnataka)	$2 \times 220 = 440$	2000, 2000
RAPS–3&4	Rawatbhata (Rajasthan)	$2 \times 220 = 440$	2000, 2000
TAPS–3&4	Tarapur (Maharashtra)	$2 \times 540 = 1080$	2005, 2006
KAIGA–3	Kaiga (Karnataka)	$1 \times 220 = 220$	2007
RAPS–5&6	Rawatbhata (Rajasthan)	$2 \times 220 = 440$	2010, 2010
KAIGA–4	Kaiga (Karnataka)	$1 \times 220 = 220$	2011
KKNPP–1	Kudankulam (Tamil Nadu)	$1 \times 1000 = 1000$	2014
Total		5780	

In addition, a number of units are under construction. These are listed in Table 1.8. It is seen from the table that the installed capacity of nuclear power will increase to 10 080 MW when the reactors under construction are completed in a few years. It is also seen that the operation of these reactors will result in the introduction of new technologies in the country. The reactor at kalpakkam is a fast breeder reactor of indigenous design using a plutonium-based fuel.

Table 1.8 Nuclear power units under construction in 2015 [36, 37]

Location	No. of units	Capacity (MW)	Type of reactor
Kakrapar (Karnataka)	2	$2 \times 700 = 1400$	PHWR
Rawatbhata (Rajasthan)	2	$2 \times 700 = 1400$	PHWR
Kudankulam (Tamil Nadu)	1	$1 \times 1000 = 1000$	PWR
Kalpakkam (Tamil Nadu)	1	$1 \times 500 = 500$	FBR
Total		4300	

Data on the electricity production from nuclear power is graphically presented in Fig. 1.13. It is seen that the electricity produced has been generally increasing over the years, as more units are getting commissioned. The highest amount, viz. 37 835 GWh was produced in 2014. The fall in certain years is because of some units being down for maintenance.

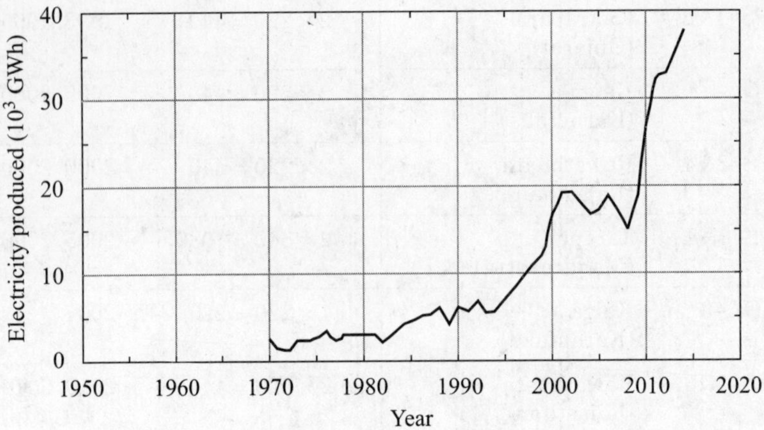

Fig. 1.13 *Electricity generation from nuclear power in India [36, 37]*

India's resources of uranium are not extensive. It is estimated that reserves available are about 119 000 t [21]. It is easy to show that the reserves would only be adequate for providing the requirements of an installed capacity of 10 000 MW for about 45 years if the country continues to use natural uranium as a fuel in PHWRs.

1.3.4 Other Sources

We now discuss the contribution of some other sources for generating electrical energy. In India, these sources are renewable sources like wind energy, solar photovoltaic power, biomass, small hydro-power, etc.

As was the case for the world, in India also, wind energy is the main contributor. The growth in installed capacity for wind energy is, therefore, plotted in Fig. 1.14 along with data on the electricity produced from the wind. The growth of installed capacity for wind energy in India has been very impressive. At the end of 1990, the capacity was only 37 MW. Twenty-five years later, at the end of 2015, it was 25 088 MW and India now ranks fifth in the world in terms of wind power installed capacity. There is considerable scope for increasing this capacity. If sites having wind power density in excess of 200 W/m^2 at 50 m hub height with 2 per cent land availability are considered, the potential for wind power is estimated to be 49 130 MW. At 80 m hub height, the potential is 102 788 MW [38,40]. Phadke *et al.* [41] have adopted a more refined approach for calculating the wind potential in India and shown that it is 2006 GW. Apart from the potential on land, a significant potential is available off-shore in India. Lu *et al.* [42] estimated that this could be as high as 1100 TWh. More details are given in Chapter 9.

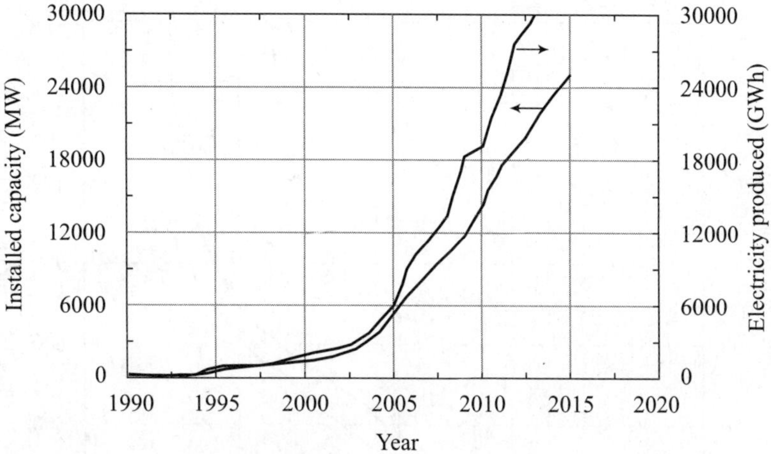

Fig. 1.14 *Installed capacity and electricity generation from wind power in India [38, 39]*

The contributions of small hydro-power, biomass and solar power are also significant. Table 1.9 presents data on the installed capacity of small hydro-power, biomass and solar (PV and thermal) power in the country.

Table 1.9 Installed capacity of small hydro-power, biomass power and solar power in India [38]

Year	Capacity (MW)		
	Small hydro-power	Biomass power	Solar power
Up to 2001	3300	3095	513
2012	3552	3565	1446
2013	3804	4013	2647
2014	3991	4165	3062
2015	4177	4678	4879

1.3.5 Electricity Production in India

Data on the installed capacity for generating electrical power from all commercial sources and the energy produced is plotted in Fig. 1.15. The data show that the installed capacity has increased from 1362 MW in 1947 at the time of independence to 16 664 MW in 1973, to 124 287 MW in 2005 and to 298 060 MW in 2015 [43–46]. These correspond to an impressive average annual growth rate of 10.1 per cent from 1947 to 1973 and to a rate of 7.1 per cent from 1973 to 2015. The corresponding figures for the electricity produced were 4073, 66 689 and 1 107 385 GWh respectively [43 – 46].

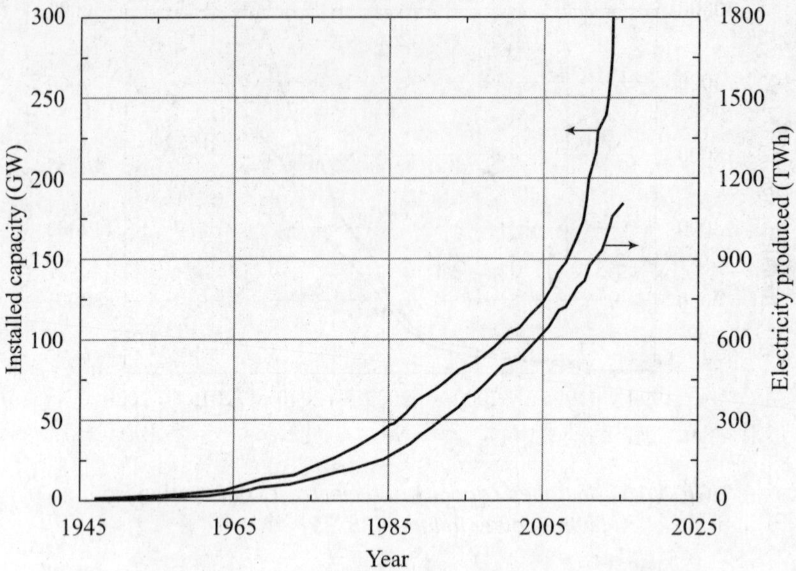

Fig. 1.15 *Total installed capacity and electricity generation in India from all commercial sources [43–46]*

The percentage contribution of various energy sources to the installed capacity is shown in Table 1.10 for the four years, viz. 1947, 1973, 2005 and 2015. The table shows that the contribution from fossil fuels is the highest. The contribution was 62.7 per cent in 1947, it declined to 54.4 per cent in 1973 and subsequently has risen again to 70.7 per cent in 2015. Coal is the dominant contributor, while the contribution of oil is now negligible. For example in 2015, the contributions of coal, oil and natural gas were as follows:

Table 1.10 Installed capacity for electrical power in India [43,44]

Year		Installed capacity				
		Fossil fuels	Hydro-power	Nuclear	Renewable	Total
1947	MW	854	508	0	0	1362
	per cent	62.7	37.3	0	0	100.0
1973	MW	9058	6966	640	0	16 664
	per cent	54.4	41.8	3.8	0	100.0
2005	MW	82 410	31 277	3360	7240	124 287
	per cent	66.3	25.2	2.7	5.8	100.0
2015	MW	210 675	42 783	5780	38 822	298 060
	per cent	70.7	14.4	1.9	13.0	100.0

Coal	185 173 MW (61.3 per cent)
Oil	993 MW (0.3 per cent)
Natural gas	24 509 MW (8.1 per cent)

The contribution of hydroelectric power to the installed capacity was around 40 per cent for the first 25 years after independence. However subsequently, it has gradually declined to 14.4 per cent in 2015. The contribution from nuclear power is 1.9 per cent. A significant contribution (13 per cent) is made by renewable sources.

It is interesting to examine the percentage contribution of various energy sources to electricity generated (Table 1.11). We see that the contribution from fossil fuels was 46.1 per cent in 1947, it increased to 53 per cent in 1973 and has been increasing over the years, reaching 79.4 per cent in 2014. The contribution of hydro-power to the electricity generation was around 54 percent in 1947 and since then its share has been declining reaching 11.7 percent in 2014. Small but significant contributions (8.9 per cent) are now being made by nuclear power and renewable sources.

Table 1.11 Contribution of various energy sources to electricity produced in India [43,44]

Year		Electricity generation				Total
		Fossil fuels	Hydro-power	Nuclear	Renewable	
1947	GWh	1877	2195	0	0	4073
	per cent	46.1	53.9	0	0	100.0
1973	GWh	35 321	28 972	2396	0	66 689
	per cent	53.0	43.4	3.6	0	100.0
1989	GWh	178 690	62 116	4625	6	245 438
	per cent	72.8	25.3	1.9	0	100.0
2006	GWh	528 490	113 502	18 802	9860	670 654
	per cent	78.8	16.9	2.8	1.5	100.0
2011	GWh	708 427	130 511	32 287	51 226	922 451
	per cent	76.8	14.1	3.5	5.6	100.0
2014	GWh	878 320	129 244	36 102	61 780	1 105 446
	per cent	79.4	11.7	3.3	5.6	100.0

In spite of the good growth in capacity for generating electrical power, the per capita consumption in India is still rather low. In 2013, the consumption was only around 957 kWh per person [44]. In contrast, the consumption in economically developed countries was 10 to 15 times larger.

The data presented in Fig. 1.15 excludes data on the capacity of captive power units in industry and the energy produced by them. Captive power units are set up by industries for their own consumption primarily because of the non-availability of adequate grid supply and the lack of reliability of this supply.

Captive power stations range in capacity from a megawatt or two to a couple of hundred megawatts. Estimates of the captive power capacity in the country vary. The Central Electricity Authority has given an estimate of about 47 082 MW in 2014 [44].

1.3.6 Estimating India's Future Needs of Electricity

In 2011–12, India consumed close to 1000 TWh of electricity equivalent to a mean value of about 900 kWh per person. In contrast, USA consumed about 12 000 kWh per person, while other economically developed countries like Germany and France had mean per capita values between 7000 and 8000 kWh/year. It is agreed that the present consumption in India is low. It is also generally accepted that the consumption in USA is high and wasteful. High consumption is harmful because one is using up non-renewable energy resources unnecessarily and also because of the consequent increase in greenhouse gas emissions. Thus, the question which is often raised from the point of view of energy policy and planning is: what per capita mean value should India aim for in the future? The question is particularly appropriate when one recognizes that on a per capita basis, India is not that well endowed with natural resources. The word 'future' as used here refers to a date about fifty or sixty years from today when the country will be economically developed and the *mean annual per capita consumption/need for electricity* (MACE) would have more or less stabilized. Obviously the value should be adequate for providing a good standard of living for most people in India. At the same time, it should not lead to unnecessary waste.

Mean Annual Per Capita Needs of Electricity (MACE) in the Future

Two approaches have been adopted in the literature for estimating the value of MACE required in a country as it seeks to become a developed country in the future. These are described by Sukhatme [47, 48].

In the first approach, one considers the electrical energy inputs required in various sectors like manufacturing, transportation, mining, agriculture, commercial, construction, etc., which provide the goods, services, food, infrastructure, etc., needed in daily life. Data on these inputs are collected in numerous locations and countries. These are averaged and converted into per capita values. To these values are added the electricity needs of the residential sector. Goldemberg *et al.* [49, 50] carried out such pioneering studies in the 1980s by collecting and analyzing data obtained by four groups in Brazil, Sweden, India and USA. The group in India was led by Professor Amulya Reddy of the Indian Institute of Science (IISc). The main result obtained from the studies was that in a country like India, a mean value of 1840 kWh per person per year would be adequate for providing a good, but frugal standard of living. The estimate was based on the assumption that sound energy conservation measures would be adopted and energy efficient devices would be used. Since the work of Goldemberg and his co-workers was conducted more than 20 years ago, it can be criticized on the grounds that with the availability of more efficient electrical devices today the required value of MACE should have decreased a little. On the other hand, it can also be argued that with increased use of electricity in more applications today than earlier, the required value of MACE consequently should have increased. Since there is no better estimate available in the literature based on the procedure adopted by Goldemberg *et al.* [49, 50], 1840 kWh per person per year, as a credible value, is acceptable.

A second approach adopted for determining the desirable value of MACE is a comparison plot of the *Human Development Index* (HDI) of a country against the value of MACE of the country. HDI of a country is a composite index which combines indices regarding life expectancy, school education and gross national income in that country. By definition, the value of HDI ranges between 0 and 1. The plot obtained by using data for all countries of the world shows that there is a fairly good correlation between the two quantities. It is seen that the value of HDI increases as the value of MACE increases and that HDI reaches a value around 0.9 when MACE is about 4000 kWh per person per year. Thereafter, the value of HDI levels off around 0.9 even if the electricity use goes on increasing. Based on this information, it has been argued that it would be desirable to provide an average of 4000 kWh per person per year to meet the future needs of electricity in India.

Thus we have two mean annual values, 1840 and 4000 kWh per person, for consideration. Keeping in mind the need for frugality in energy usage, we should opt for a value of 1840 kWh per person. At the same time, taking into consideration the aspirations of the people, aiming for an average value of 4000 kWh per person is also not unreasonable.

Demographic studies indicate that based on the latest census trends (2011), the population of India may stabilize around 1700 million in the period 2071–80. Using this value, we can obtain India's total need of electricity around 50 years from today when the nation would be economically developed with a stabilized population. This data is presented in Table 1.12.

Table 1.12 India's total need for electricity in the future [47, 48]

Mean per capita need (kWh/year)	Population (million)	Total need (TWh/year)
1840	1700	3128
4000	1700	6800

Figure 1.16 is an approximate graphical description of the variation of India's total need for electricity in the future and the contribution of fossil fuels. Curve I shows that the total production will have to increase by a factor of 6 to 7 from its value of about 1000 TWh/year in 2010 (point A) to a value around 6800 TWh/year by 2070 (point B) when it can be expected to stabilize. This is a steep increase but is achievable with good economic growth. Curve II shows the contribution of fossil fuels. As is to be expected, the shape of this curve is

similar to the curve drawn in Fig. 1.2 which shows the world's rate of production of coal. In 2010, the contribution of fossil fuels was about 80 per cent of the total (point C). As the years go by, the actual contribution to electricity production will continue to increase, pass through a maximum (point D), gradually decrease over the years and asymptotically approach zero (point E). However, the percentage contribution to the total will go on decreasing. It should be noted that apart from the location of point C, only the shape of Curve II is known. The actual values of electricity production at point D and the year in which it may occur are not known. Similarly, the year in which the production of electricity from fossil fuels is likely to become negligible (point E) is not known. The space between Curves I and II indicates the electricity produced by renewable energy sources and nuclear energy. Today, their contribution is only about 20 per cent, but eventually (perhaps after two hundred years or so) these will have to be the sources from which India will have to derive all its electricity.

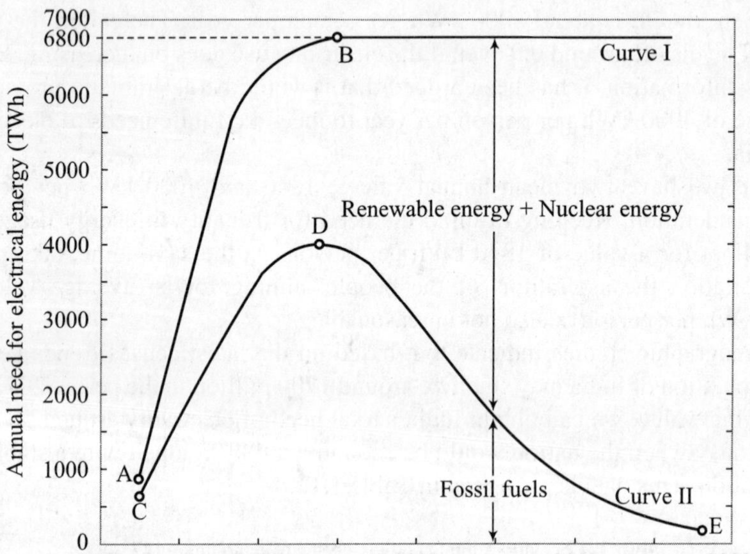

Fig. 1.16 *Variation of India's total need for electricity over the years (Curve I) if individual need is assumed to be 4000 kWh/year. Contribution of fossil fuels is also indicated (Curve II)*

Distribution of Annual Per Capita Value

Apart from increasing the value of the mean in India, it is necessary to ensure that the distribution of the annual per capita value is reasonably equitable with respect to the population. Sukhatme [48] has studied this aspect and shown that the distribution in India can be approximated by a Weibull distribution[§]. The density function for the Weibull distribution is given by

[§]The Weibull distribution has also been used later in Chapter 9 for analyzing wind speed data

$$f(x) = \left(\frac{k}{c}\right)\left(\frac{x}{c}\right)^{(k-1)} \exp\left[-\left(\frac{x}{c}\right)^{k}\right] \tag{1.2}$$

where x is the electricity consumption/future need per person per year (ACE). k and c are parameters in the distribution, k being a shape parameter and c, a scale parameter. The Weibull distribution has the advantage of being flexible in its shape. For values of k between 0 and 1, it has a Poisson-like shape, while for values of k greater than 1, it has a Gaussian-like shape. The fraction of the population, $F(x)$, covered in the distribution from 0 to x is obtained by integrating Eq. (1.2) from 0 to x and yields the expression

$$F(x) = 1 - \exp\left[-\left(\frac{x}{c}\right)^{k}\right] \tag{1.3}$$

It has been shown that the present distribution (around the year 2015) is a Weibull distribution with a Poisson-like shape (Fig. 1.17). It is highly skewed; 65.5 per cent of the population have access to a per capita value less than the mean and 15.0 per cent consume less than 100 kWh/year. Thus in India, today, in addition to the mean value being low, the distribution is highly inequitable.

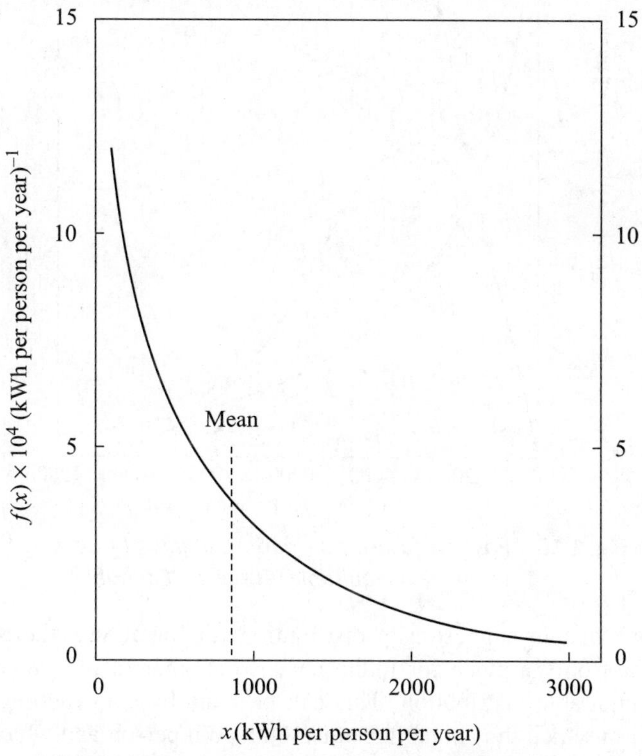

Fig. 1.17 *Present distribution in India. (k = 0.870, c = 810 kWh per person per year and mean = 868.8 kWh per person per year) [49]*

What does the future hold? If no particular steps are taken, the shape of the distribution is likely to stay the same even as the annual mean increases in the future to 4000 kWh per person. This distribution can be obtained by simply scaling up the present distribution from an annual mean value of 868.8 to 4000 kWh per person. This is shown as Curve 1 ($k = 0.87$; $c = 3729$ kWh per person per year) in Fig. 1.18. Since it is a scaled version of Fig. 1.17, again 65.5 per cent of the population will have access to per capita values less than the mean of 4000 kWh/person. In addition, calculations show that for $x = 1840$ kWh/person, $F(x) = 0.418$. Thus, 41.8 per cent of the population will consume less than 1840 kWh/person, which has been identified as a desirable frugal value. This is a large percentage and is indicative of an inequitable and unacceptable situation.

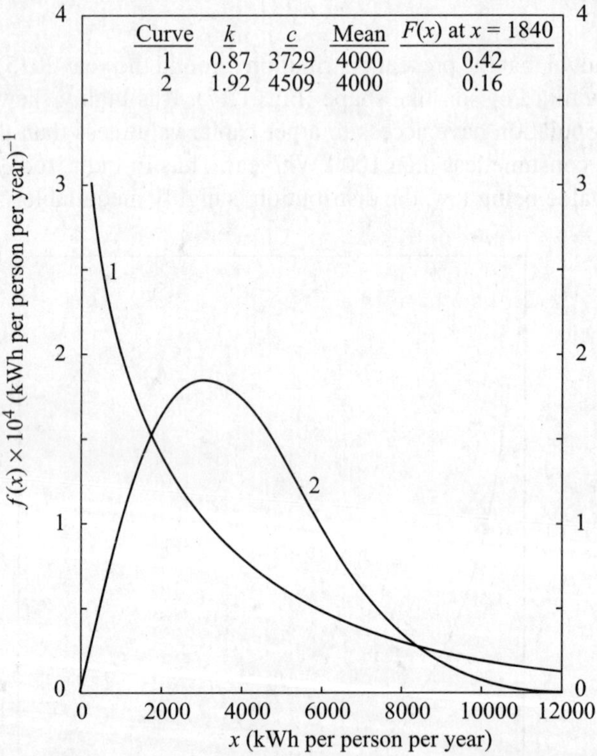

Fig. 1.18 *Possible future distributions in India [49]*
Curve 1 – Inequitable, Curve 2 – Equitable

It follows that in suggesting a distribution for the future, there is a need to specify not only a mean for India, but also another factor which will lead to a more equitable distribution. This can be done by constructing a Weibull distribution in which the annual mean is 4000 kWh/person and which specifies that the percentage of the population consuming less than 1840 kWh/person lies between 10 and 20 per cent, say around 15 per cent. The distribution thus obtained is Curve 2 in Fig. 1.18, the values of the parameters, k and c, being $k = 1.92$ and $c = 4509$ kWh per person per year. It will be noted that unlike the distribution shown in Curve 1, the value of k is now greater than 1 and the distribution is more Gaussian-like in its shape. As expected, this distribution is much more equitable. The percentage of the population consuming less than 1840 kWh per person per year comes out to be 16.4 per cent, which is acceptable. Even the percentage of the population below the mean is reduced from the value of 65.5 percent in Curve 1 to 54.8 per cent in Curve 2. Curve 2 is an indication of the shape of the distribution India should strive for in the future while simultaneously raising the value of the mean.

1.3.7 Annual Production of Energy

Following the procedure described in Section 1.2.6, let us calculate the energy content associated with the production or consumption figures for different sources of commercial energy used in India. The calculations are performed for the year 2013 and are presented in Table 1.13. It is seen that the total energy production is 28 074 × 10^{15} J. Once again the dominant role played by fossil fuels in the energy sector is apparent. Ninety-one per cent of India's requirement of commercial energy is being met by fossil fuels, with coal contributing 53 per cent, and oil and natural gas contributing 38 per cent. Water-power and nuclear power contribute only about 9 per cent to the total energy production. Comparing the total energy production in India from commercial sources with that of the world, it is seen to be only 4.7 per cent of the total world production.

Table 1.13 Energy production from commercial energy sources in India – Year 2013

Energy source	Production/ Consumption	Energy equivalent (in 10^{15} J)	Per cent contribution
Coal	565.77 Mt	14 923	53.16
Crude oil	220.76 Mt	9243	32.92
Natural gas	35.407×10^9 m^3	1382	4.92
Water-power	134 848 GWh	1387	4.94
Nuclear power	35 333 GWh	424	1.51
Other sources	59 615 GWh	715	2.55
Total		28 074	100.00

ENERGY ALTERNATIVES 1.4

1.4.1 The Need for Alternatives

Based on the preceding survey in Sections 1.2 and 1.3, it will now be useful to repeat some observations made earlier, to make some new observations and to draw some conclusions for the world as a whole.

Fossil Fuels

1. The average rate of increase of oil production in the world is declining and a peak in production may be reached around 2021 (if no significant addition occurs to the proved reserves). The serious nature of this observation is apparent when one notes that oil provides nearly 30 per cent of the world's need for energy from commercial sources and that oil is the fuel used in most of the world's transportation systems.
2. The production of natural gas is continuing to increase at a rate of about 3.5 per cent every year. Unlike oil, there has been no significant slowdown in the rate of increase of production. Present indications are that a peak in gas production may occur between 2024 and 2046.
3. As oil and natural gas become scarcer, a greater burden will fall on coal. It is likely that the production of coal will touch a maximum somewhere between 2040 and 2060.
4. Finally, it should be noted that in addition to supplying energy, fossil fuels are used extensively as feedstock material for the manufacture of organic chemicals. As resources deplete, the need for using fossil fuels exclusively for such purposes may become greater.

Water-power

Water-power is indirectly obtained from solar energy and has the advantage of being a renewable source of energy. Estimates indicate that there is considerable scope for increasing the installed capacity in the world, particularly in the developing countries. However many environmental concerns are now being raised when big projects involving the storing of large amounts of water are being considered. The impact of these concerns is reflected in the fact that the average annual increase in capacity has declined.

Nuclear Power

Nuclear power is obtained from fission reactions in pressurized water reactors or pressurized heavy water reactors which utilise only the U^{235} isotope. This isotope is available to the extent of less than 1 per cent in naturally occurring uranium. The other isotope U^{238}, which is available to the extent of more than 99 per cent, is wasted. Present indications are that the world's assured reserves of 4.59 Mt of uranium (< US $260 per kg of uranium) will be adequate for the current installed capacity of 376 200 MW for only about 70 years if the present technology utilizing U^{235} alone continues to be used.

It is thus evident that a need for developing alternative energy sources does exist. The immediate need would be to alleviate the problems caused by the depletion of oil and then natural gas, while the long-term need would be to develop means to replace U^{235} based fission technology and then coal. These conclusions are applicable for India also.

The primary sources of alternative energy which hold potential for the future can be broadly classified under four categories. These are

1. The solar energy option,
2. Other renewable energy sources,
3. The nuclear energy option, and
4. Miscellaneous sources – Tar sands, oil shale, unconventional hydrocarbons, geothermal energy

Work is in progress in many parts of the world on all these primary alternatives, which are in different stages of utilization. Some are in initial stage of exploration, appraisal, research and development. Some are in the pilot plant or prototype stage and others are already commercially available, but on a relatively small scale. In the following sections, we shall briefly describe these energy alternatives. These preliminary descriptions will hopefully help the reader to acquire a broad perspective of the energy problem, before we focus on the solar energy option and, more specifically, direct thermal methods for utilising solar energy.

In addition to the primary sources listed above, there are some secondary alternatives. These sources are different in the sense that they require a primary source or one of the existing sources of energy for their production. A good example of a secondary source is hydrogen, which is discussed later in Section 1.4.6.

1.4.2 The Solar Energy Option

Solar energy is a very large, inexhaustible source of energy. The power from the sun intercepted by the earth is approximately 1.8×10^{11} MW which is many thousands of times larger than the present consumption rate on the earth of all commercial energy sources. In principle, solar energy could supply all the present and future energy needs of the world on a continuing basis, thus making it one of the most promising of the unconventional energy sources.

In addition to its size, solar energy has two other factors in its favour. First unlike fossil fuels and nuclear power, it is an environmentally clean source of energy. Second, it is free and available in adequate quantities in almost all habitable parts of the world.

However, there are many problems associated with its use. The main problem is that it is a dilute source of energy. Even in the hottest regions on earth, the solar radiation flux available rarely exceeds 1 kW/m^2 and the total radiation over a day is at best about 7 kWh/m^2. These are low values from the point of view of technological utilisation. Consequently, large collecting areas are required in many applications and these result in excessive costs.

A second problem associated with the use of solar energy is that its availability varies widely with time. The variation in availability occurs daily because of the day-night cycle and, also, seasonally because of the earth's orbit around the sun. In addition, variations occur at a specific location because of local weather conditions. Consequently, the energy collected when the sun is shining must be stored for use during periods when it is not available. The need for storage also adds significantly to the cost of any system. Thus, the real challenge in utilising solar energy as an energy alternative is of an economic nature. One has to strive for the development of cheaper methods of collection and storage so that the large initial investments required at present in most applications are reduced.

The energy from the sun is utilized directly in two ways. The first is the thermal route in which devices are used to collect the sun's energy in the form of heat. This thermal energy is either used straightway for applications such as water heating, space heating, drying, etc. or converted into electricity through energy conversion cycles. The second way of using solar energy directly is by photovoltaic conversion, the devices used for the purpose being called solar cells. This book is essentially devoted to these two aspects through Chapters 2 to 8.

1.4.3 Other Renewable Energy Sources

The second category 'Other renewable energy sources' includes the use of water (hydroelectric) power, wind energy, biomass, tidal energy, wave energy, energy derived from temperature differences in the oceans and marine currents. All these sources (except tidal energy) derive their energy indirectly from the sun. The last four, viz. tidal energy, wave energy, the energy available from temperature differences in the ocean and energy in marine currents are often grouped under the broad heading of *ocean energy*. Excepting large hydroelectric power systems[¶], these sources will be described in Chapter 9.

1.4.4 The Nuclear Energy Option

Under the nuclear option, the two alternatives under study are

1. The breeder reactor, and 2. Nuclear fusion.

The Breeder Reactor

In order to understand the working of a breeder reactor, it is necessary to describe the fission reactions in a little detail.

Naturally occurring uranium contains three isotopes, U^{234}, U^{235} and U^{238}. The relative percentages of these isotopes are as follows:

U^{234} 0.006% U^{235} 0.711% U^{238} 99.283%

Of these isotopes, only U^{235} undergoes spontaneous fission when subjected to bombardment by slow neutrons. It is in fact the only naturally occurring fissile material. The fission reaction which occurs is shown in Fig. 1.19. The break-up of U^{235} when subjected to neutron bombardment yields fission products, neutrons and the release of a large amount of energy as heat (8.2×10^7 kJ per g of U^{235}). The neutrons are slowed down by a moderator, and used to bombard the U^{235} nucleus again, thereby setting up a controlled chain reaction.

[¶]Hydroelectric power using large units is being produced commercially since a long time. It will not be discussed in this book because the technology for its use is relatively conventional and well established.

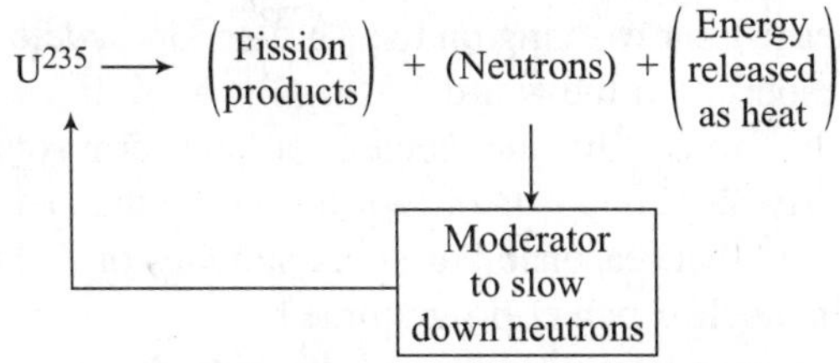

Fig. 1.19 *Fission reaction of U^{235}*

Almost all the nuclear reactors in the world today, which are used for generating electrical power, work on the above reaction and use only U^{235}. The fuel used is either natural uranium having U^{235} with a percentage of 0.711 or slightly enriched uranium having 3 to 4 per cent U^{235}. Thus, in all these reactors, the abundant U^{238} isotope is essentially wasted.

Although U^{238} is not a fissile material, it is a 'fertile' material, i.e. it can be converted by neutron bombardment into a fissile material, plutonium-239. Similarly, naturally occurring thorium-232 is also a fertile material. It can be converted into U^{233} which is a fissile material. Both these reactions are shown in Fig. 1.20.

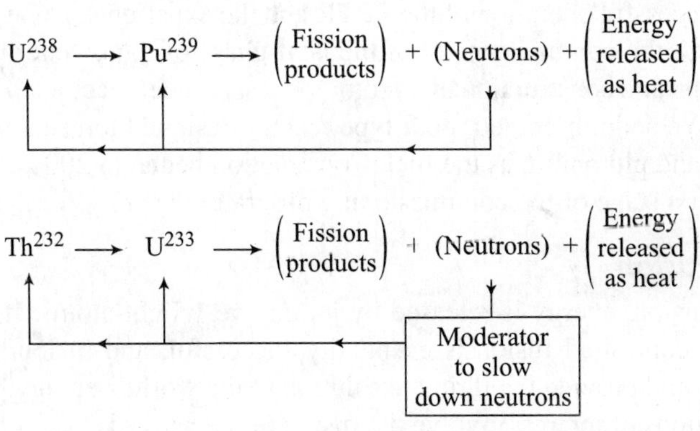

Fig. 1.20 *Breeder reactions for U^{238} and Th^{232}*

It will be seen that the neutrons generated by the fission reaction serve two purposes. They help in converting a fertile material to a fissile material and also sustain the fission reaction for the fissile material formed. The above reactions are called *breeder reactions* if they produce more fissile material than they consume and the nuclear reactor in which they are caused to occur is called a *breeder reactor*. Breeding is achieved by having both fissile and fertile materials in the reactor core under conditions which provide enough neutrons to propagate a chain reaction in the fissile material as well as to convert more fertile material into fissile material than was originally present. Thus, a breeder reactor working on the U^{238} to Pu^{239} cycle utilises naturally-occurring uranium almost completely and thereby helps to extend the supply of uranium by a factor of about 100.

Similarly, a breeder reactor working on the Th^{232} to U^{233} cycle helps in utilising the vast thorium resources of the world.

For India, the Th^{232} to U^{233} breeder cycle is of long-term significance because the country has fairly large deposits of thorium in the monazite beach sands in Kerala. It is estimated that reasonably assured supplies of 846 000 t of thorium are available for the nuclear power programme [51].

Reactors working on various breeder cycles have been built. However, the major effort has been on liquid-metal cooled, fast breeder reactors working on the U^{238} to Pu^{239} cycle. In the 1970s, it appeared that breeder reactors would be in commercial operation by the turn of the century. However, fears of nuclear accidents, difficulties associated with radioactive waste disposal and the possibility of plutonium being misused for weapons have caused the breeder development programme to be closed or to be slowed down in most countries.

In India, a *fast breeder test reactor* (FBTR) working on the U^{238}-Pu^{239} cycle has been set up and operated at Kalpakkam. The reactor became critical for the first time in 1985 and has been commissioned in stages. The steam generator was commissioned in 1993 and the turbo-generator synchronized to the grid in 1997. The reactor has operated up to a power level of 17.4 MWt/2.2 MWe. It is cooled by liquid sodium and uses a mixture of carbides of uranium and plutonium as the fuel. The successful operation of the FBTR and the experience gained in running it, have led to the commencement of the next phase of the breeder programme, viz. the design and construction of a prototype fast breeder reactor (PFBR). PFBR is a 500 MWe sodium-cooled, pool type reactor designed to use a mixed oxide of uranium and plutonium as the fuel. Construction began in 2004 and currently it is in the first stage of the commissioning programme [52].

Nuclear Fusion

In nuclear fusion, energy is released by joining very light atoms. If the current research on controlled fusion is eventually successful, and fusion reactors are built, they could provide the ultimate solution to the world's energy problem.

The reactions of interest involve the fusing of the heavy isotopes of hydrogen (deuterium D and tritium T) into the next heavier element, viz. helium. They are as follows

$$D + D \rightarrow T + p + 4.0 \text{ MeV} \quad (1.4a)$$
$$D + D \rightarrow He^3 + n + 3.2 \text{ MeV} \quad (1.4b)$$
$$D + T \rightarrow He^4 + n + 17.6 \text{ MeV} \quad (1.4c)$$
$$D + He^3 \rightarrow He^4 + p + 18.3 \text{ MeV} \quad (1.4d)$$
$$6D \rightarrow 2\,He^4 + 2p + 2n + 43.1 \text{ MeV} \quad (1.4e)$$

Equations (1.4a) and (1.4b) show that two nuclei of deuterium can fuse in two ways. Both ways are equally probable. In the first, tritium and one proton are formed, while in the second, helium-3 and one neutron are formed. The energy released by the fusion reaction is as indicated. Tritium is unstable and combines with deuterium to form helium-4 and one neutron, Eq. (1.4c), while helium-3 combines with deuterium to form helium-4 and one proton, Eq. (1.4d). The net result, Eq. (1.4e), is the addition of all the four reactions. It indicates that six deuterium nuclei are converted to two helium-4 nuclei, two protons and two neutrons with an energy release of 43.1 MeV. Equations (1.4a) and (1.4b) are referred to as the D-D reactions, Eq. (1.4c) as the D-T reaction and Eq. (1.4d) as the D-He^3 reaction. It is to be noted that the above reactions occur at very high temperatures when matter is in the form of plasma.

Deuterium occurs naturally in sea water and it is estimated that the fusion of all the deuterium in just one cubic metre of sea water would yield an energy of 12×10^9 kJ. Thus the fusion of deuterium represents an essentially inexhaustible source of energy.

A fusion reaction can be self-sustaining only if the rate at which energy is supplied to the plasma is less than the rate at which energy is generated by the fusion reaction. This is only possible above a certain critical temperature which is estimated to be of the order of 10^8 K for the above mentioned reactions. Thus the key problems in the development of a nuclear fusion reactor are the attainment of the required high temperature by initially heating the fuel charge and the confinement of the heated fuel for a long enough time for the reaction to become self-sustaining.

The research being conducted to solve these problems is proceeding broadly along two conceptual directions—*magnetic confinement* and *laser-induced fusion*. In the first concept, the fuel charge (in the form of a charged particle gas composed of positively charged nuclei and free electrons) is contained in a hermetically sealed vacuum chamber and is heated to the required high temperature by passing an electric current through it. At this temperature, the fusion reactions can take place successfully only if the gas is confined within a certain volume for a specified time and not allowed to come into contact with the containing chamber's walls. This confinement is achieved by the application of a very strong, specially shaped magnetic field. In the second concept, the fuel charge is in the form of very small pellets. These are positioned one by one at a specific location and subjected to intense focussed laser beams which heat the pellets to the required temperature and cause fusion to occur. By adopting this method, the confinement time is substantially reduced and the need for a magnetic field is removed.

Of the reactions given in Eqs (1.4a)–(1.4d), the D-T reaction takes place at the lowest temperature, about 10^7 K. It is, therefore, the least demanding in terms of its heating and confinement requirements. Attention has been focussed on using this fact. It has been suggested that since tritium does not occur naturally, the D-T reaction should be supplemented by one using lithium as follows.

$$D + T \rightarrow He^4 + n + 17.6 \text{ MeV} \tag{1.5a}$$

$$n + Li^6 \rightarrow He^4 + T + 4.8 \text{ MeV} \tag{1.5b}$$

$$D + Li^6 \rightarrow 2\,He^4 + 22.4 \text{ MeV} \tag{1.5c}$$

Research in nuclear fusion is in progress in all the technologically advanced nations. Although no research group has as yet come close to achieving the conditions mentioned, significant advances have been made in the last few years. In this context, it is worth mentioning that a number of countries have come together to launch a joint programme for building and operating a fusion power reactor. The joint programme is called the *International Thermonuclear Energy Reactor* (ITER) Project [53]. The participating countries are the European Union, Japan, China, India, Korea, Russia and USA. The aim of ITER is to establish the scientific and technological feasibility of building a controlled fusion power reactor which produces many times more power from fusion reactions than the power consumed in heating the plasma. Further, it is planned to gain the necessary experience for eventually designing, constructing and operating the first electricity-producing plant based on nuclear fusion. ITER will be located in Cadarache in France. The engineering design of the reactor was completed in 2001 and construction of buildings began in 2010.

1.4.5 Miscellaneous Sources

Under the category of 'Miscellaneous Sources', we take up for consideration four sources – tar sands, oil shale, unconventional hydrocarbons and geothermal energy. The first three in this listing are essentially fossil fuels mixed with some other naturally occurring products.

Tar Sands and Oil Shale

Tar sands is an expression used to describe porous sandstone deposits impregnated with heavy viscous oils called bitumen or simply deposits of heavy oils. There are no major technical problems involved in the production of synthetic crude oil from these sands. Some production has been going on for years and commercial production on a large scale may begin as supplies of naturally occurring crude oil dwindle.

A schematic diagram indicating the processes involved in producing synthetic crude oil from tar sands made up of sandstone deposits containing bitumen is shown in Fig. 1.21. The sands obtained from surface mining are first passed through a conditioning drum where water, steam and caustic soda are added and a slurry is formed. The slurry passes into a separation tank where the coarse sand settles at the bottom and a froth of bitumen, water and fine mineral matter forms on the top. The froth is diluted with naphtha and subjected to centrifugal action. As a result, fine mineral matter and water is removed. After this, the naphtha is recovered and recycled, and the bitumen obtained is subjected to hydroprocessing and desulphurisation to produce synthetic crude oil.

Deposits of tar sands are found in seventy countries. The resources are estimated to contain 5920 billion barrels of synthetic oil, with more than 80 per cent being located in Canada, Venezuela and U.S.A. The reserves are estimated to be 1085 billion barrels, with 434.3 billion barrels of heavy oil and 650.7 billion barrels of bitumen [54]. Though this value is comparable to the currently estimated crude oil reserves of 1653 billion barrels, it should be noted that there are many issues involved in handling tar sands. The hydrocarbons present are denser than conventional crude oil and are more viscous. Consequently, they are difficult to transport and refine. They also require a large amount of energy for the production of synthetic crude and their productivity rate is less than that obtained with conventional crude. Plants producing significant amounts of synthetic crude oil have been in operation in Canada and Venezuela for some time. In 2014, the annual output was 841 million barrels in Canada and 233.6 million barrels in Venezuela [54]. Unfortunately, no significant deposits of tar sands have been located in India.

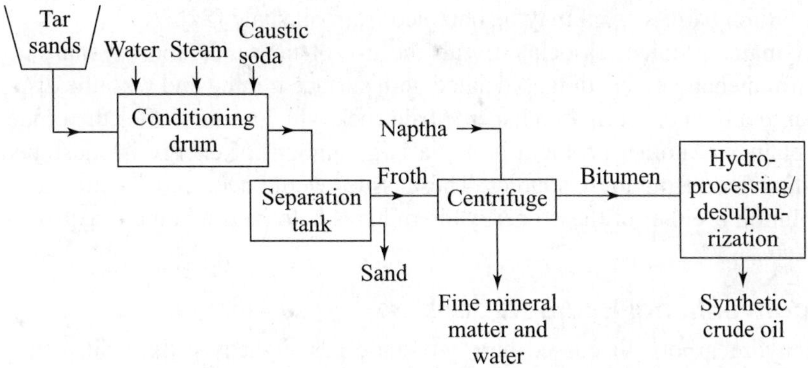

Fig. 1.21 *Production of synthetic crude oil from tar sands*

Oil shale refers to a finely textured rock mixed with a solid organic material called kerogen. When crushed, it can be burnt directly (like coal) and has a heating value ranging from 2000 to 17 000 kJ/kg. It is used in this manner for generating electricity and supplying heat. Alternatively, the oil shale can be converted to oil. This is done by heating crushed oil shale to about 500°C in the absence of air. Under these conditions, pyrolysis occurs and the kerogen is converted to oil. A schematic diagram of the process is shown in Fig. 1.22.

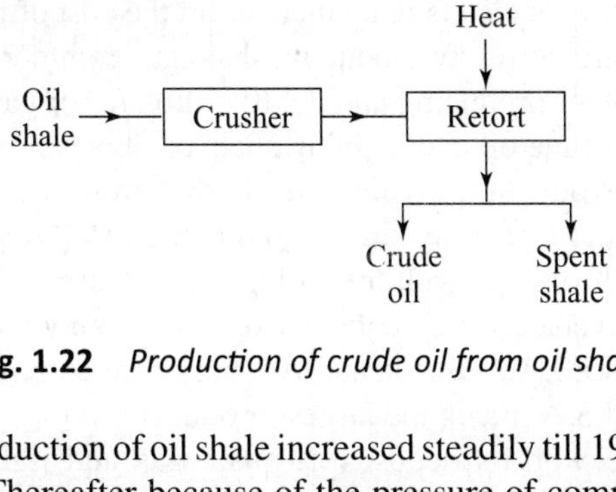

Fig. 1.22 *Production of crude oil from oil shale*

The world's production of oil shale increased steadily till 1980 when it touched a peak of 43 Mt. Thereafter because of the pressure of competition, it declined to 16 Mt in 2000. Since then, the production has been increasing gradually.

In 2014, the production was about 33 Mt [55]. Major producers of oil from oil shale are China, Estonia and Brazil which produced 16000, 14000 and 4000 barrels of oil per day respectively [55]. Though most of the attention is focussed on using oil shale as a source of oil, about 66 per cent of it is used like coal for power generation and heating, primarily in Estonia [56].

The world's deposits of oil shale are very large and are well distributed worldwide in about forty countries. It is estimated that the resources are around 7500 billion barrels of oil, more than half being in U.S.A. Significant deposits have been located in Russia, China, Israel, Estonia, Jordan and the Democratic Republic of Congo [55]. India has significant resources of oil shale in the northeast. Various seismic survey works are in progress indicating that at least 100 billion barrels of oil may be obtained from oil shale [57].

A major problem associated with the use of tar sands and oil shale is the environmental degradation associated with surface mining and with the disposal of large amounts of sand and spent shale rock which remain after the crude oil is obtained. Another problem is that a large amount of energy is consumed in producing oil from these sources. These issues would need careful attention and resolution if either of these energy alternatives is to be used on a larger scale in the future.

Unconventional Hydrocarbons

The hydrocarbons discussed here are shale gas, tight gas, tight oil, coal bed methane and gas hydrates.

Shale gas refers to natural gas trapped underground in organic-rich shales, while *tight gas* is natural gas trapped in sandstone or limestone rock formations. Unlike natural gas reservoirs, shale and tight gas reservoirs have very low permeability and low porosity. Thus the reservoirs have to be mechanically stimulated to increase the permeability and free the natural gas for collection. This is done by drilling wells and pumping water at high pressure underground to break the rock in which the gas is trapped. The process is called hydraulic fracturing. Its effectiveness is enhanced by mixing the water with sand and certain chemicals. Sometimes horizontal drilling is also done to enhance the productivity. Needless to say, the whole process results in an increase in the cost of production. U.S.A., Canada and China are currently producing shale gas commercially. U.S.A. is by far the largest producer producing about 440 billion m^3 of gas from shales [58].

Tight oil is light crude oil contained in shale or tight sandstone formations of low permeability. Production involves separation from the shale or sandstone and requires hydraulic fracturing similar to that used for the production of shale gas and tight gas. Horizontal well technology is also used. Tight oil should not be confused with oil shale which contains kerogen from which crude oil can be made by pyrolysis. U.S.A., Canada and Argentina are currently producing tight oil commercially, U.S.A. being the largest producer.

Estimates of the world's deposits of shale gas and tight gas/oil are very complex due to the heterogeneous structure of the reservoirs. The latest estimates say that the recoverable reserves of shale gas from 46 countries in the world are 214.5 trillion m^3 and of tight oil are 418.9 billion barrels [59]. It is expected that the production in some countries will pick up as the experience gained in North America is transferred to other regions.

In India, several shale formations exist in different basins such as Cambay, Gondawana, Krishna-Godavari and Cauvery. ONGC has commenced drilling of one well in the Cambay basin. Currently, there is no commercial production. The estimates are that India has 2728 billion m^3 of shale gas and 3.8 billion barrels of tight oil in different basins [59].

Coal bed methane (CBM) refers to natural gas adsorbed onto the matrix of coal in coal seams. Commercial production of CBM began in U.S.A. in the late 1980s. With technological advancements, commercial exploitation has been established. Other countries, viz. Canada, Australia, China, Russia, Indonesia and India have also taken up the production of CBM.

The world's CBM resources are estimated to be 235 trillion m^3 [60]. India's resources are estimated to be 2600 billion m^3 with recoverable sources being 280.3 billion m^3 [61]. India's production rate is only about 1.1 million m^3 per day [61]. However, efforts have been initiated for more exploration and production.

Gas hydrates (or methane hydrates) are solid ice-like forms of water containing methane molecules held within a lattice structure. They are stable on the sea floor at water depths around 500 m under certain conditions of pressure and temperature. If the conditions are altered, the hydrates quickly dissociate into water and methane gas. 1 m^3 of hydrates contains about 164 m^3 of methane gas under standard conditions.

Gas hydrates exist in permafrost Arctic regions as well as on the ocean floor. Estimates of gas hydrates vary widely from 283 000 billion m^3 to 2 830 000 billion m^3 of natural gas [62]. Tapping such sources is very complex and requires many technological developments. Though research and developmental work is in progress in a few countries, commercial production will take a significant amount of time. As far as India is concerned, it has initiated a National Gas Hydrate Programme and signed memoranda of understanding with various foreign agencies for sharing knowledge and scientific data. Initial exploration works are in progress and the results are encouraging [61].

Geothermal Energy

Geothermal energy is energy coming out of the molten interior of the earth towards the surface. The average rate at which this heat emerges is about 0.05 W/m^2, while the radial temperature gradient that causes this heat flow is about 0.03°C per metre. Thus, on an average, the temperature of the earth increases by 30°C per kilometre as one moves inwards.

Because of non-homogeneities in the earth's crust, there are numerous local hot spots just below the surface where the temperature is in fact much higher than the average value expected. Ground water comes into contact with the hot rocks in some of these locations and as a result, dry steam, wet steam and hot water or hot water alone are formed. A well drilled to these locations causes the steam/water to emerge at the surface where its energy can be utilised either for generating electricity or for space heating.

The first geothermal power plant was erected at Larderello in Italy in 1904. Over the years, more plants have been installed in the region, raising the installed capacity in Italy in 2015 to 916 MW. Other countries having high installed capacities are the USA (3450 MW) Philippines (1870 MW), Indonesia (1340 MW), Mexico (1017 MW) and New Zealand (1005 MW). For the world as a whole, the total installed capacity in 2015 was 12 636 MW [63]. In addition, a fair amount of geothermal energy is directly used as heat. 587 786 TJ of energy was used in this manner in the world during 2015 [64].

The technical potential of geothermal resources suitable for electricity generation is 210 GWe and for use in the form of heat is about 140×10^6 TJ/year [8]. Although this is a sizeable amount, it must be remembered that it is available only in certain regions of the world and can at best be expected to meet the energy needs of these regions.

The geothermal resource, which is really large, is the heat stored locally at depths of 1 or 2 km in the earth's mantle in hot dry rocks with which water has not come into contact. Utilisation of these sources requires development of techniques for artificial fracture of the rocks, the injection of water into the fractures and subsequent recovery of the steam for driving turbines.

In India, geothermal sources in the form of steam and hot water exist along the west coast, in Ladakh, and in parts of Himachal Pradesh. However, they do not seem to be suitable for generating electricity and are being used only for tapping small amounts of heat.

1.4.6 Hydrogen Energy

We will conclude this chapter with a discussion on hydrogen energy, which is being considered widely as a secondary energy alternative. As an energy source, hydrogen has a number of advantages [65]. It is a non-toxic, clean fuel and has a high specific energy on a mass basis. For example, the energy content of 9.5 kg of hydrogen is equivalent to that of 25 kg of gasoline. In addition, hydrogen can be stored for long periods of time and can be easily transported. The disadvantages are that it is highly inflamable when mixed with air, causing safety problems and it cannot be stored easily in liquid form due to requirements of a very low temperature.

Unfortunately, hydrogen is not available by itself in nature, but in compound form, e.g. water. Thus it has to be produced by some process, which requires an input of energy. Currently, almost all the world's hydrogen is produced by steam reforming of natural gas. At a high temperature between 700 and 1100°C, steam reacts with the methane in the natural gas to yield syngas, a mixture of carbon monoxide and hydrogen

$$CH_4 + H_2O \rightarrow CO + 3H_2 \qquad (1.4.9)$$

The heat energy required for the process is supplied by burning a part of the natural gas. It is worth noting that this method for hydrogen production is not sustainable, because the energy available from hydrogen produced is less than the energy put in for producing it [66]. For this reason, hydrogen is sometimes referred to as an energy carrier rather than an energy source.

Apart from the steam reforming process for methane, it is possible to produce hydrogen from other hydrocarbons and from water by many other techniques. These are listed in the block diagram in Fig. 1.23. The energy required for these processes may come from any of the existing primary sources or the alternatives suggested. Efforts are also in progress for the production of hydrogen through thermo-chemical, photo-chemical, photo-electrochemical and biological routes. It is to be noted that the production process would be sustainable only if it is based on the use of a renewable energy source.

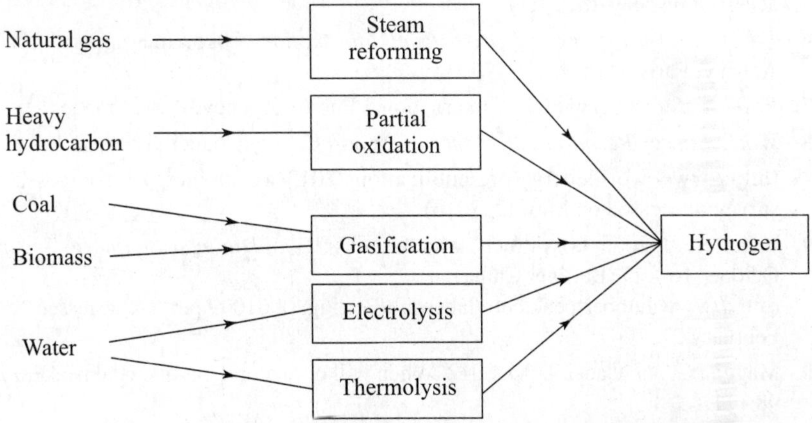

Fig. 1.23 *Various routes for production of hydrogen*

Hydrogen can be stored in gaseous form in pressure vessels, in liquid form in cryogenic containers or in the form of metallic hydrides [67]. There are a number of ways in which hydrogen can be used. It can be used to generate electricity through fuel cells, engine-generator sets or gas-fired steam turbines. It can also be used in internal combustion engines to obtain work. Indeed, hydrogen may well be the fuel of the future for transportation purposes!

REFERENCES

1. *Climate Change 2007: The Physical Science Basis.* Contribution of Working Group I to the 4th Assessment Report, Intergovernmental Panel on Climate Change, Solomon, S., Qin, D., Manning, M., Chen, Z., Marquis, M., Averyt, K.B., Tignor, M. and Miller, H.L. (eds), Cambridge University Press, Cambridge, U.K.

2. *Climate Change 2013: The Physical Science Basis.* Contribution of Working Group I to the 5th Assessment Report, Intergovernmental Panel on Climate Change, Stocker, T.F., Qin, D., Plattner, G.K., Tignor, M., Allen, S.K., Boschung, J., Nauels, A., Xia, Y., Bex, V. and Midgley, P.M. (eds), Cambridge University Press, Cambridge, U.K.

3. Karl, T.R., Arguez, A., Huang, B., Lawrimore, J.H., McMahon, J.R., Menne, M.J., Peterson, T.C., Vose, R.S. and Zhang, H. 2015. Possible artifacts of data biases in the recent global surface warming hiatus. *Science,* 348:1469.

4. Hubbert, M.K. 1974. *U.S. energy resources: a review as of 1972.* U.S. 93rd Congress, 2nd Session, Senate Committee on Interior and Insular Affairs, Ser 93-40 (92-75).

5. *International Energy Statistics.* Energy Information Administration, Department of Energy, Washington, U.S.A.
(http://www.eia.gov/cfapps/ipdbproject/IEDIndex3.cfm?tid=l&pid=7&aid=1; accessed on May 28, 2016)

6. *Key Coal Trends: excerpt from Coal Information.* 2015. International Energy Agency, Paris, France.

7. *Resources to Reserves 2013.* International Energy Agency, Paris, France.

8. *World Energy Resources 2013 Survey.* World Energy Council, London, U.K.
(http://www.worldenergy.org/publications/2013/world-energy-resources-2013-survey/; accessed on May 17, 2016)

9. Laherrere, J. 2006. Fossil fuels: what future? *Global Dialogue on Energy Security,* October 16 – 17, Beijing, China.
(http://www.hubbertpeak.com/laherrere/Beijing20061009.pdf; accessed on February 27, 2007)

10. Maggis, G. and Cacciola, G. 2012. When will oil, natural gas and coal peak?. *Fuel,* 98:111.

11. *International Energy Statistics.* Energy Information Administration, Department of Energy, Washington, U.S.A.
(http://www.eia.gov/beta/international; accessed on May 17, 2016)

12. Considine, D.M. (ed.) 1977. *Energy Technology Handbook,* McGraw Hill, New York.

13. *International Petroleum Encyclopedia.* 1981. Penn Well Publishing Co., Tulsa, U.S.A.

14. Darmstadter, J. 1970. *Energy in the World Economy.* Johns Hopkins, Baltimore, U.S.A.

15. *Encyclopedia of Energy.* 1973. McGraw Hill, New York, U.S.A.

16. *World Energy Supplies.* 1973. United Nations J. Series, New York, U.S.A.

17. *2015 Key Trends in Hydropower.* 2015. International Hydropower Association, London, U.K.

18. *Hydropower Status Report.* 2016. International Hydropower Association, London, U.K.

19. *Survey of Energy Resources.* 1974. US National Committee of the World Energy Conference, 345E, 47th Street, New York, U.S.A. p – 156.

20. *World Energy Resources – Charting the Upsurge in Hydropower Development.* 2015. World Energy Council, London, U.K.

21. *Energy Electricity and Nuclear Power Estimates for the Period upto 2050.* 2015. Reference Data Series No. 1, International Atomic Energy, Vienna, Austria.

22. *Uranium 2014: Resources, Production and Demand.* 2014. NEA No. 7209, Nuclear Energy Agency, Organisation for Economic Co-operation and Development, Paris, France.

23. *Supply of Uranium.* 2015. World Nuclear Association, London, U.K.

24. *Global Wind Report.* 2015. Global Wind Energy Council, Brussels, Belgium.

25. *World Wind Resource Assessment Report.* 2014. World Wind Energy Association, Bonn, Germany.

26. *Key World Energy Statistics 2015.* International Energy Agency, Paris, France.

27. *Geothermal Energy.* REVE, Madrid, Spain. (http://www.evwund.es/2014/06/08/530-mw-of-new-geothermal-energy-capacity-came-on-line-in-2013-bringing-total-global-capacity-to-12-gw/45783; accessed on June 07, 2016)

28. *Annual Reports.* Ministry of Coal and Mines, Government of India, New Delhi, India.

29. *Production & Supplies.* Ministry of Coal and Mines, Government of India, New Delhi, India. (http://www.coal.nic.in/content/production-supplies; accessed on May 20, 2016)

30. *Export and Import.* Ministry of Coal and Mines, Government of India, New Delhi, India. (http://www.coal.nic.in/content/export-import-coal/Exportimport0.pdf; accessed on June 06, 2016)

31. *Annual Reports.* Ministry of Petroleum and Natural Gas, Government of India, New Delhi, India.

32. *LNG: Open Government Data Platform India.* Ministry of Petroleum and Natural Gas, Government of India, New Delhi, India. (http://data.gov.in/keywords/lng; accessed on May 27, 2016)

33. *India 1991.* Research and Reference Division, Ministry of Information and Broadcasting, Government of India, New Delhi, India.

34. *Natural Gas Scenario in India.* Ministry of Petroleum and Natural Gas, Government of India, New Delhi, India. (http://www.petroleum.nic.in/docs/abtng.pdf; accessed on May 21, 2016)

35. *Review of Performance of Hydro Power Stations 2014-15.* 2015. Central Electricity Authority, Ministry of Power, Government of India, New Delhi, India.

36. *Annual Reports.* Department of Atomic Energy, Government of India, New Delhi, India.

37. *Plants under Operation.* Nuclear Power Corporation of India Limited, Mumbai, India, (http://www.npcil.nic.in; accessed on May 21, 2016)

38. *Annual report 2015-16.* Ministry of New and Renewable Energy, Government of India. New Delhi, India.

39. *Wind Energy Programme in India.* Indian Wind Energy Association, New Delhi, India.

(http://www.inwea.org/aboutwindenergy.htm; accessed on May 22, 2016)

40. *Wind Power Programme.* Ministry of New and Renewable Energy, Government of India, New Delhi, India.

(http://www.mnre.gov.in.schemes/grid-connected/solar-thermal-2/; accessed on May 30, 2016)

41. Phadke, A., Bharvirkar, R. and Khangura, J. 2012. Reassessing wind potential estimates for India: economic and policy implications. Report no. LBNL-5077E Revision I, Lawrence Berkeley National laboratory, U.S.A.

(http://hindi.indiawaterportal.org/sites/hindi.indiawaterportal.org/files/India Wind PotentialAssessmentrevisedFinal03202012[1].pdf; accessed on June 09, 2016)

42. Lu, X., McElroy, M.B. and Kiviluoma, J. 2009. Global potential for wind generated electricity. Proc. National Academy of Sciences (PNAS).

(http://www.pnas.org/content/106/27/10933.full; accessed on June 09, 2016)

43. *Monthly Report.* 2016. Central Electricity Authority, Ministry of Power, government of India, New Delhi, India.

(http://www.cea.nic.in/reports/monthly/installedcapacity/2016/installed-capacity-03.pdf; accessed on June 07, 2016)

44. *Growth of Electricity Sector in India from 1947–2015.* Central Electricity Authority, Ministry of Power, Government of India, New Delhi, India.

45. *Annual Report.* 2015. Central Electricity Authority, Ministry of Power, Government of India, New Delhi, India.

46. *Annual Reports 2015–2016.* Ministry of Power, Government of India, New Delhi, India.

47. Sukhatme, S.P. 2012. Can India's future needs of electricity be met by renewable energy sources: A revised assessment. *Current Science,* 103:1153.

48. Sukhatme, S.P. 2014. Distribution functions to describe the variation of the present per capita electricity consumption in India and future needs with respect to the population. *Current Science,* 107:1806.

49. Goldemberg, J., Johansson, T.B., Reddy, A.K.N., and Williams, R.H. 1985. Basic needs and much more with one kilowatt per capita. *Ambio,* 14:190

50. Goldemberg, J., Johansson, T.B., Reddy, A.K.N., and Williams, R.H. 1988. Energy for a sustainable world. Wiley Eastern Ltd., New Delhi, India.

51. *Thorium.* World Nuclear Association, London, U.K.

(http://www.worldnuclear.org/information-library/current-and-future-generation/thorium.aspx; accessed on June 02, 2016)

52. *Annual Report 2015 – 2016.* Department of Atomic Energy, Government of India, New Delhi, India.

53. International Thermonuclear Energy Reactor (ITER) Project.

(http://www.iter.org; accessed on June 03, 2016)

54. Schamel, S., Overland, S. and Ibatullin, R. 2015. Bitumen and heavy oil, Unconventional Energy Sources: 2015 Review. *Natural Resources Research,* 24:443.

55. Brunham, A.K. 2015. Oil shale. Unconventional Energy Sources: 2015 Review. *Natural Resources Research,* 24:443.

56. Brendow, K. 2009. Oil shale – a local asset under global constraint. *Oil Shale,* 26:357

57. *Energy.* The 11th Five Year Plan, vol.3. Planning Commission, Government of India, New Delhi.

(http://www.planningcommission.nic.in/plans/planrel/fiveyr/11th/11_v3/11v3_ch10.pdf; accessed on June 04, 2016)

58. *Shale in the United States.* Energy Information Administration, Department of Energy, Washington, U.S.A.

(http://www.eia.gov/energy_in_brief/article/shale_in_the_united_states. cfm; accessed on June 05, 2016)

59. *World Shale Resource Assessments.* Energy Information Administration, Department of Energy, Washington, U.S.A.

(http://www.eia.gov/analysis/studies/worldshalegas/; accessed on May 23, 2016)

60. Cardott, B.J., Levine, J.R., Pashin, J.C. and Tabet, D.E. 2015. Coal bed methane. Unconventional Energy Sources; 2015 Review. *Natural Resources Reasearch,* 24:443.

61. *Annual Report 2015–2016.* Ministry of Petroleum and Natural Gas, Government of India, New Delhi., India.

62. *Potential of Gas Hydrates is Great, but Practical Development is far off.* Energy Information Administration, Department of Energy, Washington, U.S.A.

(http://www.eia.gov/todayinenergy/detail.cfm?id=8690; accessed on June 05, 2016)

63. *Installed Generating Capacity.* 2016. International Geothermal Association, Bochum, Germany.

64. *Direct Uses.* 2016. International Geothermal Association, Bochum, Germany.

65. Midillia, A., Ay, M., Dineer, I. and Rosen, M.A. 2005. On hydrogen energy strategies I: Current status and needs. *Renewable and Sustainable Energy Reviews,* 9:255.

66. Manish, S., Pillai, I.R. and Banerjee, Rangan, 2006. Sustainability analysis of renewables for climate change mitigation. *Energy for Sustainable Development,* 10:25.

67. Sastri, M.V.C., Viswanathan, B.V. and Srinivasa Murthy, S. 1998. *Metal Hydrides: Fundamentals and Applications,* Narosa Publications, New Delhi, India.

The Solar Energy Option: An Overview of Thermal Applications

CHAPTER 2

INTRODUCTION 2.1

In Chapter 1, the solar energy option has been identified as one of the promising future alternative energy sources. The nature of this source, its magnitude and its characteristics have been described. Also the two ways in which solar energy can be used directly have been mentioned. The first is the thermal approach and the other, the photovoltaic conversion approach. In this chapter, the focus is on the thermal approach. Various devices used for thermal collection and storage will be described in brief. This will be followed by an overview of a number of solar thermal applications in many of which these devices are used. The overview will help the reader to acquire familiarity with solar thermal systems – their configurations, capacities and costs. It will also familiarize the reader with the history of development of the systems.

Photovoltaic conversion will be discussed in Chapter 8, while other renewable energy sources will be considered in Chapter 9.

DEVICES FOR THERMAL COLLECTION AND STORAGE 2.2

2.2.1 Flat-Plate Collectors

In any collection device, the principle usually followed is to expose a dark surface to solar radiation so that the radiation is absorbed. A part of the absorbed radiation is then transferred to a fluid like air or water. When no optical concentration is done, the device in which the collection is achieved is called a *flat-plate collector* (FPC). The flat-plate collector is the most important type of solar collector because it is simple in design, has no moving parts and requires little maintenance. It can be used for a variety of applications in which the required temperature ranges from 40°C to about 100°C.

A sketch of a liquid flat-plate collector is shown in Fig. 2.1. As stated earlier, it consists of an absorber plate on which the solar radiation falls after coming through a transparent cover (usually made of glass). The absorbed radiation is partly transferred to a liquid flowing through tubes which are fixed to the absorber plate or are integral with it. This energy transfer is the useful gain. The remaining part of the radiation absorbed in the absorber plate is lost by convection and re-radiation to the surroundings from the top surface, and by conduction through the back and the edges. The transparent cover helps in reducing the losses by convection and re-radiation, while thermal insulation on the back and the edges helps in reducing the conduction heat loss. The most commonly used liquid is water. A liquid flat-plate collector is usually held tilted in a fixed position on a supporting structure, facing south if located in the northern hemisphere. It is discussed in detail in Chapter 4.

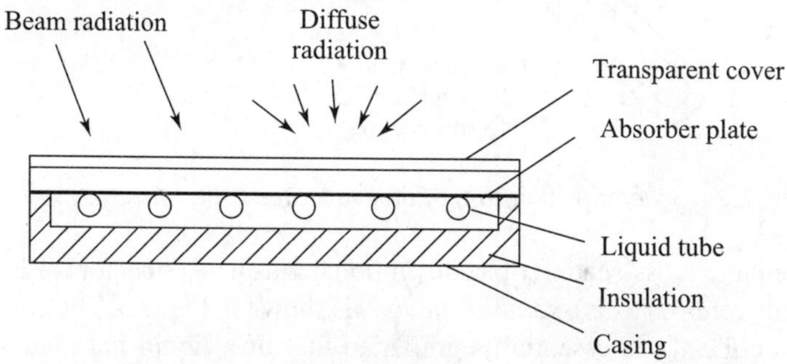

Fig. 2.1 *Liquid flat-plate collector*

In order to reduce the heat lost by re-radiation from the top of the absorber plate of a flat-plate collector, it is usual to put a selective coating on the plate. The selective coating exhibits the characteristic of a high value of absorptivity for incoming solar radiation and a low value of emissivity for out-going re-radiation. As a result, the collection efficiency of the flat-plate collector is improved. Further improvement in the collection efficiency (or in the operating temperature) is obtained by evacuating the space above the absorber plate and leads to the design of an evacuated tube collector.

An *evacuated tube collector* (ETC) consists of a number of cylindrical modules mounted side-by-side on a common frame. Figure 2.2 shows a schematic diagram of one type of evacuated tube module. It consists of two concentric glass tubes, with the annular space between them being evacuated. The outer surface of the inner glass tube is selectively coated. The incoming solar radiation is absorbed on this surface and partly conducted inwards through the tube wall. The inner tube is filled with water, and the heat is transferred to the water by thermosyphon circulation. It is to be noted that the heat loss by convection to the surroundings is reduced significantly due to the vacuum in the annular space. This results in an improvement in the collection efficiency. The evacuated tube collector is discussed further in Chapter 4.

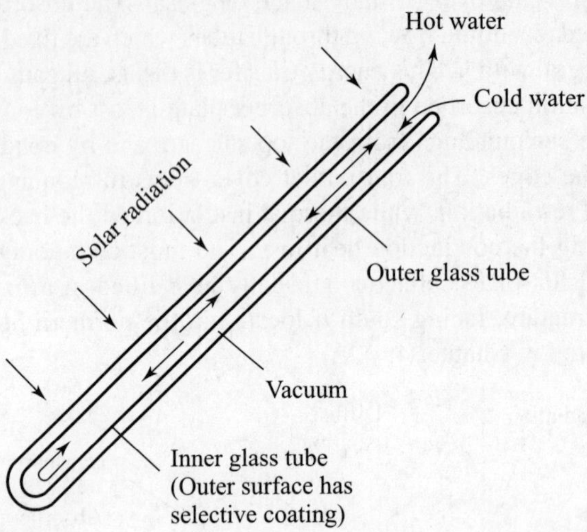

Fig. 2.2 *Schematic diagram of an evacuated tube collector module*

A schematic cross-section of a conventional flat-plate collector for heating air (commonly referred to as a *solar air heater)* is shown in Fig. 2.3. The construction of such a collector is essentially similar to that of a liquid flat-plate collector except for the passages through which the air flows. These passages have to be made larger in order to keep the pressure drop across the collector within manageable limits. In the diagram shown, the air passage is simply a parallel plate duct. The solar air heater is discussed further in Chapter 5.

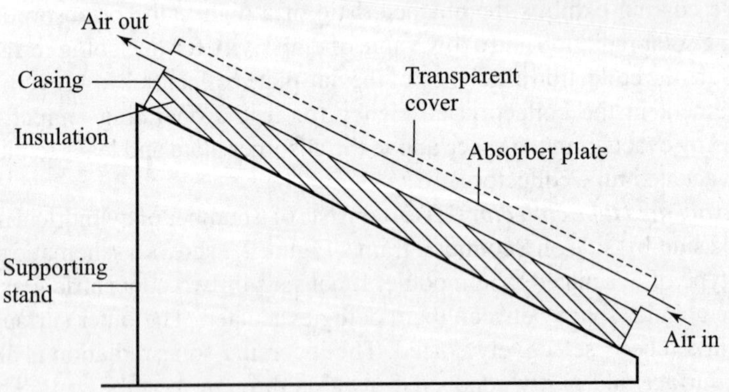

Fig. 2.3 *Solar air heater*

2.2.2 Concentrating Collectors

When higher temperatures are required, it becomes necessary to concentrate the radiation. This is achieved using *focussing* or *concentrating collectors.* A schematic diagram of a typical line-focussing concentrating collector is shown in Fig. 2.4. The collector consists of a concentrator and a receiver. The concentrator shown is a mirror reflector having the shape of a cylindrical parabola. It focusses the sunlight on to its axis, where it is absorbed on the surface of the absorber tube and transferred to the fluid flowing through it. A concentric glass cover around the absorber tube helps in reducing the convective and radiative losses to the surroundings. In order that the sun's rays should always be focussed on to the absorber tube, the concentrator has to be rotated. This movement is called *tracking.* In the case of cylindrical parabolic concentrators, rotation about a single axis is generally required. Fluid temperatures up to 400°C can be achieved in cylindrical parabolic focussing collector systems.

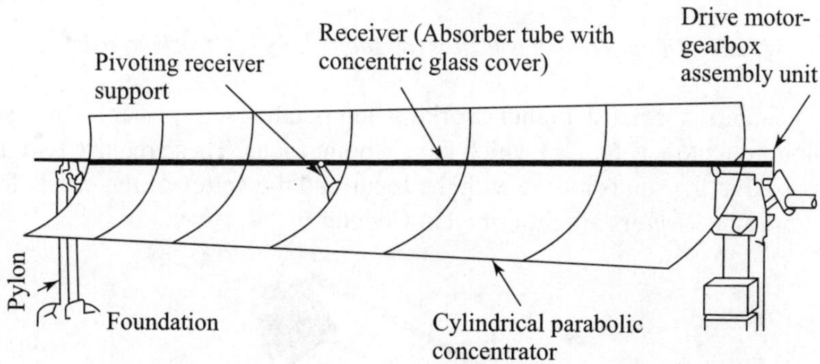

Fig. 2.4 *Cylindrical parabolic concentrating collector*

The other type of line-focussing collector is the *linear Fresnel collector* (LFR) shown in Fig. 2.5. In this collector, the reflector is made up of a field of long mirror strips held side-by-side and oriented usually in a north-south direction in a horizontal plane close to ground level. Each mirror rotates independently through the day about a north-south horizontal axis. It tracks the sun and reflects the sun rays onto a long horizontal receiver which is also oriented in the north-south direction. The receiver is held a few metres above the mirrors. A heat transfer fluid flowing through the absorber tube of the receiver absorbs the heat. Sometimes in the interest of compactness, more than one receiver can be used above the mirror field. The collector is then referred to as a compact linear Fresnel reflector (CLFR).

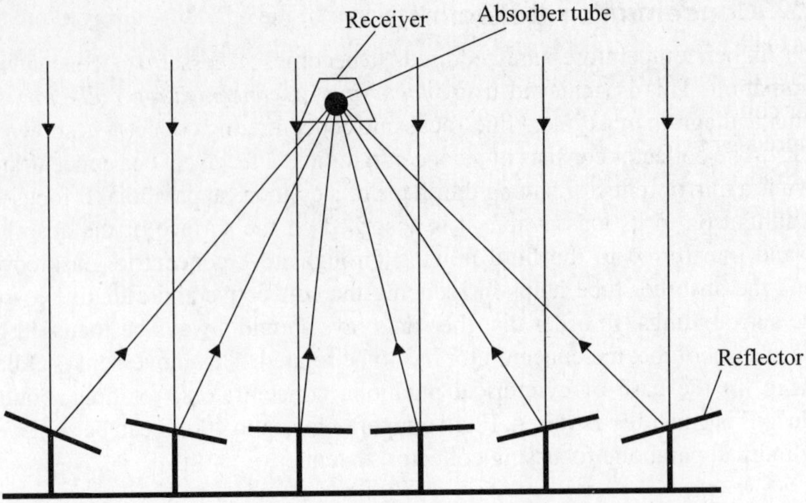

Fig. 2.5 *Linear Fresnel collector (as seen along the tracking axis)*

The generation of still higher working temperatures is possible by using paraboloid reflectors (Fig. 2.6) which have a point focus. These require two-axis tracking so that the sun is in line with the focus and the vertex of the paraboloid. Concentrating collectors are described in Chapter 6.

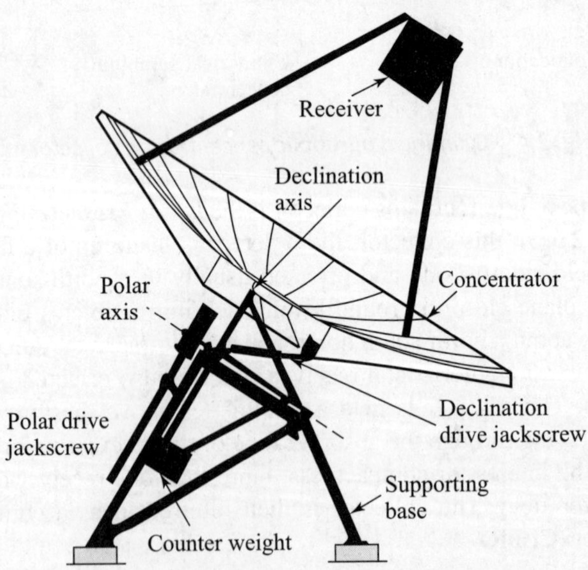

Fig. 2.6 *Paraboloid concentrating collector*

Typical values of efficiency obtained from flat-plate collectors, evacuated tube collectors, line-focussing collectors and paraboloid dish collectors are shown in

Fig. 2.7. The efficiency is plotted as a function of the operating temperature and as is to be expected, it decreases with increasing temperature. The data presented in Fig. 2.7 is useful for doing approximate calculations of thermal systems using solar collectors. Such calculations help in giving estimates of the collector area required for a given performance. Alternatively, they can be used for obtaining a rough idea of the performance of a solar thermal system using a certain number of collectors of a specified type.

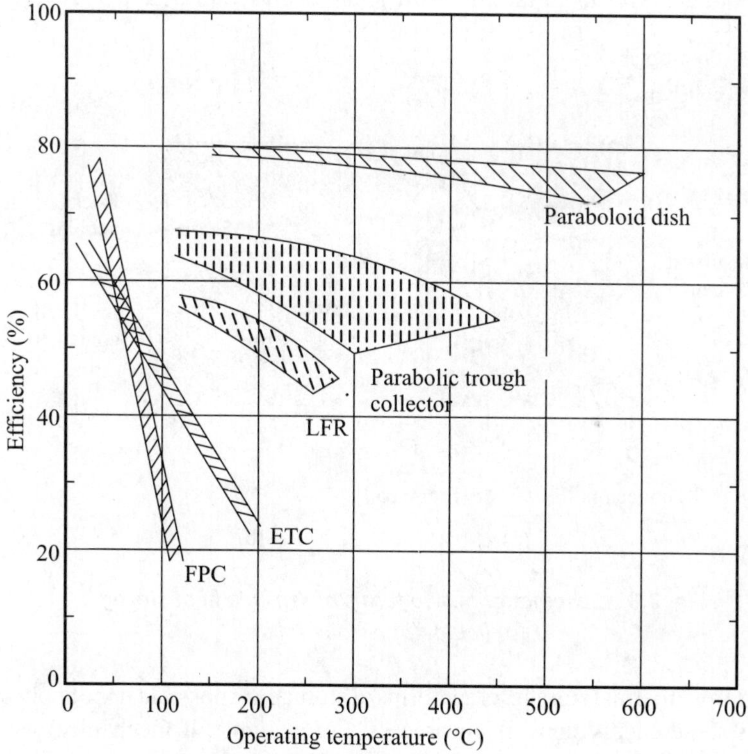

Fig. 2.7 *Approximate efficiency of various types of collectors as a function of operating temperature (Adapted partly from Gehlisch et al. [1] and Rabl [2])*

2.2.3 Energy Storage Systems

As stated earlier, one of the major problems associated with the utilization of solar energy is its variability. For this reason, most applications require some type of *energy storage* system. The purpose of such a system is to store energy when it is in excess of the requirement of an application and to make it available for extraction when the supply of solar energy is absent or inadequate. Energy storage can be in various forms–thermal, electrical, mechanical or chemical. Thermal energy can be stored as sensible heat or as latent heat.

Sensible heat storage is usually done in an insulated container containing a liquid like water or a porous solid in the form of pebbles or rocks (Fig. 2.8). The first type is preferred with liquid collectors, while the second type is compatible with air heaters. In the case of latent heat storage, heat is stored in a substance when it melts and extracted when the substance freezes. Sensible heat storage systems operate over a range of temperatures, while latent heat storage systems operate essentially at the temperature at which the phase change takes place. Both these forms of thermal storage are analysed in Chapter 7.

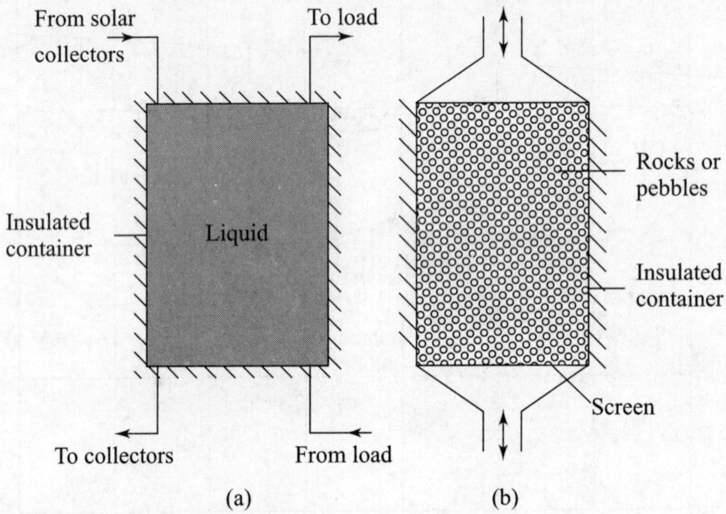

Fig. 2.8 *Schematics of two forms of sensible heat storage: (a) liquid, (b) porous solid*

Apart from thermal storage, other forms of storage are possible if the application involves some energy conversion process. For example, if mechanical work is being obtained after conversion, storage could be done in large-sized flywheels or in compressed air. It has been suggested that compressed air could be stored in large underground chambers. Similarly, if electrical power is being obtained after conversion, storage could be in the form of electric batteries. Many types of electric batteries are used for the purpose, the most common being lead-acid batteries.

2.2.4 Solar Pond

A device which combines the function of both collection and long-term storage is the *solar pond*. It consists of an expanse of water, a few metres deep, in which salts like sodium or magnesium chloride are dissolved. The concentration of the salt is more at the bottom and less at the top. Because of this, the bottom layers of water are denser than the surface layers even though they are hotter and natural convection does not occur. Thus, the heat from the sun's rays absorbed at the bottom of the pond is retained in the lower depths, and the upper layers of water act like a thermal insulator.

A schematic diagram of a solar pond is shown in Fig. 2.9. Typically, it is about 2 or 3 metres deep with a thick durable liner laid at the bottom and on the sides. Materials used for the liner include low density polyethylene, high-density polyethylene, woven polyester yarn, and hypalon reinforced with nylon mesh.

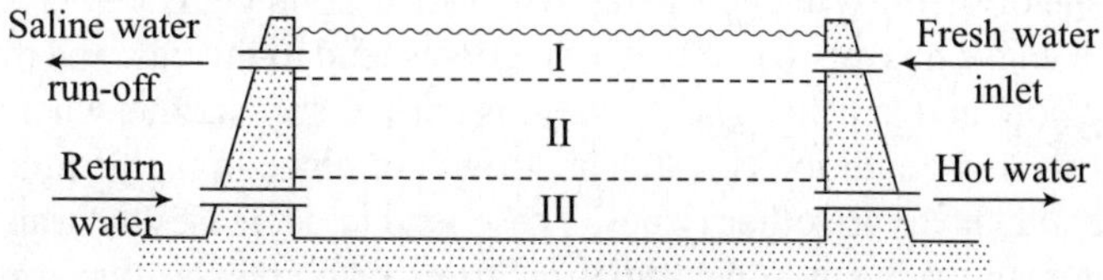

I – Surface convective zone
II – Concentration gradient zone
III – Lower convective zone

Fig. 2.9 *Schematic diagram of a solar pond*

In order to extract the energy stored, hot water is removed continuously from the bottom, passed through a heat exchanger and returned to the bottom. Because of movement and mixing of fluid, both at the top and bottom, the solar pond is characterized by three zones: a surface convective zone, a non-convective concentration gradient zone, and a lower convective zone (Fig. 2.9). The *surface convective zone* usually has a small thickness, around 10 to 20 cm. It has a low, uniform concentration, which is close to zero, as well as a fairly uniform temperature, which is close to the ambient temperature. The *non-convective zone* occupies more than half the depth of the pond. Both concentration and temperature increase with depth in this zone. It serves principally as an insulating layer and reduces heat losses in the upward direction. Some of the heat collection takes place in this zone and it serves also as part of the thermal storage. The *lower convective zone* is comparable in thickness to the non-convective zone. It serves as the main heat-collection as well as thermal storage medium. Typically, the temperature in the lower convective zone of a large pond operating in India might fluctuate cyclically between a maximum value of 85 to 95°C in summer and a minimum of 50 to 60°C in winter.

The first solar ponds were constructed in Israel in the early 1960s. Although the principle of working was demonstrated effectively and temperatures close to 90°C were obtained at the bottom, many practical difficulties were encountered and the work was abandoned. Since then, a number of solar ponds have been built all around the world and the heat energy stored in them has been used both for providing process heat as well as generating power. One of the largest solar ponds built so far was the 250 000 m^2 pond at Beit Ha'aravah in Israel. The heat collected in this pond was used to generate 800 kW of electrical power on a continuous basis using an organic fluid Rankine cycle [3]. Solar ponds were also built in USA and Australia. In India, a few small solar ponds were built and operated for various experimental investigations. The largest pond (about 6000 m^2 surface area) built was located at Bhuj, Gujarat. It was used to supply 20 000 m^3 of hot water at an average temperature of 75°C to a nearby dairy.

In spite of the seeming simplicity of solar ponds, a number of problems have been encountered in connection with their long-term operation and maintenance. Establishing and maintaining the salt concentration gradient from top to bottom has been one of them. Winds blowing over a solar pond generate waves and rain falling on it cause mixing to occur. These effects tend to increase the thickness of the surface convective zone. The problem is particularly serious when the rain is heavy and stormy conditions exist. The growth of algae in the pond and fouling due to dirt and leaves are other issues. These tend to decreases the transmissivity of the water and prevent solar radiation from reaching the two lower zones. In the long term, a problem of major concern is the degradation of the plastic liners. These problems have not lent themselves to relatively easy and economic solutions. For these reasons, there has been a declining interest in solar ponds all over the world.

THERMAL APPLICATIONS 2.3

We will now survey a number of thermal applications. These are

1. water heating
2. space heating
3. space cooling and refrigeration
4. power generation
5. distillation
6. drying, and
7. cooking.

For each application, a variety of systems in use are described and specifications and costs of typical systems available in India are given. A few numerical examples are also solved to show how systems can be approximately sized or alternatively how the performance of installed systems can be estimated.

It may be noted that most of the above applications use one or more of the devices described in Section 2.2. It may also be noted that applications 1, 2, 5, 6 and 7 use the heat energy collected from solar radiation directly, while applications 3 and 4 use the collected heat in thermodynamic cycles to obtain cooling in one case and work in the other. From an economic stand-point, applications which use the collected heat directly are obviously more attractive.

2.3.1 Water Heating

We begin with descriptions of water-heating systems because solar water heating is one of the most attractive solar thermal applications from an economic standpoint. In many countries of the world, it is already competing on equal terms with systems using other energy sources. In India, a large number of water-heating systems have been installed and they have made an encouraging impact on the Indian energy scene. Most of the systems use flat-plate collectors. It is cumulatively estimated that the total area of the flat-plate collectors in the systems installed in India till December 2015 is about 8.90×10^6 m^2. The hot water is used for domestic purposes or for meeting the needs of industries and commercial establishments.

Solar water-heating systems can be classified into two categories:

1. natural circulation (thermosyphon) systems and
2. forced circulation systems.

❑ *Natural Circulation Systems* A simple, small-capacity natural circulation system, suitable for domestic purposes, is shown in Fig. 2.10 and in Plate 2.1. The two main components of the system are the liquid flat-plate collector and the storage tank, the tank being located above the level of the collector. As water in the collector is heated by solar energy, it flows automatically to the top of the water tank and it is replaced by cold water from the bottom of the tank. Hot water for use is withdrawn from the top of the tank. Whenever this is done, cold water automatically enters at the bottom. An auxiliary-heating system is sometimes provided for use on cloudy or rainy days.

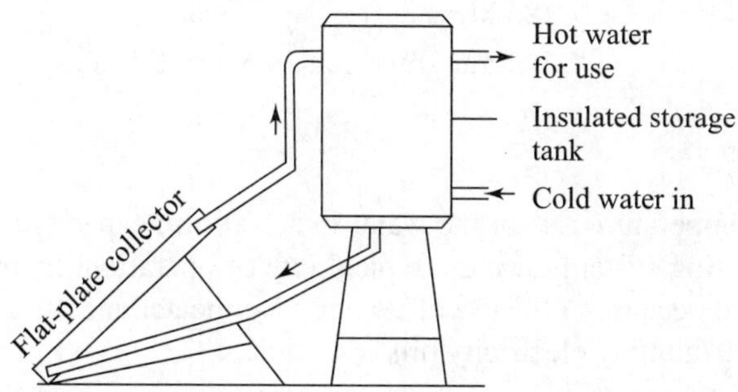

Fig. 2.10 *Small capacity natural circulation water heating system*

Solar water heating is a good example to illustrate one of the assets of the direct use of solar energy which has not been mentioned so far. This is the possibility of matching the temperature achieved in the heating device (in this case the flat-plate collector) with the temperature required for end use. As a result of this matching, the thermodynamic efficiency based on considerations of availability of energy can be shown to be higher in the case of a solar water heating system than a water heating system using natural gas or electricity.

It is of interest to note that solar water heaters of the natural circulation (thermosyphon) type were used fairly widely in many countries from the beginning of the 20th century, till about 1940, until cheap oil and natural gas became available. Now, they are being installed again in large numbers.

In India, the market for domestic solar water heating systems has been growing steadily. Most of the systems have capacities in the range of 100 to 200 litres per day, use one or two flat-plate collectors having a face area of about 2 m^2 each, and deliver hot water in the temperature range of 50 to 70°C. The cost of a domestic water heating system of 125 litres storage capacity was Rs 25 000 in 2015.

The specifications of a typical 125 litre capacity thermosiphon system are given below:

Flat-plate collector

Gross area	:	2.23 m^2
Cover	:	Toughened glass, single, 4 mm thick
Absorber plate	:	Copper sheet (0.20 mm thick) with selective coating
Risers for water flow	:	Copper tubes (12 mm O.D.); 9 numbers
Casing	:	Extruded aluminium section with aluminium sheet on back side
Side insulation	:	25 mm thick glass wool
Bottom insulation	:	50 mm thick glass wool

Storage tank

Shape	:	Cylindrical shell (with axis horizontal) made from SS 304 (1.2 mm thick)
Capacity	:	125 litres
Insulation	:	100 mm thick glass wool on all sides

Example 2.1

Estimate the temperature rise of the water in a 100-litre capacity thermosyphon solar water-heating system during a typical day of operation. Estimate also the electricity saved because of the use of a solar water heater and the corresponding reduction in the monthly electricity bill.

Solution We will perform an approximate calculation based on the following assumptions:

1. The system has one flat-plate collector with an absorber plate area of 2 m^2.
2. The solar radiation falling on the collector during a typical day is 5 kWh/m^2. (This is a good assumption for many locations in India for 250 to 300 days in a year.)
3. The collector efficiency is 50 per cent. From Fig. 2.7, this corresponds to an operating temperature of 70°C.
4. No water is withdrawn for use during the day.

Equating the energy absorbed in the collector during the day to the enthalpy change of the water, we have

Solar radiation incident on the collector per unit area per day × Collector area × Collector efficiency = Mass of water to be heated per day × Specific heat of water × Temperature rise of water

$$(5 \times 3600) \times 2 \times 0.50 = 100 \times 4.2 \times \Delta T$$

where ΔT = temperature rise of water.

Thus ΔT = 43°C.

Plate 2.1 *Natural circulation domestic solar water heating system. (Capacity: 125 litres per day)*

Plate 2.2 *Natural circulation domestic solar water heating system based on evacuated tube collectors. (Courtesy: Regional Test Centre, Devi Ahilya Vishwa Vidyalaya, Indore).*

Plate 2.3 *Forced circulation solar water heating system. Capacity: 6000 litres per day. (Courtesy: Vishwakarma International Centre, VIT, Pune).*

Assuming that the water temperature in the morning was 20°C, the temperature around 4 or 5 p.m. will be 63°C. Thus the value assumed for the efficiency of the collector is justified.

In order to calculate the electrical energy saved, we assume that an electric geyser has an efficiency of 95 per cent. Thus, electricity saved every day

$$= \frac{5 \times 2 \times 0.50}{0.95} = 5.26 \text{ kWh}$$

Taking the cost of electricity to be Rs 3.50 per unit, money saved per day

$$= \text{Rs } 5.26 \times 3.50 = \text{Rs } 18.42$$

Reduction in monthly electricity bill

$$= \text{Rs } 18.42 \times 30 = \text{Rs } 553$$

This is a significant reduction.

A domestic water heating system based on evacuated tube collectors is shown in Plate 2.2. The system consists of a number of evacuated tube modules connected directly to a storage tank. The inside of each evacuated tube module is filled with water, which gets heated up and a thermosyphon circulation is established. Hot water flows out of each module and cold water from the storage tank takes its place. Such domestic water heating systems are becoming popular in India. A commercially available system has 15 modules with a storage tank of 100 litres capacity. Each module made of borosilicate glass is 1.5 m long and the diameters of the outer and inner tubes are 47 and 37 mm, respectively. The cost of such a system presently is about Rs 18 000.

❑ *Forced Circulation Systems* When a large amount of hot water is required for supplying process heat in an industry or in a commercial establishment, a natural circulation system is not suitable. Large arrays of flat-plate collectors are then used and forced circulation is maintained with a water pump (Plate 2.3). The restriction that the storage tank should be at a higher level is thus ruled out. A schematic diagram of a typical closed loop system is shown in Fig. 2.11. Water from a storage tank is pumped through a collector array, where it is heated and then made to flow back into the storage tank. Whenever hot water is withdrawn for use, cold make-up water takes its place because of the ball-float control shown. The pump for maintaining the forced circulation is operated by an on-off controller which senses the difference between the temperature of the water at the exit of the collectors and a suitable location inside the storage tank. The pump is switched 'on' whenever this difference exceeds a certain value and 'off' when it falls below a certain value. Provision is also usually made for an auxiliary heater. In Fig. 2.11, the auxiliary heater is shown as an immersion heater located in the storage tank. Often it is located on the hot water line leading out of the storage tank. Solar water systems of the type shown in Fig. 2.11 are well suited for factories, hospitals, hotels, offices, etc. Typically, these systems cost about Rs 120 per litre/day of hot water.

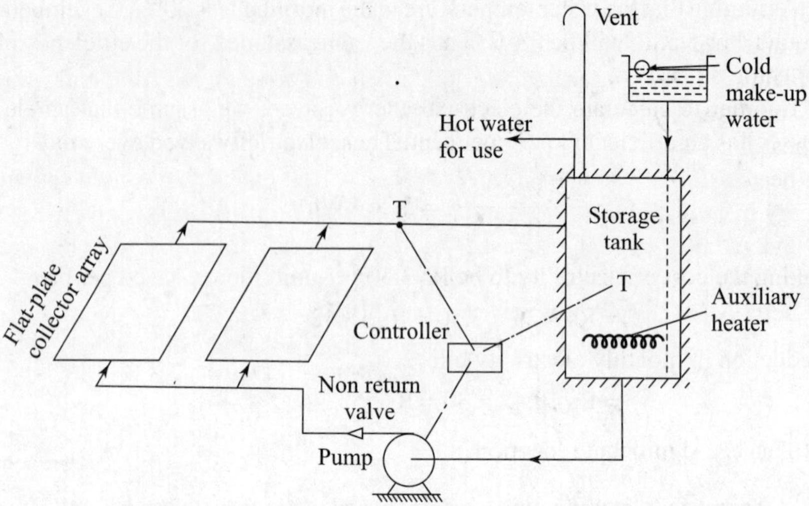

Fig. 2.11 *Forced circulation water heating system – closed loop configuration*

Many variations are possible in the configuration shown in Fig. 2.11. A popular variation is one with an open loop configuration shown in Fig. 2.12. In such a configuration, the water flows from a cold water tank through the collectors to an insulated hot water storage tank from where it is withdrawn for use. The control of such a system is usually designed so that hot water at the required temperature flows out of the collector array, the flow rate being adjusted to meet this temperature requirement. The specifications of a large open-loop system installed at a textile mill in Mumbai are as follows:

System capacity	:	100 000 litres of hot water per day
Average temperature of inlet feed water	:	25°C
Required outlet temperature	:	80°C
Number of flat-plate collectors (each about 2 m^2 in area)	:	1021
Capacity of cold water tanks	:	100 000 litres
Capacity of hot water tanks	:	175 000 litres

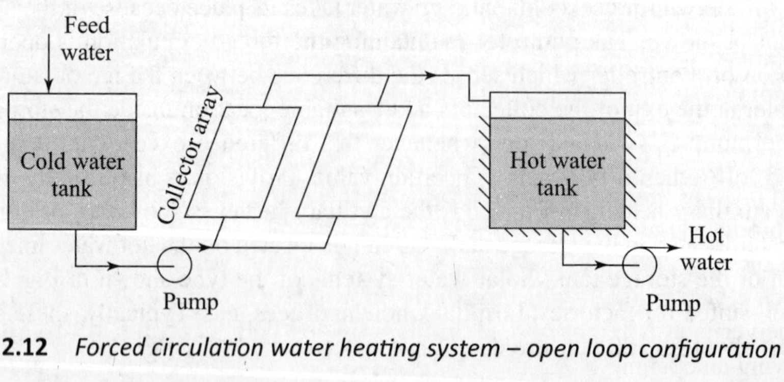

Fig. 2.12 *Forced circulation water heating system – open loop configuration*

It is claimed that this system meets the design specification for 300 days in the year and helps to save about 200 000 litres of oil annually.

Figure 2.13 shows a schematic diagram of one more forced circulation system for supplying process heat. In this case, a concentrating collector array is used to heat pressurized water or a thermic fluid to a temperature above 100°C and the heated liquid is kept in a storage tank. A heat exchanger is used to extract energy from the heated liquid and obtain either hot water or low pressure steam for industrial or commercial use. An auxiliary heater is provided in the line in order to supply make-up energy is case the solar energy is inadequate. A number of such systems using LFRs or paraboloid dishes have been installed in India in recent years. Three of them are described in the following paragraphs.

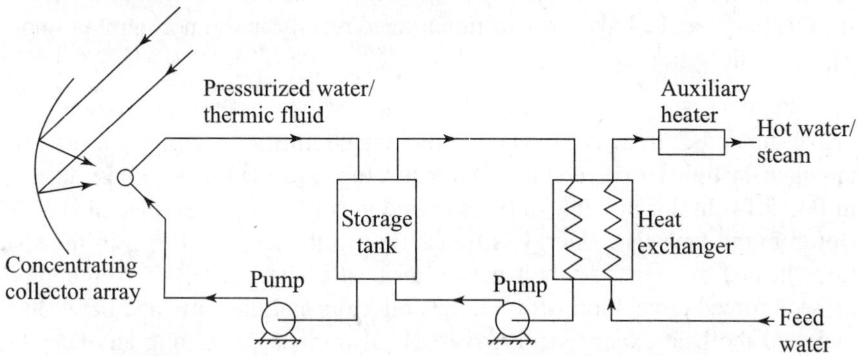

Fig. 2.13 *Solar industrial process heating system using a concentrating collector array*

A solar steam generating system based on LFR collectors was installed in 2013 in a coastal village of Ramanathapuram district in Tamil Nadu for providing saturated steam at 800 kg/h and 21 bar [4,5]. The steam is used as a heat source for a multi-effect distillation system producing 6000 litres/h of desalinated water. The solar system has two LFR collectors having an aperture area of 1404 m^2. Each collector has a linear cavity receiver with multiple absorber tubes and water is used as the working fluid. The solar system does not have any storage. A biomass boiler provides heat during periods of insufficient solar radiation. A photograph of the plant is shown in Plate 2.4.

A demonstration plant based on LFR collectors was set up by M/s Thermax Limited, Pune [6] in June 2015. The plant is used for preheating water fed to the deaerator of a heavy water plant in Manuguru, Telangana. Therma B is used as the heat transfer fluid in the collector. The system is designed to heat oil up to 280°C and provide 250 kWth peak. Currently, the heat is used at a lower temperature to produce hot water at 95°C and 6 bar which is used in the deaerator. There are 4 LFR collectors connected in series, each having an aperture area of 120 m^2. Each collector has 12 reflector-rows, the width of each row being 0.625 m. The receiver is at a height of 4 m and is provided with a secondary CPC reflector having an opening of 0.3 m.

An industrial process heat system using six paraboloid dishes has been installed by M/s Megawatt Solutions in Hyderabad. It provides 30 000 litres of hot water daily at 90°C and 3 bar as process heat for preparing various chemical compounds [7]. The dishes, each having 90 m^2 aperture area are installed in a parallel configuration (Plate 2.5). The unit is in operation since August 2014. Each dish heats up Therminol 55 and delivers the fluid at about 180°C. About 6×10^4 kJ of heat energy are delivered on a typical day by each dish. As a result, the company is able to save around 200 litres of furnace oil per day

2.3.2 Space Heating

We now turn our attention to space-heating systems. Space heating is of particular relevance in colder countries where a significant amount of energy is required for this purpose. In India, it is of importance mainly in the northern and north-eastern regions in winter.

❑ *Active Methods* We will first describe space-heating systems using active methods. An active method is one which utilizes a pump or a blower to circulate the fluids involved in the space-heating system. One system is illustrated in Fig. 2.14. In this system, water is heated in solar flat-plate collectors (A) and stored in the tank (B). Energy is transferred to the air circulating in the space to be heated by means of the water-to-air heat exchanger (E). Two pumps (C) provide forced circulation between the collectors and the tank, and between the tank and the heat exchanger. Provision is also made for adding auxiliary heat (D). Since the solar energy is first being used to heat water, the system shown in Fig. 2.14 can be easily modified to be a two-in-one system supplying hot water as well as hot air for space heating.

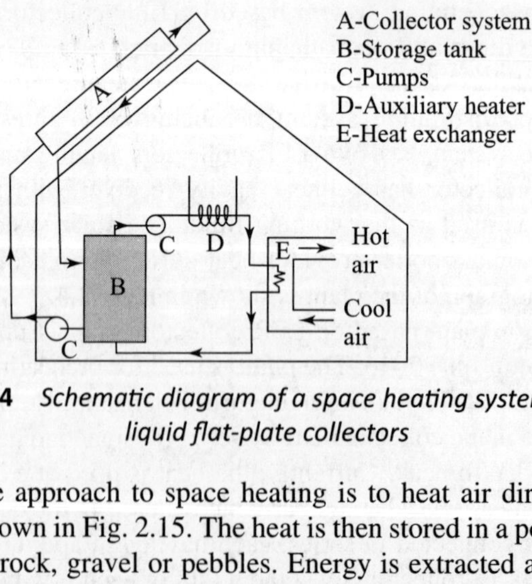

Fig. 2.14 *Schematic diagram of a space heating system using liquid flat-plate collectors*

An alternative approach to space heating is to heat air directly in solar air heaters (*A*), as shown in Fig. 2.15. The heat is then stored in a porous bed storage (*B*) packed with rock, gravel or pebbles. Energy is extracted and transferred to the space to be heated by blowing cool air through the porous bed. Once again an auxiliary heater (*D*) is provided for supplying make-up heat.

Plate 2.4 *LFR steam generating system installed in the coastal district of Ramanathapuram (Courtesy: M/s KGDS Renewable Energy Pvt. Ltd., Coimbatore, Tamil Nadu)*

Plate 2.5 *Process heat system using paraboloid dishes. (Courtesy: Megawatt Solutions, New Delhi)*

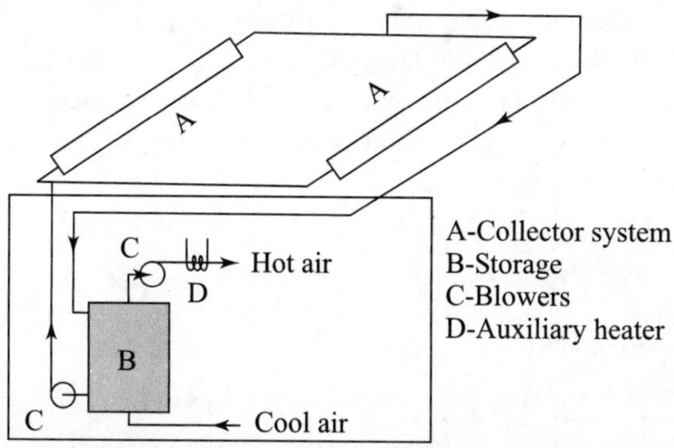

Fig. 2.15 *Schematic diagram of a space heating system using solar air heaters*

Another type of space heating system which is being adopted in cold countries consists of an unglazed transpired collector (UTC) (*A*), fixed in front of a sun-facing wall (*B*), a fan unit (*C*) and an air-distributing duct (*D*) (Fig. 2.16). A UTC (details are given in Chapter 5) is usually a perforated dark-coloured metallic plate and gets heated up by the absorption of solar radiation. Ambient air, drawn through the holes of the UTC by the fan, gets heated as it flows in the space between the UTC and the wall, and is subsequently distributed through the duct to the living space. Over 70 such systems, with collector areas between 500 and 10 000 m^2 have been installed in Canada, USA, Germany and Japan [8].

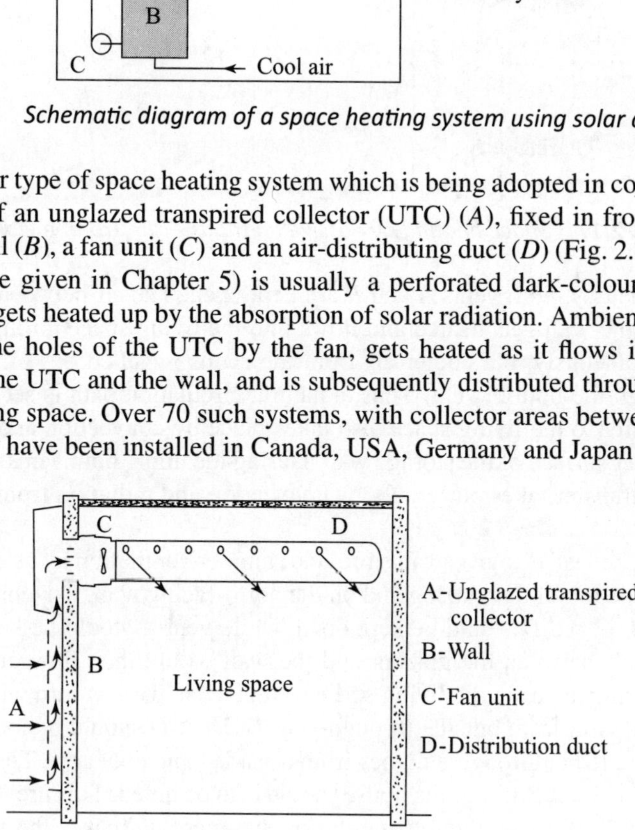

Fig. 2.16 *Schematic diagram of a UTC-based space heating system*

❑ *Passive Methods* In contrast to the active methods, space heating giving a fair degree of comfort can also be done by adopting passive methods. A passive method is one in which thermal energy flows through a living space by natural means without the help of a mechanical device like a pump or a blower.

A schematic diagram of a passive space-heating system designed by Professor Trombe is shown in Fig. 2.17. The south face* of the house to be heated is

*The south face is for locations in the northern hemisphere. The north face would be used in the southern hemisphere.

provided with a single or double glazing. Behind it is a thick, "black", concrete wall, which absorbs the sun's radiation and serves as a thermal storage. Vents (*A* and *B*), which can be kept open or closed, are provided near the top and bottom of the storage wall. The whole unit consisting of the storage wall with vents and the glazing is referred to as a *Trombe wall*.

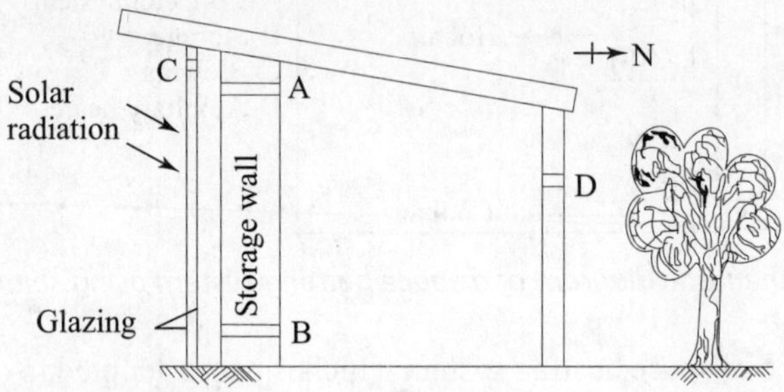

Fig. 2.17 *Space heating by passive methods – the Trombe wall*

During the day, both vents *A* and *B* are kept open. The air between the inner glazing and the wall gets heated and flows into the living space through the top vent. Simultaneously, the cooler air from the room is pulled out of the living space through the bottom vent. Thus, a natural circulation path is set up. Some energy transfer to the living space also takes place by convection and radiation from the inner surface of the storage wall. During the night, both vents are closed and energy transfer takes place only by convection and radiation from the inner surface.

The Trombe wall design can also provide summer ventilation by using vents *C* and *D* near the top of the glazing and on the north-facing wall. On a hot summer day, vents *B, C* and *D* would be kept open, while vent *A* would be kept closed. The heated air between the glazing and the wall would then flow out through vent *C,* drawing air from the living space to replace it. This in turn would cause air to be pulled in from outside through vent *D*. Vent *D* should be located such that the air pulled in through it comes from a shaded and cool area. The presence of an overhang on the roof of the house should also be noted. This prevents direct radiation from falling on the glazing during summer and makes the ventilation more effective.

Another passive design approach for space heating is to construct a sunspace or greenhouse next to the living space to be heated. The sunspace is located on the south side of the building. It has a large glass roof and a thick wall with vents at the top and bottom linking it with the living space (Fig. 2.18). Thus the sunspace acts as a buffer zone between the living space and the outdoor conditions. During the day, the sunspace gets heated up and warm air enters the living space through the top opening *A*. Cooler air from the living space is pulled out through the bottom opening *B* thereby establishing a natural circulation flow. In addition, energy is stored in the link wall. During the night, the openings are closed and the energy stored in the link wall is conducted through it and transferred to the living space by convection and radiation from the inner surface.

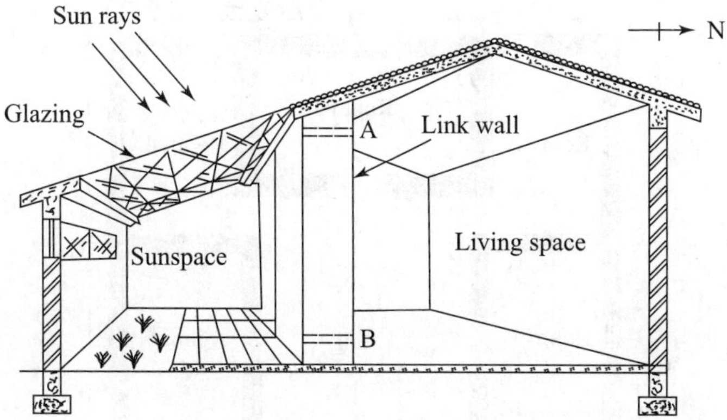

Fig. 2.18 *Schematic diagram of a sunspace for passive space heating*

Many variations are possible in the design described.

1. In order to increase the energy storage capacity in the sunspace room, concrete benches, water drums or rock beds on the floor can be provided.
2. In order to increase the solar radiation entering the sunspace, the side walls may also be provided with glazing.
3. Since the area of the glass in the sunspace is large, the thermal loss during the night can be excessive. This can be reduced by providing a moveable cover on the glass during the night. The moveable cover can also be used for shading the glazing in summer during the day to prevent overheating.

Generally, a building may incorporate more than one passive heating feature. Figure 2.19 shows the plan of one such building at Choglamsar, Leh [9]. The building has an attached greenhouse and also Trombe walls. The floor of the greenhouse consists of solid masonry to provide thermal storage and it is fitted with a moveable internal shade. The performance of the building was monitored. It was seen that in the month of December when the maximum and minimum ambient temperatures at Choglamsar were 2.8° and –9.4°C, respectively, the corresponding temperatures of Room 5 were 21.6° and 6.8°C, and in the hall were 17.0° and 4.6°C respectively. Thus because of the passive heating features, the temperatures inside were about 14° to 18°C higher than the temperatures outside. Clearly, space heating by passive methods integrated into the initial design of a building can be beneficial in the northern parts of India [10,11].

The use of passive techniques for heating, as well as ventilation and cooling, is not new. They were adopted by almost all the ancient civilisations. Excavations at Mohenjodaro and Harappa have shown that the buildings had thick mud brick walls with no openings on the north side so as to keep away the winter winds. Also, many of the building designs were of the courtyard type, which is a good layout for regions having climatic extremes and large temperature variations during the day.

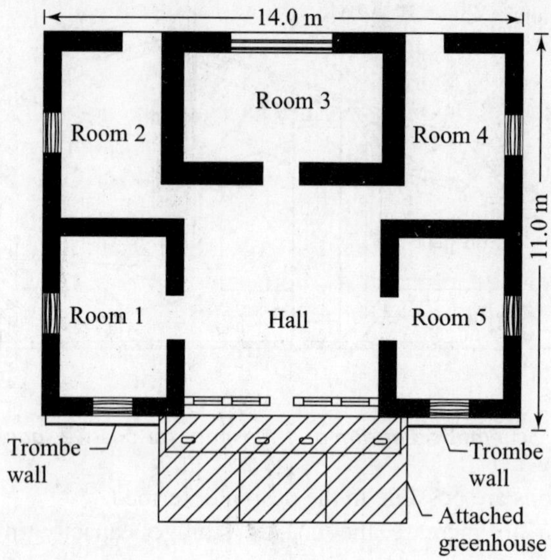

Fig. 2.19 *Plan of S.O.S. dormitory building, Choglamsar, Leh [9]*

The question which may well be asked is why passive methods are not used more extensively. The answer lies in the fact that while a passive method does provide comfort, it does not provide the same degree of comfort as an active heating system. For example, a passive method might result in a daily temperature fluctuation of 10°C in a living space and might not be able to take care of an extreme climatic condition adequately. On the other hand, a properly designed active method would probably result in a smaller variation of 3° or 4°C, and would take care of all situations. For this reason, with increased expectations of comfort, passive methods have been discarded. Now, however, with rising fossil fuel costs, people are again realising the virtues of passive methods and they are being slowly "rediscovered". In many new buildings, hybrid systems using both passive and active methods are being considered. In this way, the size of the active system is considerably reduced.

2.3.3 Space Cooling and Refrigeration

One of the interesting thermal applications of solar energy is for the purpose of cooling. Space cooling may be done with the objective of providing comfortable living conditions (air-conditioning) or of keeping a food product cold (refrigeration). Since the energy of the sun is being received as heat, the obvious choice is a system working on the absorption refrigeration cycle which requires most of its energy input as heat. Cooling is required most in summer. Hence, in this case, there is a seasonal matching between the energy needs of the space cooling system and the availability of solar radiation.

A diagram of a simple solar operated absorption system is shown in Fig. 2.20. Water heated in a flat-plate collector array is passed through a heat exchanger called the generator, where it transfers heat to a solution mixture of the absorbent and refrigerant, which is rich in the refrigerant. Refrigerant vapour is boiled off at a high pressure and goes to the condenser where it is condensed into a high pressure liquid. The high pressure liquid is throttled to a low pressure and temperature in an expansion valve, and passes through the evaporator coil. Here, the refrigerant vapour absorbs heat and cooling is therefore obtained in the space surrounding this coil. The refrigerant vapour is now absorbed into a solution mixture withdrawn from the generator, which is weak in refrigerant concentration. This yields a rich solution which is pumped back to the generator, thereby completing the cycle. The rich solution flowing from the absorber to the generator is usually heated in a heat exchanger by the weak solution withdrawn from the generator. This helps to improve the performance of the cycle. Some of the common refrigerant-absorbent combinations used are ammonia-water and water-lithium bromide, the latter being used essentially for air-conditioning purposes. Typical values for the coefficient of performance (the ratio of the refrigerating effect to the heat supplied in the generator) range between 0.5 and 0.8.

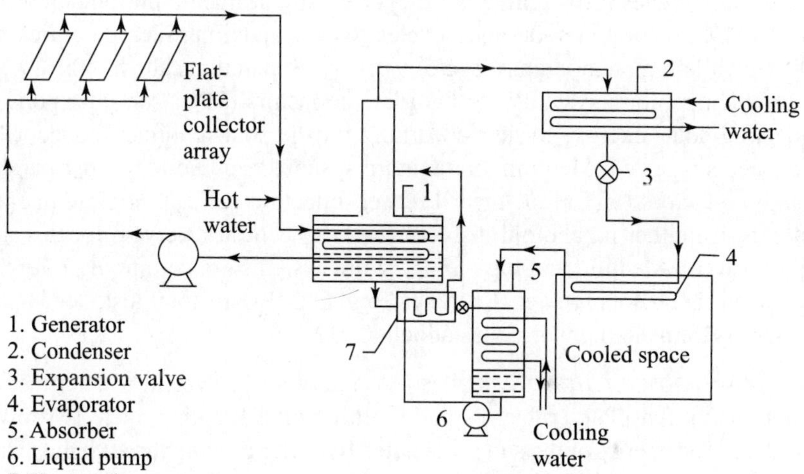

Fig. 2.20 *Solar absorption refrigeration system*

Various types of solar cooling systems have been reported in the literature. Al-Alil *et al.* [12] and Nkwetta and Sandercock [13] have carried out a review of such technologies and presented their features. Unfortunately, the installation cost of a solar absorption refrigeration system is high because of the cost of the large collector array required. Thus commercialization has not taken place although a few demonstrations units have been set up. As an example, the specifications and cost of a solar air-conditioning plant installed near Ahmedabad are given below [14]:

Cooling cycle	:	Water-lithium bromide absorption cycle
Capacity	:	25 tonnes of refrigeration
Solar collectors	:	Heat pipe evacuated tube collectors
Collector area	:	280 m^2
Heat storage tank	:	5000 litres
Area covered for cooling	:	227 m^2
Indoor air temperature	:	18–19°C
Hours of operation per day	:	9
Cost of plant	:	Rs 58 lakh (Solar unit Rs 40 lakh. Air-conditioning unit Rs 18 lakh)

2.3.4 Power Generation

The generation of electrical power is one of the most important applications of an energy source. Here we shall discuss only methods for solar thermal power generation.† Other methods using solar energy for generating electrical power (either directly or indirectly) will be described in Chapters 8 and 9.

Solar thermal power cycles can be classified as low, medium and high temperature cycles. Low temperature cycles work at maximum temperatures of about 100°C, medium temperature cycles work at maximum temperatures up to 400°C, while high temperature cycles work at temperatures above 400°C.

Low temperature systems use flat-plate collectors (FPCs) or solar ponds for collecting solar energy. Systems working on the solar chimney concept have also been suggested. Medium temperature systems use line-focussing parabolic trough collectors (PTCs) or linear Fresnel reflectors (LFRs). High temperature systems use either paraboloid dish collectors or central receivers located at the top of towers. Medium and high temperature systems are commonly referred to as concentrating solar power (CSP) systems. In 2014, the total installed capacity of these systems in the world was about 4000 MW.

❑ *Low Temperature Systems* A diagram of a typical low temperature system using flat-plate collectors and working on a Rankine cycle is shown in Fig. 2.21. The energy of the sun is collected by water flowing through the array of flat-plate collectors. In order to get the maximum possible temperature, booster mirrors which reflect radiation on to the flat-plate collectors are sometimes

---

†From a historical viewpoint, it is of interest to note that the first experiments on the generation of solar thermal power in India were conducted by an Englishman, William Adams, about one hundred years ago. Adams stayed in Colaba, Mumbai and performed his experiments in the compound of his bungalow. He used a spherical reflector 12 m in diameter, made from sheets of glass mirror. The sun's rays were focussed on a boiler having a capacity of about 60 litres and the steam generated was used to drive a 2.5 HP steam pump. Adams' work is described in a book written by him entitled *Solar Heat—A Substitute for Fuel in Tropical Countries for Heating Steam Boilers and Other Purposes* (Education Society's Press, Byculla, Bombay, 1878).

used. The hot water at temperatures close to 100°C is stored in a well-insulated thermal storage tank. From here, it flows through a vapour generator through which the working fluid of the Rankine cycle is also passed. The working fluid has a low boiling point. Consequently, vapour at about 90°C and a pressure of a few atmospheres leaves the vapour generator. This vapour then executes a regular Rankine cycle by flowing through a prime mover, a condenser and a liquid pump. The working fluids normally used are organic fluids like methyl chloride and toluene, and refrigerants like R-11, R-113 and R-114.

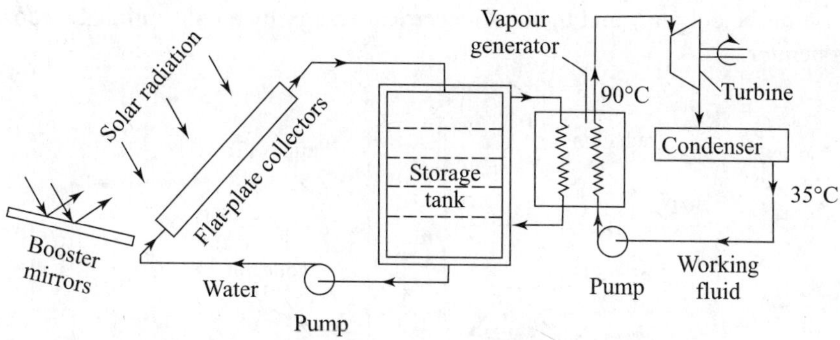

Fig. 2.21 *Low temperature power generation cycle using liquid flat-plate collectors*

It has to be noted that the overall efficiency of this system is rather low, because the temperature difference between the vapour leaving the generator and the condensed liquid leaving the condenser is small. For the cycle, shown in Fig. 2.21, the temperature difference is only 55°C. This leads to a Rankine cycle efficiency of 7 to 8 per cent. The efficiency of the collector system is of the order of 25 per cent. Hence an overall efficiency of only about 2 per cent is obtained.

Plants of this type, of French design, having generation capacities up to about 50 kW were installed in many parts of the world, particularly Africa, in the 1970s. A 10 kW plant was also installed at IIT Madras in 1979–80 under an Indo-German collaboration agreement. However, such plants have been found to be very costly because of the large collector areas involved. Typically, the installed cost was about Rs 300 000 per kW for 6 to 8 hours of daily operation, the main component of cost being the collectors.

In order to reduce the cost, solar ponds have been used instead of flat-plate collectors. The first two solar pond power plants having capacities of 6 kW and 150 kW were constructed in Israel about 40 years ago. These were followed in 1984 by the largest plant in the world at Beit Ha'aravah generating about 800 kW on a continuous basis. The working of these plants established the technical viability of solar pond power plants. However, they were also not found to be economically lucrative in spite of being less costly than plants using flat-plate collector systems.

The concept of a *solar chimney power plant* was suggested in the 1970s. It is also called a solar updraft tower power plant. In such a plant, a tall central chimney is surrounded at its base by a circular greenhouse consisting of a transparent cover supported a few metres above the ground by a metal frame (Fig. 2.22). Sunlight passing through the transparent cover causes the air trapped in the greenhouse to heat up by 10° to 20°C. Thereby a convection system is set up in which the hot air is drawn up through the central chimney, and is continuously replenished by fresh air drawn in at the periphery of the greenhouse. The energy contained in the updraft air is converted into mechanical energy by a turbine located at the base of the chimney and then into electrical energy by a conventional electrical generator.

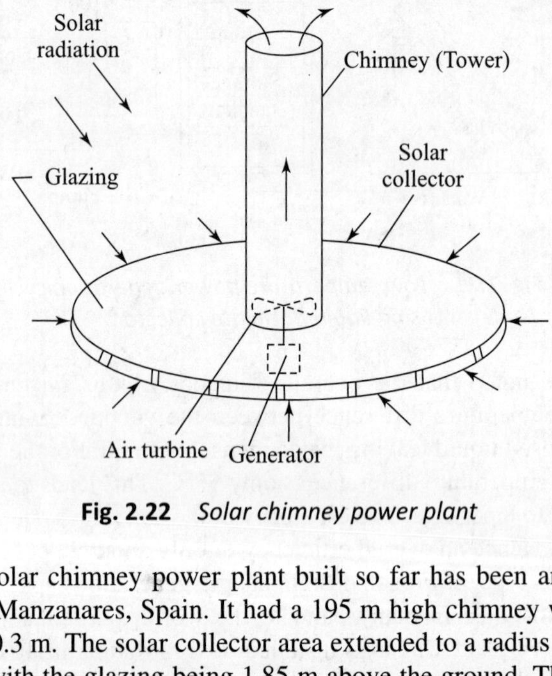

Fig. 2.22 *Solar chimney power plant*

The only solar chimney power plant built so far has been an experimental pilot plant in Manzanares, Spain. It had a 195 m high chimney with a constant diameter of 10.3 m. The solar collector area extended to a radius of 122 m from the chimney with the glazing being 1.85 m above the ground. The total area of the glazing was 46 000 m^2, 40 000 m^2 being plastic membrane and 6000 m^2 being glass. The turbine, housed at the base of the chimney, had four 5 m long blades and rotated at 1500 rpm to produce a peak output of 50 kW. Construction of the plant was completed in 1982 and it was operated up to 1989. The purpose was to demonstrate the operating principle and obtain operational data and experience on a long-term basis. From mid-1986 to early 1989 for a period of 32 months, the plant was run regularly on a daily basis. During this period, it operated fully automatically for an average of 8.9 hours every day and generated 44 MWh annually. The results obtained showed that the plant as a whole was capable of reliable operation over extended periods of time.

The energy conversion efficiency of a solar chimney is inherently low. It can be shown that the maximum possible efficiency is given by the expression [15]:

$$\eta_{max} = \frac{g\,H}{C_p T_a} \qquad (2.1)$$

where H = height of the chimney tower, and
T_a = temperature of the ambient air.

Equation (2.1) displays a basic characteristic of a solar chimney, viz. that the efficiency is directly proportional to the height. The equation is also useful for estimating the overall efficiency of a solar chimney power plant.

Example 2.2

It is proposed to set up a solar chimney power plant in Rajasthan with a chimney 300 m high. Calculate the maximum possible conversion efficiency obtainable with the chimney. Also estimate the efficiency of the plant as a whole and the daily electrical output in a typical summer month (in kWh), if the solar collection area of the greenhouse is 50 000 m^2.

Solution We will use Eq. (2.1). Substituting H = 300 m, C_p = 1005 J/kg-K and T_a = 305 K, we get

$$\eta_{max} = \frac{9.81 \times 300}{1005 \times 305} = 0.00960, \text{ i.e. } 0.96\ \%.$$

Assume:

1. The turbine-generator set converts only 50 per cent out of the maximum available energy into electrical energy, and
2. The collection efficiency of the greenhouse is 25 per cent

Then

$$\eta_{overall} = 0.25 \times 0.00960 \times 0.50 = 0.0012, \text{ i.e. } 0.12\ \%.$$

The solar radiation on a typical summer day in May or June in Rajasthan may be taken as 6.5 kWh/m^2. Therefore, the daily electrical output of the plant

$$= 6.5 \times 50\,000 \times 0.0012 = 390 \text{ kWh}$$

Example 2.2 shows that the overall energy conversion efficiency of a solar chimney power plant is low. Because of this fact, a small-capacity plant (in the kilowatt range) is not economically viable. One needs to build large capacity plants with very tall chimneys and large collector areas at the ground level to achieve economy with scale-up. Plans had been developed for building a 200 MWe plant in Australia. The design concept visualized a chimney 1 km high and a greenhouse collector 7 km in diameter at the base [16]. The plant was not built.

Based on the above review of low temperature thermal systems for generating electricity, it is clear that the energy conversion efficiency is very low and the initial capital cost is too high. This is the case regardless of whether one uses flat-plate collectors, a solar pond or a solar chimney. For these reasons, such systems are not being considered for commercialization any more.

❑ Medium Temperature Systems

Parabolic Trough Collector Plants Among solar thermal-electric power plants, those operating on medium temperature cycles and using the line-focussing parabolic collector technology at a temperature close to 400°C, have proved to be the most cost effective and successful so far. A schematic diagram of a typical plant is shown in Fig. 2.23. An array of long parabolic concentrating collectors is used to collect the incident solar radiation. The axis of the mirrors is usually horizontal and in a north-south direction. During the day, they track the sun by slowly rotating from east to west and thereby focus the solar beam radiation along the focal line. Receiver tubes with a fluid flowing through are located on the focal axis and pick up the heat energy. The fluid is a mineral oil, a synthetic oil or a molten salt and typically, it is heated to a temperature around 400°C. The heat is transferred to water in heat exchangers (preheater, steam generator and superheater) and the resulting high pressure superheated steam is expanded through a turbine and used in a closed Rankine cycle to generate electricity. Solar power plants are usually located in arid areas where water is not readily available. The heat energy in the condenser is, therefore, rejected to the atmosphere through cooling towers.

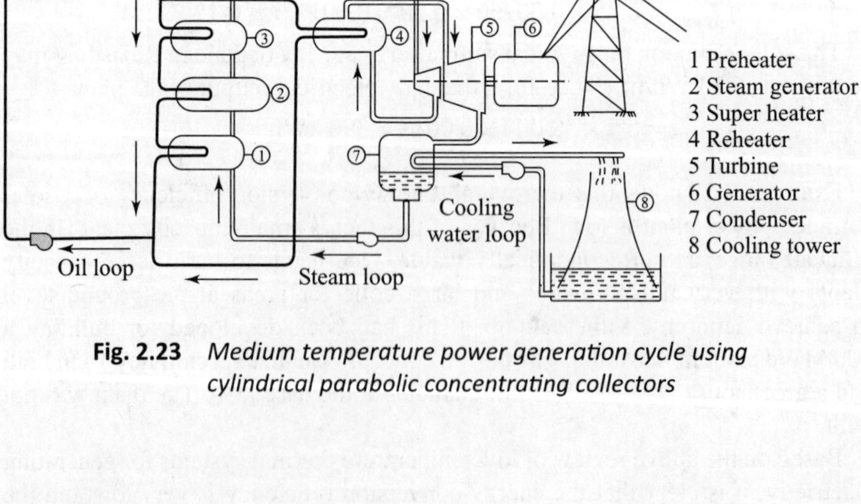

Fig. 2.23 *Medium temperature power generation cycle using cylindrical parabolic concentrating collectors*

One of the advantages which solar thermal power plants enjoy over solar photovoltaic plants is that heat energy can be more easily stored compared to electrical energy. Thus, very often thermal energy storage tanks are used as a buffer between the solar collector array and the steam generation system. The use of a thermal energy storage system helps the plant to generate electricity even when solar radiation is not available. This advantage will gain in importance as renewable energy sources increase their share of the electricity produced in a country.

The first commercial PTC plant (Solar Electric Generating System – SEGS I) having a capacity of 14 MW was set up in 1984. Since then six plants of 30 MW capacity each (SEGS II to VII), followed by two plants of 80 MW each (SEGS VIII and IX), were installed and commissioned. All these plants set up by LUZ International in California had a total installed capacity of 354 MW.

The collector array for SEGS IX has an area of 483 960 m^2. The cylindrical parabolic collectors used have their axes oriented north-south. The absorber tube is made of stainless steel and has a specially developed selective coating. It is surrounded by a glass cover with a vacuum. The collectors heat a synthetic oil (Therminol) to a temperature of 390°C with a collector efficiency of about 0.7 for beam radiation. The synthetic oil is used for generating superheated high pressure steam which executes a Rankine cycle with an efficiency of 38 per cent. The plant has no thermal storage and generally produces electricity for about eight hours a day. It is coupled with natural gas for continuous operation. The installed cost of this type of plant has reduced over the years because of the increasing installed capacity. However it is still very high. SEGS VIII which commenced operation in 1990 is reported to have cost $ 4000 per kW. SEGS I – IX have continued to operate over the years and valuable operating experience has been obtained.

After a break of about 15 years, a very large number of PTC plants have been commissioned from 2008 onwards. At the end of 2015, there were more than 50 operating units having capacities of 50 MW or more. A large number are installed in Spain. Table 2.1 gives details of a few plants. The table includes data on Andasol-1, the first commercial plant in Europe, and Godawari, the first and largest plant in India located in Rajasthan. A view of the 50 MW Godawari CSP plant showing the parabolic collector field is given in Plate 2.6. The table also includes Solana, the largest plant in the world in 2015. It may be noted that the plants Andasol-1 and Solana have thermal storage systems. Thus the heat absorbed in the receiver fluid is used for converting water to high pressure steam as well as heating molten salt stored in large tanks. The stored thermal energy is used to generate electricity when there is no sunlight.

Table 2.1 Solar parabolic trough power plants

Plant name	SEGS IX	Andasol-1	Solana	Solaben 6	Godawari	Genesis
Country	USA	Spain	USA	Spain	India	USA
Output (MWe)	89 Gross 80 Net	50 Gross 49.9 Net	280 Gross 250 Net	50 Gross 50 Net	50 Gross 50 Net	250 Gross 250 Net
Reflector aperture area (m^2)	483 960	510 120	2 200 000	300 000	392 400	1 928 320
Receiver fluid	Therminol	Dowtherm A	Therminol VP-1	Therminol VP-1	Dowtherm A	Therminol VP-1
Outlet temperature (°C)	390	393	393	393	385	393
Thermal storage:						
Fluid	-	Molten salt	Molten salt	-	-	-
Capacity (h)	-	7.5	6	-	-	-
Back-up	Natural gas	-	Natural gas	-	-	-
Year of operation	1990	2008	2013	2013	2013	2014

Example 2.3

Make suitable assumptions and estimate the collector area required for a 80 MWe line-focussing solar thermal power plant producing electricity for 8 hours every day.

Solution We will make the following assumptions:

1. The collector is operating at a temperature of 400°C. At this temperature, the collector efficiency may be taken as 0.6 (refer to Fig. 2.7).
2. The Rankine cycle has an efficiency of 0.36.
3. The electrical generator efficiency is 0.96.
4. The solar insolation during a typical day is 6 kWh/m^2.

An energy balance over a day yields the following:
Solar radiation incident on the collectors per unit area per day × Collector area × Collector efficiency × Rankine cycle efficiency × Generator efficiency = Electricity production × Hours per day

$$6 \times A_c \times 0.6 \times 0.36 \times 0.96 = 80\,000 \times 8$$

$$A_c = 514\,403 \text{ m}^2$$

(Note: This value agrees reasonably well with the area of 483 960 m^2 used in SEGS IX.)

Linear Fresnel Reflector Plants Linear Fresnel reflector plants are of recent origin. One of the first commercial plants, Kimberlina, was commissioned in Bakersfield, California in 2008. It has a capacity of 5 MW and a solar field of 25 988 m^2. The reflectors are in three rows, 385 m long. In this plant, the heat transfer fluid is water which is converted to steam at 40 bar and 300°C. There is no thermal storage.

A plant (PE1) having a capacity of 1.4 MW was connected to the grid in 2009. It is located in southern Spain. The total area of the mirror strips is 18 000 m^2 and they are placed in two rows, 806 m long. The linear receiver is located 7.4 m above the reflector strips. Water flows through the absorber tube located at the focal line of the mirror field. The water is heated from 140°C and converted into saturated steam at 270°C and 55 bar. The steam runs a turbine connected to a power generator. After demonstrating the successful operation of PE1, a second plant PE2 has also been commissioned in Spain in 2012. It has a capacity of 30 MW. This plant has a mirror surface area of 300 000 m^2.

LFR technology was introduced in India when construction of a 125 MW plant was begun in Dhursar, Rajasthan. The estimated cost is Rs 2100 crores. The plant is expected to generate 280 000 MWh per year when it is fully operational.

It should be noted that the efficiencies of current LFR plants are lower than the values obtained in parabolic trough plants. However, they require less land and it is generally expected that LFR plants will be cost competitive compared to other solar power plants.

It may be mentioned that a combination of parabolic trough and LFR collectors can be used as a solar field for power generation [17]. A 1 MWe demonstration power plant using this concept has recently been commissioned by IIT Bombay on the campus of the National Institute of Solar Energy of MNRE at Gwalpahari, Haryana [18,19].

❑ *High Temperature Systems* Two concepts have been experimented with in the case of high temperature systems. These are the paraboloid dish concept and the central receiver concept.

Paraboloid Dish Systems In the paraboloid dish concept, the concentrator tracks the sun by rotating about two axes and the sun's rays are brought to a point focus. A fluid flowing through a receiver at the focus is heated and this heat is used to drive a prime mover. Typically, Stirling engines have been favoured as the prime movers. For this reason, such systems are referred to as *Dish-Stirling Systems*. Because of the limitations on the size of the concentrator, paraboloid dish systems can be expected to generate power in kilowatts rather than megawatts. Thus they can be expected to meet the local power needs of communities, particularly in rural areas.

A few details about four systems developed for commercial applications are given in Table 2.2. It will be seen that the peak outputs range from 8.5 to 25.3 kW, while the peak net efficiencies range from 19 to 29.4 per cent.

Table 2.2 Paraboloid dish systems⁺

System name	Sun Dish	Euro Dish	SES	WGA
Location	USA	Spain	USA	USA
Rated output (kW)	22	10	25	9.5
Peak output (kW)	22.9	8.5	25.3	11.0
Peak net efficiency (per cent)	20	19	29.4	24.5
Area of reflecting glass (m^2)	117.2	60.0	91.0	42.9
Receiver aperture diameter (m)	0.38	0.15	0.20	0.14
Power cycle	Stirling	Stirling	Stirling	Stirling
Working fluid	Hydrogen	Helium	Hydrogen	Hydrogen

⁺Adapted from Mancini *et al.* [20].

Two important issues for commercializing Dish-Stirling systems are cost and reliability. Initial costs of such systems are very high, at around $ 10 000 per kW. Consequently, very few systems have been built. In 2013, the global installed capacity was only about 3 MW.

Central Receiver Power Plants In a central receiver system, solar radiation reflected from an array of large mirrors is concentrated on a receiver situated at the top of a supporting tower. The mirrors are called *heliostats* and they are placed on the ground around the tower. Their orientation is individually controlled so that throughout the day they reflect beam radiation on the receiver. A fluid flowing through the receiver absorbs the concentrated radiation and transports the heat to the ground level where it is used to operate a thermodynamic cycle like the Rankine or the Brayton cycle. Molten salts, water (converted to steam) and air have been used as the heat transfer fluids. Because of the use of a receiver placed at the top of a tower, a central receiver system is also referred to as a power tower.

A schematic diagram of a typical central receiver system using a molten salt as the heat transfer fluid is shown in Fig. 2.24. The molten salt used frequently is a mixture of 60 per cent sodium nitrate and 40 per cent potassium nitrate. Cold salt at 290°C is pumped from a tank at ground level to the receiver at the top of a tower where it is heated by the concentrated radiation to a temperature of 565°C. The salt flows back to another tank at ground level. In order to generate electricity, hot salt is pumped from the hot tank through a steam generator where superheated steam is produced. The superheated steam then goes through a Rankine cycle to produce mechanical work and then electricity. The heliostat array can be sized to collect more power than is required by the electricity generation system. In that case, the excess thermal energy in the form of excess salt at 565°C accumulates in the hot tank and serves as a thermal storage.

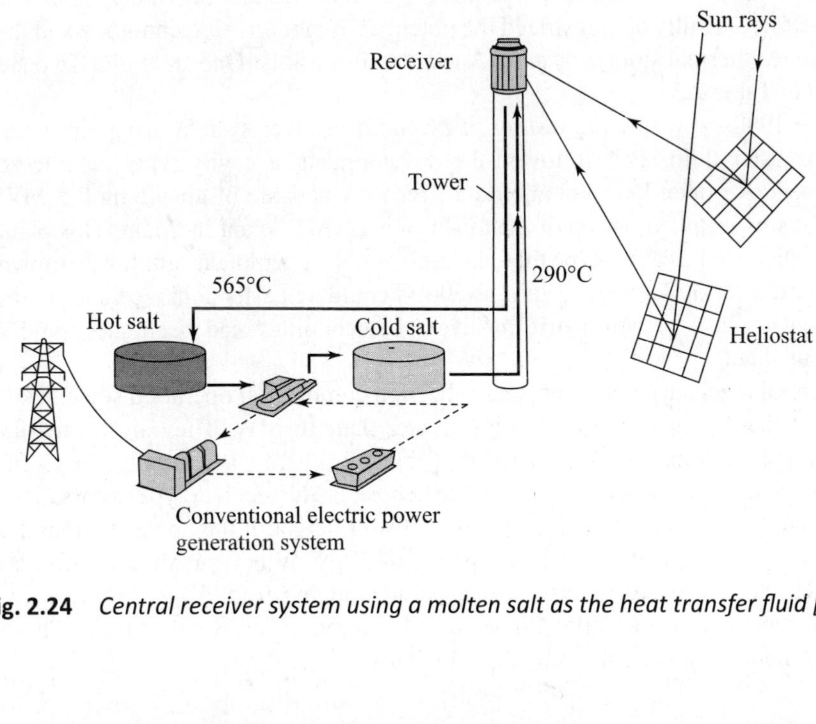

Fig. 2.24 *Central receiver system using a molten salt as the heat transfer fluid [21]*

From 1981 to 1996, a number of central receiver systems were built to demonstrate the viability of the concepts involved. These systems were pilot plants having electrical outputs from 0.5 to 10 MW. The largest plant was Solar One at Barstow, California. Solar One had an external type, tubular receiver with sub-cooled water entering at the inlet and superheated steam at 102 bar and 510°C being delivered at the outlet. Most of the time, the steam was directly fed to a turbine which was part of a Rankine cycle. The plant was operated for six years from 1982 to 1988 and the feasibility of the power tower concept was successfully demonstrated. However, Solar One suffered from some problems. The receiver had a low collection efficiency and the tubes failed frequently because of the thermal stresses which they had to encounter in handling single phase liquid flow, two-phase flow and single phase vapour flow. Also, the thermal storage system (which was an oil-rock system) was of inadequate capacity and operated only in a temperature range from 220° to 305°C, which was too low.

Because of the problems encountered during the operation of Solar One, it was decided to make modifications in some of the components. Molten salt was used as the heat transfer fluid instead of water/steam so that only single-phase liquid flow occurred in the tubes of the receiver. In addition, a new molten salt thermal storage system with a larger capacity was installed instead of the oil-rock storage system and 108 heliostats (each having an area of 95 m^2) were added. The other hardware of Solar One was essentially reused. The modified plant was called Solar Two. It began operation in 1996 and was run for three years. The project successfully demonstrated the potential of molten salt technology and the associated thermal storage system. A few details of Solar One and Solar Two are given in Table 2.3.

The 1990s also saw the testing of a central receiver system using air as the heat transfer fluid. This followed the development of cavity type, volumetric receivers. An open-loop volumetric air receiver capable of absorbing 2.5 MW of heat was installed on top of the tower of a 1.2 MW plant in Spain. This plant had earlier used a cavity-type tubular receiver to heat a molten salt. It was shown that a receiver outlet temperature of 700°C could be easily achieved and it was felt that the use of volumetric receiver air technology had been successfully demonstrated.

Central receiver power plants have been in commercial operation since 2007. Five of these plants have capacities greater than 10 MW. They are located in Spain and USA and are also listed in Table 2.3 where some details are given. These include the number and size of the heliostats, the receiver type, the receiver fluid, the thermal storage medium and the tower height. It may be noted that the largest plant having an aggregate capacity of 377 MW is Ivanpah in California, USA. In addition, five plants with capacities ranging from 50 to 150 MW are under construction in Chile, China, Israel, Morocco and South Africa. These plants should be in operation by 2017 or 2018.

Table 2.3 Data on solar central receiver plants

Plant name	Solar One	Solar Two	PS 10	PS 20	Gemasolar	Ivanpah	Crescent Dunes
Country	USA	USA	Spain	Spain	Spain	USA	USA
Net output (MWe)	10.0	10.0	11.0	20.0	19.9	377.0	110.0
Number of heliostats	1818	1818, 108	624	1255	2650	173 500	10 347
Area of heliostats (m^2)	39.3	39.3, 95.0	120.0	120.0	120.0	14.05	115.7
Total reflecting area (m^2)	71 447	81 707	74 880	150 660	318 000	2.438×10^6	1.197×10^6
Receiver type	External	External	Cavity	Cavity	External	External	External
Receiver fluid	Water/steam	Molten salt	Water/steam	Water steam	Molten salt	Water/ steam	Molten salt
Thermal storage medium	Oil/rock	Molten salt	Saturated steam	Saturated steam	Molten salt	None	Molten salt
Tower height (m)	80	80	115	165	140	137	195
Start of operation	1982	1996	2007	2009	2011	2014	2015

In India, the only central receiver power plant is a 2.5 MW pilot plant in Bikaner, Rajasthan. The plant is operational since 2011. Details of the plant are as follows [22]:

Area of one heliostat	: 1.136 m^2
Number of heliostats	: 14 280
Total reflecting area	: 16 122 m^2
Receiver type	: Cavity
Receiver fluid	: Water/Steam
Receiver inlet temperature	: 218°C
Receiver outlet temperature	: 440°C
Power cycle	: Rankine
Inlet pressure and temperature	: 60 bar, 400°C

It will be noted that the area of one heliostat is very small compared to the heliostats in use in other plants today. Since each heliostat has to be individually installed, this would have led to an increase in the initial capital expenditure. However, this increase would have been off-set to some extent by the less expensive support and foundation requirements per heliostat.

Considerable interest is focused on the plant at Ivanpah in California because it is currently the largest in the world with a gross output of 392 MW and a net output of 377 MW. The plant consists of three units, each with its own power tower. Each unit is designed to produce about 130 MW. The receiver fluid is water which enters the receiver at 250°C and leaves as steam at 160 bar and 565°C. The system operates on a Rankine cycle having a conversion efficiency of 28.7 per cent. The annual solar resource at Ivanpah is estimated to be 2.717 MWh/m^2. Based on this input of solar radiation on the heliostats and considering optical and other losses, it is calculated that the heat energy absorbed in the receiver fluid would be 3.642 TWh/year and that the electrical energy output would be 1.079 TWh/year. Based on the gross power output of 392 MW, this is equivalent to a capacity factor of 0.31. The Ivanpah plant has no thermal storage. It generates power only during the day and has a fossil back-up of natural gas which is needed during certain periods every day.

The commercial operation of three units started at different times in 2014 and the electrical energy output was about 419 GWh. In 2015, the output was about 645 GWh [23-25], which is about 60 per cent of the expected value. The performance is expected to improve in the coming years.

As mentioned, the Ivanpah plant does not have a thermal storage system. The operation of the plant has brought to the fore the need for thermal storage in any CSP plant. CSP plants are now competing against PV installations. One advantage which CSP plants have is that solar energy can be stored as heat in a thermal storage. This energy can be used in a controlled fashion to give a desired electrical output not only through the day but even through the night. In contrast, PV installations produce an electrical output which cannot be stored so readily, particularly in large quantities.

Although the central receiver systems described in Table 2.3 are operating successfully, the available data indicate that the construction costs have been high. For example, Solar One costed approximately \$14 000 per kW. However, costs have been reduced significantly because of improvements in design. The costs of the plants PS 10 and Ivanpah are reported to be 3500 Euro (\$ 3815) per kW and \$ 5800 per kW respectively. Initial costs are likely to reduce further with more operational and design experience.

❑ *Remarks* It needs to be reiterated that solar thermal energy can make a real impact if it leads to large scale cost-effective electrical power generation. Our survey of the world scene shows that as yet, this is far from being the case. However, impressive developments have taken place in the last two or three decades. Many CSP systems using line-focussing parabolic collector technology have been commercialized. A few systems involving LFR technology have also been commissioned. Similarly, some central receiver plants have been commissioned. All these achievements indicate that these systems show promise for the future, particularly if they are coupled with thermal storage. Our survey also shows that India is lagging behind compared to other sun-rich countries in utilizing solar thermal energy for power generation.

2.3.5 Distillation

In many small communities, the natural supply of fresh water is inadequate in comparison to the availability of brackish or saline water. Solar distillation can prove to be an effective way of supplying drinking water to such communities.

The principle of solar distillation is simple and can be explained with reference to Fig. 2.25, in which a conventional basin-type solar still is shown. The still consists of a shallow air-tight basin lined with a black, impervious material which contains the saline water. A sloping transparent cover is provided at the top. Solar radiation is transmitted through the cover and is absorbed in the black lining. It thus heats up the water by about 10° to 20°C and causes it to evaporate. The resulting vapour rises, condenses as pure water on the underside of the cover and flows into condensate collection channels on the sides. An output of about

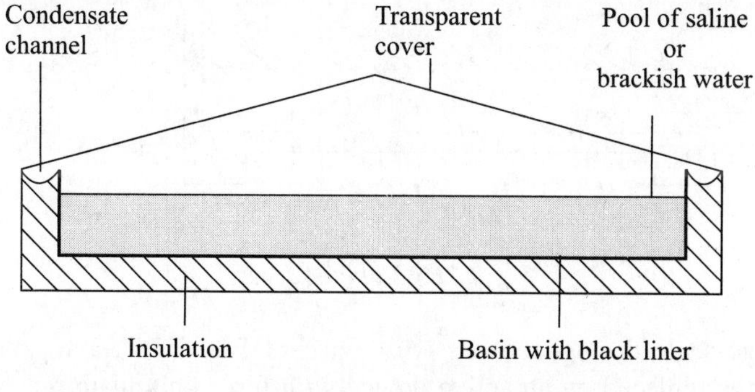

Fig. 2.25 *Solar still*

3 litres/m^2 with an associated efficiency of 30 to 35 per cent can be obtained in a well-designed still on a good sunny day.

A number of basin-type solar-still plants having areas greater than 100 m^2 are in operation in many parts of Africa and the West Indies.

2.3.6 Drying

One of the traditional uses of solar energy has been for drying of agricultural products. The drying process removes moisture and helps in the preservation of the product. Traditionally, drying is done on open ground. The disadvantages associated with this are that the process is slow and that insects and dust get mixed with the product. The use of dryers helps to eliminate these disadvantages. Drying can then be done faster and in a controlled fashion. In addition, a better-quality product is obtained. A few types of solar dryers will be discussed here.

A cabinet-type solar dryer, suitable for small-scale use, is shown in Fig. 2.26. The dryer consists of an enclosure with a transparent cover. The material to be dried is placed on perforated trays. Solar radiation entering the enclosure is absorbed in the product itself and the surrounding internal surfaces of the enclosure. As a result, moisture is removed from the product and the air inside is heated. Suitable openings at the bottom and top ensure a natural circulation. Temperatures ranging from 50°C to 80°C are usually attained and the drying time ranges from 2 to 4 days. Typical products which can be dried in such devices are dates, apricots, chillies, grapes, etc.

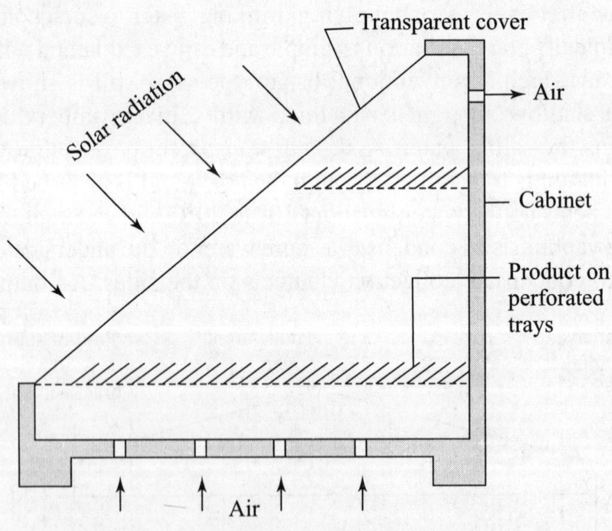

Fig. 2.26 *Cabinet dryer*

For large-scale drying, the passive device of Fig. 2.26 relying on natural circulation is replaced by an active device with forced circulation as shown in Fig. 2.27. Systems of this type have been used for drying timber.

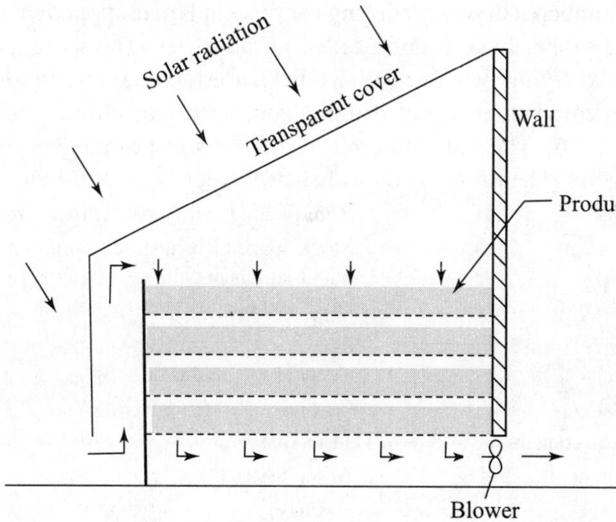

Fig. 2.27 *Forced circulation dryer (direct gain)*

An indirect type of active device is used when the solar radiation falling directly on the product (as in Figs 2.26 and 2.27) is not adequate, or the temperature of the product needs to be controlled. One such system is shown in Fig. 2.28. Here, the air is heated separately in an array of solar air heaters and then ducted to the chamber in which the product to be dried is stored. Such dryers are suitable for food grains, tea, spices, etc. and for products like leather and ceramics. An 11 m^2 solar air heating system installed in Cuddallore, Tamil Nadu for drying fish is shown in Plate 2.7.

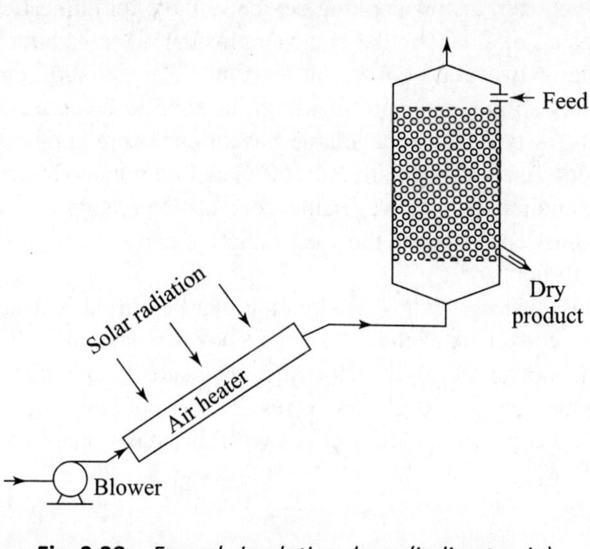

Fig. 2.28 *Forced circulation dryer (indirect gain)*

In India, a number of dryers for drying various kinds of crops and food products, and for drying timber have been installed in industries. The specifications of a typical industrial drying system which has been used since 1995 in a spice powder manufacturing company in Theni, Tamil Nadu, for drying chillies and coriander seeds are given [26]. The solar system is an add-on to an existing dryer, which is run with firewood. The firewood dryer has been kept as a stand-by unit.

Collector array	: Roof mounted air heaters with an area of 290 m^2
Collector	: Single glazed, corrugated aluminium absorber with matt black paint (Air flows between the cover and the absorber.)
Typical temperature rise	: 30°C
Type of dryer	: Tunnel (continuous drying)
Material dried	: Chillies
Initial moisture content	: 11 to 14%
Moisture content after drying	: About 5%
Capacity	: 10 t/day

It may be noted that the dryer described here has no thermal storage. Thermal storage would provide heat for drying during periods when solar radiation is not available. Dryers with thermal storage have also been developed [27,28].

2.3.7 Cooking

An important domestic thermal application is that of cooking. Over the past 40 years, a number of designs of solar cookers have been developed. Panwar *et al.* [29] and Cuce and Cuce [30] have presented a review of solar cookers. A few type of cookers are described here.

Solar cooker designs generally fall into one of two categories. The first category is the box-type cooker, a slow cooking device suitable for domestic purposes. It essentially consists of a rectangular enclosure insulated on the bottom and sides, and having one or two glass covers on the top. Solar radiation enters through the top and heats up the enclosure in which the food to be cooked is placed in shallow vessels. A typical size available has an enclosure about 50 cm square and 12 cm deep. Temperatures around 100°C can be obtained in these cookers on sunny days and pulses, rice, vegetables, etc, can be readily cooked. The time taken for cooking depends upon the solar radiation and varies from half an hour to two and a half hours.

A single glass reflector whose inclination can be varied is usually attached to the box-type cooker. A sketch of such a cooker is shown in Fig. 2.29. The addition of the mirror helps in achieving enclosure temperatures which are higher by about 15° to 20°C. As a result, the cooking time is reduced. Cookers with reflectors on all four edges with thermal storage have also been developed [30].

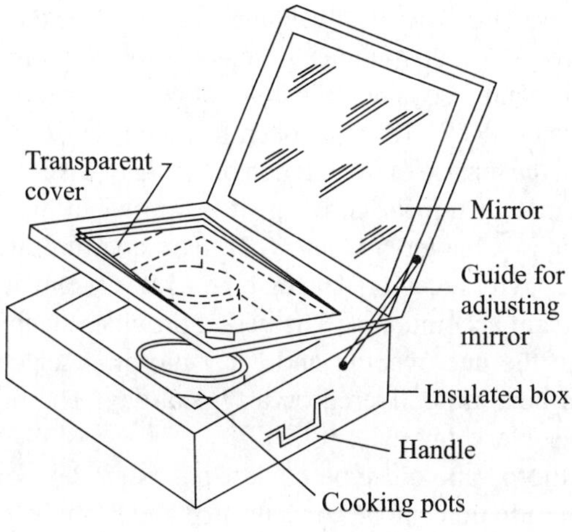

Fig. 2.29 *Box-type cooker with one reflector*

Box-type cookers with no reflector or with one reflector are simple to use and require little attention. As a result, they have found the maximum acceptance amongst all the designs developed. The item to be cooked has only to be placed inside and taken out, so that with some experience, the operator does not have to spend much time in the sun. However, the disadvantage is that they cannot be used for cooking items like *chapatis* and *purees* since these require higher temperatures.

Box-type cookers are being made by many manufacturers in India. The cost of a typical good quality cooker with a mirror varies from Rs 1000 to 3000. In most places in India, such cookers can be used for at least 250 days in the year and it is estimated that the cost of a box-type cooker is recovered within a period of three to four years. Because of the relatively low payback period, solar cookers have become fairly popular. It is estimated that more than 680 000 cookers have been sold so far and the number is growing at the rate of about 20 000 to 30 000 every year. Today, India has the distinction of having the largest number of solar cookers amongst all the countries of the world.

The second category of solar cookers developed are those in which the radiation is concentrated by a paraboloid reflecting surface. The cooking vessel is placed at the focus of the paraboloid mirror and is, thus, directly heated. The dimensions, specifications and cost of a typical cooker are as follows:

Aperture diameter	: 1.4 m
Focal length	: 0.28 m
Reflecting surface	: Anodized aluminium sheet
Cost	: Rs 3500 to 5000

This cooker is referred to as a dish solar cooker. Temperatures well above 200°C are obtained in it and it can be used for cooking food items requiring roasting, frying or boiling. The disadvantage of a dish cooker is that it requires manual tracking every 15 or 20 minutes. Also, since the cooking is done outdoors, the operator has to spend a considerable amount of time in the sun.

A paraboloid dish (ARUN 100) has been installed by Clique Developments, Mumbai in 2013 on the terrace of the kitchen building of the Ramkrishna Mission at Chennai to generate steam for cooking. It is similar in many respects to the one described later in Chapter 6 (Plate 6.2). The aperture area of the dish is 104 m^2 and it has a cavity receiver. The dish generates steam at 145°C and 6 bar and stores it in a steam accumulator. The vessel provides steam during sunshine hours as well as in the late evening and early morning for cooking purposes. About 1045 MJ of heat are delivered on a typical day. The organization saves about 36 kg of LPG every day [31].

A variation of the paraboloid-type cooker has been developed by Scheffler [32] such that solar radiation can be concentrated and brought inside the kitchen. The reflector is a small lateral section of a much larger paraboloid. The inclined cut produces the typical shape of the Scheffler reflector. The reflector usually consists of a number of mirror facets supported by a steel frame. It is mounted outside the kitchen facing south (in the northern hemisphere). It reflects the solar rays through an opening in the north wall of the kitchen on to a secondary reflector inside the kitchen. The secondary reflector further concentrates the rays on to the bottom of the cooking vessel (Fig. 2.30). These cookers require tracking such that the focus stays fixed. The daily rotation is generally carried out by a mechanical tracking device (clock work). In addition, every second or third day,

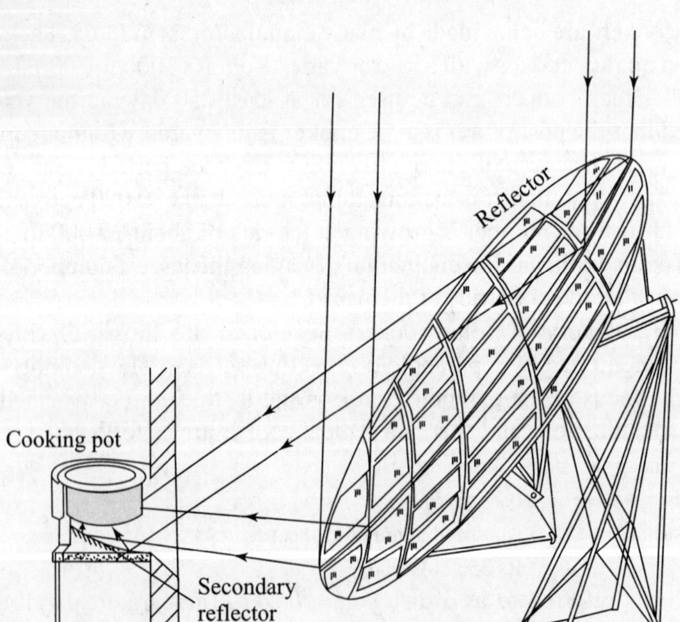

Fig. 2.30 *Schematic diagram of a Scheffler cooker [32]*

Plate 2.6 *Parabolic collector field of 50 MW Godawari CSP plant in Rajasthan. (Courtesy: Godawari Green Energy Ltd., Raipur)*

Plate 2.7 *Solar air heating system installed in Cuddalore, Tamil Nadu for fish drying. (Courtesy: Planters Energy Network, Theni, Tamil Nadu)*

Plate 2.8 *Steam cooking system based on Scheffler reflectors installed at Tirumala Tirupati Devsasthanam, Andhra Pradesh. (Courtesy: Gadhia Solar Energy Systems Pvt. Ltd., Valsad, Gujarat)*

Plate 2.9 *Spherical bowl concentrator in Auroville used for generating steam for cooking. (Courtesy: Centre for Scientific Research, Auroville, Puducherry).*

the tilt of the reflector is adjusted and the reflector is flexed so that most of the solar rays are reflected to the secondary reflector. Temperatures as high as 400°C can be attained and all types of cooking/frying are possible. One does not need to go outside in the sun unlike a normal paraboloid cooker. Cooking for about 40 to 50 persons is possible with this cooker. Two sizes (16 and 32 m^2) of Scheffler reflectors are available commercially in India.

The Scheffler cooker is useful for institutional kitchens, hotels, etc. The typical cost of such a cooker is about Rs 18 000 per m^2 of reflector [33]. A series of such cookers can also be used to generate steam which can be used for cooking. Such installations have been made at a few places in India. It may be mentioned that Scheffler reflectors are also used to provide low pressure steam for industrial applications [34, 35].

The world's largest steam cooking system based on Scheffler reflectors has been installed at Tirumala Tirupathi Devasthanam, Andhra Pradesh (Plate 2.8). The specifications of the system are as follows:

Capacity	: Food for 15 000 persons per day
Number of reflectors	: 106; each of 9.2 m^2 area
Reflector material	: Acrylic mirror
Output	: 4000 kg of steam per day at 180°C and 10 bar

A unique solar cooker using a fixed bowl with a moving receiver has been developed and installed at the Centre for Scientific Research, Puducherry [36] (Plate 2.9). It consists of a large spherical bowl concentrator integrated into the kitchen roof using prefabricated ferrocement curved channels. Laminated mirror pieces, 15 cm × 15 cm in size, have been fixed using silicone and act as the reflecting surface. The aperture area of the spherical reflector is 176 m^2. The receiver is suspended over the collector from a single support pipe and is rotated about a polar axis. It consists of a long coil of MS pipes wrapped around a steel frame. Water is pumped through the coil to generate steam which is taken to the kitchen and used for cooking purposes. The typical peak thermal energy output recorded is 63 kW. This corresponds to a steam production of 86 kg/h.

PROBLEMS

Note: *Make suitable assumptions where necessary.*

2.1 An industry requires 5000 litres of hot water per day at 75°C as process heat. The following meteorological data are available for the location:

Annual average daily global solar radiation on horizontal surface = 5.8 kWh/m^2

Average ambient temperature = 24°C

What type of solar system would you recommend? Estimate the collector area and give the approximate cost of the system.

2.2 A low temperature solar thermal power generation system using flat-plate collectors is proposed to be set up. The system is required to generate 50 kWe round the clock. Estimate the collector area required and the cost of the solar system of such a plant.

2.3 It is proposed to build a 50 MWe solar chimney plant with a chimney 1 km high. Estimate the collector area required.

2.4 It is desired to generate power through the solar thermal route at a suitable place in Rajasthan. Which type of system would you recommend? Estimate the area of solar collectors required for such a plant of 85 MWe capacity, if power is generated only for six hours during the day.

2.5 A paraboloid dish concentrator is to be deployed for generating 30 kWe in a remote location. The system does not have any storage. Estimate the size of the dish required.

2.6 Consider an ammonia–water solar refrigeration system of 2 ton cooling capacity. Estimate the collector area required for the system, if it works continuously using flat-plate collectors and a storage. What would be the approximate cost of the solar part of the refrigeration system? Comment on your result.
Given: Total solar radiation on a horizontal surface = 6 kWh/m^2-day.

2.7 A petrol pump dealer plans to sell 100 litres of distilled water per day. Estimate the area of the basin-type solar stills required for the purpose.

2.8 It is desired to use an active solar drying system to remove 200 kg of moisture per day during sunshine hours from 1000 kg of freshly harvested grain. Estimate the collector area required and the cost of the solar system.
Given: Monthly average daily global solar radiation on horizontal surface = 6 kWh/m^2.

REFERENCES

1. Gehlisch, K., Heikal, H., Mobarak, A. and Simon, M. 1982. Large parabolic dish collectors with small gas-turbine, Stirling engine or photovoltaic power conversion systems, *Proc. Intersociety Energy Conversion Engineering Conference*, Los Angeles, USA, 5: 2186.
2. Rabl, A. 1985. *Active Solar Collectors and their Applications*, Oxford University Press, New York, USA.
3. Einav, A. 2004. Solar energy ressearch and development achievements in Israel and their practical significance. *Journal of Solar Energy Engineering, Trans. ASME*, 126: 921.
4. Sureshkumar, C.S. 2016. Personal communication. KGDS Renewable Energy Pvt. Ltd., Coimbatore, Tamil Nadu, India.
5. KGDS Renewable Energy Pvt. Ltd.
(http://www.solar.kgisl.com; accessed on April 07, 2016)
6. Deshpande, K. and Joshi, Y. 2016. Personal communication. Thermax Limited, Pune, India.
7. Malik, S. 2016. Personal communication. Megawatt Solutions, New Delhi, India.

8. Decker, G.W.E., Hollands, K.G.T. and Brunger, A.P. 2001. Heat-exchange relations for unglazed transpired solar collectors with circular holes on a square or triangular pitch. *Solar Energy*, 71: 33.

9. Gupta, C.L. 1993. Solar passive buildings for developing countries. *Proc. Indian Academy of Sciences* (*Engg. Sciences*), Sadhana, Part 1, 18: 77.

10. Majumdar, M. 2001. *Energy Efficient Buildings in India*. Tata Energy Research Institute, New Delhi, India.

11. Nayak, J.K. and Prajapati, J. 2006. *Handbook on Energy Conscious Buildings*. Interactive R & D project no.3/4(03)99-SEC, Ministry of New and Renewable Energy, New Delhi, India.

12. Al-Alili, A., Hwang, Y. and Radermacher, R. 2014. Review of solar thermal airconditioning technologies. *International Journal of Refrigeration*, 39: 4.

13. Nkwetta, D.N. and Sandercock, J. 2016. A state-of-the-art review of solar air-conditioning systems. *Renewable and Sustainable Energy Reviews*, 60: 1351.

14. Kumar, A. 2006. Solar air-conditioning. *Akshay Urja,* 2: 14.

15. Schlaich, J., Bergermann, R., Schiel, W. and Weinrebe, G. 2005. Design of commercial solar updraft tower systems - Utilization of solar induced convective flows for power generation. *Journal of Heat Transfer, Trans. ASME*, 127: 117.

16. EnviroMission Solar Tower

(http://pesn.com/radio/Free_Energy_Now/shows/2006/10/28/9700213_EnviroMis sion_solar_tower; accessed on November 27, 2006)

17. Desai, N.B. and Bandyopadhyay, S. 2015. Integration of parabolic trough and linear Fresnel collectors for optimum design of concentrating solar thermal power plant. *Clean Tech Environ Policy*, 17: 1945.

18. Nayak, J.K., Kedare, S.B., Banerjee, R., Bandyopadhyay, S., Desai, N.B., Paul, S. and Kapila, A. 2015. A 1 MW national solar thermal research cum demonstration facility at Gwalpahari, Haryana, India. *Current Science*, 109: 1445.

19. Nayak, J.K., Banerjee, R., Bandyopadhyay, S. and Kedare, S.B. 2015. Development of a Mega Watt-scale National Solar Thermal Power Testing, Simulation and Research Facility, *Final report of Project no. 15/19/2008-09/ST*, Ministry of New and Renewable Energy, New Delhi, India.

20. Mancini, T., Heller, P., Butler, B., Osborn, B., Schiel, W., Goldberg, V., Buck, R., Diver, R., Andraka, C. and Moreno, J. 2003. Dish-Stirling systems: an overview of development and status. *Journal of Solar Energy Engineering, Trans*. ASME. 125: 135.

21. Romero, M., Buck, R. and Pacheco, J.E. 2002. An update on solar central receiver systems, projects and technologies. *Journal of Solar Energy Engineering, Trans. ASME*. 124: 98.

22. Jacob, J. 2016. Personal communication, ACME Cleantech Solutions Pvt. Ltd., Gurgaon, India.

23. Electricity Data Browser, U.S. Energy Information Administration, Washington, U.S.A.

(http://www.eia.gov/electricity/data/browser/#/plant/57074; accessed on February 01, 2016)

24. Electricity Data Browser, U.S. Energy Information Administration, Washington, U.S.A.

(http://www.eia.gov/electricity/data/browser/#/plant/57073; accessed on February 01, 2016)

25. Electricity Data Browser, U.S. Energy Information Administration, Washington, U.S.A.

(http://www.eia.gov/electricity/data/browser/#/plant/57075; accessed on February 01, 2016)

26. Palaniappan, C. 2007. Personal communication, Planters Energy Network, Theni, Tamil Nadu, India.

27. Belessiotis, V. and Delyannis, E. 2011. Solar drying. *Solar Energy*, 85: 1665.

28. El-Sebaii, A.A. and Salaby, S.M. 2012. Solar drying of agricultural products: a review. *Renewable and Sustainable Energy Review*, 16: 37.

29. Panwar, N.L., Kaushik, S.C. and Kothari, S. 2012. State of the art of solar cooking: an overview. *Renewable and Sustainable Energy Review*, 16: 3776.

30. Cuce, E. and Cuce, P.M. 2013. A comprehensive review on solar cookers. *Applied Energy*, 102: 1399.

31. Paranjape, A.D. 2016. Personal communication. Clique Developments Ltd., Mumbai, India.

32. The Scheffler-reflector

(http://www.Solare-Bruecke.org; accessed on November 28, 2006)

33. Fixed Focus Elliptical Solar Dish (Scheffler), UNDP-GEP Project on CSH, Ministry of New and Renewable Energy, Government of India, New Delhi, India.

(http://www.cshindia.in/images/pdf/Fixed%20Focus%20Elliptical%20Solar%20 Dish.pdf; accessed on June 20, 2016).

34. Jayasimha, K. 2006. Application of Scheffler reflectors for process industry. *International Solar Cooker Conference*, Granada, Spain, July 12-16.

(http://www.Solar-Bruecke.org)

35. Gadhia, D. and Gandhi, S. 2006. Parabolic solar concentrators for cooking, food processing and other applications. *International Solar Cooker Conference*, Granada, Spain, July 12 – 16.

(http://www.Solar-Bruecke.org)

36. Gupta, C.L. 2007. A spherical solar concentrator functioning in Auroville. Personal communication, Solar Energy Unit, Sri Aurobindo Ashram, Puducherry.

Solar Radiation

CHAPTER 3

INTRODUCTION 3.1

This chapter discusses the availability of solar radiation as an energy source. Extraterrestrial radiation, its spectral distribution and the radiation at the earth's surface are discussed first. This is followed by a description of instruments used for measuring solar radiation and methods used for presenting data. The geometry of solar radiation and the various angles used are defined so that it is possible to convert the flux on one plane to an equivalent flux on another. Finally, equations for predicting the availability of solar radiation at a location are given.

SOLAR RADIATION OUTSIDE THE EARTH'S ATMOSPHERE 3.2

The characteristics of the sun's energy available outside the earth's atmosphere are first considered.

The sun is a large sphere of very hot gases, the heat being generated by various kinds of fusion reactions. Its diameter is 1.39×10^6 km, while that of the earth is 1.27×10^4 km. The mean distance between the two is 1.496×10^8 km. Although the sun is large, it subtends an angle of only 32 minutes (0.53°) at the earth's surface. This is because the sun is also at a very large distance from the earth. Thus, the beam radiation received from the sun on the earth is almost parallel. The brightness of the sun varies from its centre to its edge. However, for engineering calculations, it is customary to assume that the brightness all over the solar disc is uniform.

Measurements indicate that the energy flux received from the sun outside the earth's atmosphere is essentially constant. The *solar constant* I_{sc} is the rate at which energy is received from the sun on a unit area perpendicular to the rays of the sun, at the mean distance of the earth from the sun. The value of the solar constant has been the subject of many experimental investigations. Based on measurements made up to 1970, a standard value of 1353 W/m^2 was adopted in 1971 [1]. However, based on subsequent measurements, a revised value of 1367 W/m^2 has been recommended [2]. The difference between the two values is only 1 per cent.

The earth revolves around the sun in an elliptical orbit having a very small eccentricity, and with the sun at one of the foci. Consequently, the distance between the earth and the sun varies a little through the year. Because of this variation, the extraterrestrial flux also varies. The value on any day can be calculated from the equation

$$I'_{sc} = I_{sc}\left(1 + 0.033 \cos \frac{360n}{365}\right) \tag{3.1}$$

where n is the day of the year. From Eq. (3.1), we see that since the cosine function varies from +1 to –1, the extraterrestrial radiation flux varies by ±3.3 per cent over a year.

It is also useful to know the spectral distribution of extraterrestrial solar radiation. Measurements corresponding to the value of the solar constant of 1367 W/m^2 are given in Table 3.1. The data shows that the spectral value first increases sharply with wavelength, passes through a maximum at a wavelength of 0.48 μm and then decreases asymptotically to zero. 99 per cent of the sun's radiation is obtained up to a wavelength of 4 μm. The percentage of radiation obtained up to a certain wavelength is also given in Table 3.1. This value is useful for calculating the solar flux emitted in any specified range of wavelengths. For easy visualisation, the spectral distribution data is also plotted in Fig. 3.1.

Table 3.1 Spectral distribution of extraterrestrial solar radiation flux at mean sun-earth distance (I_{sc} = 1367 W/m^2) [3]

λ (μm)	e_λ (W/m^2-μm)	D_λ
0.2	–	–
0.3	528	1.21
0.38	1210	6.46
0.4	1702	8.03
0.48	2055	18.85
0.5	1863	21.68
0.6	1720	35.26
0.7	1428	46.84
0.78	1195	54.47
0.8	1145	56.16
0.9	913	63.59
1.0	744	69.45

(Contd.)

2.0	119	94.10
3.0	25	98.07
4.0	8.5	99.16
5.0	3.5	99.57
6.0	1.8	99.75
7.0	0.95	99.84
8.0	0.55	99.89
10.0	0.20	99.95

$$D_\lambda = \left\{ \left[\int_0^\lambda e_\lambda \, d\lambda \right] \Big/ \left[\int_0^\infty e_\lambda \, d\lambda \right] \right\} \times 100 = \left\{ \left[\int_0^\lambda e_\lambda \, d\lambda \right] \Big/ I_{sc} \right\} \times 100$$

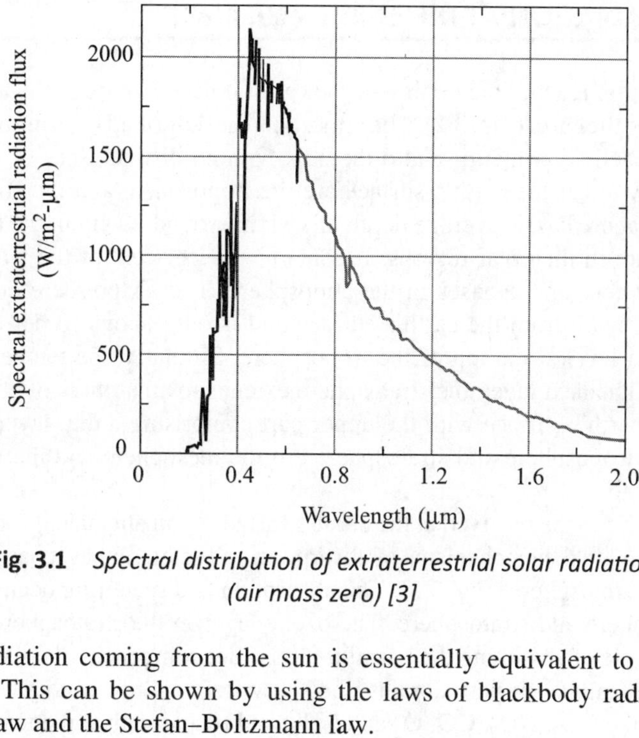

Fig. 3.1 *Spectral distribution of extraterrestrial solar radiation (air mass zero) [3]*

The radiation coming from the sun is essentially equivalent to blackbody radiation. This can be shown by using the laws of blackbody radiation, viz. Planck's law and the Stefan–Boltzmann law.

Planck's law states that the spectral emissive power of a black surface is given by

$$e_{b\lambda} = \frac{2\pi C_1}{\lambda^5 \left[\exp\left(C_2/\lambda T\right) - 1\right]} \tag{3.2}$$

where C_1 and C_2 are constants whose values are 0.596×10^{-16} W-m^2 and 0.014387 m-K respectively, λ is the wavelength and T, the temperature of the black surface (in K).

The Stefan–Boltzmann law is obtained by integrating Planck's law over all the wavelengths from 0 to ∞ and states that the emissive power of a black surface is given by

$$e_b = \sigma T^4 \tag{3.3}$$

where T is the temperature of the surface (in K) and σ is a constant called the Stefan–Boltzmann constant. $\sigma = 5.670 \times 10^{-8}$ W/m^2-K^4.

In order to show that the radiation emitted by the sun is essentially blackbody radiation, we substitute the value of the solar constant in Eq. (3.3) and obtain the equivalent blackbody temperature of the sun to be 5779 K. Substituting this value in Planck's law, Eq. (3.2), we can obtain values of $e_{b\lambda}$ for different values of λ. The magnitudes of the values obtained agree well with the values of spectral extraterrestrial radiation flux given in Table 3.1 and Fig. 3.1 (Problem 3.1).

SOLAR RADIATION AT THE EARTH'S SURFACE 3.3

Solar radiation reaches the earth's surface after it passes through the atmosphere surrounding the earth (Fig. 3.2). The atmosphere is described in terms of a number of layers, each layer merging into the next with no distinct lines of separation. The first two near the earth's surface are the troposphere and the stratosphere. The troposphere has an average depth of 15 km, around 20 km in the tropics and around 7 km in the polar regions. It contains 99 per cent of the water vapour and 75 per cent of the gases in the atmosphere. The air pressure decreases as one moves away from the earth's surface and is only about 10 per cent of the value at sea level at the top of the troposphere. Similarly, the temperature also decreases. The next layer, the stratosphere extends to an altitude of about 50 km above the earth's surface with the upper part comprising a thin layer of ozone. Above the troposphere and stratosphere are the mesophere and the ionosphere respectively.

Solar radiation is received at the earth's surface in an attenuated form because it is subjected to the mechanisms of absorption and scattering as it passes through the earth's atmosphere (Fig. 3.3). The absorption and scattering occur mainly in the troposphere and stratosphere. The ozone layer in the stratosphere plays the important role of absorbing the harmful ultraviolet component of solar radiation. Other wavelengths are partially absorbed by water vapour and to a lesser extent by gases (like CO_2, NO_2, CO, O_2 and CH_4) and by particulate matter (aerosols). Absorption results in an increase of the internal energy of the atmosphere. Scattering occurs due to the interaction of the radiation with particulate matter present in the atmosphere. The scattered radiation is redistributed in all directions, some being reflected back into space and some reaching the earth's surface.

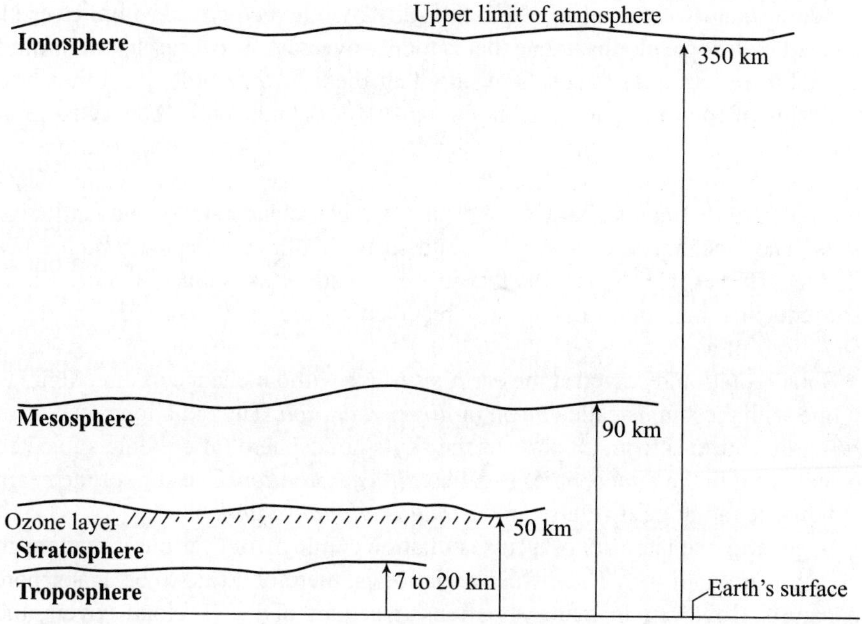

Fig. 3.2 *The earth's atmosphere*

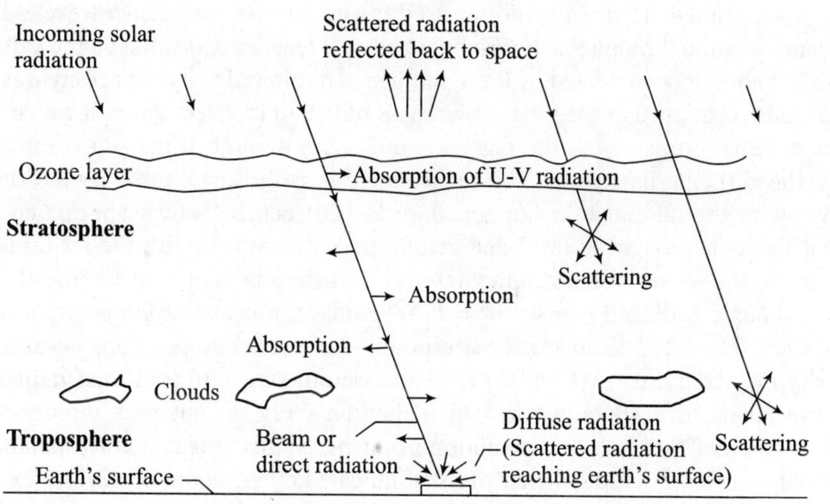

Fig. 3.3 *Schematic representation: (i) Mechanisms of absorption and scattering in the troposphere and stratosphere (ii) Beam and diffuse radiation received at the earth's surface*

At any location on the earth's surface, the terms 'clear sky' and 'cloudy sky' are often used to classify two types of sky. A *clear sky* is described as one without any clouds and with relatively 'low turbidity', while a *cloudy sky* implies a sky covered with some clouds or one that is totally overcast. The term 'low turbidity' is used to describe an atmosphere in which there is very little absorption and scattering of solar radiation. The main reason for distinguishing between a clear sky and a cloudy sky is that it is easier to analyse and develop correlations for 'clear sky' situations. Marty and Philipona [4] have defined a term called Clear Sky Index to distinguish between a clear sky and a cloudy sky on a quantitative basis. The mechanism of absorption and scattering are similar with both types of skies. However, it is obvious that less attenuation takes place in a clear sky. Consequently maximum radiation is received on the earth's surface under the conditions of a clear sky.

Solar radiation received at the earth's surface without change of direction, i.e. in line with the sun, is called *beam* or *direct radiation.* The radiation received at the earth's surface from all parts of the sky's hemisphere (after being subjected to scattering in the atmosphere) is called *diffuse radiation.* The sum of the beam and diffuse radiation is referred to as *total* or *global radiation.*

In general, the intensity of diffuse radiation coming from various directions in the sky is not uniform. The diffuse radiation is, therefore, said to be anisotropic in nature. However, in many situations (e.g. part or heavy cloud cover), the intensity from all directions tends to be reasonably uniform. It is then modelled as being perfectly uniform and is said to be isotropic in nature.

A term called the *air mass* is often used as a measure of the distance travelled by beam radiation through the atmosphere before it reaches a location on the earth's surface. It will be indicated by the symbol m and referred to by the acronym AM. Air mass is defined as the ratio of the mass of the atmosphere through which the beam radiation passes to the mass it would pass through if the sun is directly overhead (i.e. at its zenith). The *zenith angle* θ_z is the angle made by the sun's rays with the normal to a horizontal surface. It can be shown approximately that for locations at sea level and zenith angles from 0 to 70°, the air mass is equal to the secant of the zenith angle. Thus if the sun is directly overhead, the zenith angle is 0° and $m = \sec 0° = 1$. When the sun is at a zenith angle of 60°, $m = \sec 60° = 2$. The air mass corresponding to these two positions is referred to by the acronyms AM1 and AM2. The case of extra-terrestrial solar radiation corresponds to a situations in which the sun's ray do not pass through the atmosphere. Thus from the definition of air mass, this situation corresponds to $m = 0$. For this situation, the air mass is indicated by the acronym AM0.

The design of devices like concentrating collectors for solar thermal application or the determination of the area of a solar PV array at a specified location requires a knowledge of the variation with time of direct radiation and/or global radiation at the location. In the past, designers of solar equipment resorted to one of the following options:

1. Making radiation measurements over a period of time at the location in question where the solar equipment is to be installed.

2. Using measurements available for some other location where the climate is known to be reasonably similar to the location under consideration.
3. Using empirical predictive equations which link the values of solar radiation with other meteorological parameters whose values are known for the location under consideration. Examples of such parameters are the number of hours of sunshine received per day, the cloud cover, etc. (Sections 3.7.1 – 3.7.4)

However, in the past one to two decades, extensive studies have been made on the mechanisms of absorption and scattering in the atmosphere, and the ability to calculate the attenuation coefficients for different constituents more accurately has increased. This has been possible because of observations made by sensors on satellites and also because of the relative ease with which computations involving large amounts of data can now be carried out. Consequently, a number of models have been developed and these models are being increasingly used for calculating the availability of global/beam radiation at a specified location for a particular application (Section 3.7.5).

INSTRUMENTS FOR MEASURING SOLAR RADIATION AND SUNSHINE 3.4

The instruments used for measurement will now be described. Solar radiation flux is usually measured with the help of a *pyranometer* or a *pyrheliometer.*

A *pyranometer* is an instrument which measures either global or diffuse radiation falling on a horizontal surface over a hemispherical field of view (Plate 3.1). The global value is also referred as the *Global Horizontal Irradiance* (GHI). A sketch of one type of pyranometer, as installed for the measurement of global radiation, is shown in Fig. 3.4. Basically, the pyranometer consists of a 'black' surface which heats up when exposed to solar radiation. Its temperature increases until the rate of heat gain by solar radiation equals the rate of heat loss by convection, conduction and reradiation. The hot junctions of a thermopile are attached to the black surface, while the cold junctions are located under a guard plate so that they do not receive the radiation directly. As a result, an emf is generated (usually in the range of 0 to 10 mV) which can be read, recorded or integrated over a period of time and is a measure of the global radiation.

The pyranometer, shown in Fig. 3.4 is used commonly in India. It has its hot junctions arranged in the form of a horizontal circular disc of diameter 25 mm and coated with a special black lacquer having a very high absorptivity in the solar wavelength region. The disc is placed on a large diameter guard plate which may be horizontal or sloping (Plate 3.1). Two concentric hemispheres, 30 and 50 mm in diameter respectively, made of optical glass having excellent transmission characteristics, are used to protect the disc surface from weather conditions. An accuracy of about ±2 per cent can be obtained with the instrument.

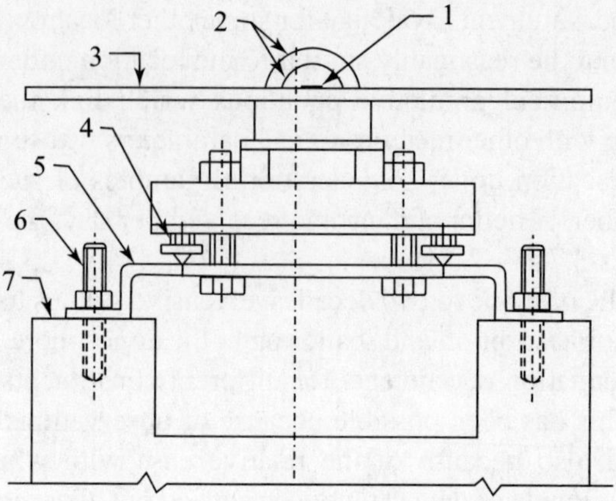

Fig. 3.4 *Pyranometer for measuring global radiation (1. Black surface, 2. glass domes, 3. guard plate, 4. leveling screws, 5. mounting plate, 6. grouted bolts, 7. platform)*

The pyranometer can also be used for the measurement of diffuse radiation. This is done by mounting it at the centre of a semicircular *shading ring*, which is fixed in such a way that its plane is parallel to the plane of the path of the sun's daily movement across the sky. Besides, it shades the thermopile element and the two glass domes of the pyranometer at all times from direct sunshine. Consequently, the pyranometer measures only the diffuse radiation received from the sky.

The construction of one type of shading ring is shown in Fig. 3.5. *ABCD* is a horizontal rectangular frame 35 cm × 80 cm with its long sides in an east-west direction. To the sides *AB* and *CD* of the frame are pivoted two angle-iron arms *EF* and *GH*, 70 cm long with slots along their length, carrying sliders, *SS*, on

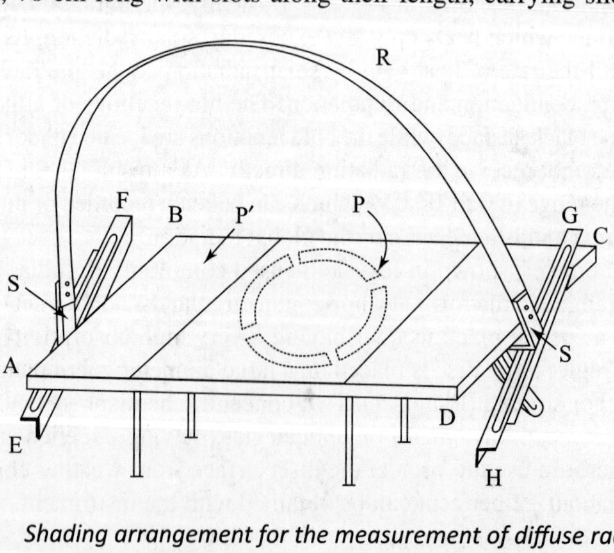

Fig. 3.5 *Shading arrangement for the measurement of diffuse radiation*

Plate 3.1 *Pyranometer for measuring global solar radiation. (Courtesy: Dynalab, Pune)*

Plate 3.2 *Sunshine recorder. (Courtesy: National Instruments Ltd., Kolkata).*

which is mounted the semicircular shading ring *R*. The arms are pivoted about a horizontal axis, which passes through the centre of the rectangular frame and can be adjusted at an angle to the horizontal equal to the latitude of the station. The movement of the ring up and down the arms allows for changes in the sun's declination. The shading ring is of aluminium, 50 mm broad, and is bent to a radius of 450 mm. The inner surface of the ring is painted dull black, while the rest of the shading ring arrangement is painted dull matt white. To the bottom of the frame *ABCD* a thick metal plate *P* with a circular slot is fixed so that the frame, when fixed on a masonry platform with nuts and bolts, can be adjusted in its proper position by rotation about a vertical axis. To the top of the frame another thick metal plate *P′* is fitted on which the pyranometer is mounted.

A *pyrheliometer* is an instrument which measures the solar beam radiation flux falling on a surface normal to the sun's rays. This is referred to as the *Direct Normal Irradiance* (DNI). In contrast to a pyranometer, the black absorber plate (with the hot junctions of a thermopile attached to it) is located at the base of a collimating tube (Fig. 3.6). The tube is aligned with the direction of the sun's rays with the help of a two-axis tracking mechanism and an alignment indicator. Thus the black plate receives only beam radiation and a small amount of diffuse radiation falling within the 'acceptance angle' of the instrument.

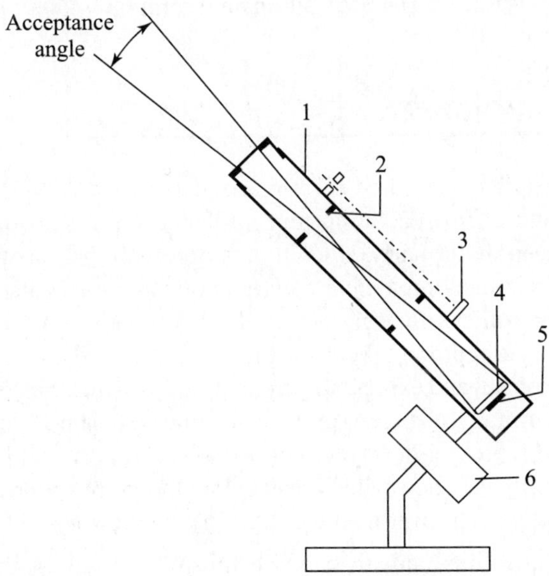

Fig. 3.6 *Pyrheliometer for measuring beam radiation (1. Tube blackened on inside surface, 2. baffle, 3. alignment indicator, 4. black absorber plate, 5. thermopile junctions, 6. two-axis tracking mechanism)*

The duration of bright sunshine in a day is measured by means of a *sunshine recorder* shown in Fig. 3.7 and in Plate 3.2. The sun's rays are focused by a glass sphere to a point on a card strip held in a groove in a spherical bowl mounted concentrically with the sphere. Whenever there is bright sunshine, the image formed is intense enough to burn a spot on the card strip. Through the day as the sun moves across the sky, the image moves along the strip. Thus, a burnt trace whose length is proportional to the duration of sunshine is obtained on the strip.

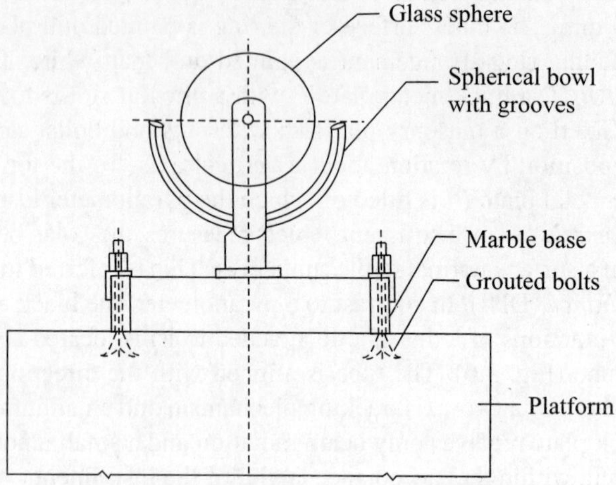

Fig. 3.7 *Sunshine recorder*

SOLAR RADIATION DATA 3.5

Most radiation data is measured for horizontal surfaces. A typical daily record of the global and diffuse radiation flux measured on a clear day is shown in Fig. 3.8. It is seen that a fairly, smooth variation with the maximum occurring around noon is obtained on a clear day. In contrast, an irregular variation with many peaks and valleys may be obtained on a cloudy day. We will use the symbols I_g and I_d to represent the instantaneous values of the global and diffuse flux plotted in Fig. 3.8 and express these quantities in W/m^2. Since solar radiation fluxes do not normally change rapidly with time, the same symbols I_g and I_d are also used to represent hourly values. These quantities will be expressed in kWh/m^2-h or kJ/m^2-h. The shaded areas below the graphs represent the global and diffuse flux incident over a whole day. The symbols H_g and H_d are used to represent these quantities and are expressed in kWh/m^2-day or kJ/m^2-day.

Solar radiation flux is sometimes reported in langleys per hour or per day (1 langley = 1 cal/cm^2 = 1.163×10^{-2} kWh/m^2). The unit 'langley' has been adopted in honour of Samuel Langley who made the first measurement of the spectral distribution of the sun.

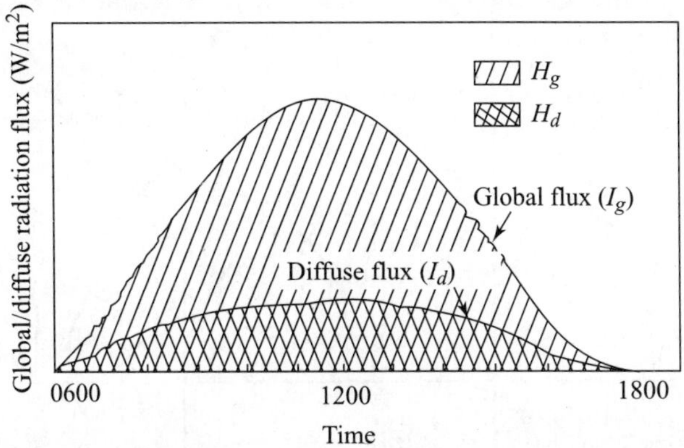

Fig. 3.8 *Record of global and diffuse radiation flux measured on a clear day*

A solar designer is primarily interested in average values for a location. The averaging is usually done over all days of a month and indicated by a bar over the symbol. Thus the monthly average of the hourly global and diffuse radiation are indicated by the symbols $\overline{I}_g$ and $\overline{I}_d$ respectively, while the monthly average of the daily global and diffuse radiation are indicated by the symbols, $\overline{H}_g$ and $\overline{H}_d$ respectively.

Solar radiation data for India is available from many sources. Data is collected at about 45 stations by the Indian Meteorological Department (IMD) and can be obtained from the department [5]. Data for a few locations is also available in the form of a handbook [6] and can be obtained from the Ministry of New and Renewable Energy (MNRE) website [7]. Based on satellite data collected from January 2002 to December 2008, solar maps containing monthly and annual values of DNI and GHI have been developed. These can also be accessed from the website of MNRE [7]. Presently, the National Institute of Wind Energy at Chennai is carrying out an extensive solar resource mapping programme by acquiring data at 111 stations across the country. This data can be obtained from the Institute [8]. The data for some stations in India is also available from the World Radiation Data Centre [9]. Data for two locations, New Delhi and Mumbai, is given in Appendix 3.

A general idea of the availability of solar radiation over different regions can be obtained by constructing solar radiation maps. Such maps have been drawn by Mani and Chacko [10] for India and are reproduced in Figs 3.9 and 3.10. Figure 3.9 shows the distribution of the average daily global radiation, while Fig. 3.10 shows the distribution of the average daily diffuse radiation.

From Fig. 3.9, it is seen that the annual average daily global radiation received over the whole country is around 450 langleys per day (5.23 kWh/m^2-day). Peak values are generally measured in April or May, with parts of Rajasthan and Gujarat, Western Madhya Pradesh and North Maharashtra receiving over 600 langleys per day. In contrast, during the monsoon and winter months, the daily global radiation decreases to about 300–400 langleys per day.

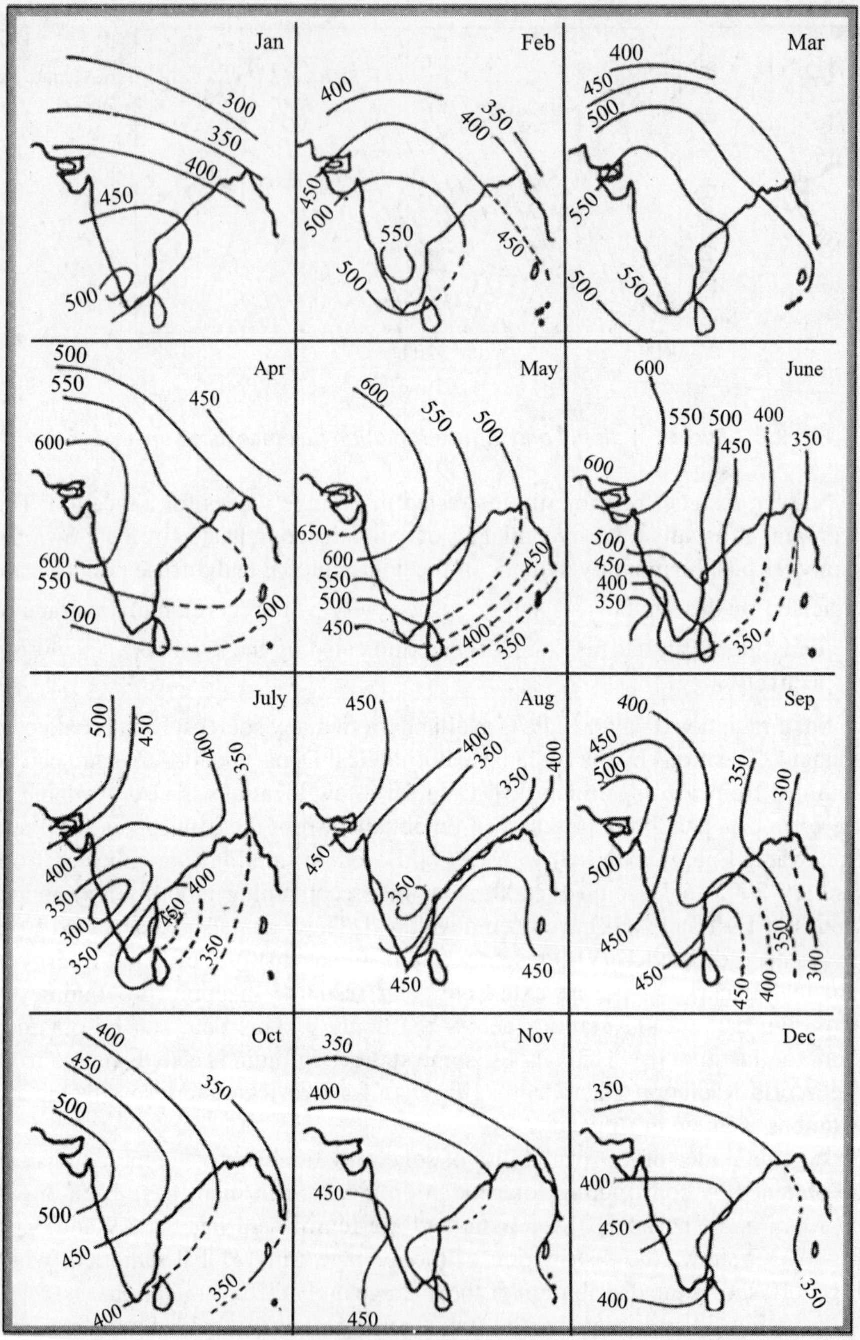

Fig. 3.9 *Average daily global radiation over India in cal/cm^2-day [10]*

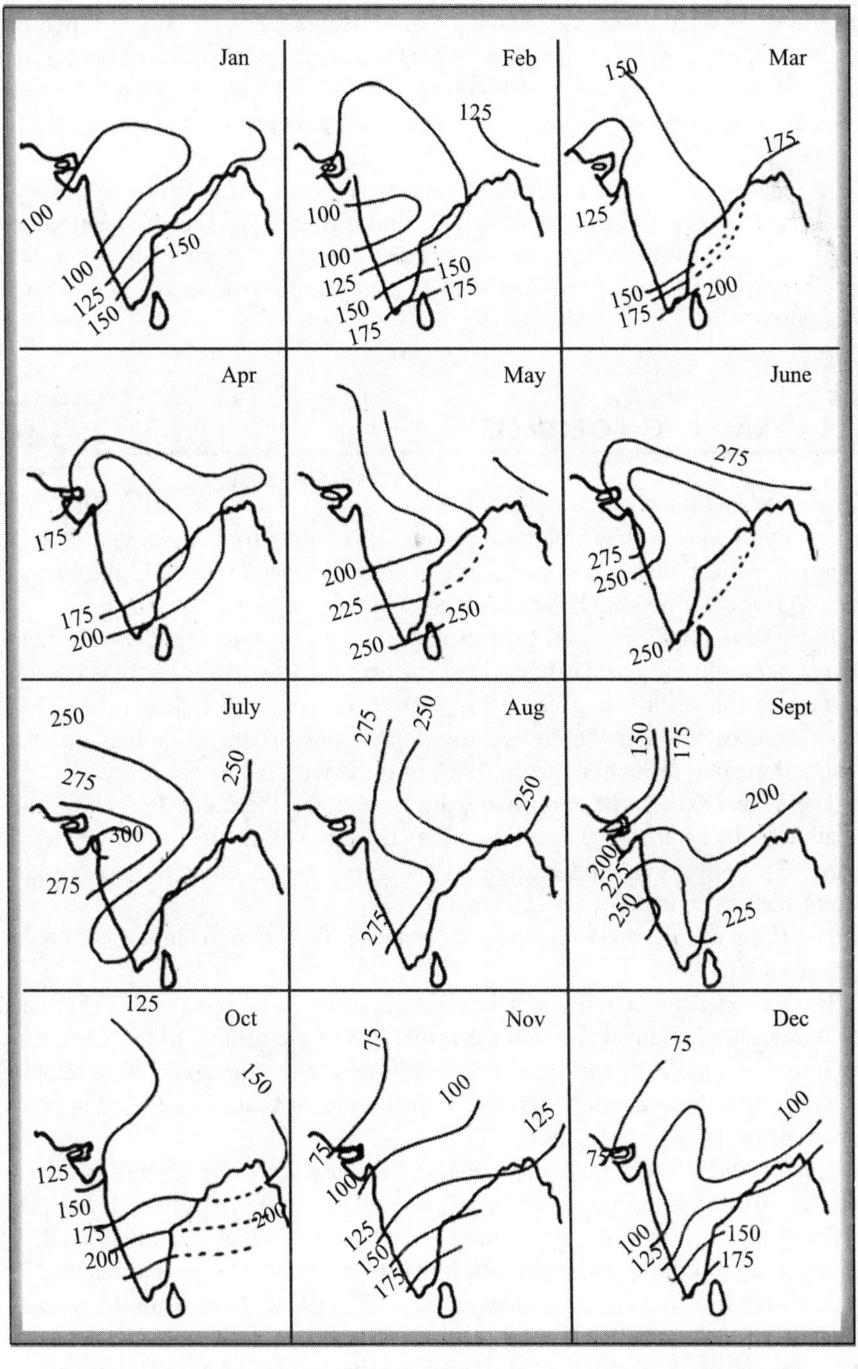

Fig. 3.10 *Average daily diffuse radiation over India in cal/cm²-day [10]*

From Fig. 3.10, it is observed that the annual average daily diffuse radiation received over the whole country is around 175 langleys per day (2.04 kWh/m^2-day). The maximum values measured over the whole country are about 300 langleys per day in Gujarat in July, while minimum values, between 75 and 100 langleys per day, are measured over many parts of the country during November and December as winter sets in.

In recent years, an issue of some concern in India is the increased presence of aerosols in the atmosphere due to pollution caused by vehicular emissions, industrialisation and biomass burning and also due to dust storms. This increase is reported to be the cause of reduction of measured values of the global horizontal irradiance [11].

SOLAR RADIATION GEOMETRY 3.6

In order to find the beam energy falling on a surface having any orientation, it is necessary to convert the value of the beam flux coming from the direction of the sun to an equivalent value corresponding to the normal direction to the surface. Relationships for making this conversion will now be given.

If θ is the angle between an incident solar beam flux I_{bn} *(Direct Normal Irradiance)* and the normal to a plane surface, then the equivalent flux falling normal to the surface is given by I_{bn} cos θ. The angle θ can be related by a general equation to ϕ the latitude, β the slope, γ the surface azimuth angle, δ the declination, and, ω the hour angle. Each of these will first be defined.

The *latitude* ϕ of a location is the angle made by the radial line joining the location to the centre of the earth with the projection of the line on the equatorial plane. By convention, the latitude is measured as positive for the northern hemisphere. It can vary from –90° to + 90°.

The *slope* β is the angle made by the plane surface with the horizontal. It can vary from 0 to 180°.

The *surface azimuth angle* γ is the angle made in the horizontal plane between the horizontal line due south and the projection of the normal to the surface on the horizontal plane. It can vary from –180° to +180°. The convention adopted here is to take the angle to be positive if the normal is east of south and negative if west of south.

The *declination* δ is the angle made by the line joining the centres of the sun and the earth with the projection of this line on the equatorial plane. It arises by virtue of the fact that the earth rotates about an axis, which makes an angle of approximately 66.5° with the plane of its rotation around the sun. The declination angle varies from a maximum value of +23.45° on June 21 to a minimum value of –23.45° on December 21. It is zero on the two equinox days of March 21 and September 22. Cooper [12] has given the following simple relation for calculating the declination

$$\delta \text{ (in degrees)} = 23.45 \sin \left[\frac{360}{365} (284 + n) \right] \tag{3.4}$$

where n is the day of the year. Equation (3.4) is plotted in Fig. 3.11. Its accuracy of prediction is adequate for engineering purposes. A more accurate empirical equation for calculating the declination has been given by Iqbal [13].

The *hour angle* ω is an angular measure of time and is equivalent to 15° per hour. It also varies from –180° to +180°. The convention adopted is to measure it from noon based on local apparent time (LAT). Further, it is taken to be positive in the morning and negative in the afternoon. (The term 'local apparent time' will be defined later.)

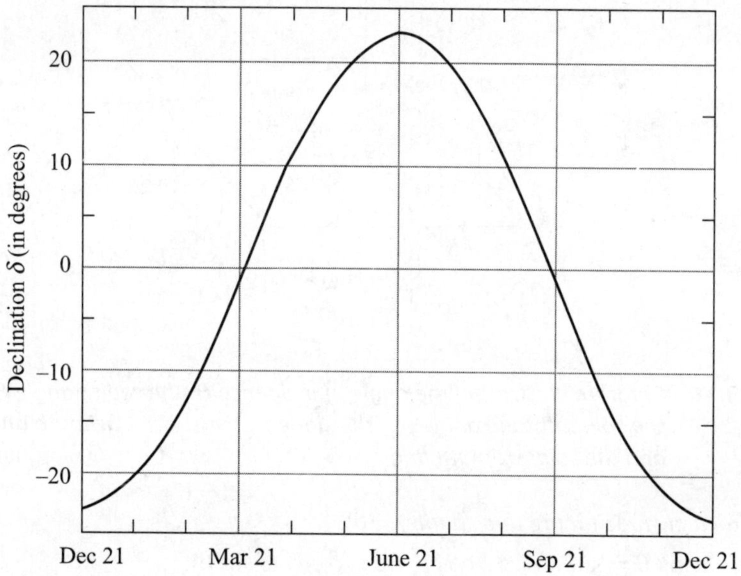

Fig. 3.11 *Variation of declination over the year – Eq. (3.4)*

Based on the preceding definitions, it can be shown that

$$\begin{aligned}\cos\theta = {} & \sin\phi\,(\sin\delta\cos\beta + \cos\delta\cos\gamma\cos\omega\sin\beta) \\ & + \cos\phi\,(\cos\delta\cos\omega\cos\beta - \sin\delta\cos\gamma\sin\beta) \\ & + \cos\delta\sin\gamma\sin\omega\sin\beta\end{aligned} \tag{3.5}$$

Special cases of Eq. (3.5) are normally required. Some of these are as follows

Vertical surface $\beta = 90°$,

$$\begin{aligned}\cos\theta = {} & \sin\phi\cos\delta\cos\gamma\cos\omega - \cos\phi\sin\delta\cos\gamma \\ & + \cos\delta\sin\gamma\sin\omega\end{aligned} \tag{3.6}$$

Horizontal surface $\beta = 0°$,

$$\cos\theta = \sin\phi \sin\delta + \cos\phi \cos\delta \cos\omega \tag{3.7}$$

The angle θ in this case is the *zenith angle* θ_z (defined earlier in Section 3.3). The complement of the zenith angle is also used quite often in calculations. It is called the *solar altitude angle* and will be denoted by the symbol α_a. Some of the above angles are shown in Fig. 3.12.

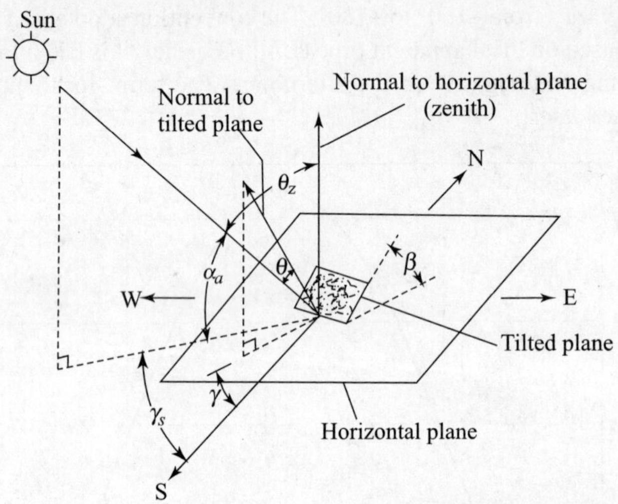

Fig. 3.12 *Diagram illustrating the angle of incidence θ, the zenith angle θ_z, the solar altitude angle α_a, the slope β, the surface azimuth angle γ and the solar azimuth angle γ_s*

Inclined surface facing due south $\gamma = 0°$,

$$\begin{aligned}\cos\theta &= \sin\phi\,(\sin\delta\cos\beta + \cos\delta\cos\omega\sin\beta)\\ &\quad + \cos\phi\,(\cos\delta\cos\omega\cos\beta - \sin\delta\sin\beta)\\ &= \sin\delta\sin(\phi-\beta) + \cos\delta\cos\omega\cos(\phi-\beta)\end{aligned} \tag{3.8}$$

Vertical surface facing due south $\beta = 90°$, $\gamma = 0°$,

$$\cos\theta = \sin\phi\cos\delta\cos\omega - \cos\phi\sin\delta \tag{3.9}$$

Inclined surface facing due north $\gamma = 180°$,

$$\cos\theta = \sin\delta\sin(\phi+\beta) + \cos\delta\cos\omega\cos(\phi+\beta) \tag{3.10}$$

The angle of incidence θ can also be expressed in terms of θ_z the zenith angle, β the slope, γ the surface azimuth angle and γ_s the solar azimuth angle. Braun and Mitchell [14] have shown that

$$\cos\theta = \cos\theta_z\cos\beta + \sin\theta_z\sin\beta\cos(\gamma_s - \gamma) \tag{3.11}$$

The *solar azimuth angle* γ_s is the angle made in the horizontal plane between the horizontal line due south and the projection of the line of sight of the sun on the horizontal plane. Thus, it gives the direction of the shadow cast in the horizontal plane by a vertical rod. By convention, the solar azimuth angle is taken to be positive if the projection of the line of sight is east of south and negative if west of south. In order to use Eq. (3.11), it is necessary to first calculate θ_z and γ_s. θ_z is obtained from Eq. (3.7), while γ_s is obtained from the expression [13]

$$\cos \gamma_s = (\cos \theta_z \sin \phi - \sin \delta)/\sin \theta_z \cos \phi \tag{3.12}$$

Example 3.1

(a) Calculate the angle made by beam radiation with the normal to a flat-plate collector on May 1 at 0900 h (local apparent time). The collector is located in New Delhi (28°35′ N, 77°12′ E). It is tilted at an angle of 36° with the horizontal and is pointing due south.

(b) Repeat the calculation for 1200 h (local apparent time).

Solution

(a) For this case, $\gamma = 0°$ and Eq. (3.8) is applicable. On May 1, $n = 121$. From Eq. (3.4),

$$\delta = 23.45 \sin \left[\frac{360}{365}(284 + 121)\right] = 14.90°$$

At 0900 h (local apparent time), $\omega = 45°$. Substituting in Eq. (3.8) we have

$$\cos \theta = \sin 14.90° \sin (28.58° - 36°) + \cos 14.90° \cos 45° \cos (28.58° - 36°)$$

$$= 0.6444$$

$$\theta = 49.9°$$

The angle θ can also be calculated using Eq. (3.11) in conjunction with Eqs (3.7) and (3.12). From Eq. (3.7),

$$\cos \theta_z = \sin 28.58° \sin 14.90° + \cos 28.58° \cos 14.90° \cos 45°$$

$$= 0.7231$$

$$\theta_z = 43.69°$$

From Eq. (3.12),

$$\cos \gamma_s = \frac{[\cos 43.69° \sin 28.58° - \sin 14.90°]}{\sin 43.69° \cos 28.58°} = 0.1464$$

$$\gamma_s = 81.58°$$

Substituting into Eq. (3.11), we get

$$\cos \theta = \cos 43.69° \cos 36° + \sin 43.69° \sin 36° \cos(81.58° - 0°)$$

$$= 0.6444$$

$$\theta = 49.9°$$

(b) At 1200 h (local apparent time), $\omega = 0°$.
Proceeding in the same manner as in part (a), we have from Eq. (3.8),

$$\cos\theta = 0.9251$$
$$\theta = 22.3°$$

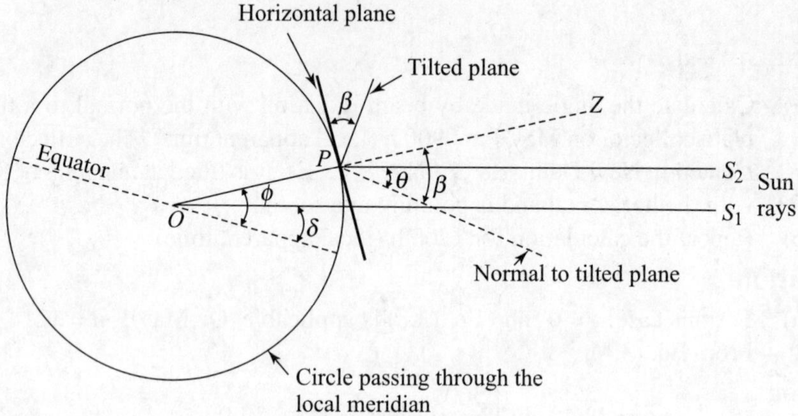

Fig. 3.13 *Example 3.1*

Alternatively, consider Fig. 3.13, which shows a cross-section through the local meridian. Since $\omega = 0°$ and $\gamma = 0°$, all the other angles involved, viz. θ, ϕ, δ and β lie in this plane. The answer for part (b) is obvious from Fig. 3.13. Considering the parallel sun rays OS_1 and PS_2 and the intersecting line OPZ, we have

$$\angle POS_1 = \angle ZPS_2$$

i.e. $\phi - \delta = \beta - \theta$

or, $\theta = \beta - \phi + \delta = 22.3°$

Apparent Motion of the Sun

It will be useful at this stage to describe the apparent motion of the sun as seen from the earth. To an observer on the earth, on any given day, the sun rises in the east, moves in a plane tilted at an angle of $(90° - \phi)$ with the horizontal, and finally sets in the west. Thus, the apparent plane in which the sun moves intersects the horizontal plane in a line pointing east-west. However, because of the declination angle this E-W line of intersection does not coincide with the E-W line passing through the observer O (Fig. 3.14). For a location in the northern hemisphere, this line is to the south in winter (declination negative), to the north in summer (declination positive) and coincides with the E-W line passing through the observer on the two equinox days of March 21 and September 22 (declination zero).

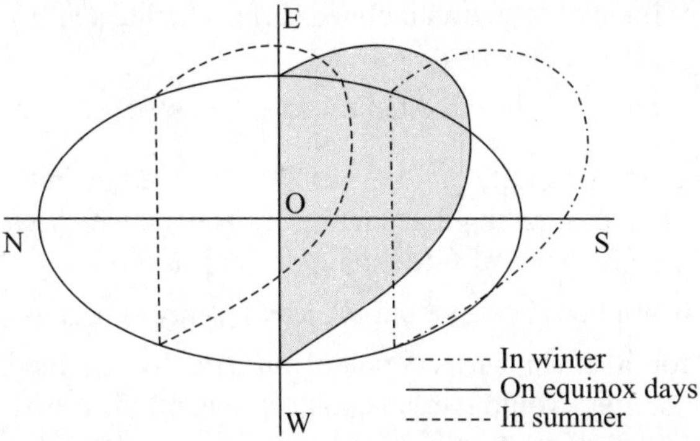

Fig. 3.14 *Apparent plane of motion of the sun for a location in the northern hemisphere*

Sunrise, Sunset and Day Length

❑ *Horizontal Surface* The hour angle corresponding to sunrise or sunset (ω_s) on a horizontal surface can be found from Eq. (3.7) if one substitutes the value of 90° for the zenith angle. We obtain

$$\cos \omega_s = -\tan \phi \tan \delta$$

$$\omega_s = \cos^{-1} (-\tan \phi \tan \delta) \tag{3.13}$$

Equation (3.13) yields a positive and a negative value for ω_s, the positive value corresponding to sunrise and the negative to sunset. Since 15° of the hour angle is equivalent to 1 hour, the corresponding *day length* (in hours)

$$S_{\max} = \frac{2}{15}\,\omega_s = \frac{2}{15}\,\cos^{-1}(-\tan \phi \tan \delta) \tag{3.14}$$

where ω_s is in degrees.

❑ *Inclined Surface Facing Due South* The hour angle at sunrise or sunset as seen by an observer on an inclined surface facing south ($\gamma = 0°$) will also be given by Eq. (3.13) if the day under consideration lies between September 22 and March 21, and the location is in the northern hemisphere. This is because during this period, the declination is negative and the apparent plane of motion of the sun intersects the horizontal plane in an E-W line which lies to the south of the E-W line passing through an observer on the inclined plane. However, if the day under consideration lies between March 21 and September 22, the hour angle at sunrise or sunset (ω_{st}) would be smaller in magnitude than the value given by Eq. (3.13) and would be obtained by substituting $\theta = 90°$ in Eq. (3.8). This yields

$$\omega_{st} = \cos^{-1} [-\tan (\phi - \beta) \tan \delta] \tag{3.15}$$

Thus the magnitude of ω_{st} for an inclined surface facing south ($\gamma = 0°$) is given by

$$|\omega_{st}| = \min[|\cos^{-1}(-\tan\phi\tan\delta)|, |\cos^{-1}\{-\tan(\phi-\beta)\tan\delta\}|] \quad (3.16)$$

❑ *Inclined Surface Facing Due North* Proceeding in the same manner as for an inclined surface facing due south, it can be shown by using Eqs (3.13) and (3.5) that for an inclined surface facing due north

$$|\omega_{st}| = \min[|\cos^{-1}(-\tan\phi\tan\delta)|, |\cos^{-1}\{-\tan(\phi+\beta)\tan\delta\}|] \quad (3.17)$$

In general, for a plane surface not symmetrically oriented, hour angles at sunrise and sunset would be unequal in magnitude apart from having opposite signs. The general procedure would be to calculate ω_{st} by substituting $\theta = 90°$ in Eq. (3.5) and by using Eq. (3.13). Depending upon the day of the year and the orientation of the surface, proper judgement would need to be exercised in selecting the correct values from the solutions thus obtained.

Example 3.2

Calculate the hour angle at sunrise and sunset on June 21 and December 21 for a surface inclined at an angle of 10° and facing due south ($\gamma = 0°$). The surface is located in Mumbai (19°07′ N, 72°51′ E).

Solution

On June 21, $\delta = 23.45°$. From Eq. (3.16),

$$|\omega_{st}| = \min[|\cos^{-1}(-\tan 19.12° \tan 23.45°)|, |\cos^{-1}\{-\tan(19.12° - 10°)\tan 23.45°\}|]$$

$$= \min[98.6°, 94.0°] = 94.0°$$

$$\omega_{st} = \pm 94.0°$$

On December 21, $\delta = -23.45°$

From Eq. (3.16),

$$|\omega_{st}| = \min[|\cos^{-1}\{-\tan 19.12° \tan(-23.45°)\}|, |\cos^{-1}\{-\tan(19.12° - 10°)\tan(-23.45°)\}|]$$

$$= \min[81.4°, 86.0°] = 81.4°$$

$$\omega_{st} = \pm 81.4°$$

Local Apparent Time

The time used for calculating the hour angle ω is the *local apparent time*. This can be obtained from the standard time observed on a clock by applying two corrections. The first correction arises because of the difference between the longitude of a location and the meridian on which the standard time is based. The correction has a magnitude of 4 minutes for every degree difference in longitude.

The second correction called the *equation of time* correction is due to the fact that the earth's orbit and rate of rotation are subject to small variations. This correction is based on experimental observations and is plotted in Fig. 3.15.

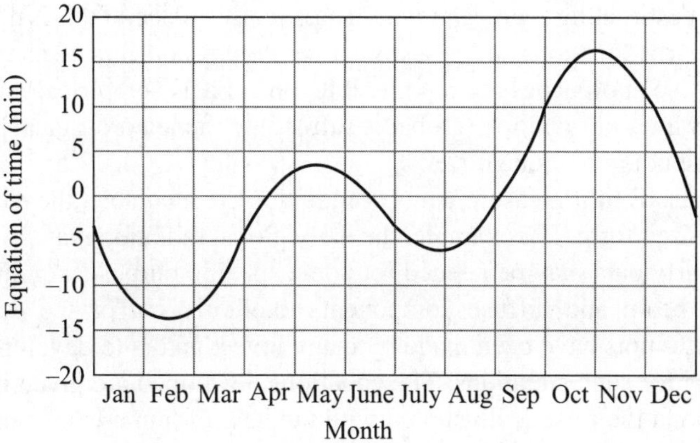

Fig. 3.15 *Equation of time correction*

Thus,

Local apparent time = Standard time

$\pm$ 4 (Standard time longitude – longitude of location)

+ (Equation of time correction)

The negative sign in the first correction is applicable for the eastern hemisphere, while the positive sign is applicable for the western hemisphere.

The equation of time correction (in minutes) can also be calculated from the following empirical relation [13]:

$$E = 229.18\,(0.000075 + 0.001868 \cos B - 0.032077 \sin B - 0.014615 \cos 2B - 0.04089 \sin 2B) \tag{3.18}$$

where $B = (n - 1)\,360/365$ and n is the day of the year.

Example 3.3

Determine the local apparent time (LAT) corresponding to 1430 h (IST) at Mumbai (19°07′ N, 72°51′ E) on July 1. In India, standard time is based on 82.50°E.

Solution

From Eq. (3.18), the equation of time correction on July 1 is (–3.5) minutes. Therefore,

$$\begin{aligned}\text{Local apparent time} &= 1430\text{ h} - 4(82.50 - 72.85)\text{ minutes} + (-3.5\text{ minutes})\\ &= 1430\text{ h} - 38.6\text{ minutes} - 3.5\text{ minutes} = 1348\text{ h}\end{aligned}$$

PREDICTING THE AVAILABILITY OF SOLAR RADIATION 3.7

The measurement of solar radiation over a period of time at a place is the best approach for estimating average radiation data for that place. If this is not possible, data from nearby locations having a similar geography and climate can be used. In case both these approaches are not possible, one can use empirical relationships linking the values of radiation (global or diffuse) with meteorological parameters like sunshine hours or cloud cover.

Sometimes, even if measurements of radiation are available, the data may not be in the desired form. For example, the daily global radiation may be measured, whereas hourly data may be needed for some specific purpose. Again, one may require the beam and diffuse components explicitly out of the daily global radiation. Attempts have been made by many investigators to develop empirical relationships for such situations. The equations developed are given in Sections 3.7.1–3.7.4 and their use is illustrated with the help of numerical examples. The relationships given are generally valid for cloudy skies.

3.7.1 Monthly Average Daily Global Radiation

The first attempt at estimating solar radiation was due to Angstrom who suggested that it could be related to the amount of sunshine by a simple linear relation of the form

$$\frac{\overline{H}_g}{\overline{H}_c} = a + b\left(\frac{\overline{S}}{\overline{S}_{\max}}\right) \tag{3.19}$$

where,

$\overline{H}_g$ = monthly average of the daily global radiation on a horizontal surface or daily global horizontal irradiance at a location (kJ/m^2-day),

$\overline{H}_c$ = monthly average of the daily global radiation on a horizontal surface at the same location on a clear day (kJ/m^2-day),

$\overline{S}$ = monthly average of the sunshine hours per day at the location (h),

$\overline{S}_{\max}$ = monthly average of the maximum possible sunshine hours per day at the location, i.e. the day length on a horizontal surface (h), and

a, b = constants obtained by fitting data.

Because of difficulties in deciding what constitutes a clear day, Page [15] suggested that $\overline{H}_c$ in Eq. (3.19) be replaced by $\overline{H}_o$, the monthly average of the daily extraterrestrial radiation which would fall on a horizontal surface at the location under consideration.

$$\frac{\overline{H}_g}{\overline{H}_o} = a + b\left(\frac{\overline{S}}{\overline{S}_{max}}\right) \tag{3.20}$$

Values of a and b have been obtained by regression analysis of measured values of global solar radiation and sunshine duration of many cities in the world by Lof *et al.* [16]. The values are given in Table 3.2. A similar set of values have also been obtained for 17 Indian cities [17] and are given in Table 3.3.

Table 3.2 Constants a and b in Eq. (3.20) for many cities of the world*

Location	Range of $\overline{S}/\overline{S}_{max}$	a	b
Albuquerque, New Mexico, USA	0.68–0.85	0.41	0.37
Atlanta, Georgia, USA	0.45–0.71	0.38	0.26
Blue Hill, Mass., USA	0.42–0.60	0.22	0.50
Brownsville, Texas, USA	0.47–0.80	0.35	0.31
Buenos Aires, Argentina	0.47–0.68	0.26	0.50
Charleston, S.C., USA	0.60–0.75	0.48	0.09
Dairen, Manchuria	0.55–0.81	0.36	0.23
El Paso, Texas, USA	0.78–0.88	0.54	0.20
Ely, Nevada, USA	0.61–0.89	0.54	0.18
Hamburg, Germany	0.11–0.49	0.22	0.57
Honolulu, Hawaii, USA	0.57–0.77	0.14	0.73
Madison, Wisconsin, USA	0.40–0.72	0.30	0.34
Malange, Angola	0.41–0.84	0.34	0.34
Miami, Florida, USA	0.56–0.71	0.42	0.22
Nice, France	0.49–0.76	0.17	0.63
Pune, India	0.25–0.49	0.30	0.51
	0.65–0.89	0.41	0.34
Stanleyville, Congo	0.34–0.56	0.28	0.39
Tamanrasset, Algeria	0.76–0.88	0.30	0.43

*Adapted with permission from Lof *et al.* [16].

Table 3.3 Constants a and b in Eq. (3.20) for Indian cities*

Location	a	b	Mean error (per cent)
Ahmedabad	0.28	0.48	3.0
Bengaluru	0.18	0.64	3.9
Bhavnagar	0.28	0.47	2.8
Kolkata	0.28	0.42	1.3
Goa	0.30	0.48	2.1
Jodhpur	0.33	0.46	2.0
Kodaikanal	0.32	0.55	2.9
Chennai	0.30	0.44	3.5
Mangalore	0.27	0.43	4.2
Minicoy	0.26	0.39	1.4
Nagpur	0.27	0.50	1.6
New Delhi	0.25	0.57	3.0
Pune	0.31	0.43	1.9
Shillong	0.22	0.57	3.0
Srinagar	0.35	0.40	4.7
Thiruvananthapuram	0.37	0.39	2.5
Vishakhapatnam	0.28	0.47	1.2

* Used with permission from Modi and Sukhatme [17].

In the above computations, the quantity $\bar{H}_o$ is the mean of the value (H_o) for each day of the month. H_o is obtained by integrating over the day length as follows:

$$H_o = I_{sc}\left(1 + 0.033 \cos \frac{360n}{365}\right)\int (\sin\phi \sin\delta + \cos\phi \cos\delta \cos\omega)\, dt \tag{3.21}$$

Now, $t = \dfrac{180\omega}{15\pi}$

where t is in hours and ω is in radians.

Hence, $dt = \dfrac{180}{15\pi} d\omega$

Substituting in Eq. (3.21)

$$H_o = \frac{12}{\pi} I_{sc}\left(1 + 0.033 \cos \frac{360n}{365}\right)\int_{-\omega_s}^{+\omega_s} (\sin\phi \sin\delta + \cos\phi \cos\delta \cos\omega)\, d\omega$$

$$= \frac{24}{\pi} I_{sc}\left(1 + 0.033 \cos \frac{360n}{365}\right)(\omega_s \sin\phi \sin\delta + \cos\phi \cos\delta \sin\omega_s) \tag{3.22}$$

The calculation of $\bar{H}_o$ has been simplified by Klein [18], who has determined the particular day in each month on which the extraterrestrial radiation is nearly equal to the monthly mean value. The dates on which the value of H_o is equal to $\bar{H}_o$ are as follows: January 17, February 16, March 16, April 15, May 15, June 11, July 17, August 16, September 15, October 15, November 14 and December 10. As expected, these dates are close to the middle of the month.

Equation (3.20) can be used for calculating the average daily global radiation at a location when data on sunshine hours is available, and values of *a* and *b* are known for a nearby location with a similar geography and climate. The procedure is illustrated in the following example.

Example 3.4

Estimate the monthly average daily global horizontal irradiance at Vadodara (22°00′ N, 73°10′ E) during the month of March if the average sunshine hours per day is 9.5.

Solution

Let us assume that the values of $a = 0.28$ and $b = 0.48$ obtained for Ahmedabad (Table 3.3) are valid for Vadodara. This is a good assumption since the two cities are fairly close to each other and their climate is similar. Based on Klein's recommendation, we will take $\bar{H}_o$ to be equal to the value of H_o on March 16. Further, we will assume that this recommendation is also valid for the calculation of $\bar{S}_{max}$.

On March 16, $n = 75$. From Eq. (3.4),

$$\delta = 23.45 \sin\left[\frac{360}{365}(284 + 75)\right] = -2.42°$$

From Eq. (3.13), the sunrise hour angle

$$\omega_s = \cos^{-1}\{-\tan 22° \tan(-2.42°)\} = 89.02° = 1.554 \text{ radians}$$

$$\text{Day length, } S_{max} = \frac{2}{15} \times 89.02 = 11.87 \text{ h}$$

From Eq. (3.22),

$$H_o = \frac{24}{\pi} \times 1.367 \times 3600 \left(1 + 0.033 \cos\frac{360}{365} \times 75\right) \{1.554 \sin 22° \sin(-2.42°) + \cos 22° \cos(-2.42°) \sin 89.02°\}$$

$$= 34\,206 \text{ kJ/m}^2\text{-day}$$

Therefore, from Eq. (3.20),

$$\bar{H}_g = \left\{0.28 + 0.48\left(\frac{9.5}{11.87}\right)\right\} \times 34\,206 = 22\,718 \text{ kJ/m}^2\text{-day}$$

Other meteorological parameters have also been used for predicting solar radiation. These include cloud cover (the amount of sky dome covered by clouds) and precipitation (the number of days in the month with precipitation greater than 0.3 mm). However, in general, the sunshine ratio parameter ($\overline{S}/\overline{S}_{max}$) has been found to be the most reliable predictor.

Correlations similar to Eq. (3.20) have been suggested by many investigators based on data for specific locations and countries. Some of these include additional parameters and help to generalize the applicability of the correlations. Gopinathan [19] has suggested the correlation

$$\frac{\overline{H}_g}{\overline{H}_o} = a_1 + b_1 \left[\frac{\overline{S}}{\overline{S}_{max}} \right] \tag{3.23}$$

Equation (3.23) is based on the data of 40 locations around the world. The constants a_1 and b_1 are related to three parameters, the latitude, the elevation and the sunshine hours as follows:

$$a_1 = -0.309 + 0.539 \cos \phi - 0.0693\, E_L + 0.290 \left(\frac{\overline{S}}{\overline{S}_{max}} \right)$$

$$b_1 = 1.527 - 1.027 \cos \phi + 0.0926\, E_L - 0.359 \left(\frac{\overline{S}}{\overline{S}_{max}} \right) \tag{3.24}$$

where ϕ = latitude (in degrees) and

E_L = elevation of the location above mean sea level (in kilometers)

Equation (3.24) is recommended for predicting the daily global radiation at locations all over the world including locations in India.

Example 3.5

Use Gopinathan's correlation to calculate the value of $\overline{H}_g$ for Vadodara for the month of March.

Given: $E_L = 34$ m

Solution

From Example 3.4, $\phi = 22.00°$, $\overline{S} = 9.5$ h, $\overline{S}_{max} = 11.87$ h and $\overline{H}_o = 34\,206$ kJ/m^2-day

Substituting in Eq. (3.24),

$$a_1 = -0.309 + 0.539 \cos 22.00° - 0.0693 \times 0.034 + 0.290 \left(\frac{9.5}{11.87} \right)$$

$$= 0.421$$

$$b_1 = 1.527 - 1.027 \cos 22.00° + 0.0926 \times 0.034 - 0.359 \left(\frac{9.5}{11.87} \right)$$

$$= 0.291$$

Therefore

$$\overline{H}_g = \left\{0.421 + 0.291\left(\frac{9.5}{11.87}\right)\right\} \times 34\,206 = 22\,367 \text{ kJ/m}^2\text{-day}$$

3.7.2 Monthly Average Daily Diffuse Radiation

Based on a study of data for a few countries, Liu and Jordan [20] showed that the daily diffuse-to-global radiation ratio could be correlated against the daily global-to-extraterrestrial radiation ratio. The correlation was expressed by the following cubic equation

$$\frac{\overline{H}_d}{\overline{H}_g} = 1.390 - 4.027\left[\frac{\overline{H}_g}{\overline{H}_o}\right] + 5.531\left[\frac{\overline{H}_g}{\overline{H}_o}\right]^2 - 3.108\left[\frac{\overline{H}_g}{\overline{H}_o}\right]^3 \tag{3.25}$$

where $\overline{H}_d$ = monthly average of the daily diffuse radiation on a horizontal surface (kJ/m^2-day).

The other symbols have the same meaning as given earlier. The ratio $(\overline{H}_g/\overline{H}_o)$ is often denoted by the symbol $\overline{K}_T$ and is called the monthly average clearness index. Kreith and Kreider [21] have pointed out that Eq. (3.25) has been obtained with a value of 1394 W/m^2 for the solar constant.

As in the case of monthly average daily global radiation, many investigators have developed empirical equations for estimating the diffuse-to-global radiation ratio for various parts of the world. Gopinathan and Soler [22] have examined radiation data for 40 widely spread locations all over the world in the latitude range 36°S to 36°N. They have proposed the following equation involving the clearness index and the sunshine ratio:

$$\frac{\overline{H}_d}{\overline{H}_g} = 0.87813 - 0.33280\,\overline{K}_T - 0.53039\left[\frac{\overline{S}}{\overline{S}_{max}}\right] \tag{3.26}$$

Equation (3.26) is based on more recent data than that available to Liu and Jordan and is recommended for use for predicting the daily diffuse radiation at locations across the world.

When available Indian data was analysed, the following linear equation was obtained by Modi and Sukhatme [17]:

$$\frac{\overline{H}_d}{\overline{H}_g} = 1.411 - 1.696\left[\frac{\overline{H}_g}{\overline{H}_o}\right] \tag{3.27}$$

Garg and Garg [23] have examined radiation data for 11 Indian cities and proposed the equation

$$\frac{\overline{H}_d}{\overline{H}_g} = 0.8677 - 0.7365\left[\frac{\overline{S}}{\overline{S}_{max}}\right] \tag{3.28}$$

Equations (3.27) and (3.28) agree well with each other. Either one of the equations may be used for Indian locations. However, when compared to Gopinathan and Soler's correlation, it is observed that there are significant differences. Equation (3.26) predicts values for the diffuse radiation which are lower than the predictions of Eqs (3.27) and (3.28). This is due to the fact that the diffuse component is in fact much larger in India.

Example 3.6

Estimate the monthly average daily diffuse radiation on a horizontal surface at Vadodara during the month of March.

Solution

We will first use Eqs (3.27) and (3.28) for making the estimation. From Example 3.4, $\overline{H}_g$ = 22 718 kJ/m^2-day and $\overline{H}_o$ = 34 206 kJ/m^2-day. Monthly average clearness index $\overline{K}_T$ = 0.6642. Also $\overline{S}$ = 9.5 h, $\overline{S}_{max}$ = 11.87 h.

From Eq. (3.27),

$$\overline{H}_d = 22\,718\,(1.411 - 1.696 \times 0.6642) = 6465 \text{ kJ/m}^2\text{-day}$$

From Eq. (3.28),

$$\overline{H}_d = 22\,718\left\{0.8677 - 0.7365 \times \left(\frac{9.5}{11.87}\right)\right\} = 6321 \text{ kJ/m}^2\text{-day}$$

Thus we see that the predictions made by Eqs (3.27) and (3.28) differ by only 2.2 per cent.

Next, we will use Eq. (3.26). We get

$$\overline{H}_d = 22\,718\left\{0.87813 - 0.33280 \times 0.6642 - 0.53039\left(\frac{9.5}{11.87}\right)\right\}$$

$$= 5284 \text{ kJ/m}^2\text{-day}$$

This value is 18 per cent less than the prediction of 6465 kJ/m^2-day made by Eq. (3.27) and 16 per cent less than the prediction of 6321 kJ/m^2-day made by Eq. (3.28). As stated earlier, Eq. (3.26) significantly underpredicts the diffuse component for Indian locations. It is therefore desirable to use either Eq. (3.27) or Eq. (3.28) for India.

3.7.3 Monthly Average Hourly Global Radiation

A number of studies have also been conducted with the objective of obtaining relations for predicting the diurnal variation of the monthly average hourly global radiation at a location. Collares-Pereira and Rabl [24] have developed the following relation

$$\frac{\overline{I}_g}{\overline{H}_g} = \frac{\overline{I}_o}{\overline{H}_o}\,(a + b\cos\omega) \qquad (3.29)$$

where $a = 0.409 + 0.5016 \sin(\omega_s - 60°)$,

$b = 0.6609 - 0.4767 \sin(\omega_s - 60°)$,

$\overline{I}_g$ = monthly average of the daily global radiation on a horizontal surface or hourly global horizontal irradiance (kJ/m^2-h),

$\overline{I}_o$ = monthly average of the hourly extraterrestrial radiation on a horizontal surface (kJ/m^2-h).

The other symbols in Eq. (3.29) have been defined earlier. Equation (3.29) has been developed on the basis of the following facts:

1. Measured data generally show a similarity between the diurnal variation of $\overline{I}_g$ and $\overline{I}_o$.
2. There is a close correlation between the values of the ratios $(\overline{I}_g/\overline{H}_g)$ and $(\overline{I}_o/\overline{H}_o)$.

It will be noted from Eq. (3.29) that the symbol I has been used for denoting an hourly value (kJ/m^2-h), while earlier it was used for denoting an instantaneous value, i.e. a flux (kW/m^2). As stated earlier in Section 3.5, for most situations involving solar radiation, transient processes occur at a slow pace. It is, therefore, not necessary to make a distinction between the two quantities. Wherever needed, an hourly value (kJ/m^2-h) can be calculated from an instantaneous value by taking the instantaneous value (kW/m^2) at the mid-point of the hour and multiplying by 3600.

Gueymard [25] modified Eq. (3.29) suggested by Collares-Pereira and Rabl by incorporating a normalizing factor f_c. Thus Eq. (3.29) becomes

$$\frac{\overline{I}_g}{\overline{H}_g} = \frac{\overline{I}_o}{\overline{H}_o}(a + b\cos\omega)/f_c \tag{3.30}$$

$$\text{where } f_c = a + 0.5\, b \left[\frac{\dfrac{\pi\omega_s}{180} - \sin\omega_s \cos\omega_s}{\sin\omega_s - \dfrac{\pi\omega_s}{180}\cos\omega_s}\right]$$

Subsequently Gueymard [26] has reviewed a number of predictive models using a large data set for 135 locations and has concluded that Eq. (3.30) is the simplest and most satisfactory correlation for predicting monthly average hourly global radiation for sites all over the world within latitudes 65°N to 65°S.

Satyamurty and Lahiri [27] have tested the predictions of Eq. (3.29) against the measured data of 14 locations in India and shown that there is a good correlation between the two sets of values. The rms difference between the predicted and measured values of $(\overline{I}_g/\overline{H}_g)$ ranged only between 2.6 and 5.5 per cent for 13 of the 14 locations. A significant difference of 10.4 per cent was obtained only at one location.

3.7.4 Monthly Average Hourly Diffuse Radiation

In a manner similar to that adopted for developing Eq. (3.29), Liu and Jordan [20] have suggested the following relation for estimating the monthly average hourly diffuse radiation

$$\frac{\bar{I}_d}{\bar{H}_d} = \frac{\bar{I}_o}{\bar{H}_o} \tag{3.31}$$

Satyamurty and Lahiri [27] have also tested the predictions of Eq. (3.31) against the measured data of 14 locations in India. In this case, relatively poor agreement is obtained, the rms difference between the predicted and measured values of ($\bar{I}_d/\bar{H}_d$) ranging between 5.7 and 13.4 per cent. Satyamurty and Lahiri have therefore suggested the following improved relation

$$\frac{\bar{I}_d}{\bar{H}_d} = (a' + b' \cos \omega) \frac{\bar{I}_o}{\bar{H}_o} \tag{3.32}$$

where $a' = 0.4922 + \{0.27/(\bar{H}_d/\bar{H}_g)\}$ for $0.1 \le (\bar{H}_d/\bar{H}_g) \le 0.7$

or $a' = 0.76 + \{0.113/(\bar{H}_d/\bar{H}_g)\}$ for $0.7 < (\bar{H}_d/\bar{H}_g) \le 0.9$

and $b' = 2(1 - a')(\sin \omega_s - \omega_s \cos \omega_s)/(\omega_s - 0.5 \sin 2\omega_s)$

Example 3.7

Use the prediction equations to calculate the monthly average of the daily and hourly global and diffuse radiation during the month of April on a horizontal surface at New Delhi (28°35′ N, 77°12′ E, elevation 216 m). Time: 0900 to 1000 h (LAT). The average number of sunshine hours per day is 8.6. Compare the predicted values with the available measurements.

Solution

Daily Radiation

The representative day in April is the 15th. On April 15, $n = 105$

From Eq. (3.4), $\delta = 23.45 \sin\left[\frac{360}{365}(284 + 105)\right] = 9.42°$

From Eq. (3.13),

$$\omega_s = \cos^{-1}(-\tan 28.58° \times \tan 9.42°)$$
$$= 95.18° = 1.661 \text{ radians}$$

$$S_{max} = \frac{2}{15} \times 95.18 = 12.69 \text{ h}$$

Substituting in Eq. (3.22), we get

$$H_o = \frac{24}{\pi} \times 1.367 \times 3600 \left(1 + 0.033 \cos \frac{360}{365} \times 105\right)$$
$$\times (1.661 \sin 28.58° \sin 9.42° + \cos 28.58° \cos 9.42° \sin 95.18°)$$
$$= 37\,034 \text{ kJ/m}^2\text{-day}$$

From Table 3.3, $a = 0.25$ and $b = 0.57$

Putting $\overline{H}_o$ equal to H_o on April 15, and substituting in Eq. (3.20), we get

$$\overline{H}_g = 37\,034[0.25 + 0.57 \times (8.6/12.69)] = 23\,564 \text{ kJ/m}^2\text{-day}$$

Alternatively using Eq. (3.24) suggested by Gopinathan, we get $a_1 = 0.3459$, $b_1 = 0.4019$ and $\overline{H}_g = 22\,895$ kJ/m^2-day which is in close agreement with the value of 23 564 kJ/m^2-day. We proceed with the value of $\overline{H}_g = 23\,564$ kJ/m^2-day.

From Eqs (3.27) and (3.28), we get

$$\overline{H}_d = 23\,564\,[1.411 - 1.696 \times (23\,564/37\,034)]$$
$$= 7820 \text{ kJ/m}^2\text{-day}$$

and $$\overline{H}_d = 23\,564\,[0.8677 - 0.7365 \times (8.6/12.69)] = 8685 \text{ kJ/m}^2\text{-day}$$

Hourly Radiation

The hourly extraterrestrial radiation on a horizontal surface on April 15 between 0900 and 1000 h is obtained by calculating the instantaneous value at 0930 h and multiplying by 3600 s.

From Eq. (3.7),

$$I_o = 1.367\left(1 + 0.033\cos\frac{360}{365}\times 105\right)$$
$$\times (\sin 28.58° \sin 9.42° + \cos 28.58° \cos 9.42° \cos 37.5°)$$
$$= 1.0384 \text{ kW/m}^2 = 3738 \text{ kJ/m}^2\text{-h}$$

Substituting in Eq. (3.30),

$$a = 0.409 + 0.5016 \sin (95.18° - 60°) = 0.69800$$
$$b = 0.6609 - 0.4767 \sin (95.18° - 60°) = 0.38625$$
$$f_c = 0.9931$$

Once again putting $\overline{I}_o$ equal to I_o on April 15, we get

$$\frac{\overline{I}_g}{23\,564} = \frac{3738}{37034}\,(0.69800 + 0.38625 \cos 37.5°)$$

$$\overline{I}_g = 2406 \text{ kJ/m}^2\text{-h}$$

Finally, from Eq. (3.31), with $\overline{H}_d = 7820$ kJ/m^2-day

$$\overline{I}_d = 7820 \times (3738/37\,034) = 789 \text{ kJ/m}^2\text{-h}$$

Instead of Eq. (3.31), if we use Eq. (3.32), a slightly different value is obtained. We get

$$\overline{I}_d = 780 \text{ kJ/m}^2\text{-h}$$

Alternatively, putting $\overline{H}_d = 8685$ kJ/m^2-day, we have

$$\overline{I}_d = 866 \text{ kJ/m}^2\text{-h}$$

Comparison between Predicted and Measured Values

Measured values of $\overline{H}_g$, $\overline{H}_d$, $\overline{I}_g$ and $\overline{I}_d$ for New Delhi are given in Tables A 3.1 and A 3.2. Table 3.4 gives a comparison between the predicted and measured values. It is seen that in all the cases the predicted values are in reasonably good agreement with the measured values.

Table 3.4 Comparison between the predicted and measured values of solar radiation

Location: New Delhi

Month : April Time: 0900–1000 h (LAT)

	Predicted		Measured	Difference (%)	
$\overline{H}_g$ (kJ/m^2-day)	23 564		24 966	5.6	
$\overline{H}_d$ (kJ/m^2-day)	7820	8685	8902	12.2	2.4
$\overline{I}_g$ (kJ/m^2-h)	2406		2617	8.1	
$\overline{I}_d$ (kJ/m^2-h)	780	866	850	8.2	–1.9

3.7.5 Hourly Global, Beam and Diffuse Radiation Under Clear Skies

We now turn our attention to the prediction of radiation under clear skies. Based on an analysis of US data, ASHRAE [28] developed a method for estimating the hourly values of the global horizontal irradiance and direct normal irradiance. The equations are based on an exponential decay model in which the beam radiation decreases with increase in the distance traversed through the atmosphere. The global horizontal irradiance I_g on the earth is given by

$$I_g = I_b + I_d$$

where I_g = hourly global horizontal irradiance

I_b = hourly beam component

I_d = hourly diffuse component

Now, $$I_b = I_{bn} \cos \theta_z \tag{3.33}$$

where I_{bn} = direct normal irradiance

and θ_z = zenith angle

Thus, $$I_g = I_{bn} \cos \theta_z + I_d \tag{3.34}$$

In the ASHRAE-1972 model, it is postulated that

$$I_{bn} = A \exp [-B/\cos \theta_z] \tag{3.35}$$

and $$I_d = C\, I_{bn} \tag{3.36}$$

where *A*, *B* and *C* are constants whose values have been determined on a monthwise basis. These constants change during the year because of seasonal changes in the dust and water vapour content of the atmosphere, and also because of the changing earth-sun distance. The values of *A*, *B* and *C* were initially given by Threlkeld and Jordan [29] and subsequently revised by Iqbal [13]. The revised values are given in Table 3.5.

Table 3.5 Values of the constants *A*, *B* and *C* for predicting hourly solar radiation on clear days in ASHRAE-1972 method

	A (W/m^2)	*B*	*C*
January 21	1202	0.141	0.103
February 21	1187	0.142	0.104
March 21	1164	0.149	0.109
April 21	1130	0.164	0.120
May 21	1106	0.177	0.130
June 21	1092	0.185	0.137
July 21	1093	0.186	0.138
August 21	1107	0.182	0.134
September 21	1136	0.165	0.121
October 21	1136	0.152	0.111
November 21	1190	0.144	0.106
December 21	1204	0.141	0.103

Example 3.8

The following hourly values are measured around mid-day on a clear day at Nagpur (21°06′N, 79°03′E).

Date: June 20, 1968.

Local apparent time	0900–1000	1000–1100	1100–1200	1200–1300	1300–1400	1400–1500
Global radiation (kJ/m^2-h)	2830	3224	3320	3538	3329	2680
Diffuse radiation (kJ/m^2-h)	716	636	737	779	708	737

Compare these values with values which would be predicted by the ASHRAE-1972 method.

Solution

On June 20, 1968, $n = 171$. From Eq. (3.4),

$$\delta = 23.45 \sin\left\{\frac{360}{365}(284 + 171)\right\} = 23.44°$$

Therefore from Eq. (3.7),

$$\cos\theta_z = \sin 21.10° \sin 23.44° + \cos 21.10° \cos 23.44° \cos\omega$$
$$= 0.1432 + 0.8560 \cos\omega$$

From Table 3.5, the values of *A*, *B* and *C* are

$$A = 1092 \text{ W/m}^2 = 3931.2 \text{ kJ/m}^2\text{-h}$$

$$B = 0.185$$
$$C = 0.137$$

The remaining calculation is conveniently done in tabular form. The values obtained for I_b, I_d and I_g from Eqs (3.33)–(3.36) are given in Table 3.6 and are compared with the measured data. It is seen that the values of *A, B, C* obtained on the basis of US data predict high values of the beam radiation and low values of the diffuse radiation, when used to predict radiation under Indian conditions. The values of beam radiation are 15 to 33 per cent higher, while the values of diffuse radiation are 31 to 43 per cent lower. The two effects tend to balance each other to some extent. As a result, the values of global horizontal irradiance are predicted to a reasonable degree of accuracy.

Table 3.6 Comparison of predicted and measured values of hourly radiation using ASHRAE-1972 method

LAT	ω (deg)	I_b (kJ/m^2-h)		I_d (kJ/m^2-h)		I_g (kJ/m^2-h)	
		Predicted	Measured	Predicted	Measured	Predicted	Measured
0900–1000	37.5	2581	2114	430	716	3011	2830
1000–1100	22.5	3012	2588	442	636	3454	3224
1100–1200	7.5	3236	2583	447	737	3683	3320
1200–1300	–7.5	3236	2759	447	779	3683	3538
1300–1400	–22.5	3012	2621	442	708	3454	3329
1400–1500	–37.5	2581	1943	430	737	3011	2680

Based on recent solar radiation data, the ASHRAE-1972 method has been updated. In the new method (ASHRAE-2013), it is postulated that both I_{bn} and I_d are subjected to an exponential decay. The direct normal irradiance is given by [30]

$$I_{bn} = I'_{sc} \exp\left[-\tau_b\, m^{ab}\right] \tag{3.37}$$

and the diffuse component by

$$I_d = I'_{sc} \exp\left[-\tau_d\, m^{ad}\right] \tag{3.38}$$

where τ_b and τ_d are beam and diffuse optical depths respectively, *ab* and *ad* are beam and diffuse air mass exponents. Values of τ_b and τ_d depend on the atmospheric conditions of a location and hence vary during the year. Their average values have been determined for a number of locations of the world for the 21st day of each month [30]. Table 3.7 lists these values for a few locations in India.

Table 3.7 Values of τ_b and τ_d for a few locations in India [30]

Location	Para-meters	January	February	March	April	May	June	July	August	September	October	November	December
New Delhi (Palam)	τ_b	0.544	0.527	0.545	0.626	0.793	0.900	0.846	0.677	0.615	0.640	0.624	0.572
	τ_d	1.908	1.930	1.877	1.679	1.406	1.310	1.399	1.672	1.814	1.697	1.729	1.849
Mumbai (Santa-cruz)	τ_b	0.484	0.477	0.512	0.556	0.608	0.744	0.858	0.695	0.564	0.539	0.504	0.497
	τ_d	2.194	2.184	2.046	1.911	1.791	1.563	1.411	1.669	1.991	2.086	2.177	2.172
Nagpur	τ_b	0.472	0.451	0.486	0.501	0.545	0.623	0.652	0.556	0.507	0.475	0.468	0.456
	τ_d	2.200	2.244	2.088	2.019	1.898	1.747	1.686	1.928	2.118	2.222	2.223	2.243
Kolkata	τ_b	0.675	0.632	0.615	0.666	0.733	0.750	0.636	0.605	0.586	0.571	0.605	0.648
	τ_d	1.712	1.777	1.789	1.674	1.568	1.533	1.752	1.846	1.929	1.993	1.877	1.763
Chennai	τ_b	0.477	0.455	0.483	0.514	0.582	0.568	0.557	0.527	0.508	0.521	0.511	0.519
	τ_d	2.275	2.321	2.187	2.076	1.880	1.906	1.941	2.059	2.173	2.184	2.237	2.185

Air mass exponents *ab* and *ad* are correlated to τ_b and τ_d as follows

$$ab = 1.454 - 0.406\ \tau_b - 0.268\ \tau_d + 0.021\ \tau_b\ \tau_d \tag{3.39}$$

and

$$ad = 0.507 + 0.205\ \tau_b - 0.080\ \tau_d - 0.190\ \tau_b\ \tau_d \tag{3.40}$$

Example 3.9

Repeat the calculations of Example 3.8 using the ASHRAE-2013 method.

Solution

The calculations of Example 3.8 are repeated using Eqs (3.37)–(3.40) and the results are presented in Table 3.8. It is seen that there is a marked improvement in the prediction of hourly values of beam and diffuse radiation by using the ASHRAE-2013 method. The predicted values of I_b are less than the measured values, the variation ranging from –2% to –12%. On the other hand, the predicted values of I_d are slightly higher than the measured values, the variation ranging from +2% to +26%.

Table 3.8 Comparison of predicted and measured values of hourly radiation using ASHRAE-2013 method

LAT	ω (deg)	I_b (kJ/m^2-h)		I_d (kJ/m^2-h)		I_g (kJ/m^2-h)	
		Predicted	Measured	Predicted	Measured	Predicted	Measured
0900–1000	37.5	1901	2114	750	716	2651	2830
1000–1100	22.5	2308	2588	802	636	3110	3224
1100–1200	7.5	2523	2583	826	737	3349	3320
1200–1300	–7.5	2523	2759	826	779	3349	3538
1300–1400	–22.5	2308	2621	802	708	3110	3329
1400–1500	–37.5	1901	1943	750	737	2651	2680

As described earlier in Section 3.3, the earth's atmosphere consists of various gases, water vapour and particulate matter (called aerosols). Solar radiation entering the earth's atmosphere is scattered and absorbed in varying degrees by each of these constituents. Thus, their composition and distribution plays a vital role in the availability of solar radiation on the ground under clear sky conditions. The ASHRAE methods just described use simplified models which take an integrated view of the attenuation process rather than considering the attenuation caused by the individual constituents. Consequently, they are limited in their accuracy and applicability. A number of sophisticated models have been developed over the years to estimate clear sky radiation more accurately. These models are based on satellite observations and consider the scattering and absorption caused by different atmospheric constituents individually. Ineichen [31] has evaluated eight such models under a wide range of conditions and has recommended the SOLIS model [32] as the best. More recently, Sengupta and Goteseff [33] have compared the result of three satellite-based models with measured clear sky data. The models are the Simplified Solis Broadband model [34], the Bird model [35] and the REST2 model [36]. The comparison shows that the Bird model is the best from the point of view of accuracy. The details are beyond the scope of this book.

SOLAR RADIATION ON TILTED SURFACES 3.8

From the preceding sections, it is seen that very often measuring instruments give the values of solar radiation falling on a horizontal surface. The same is also true of the empirical equations given in Section 3.7. However, most solar equipment (e.g. flat-plate collectors or PV modules) for absorbing radiation, are tilted at an angle to the horizontal. It therefore becomes necessary to calculate the flux which falls on a tilted surface. This flux is the sum of the beam and diffuse radiation falling directly on the surface and the radiation reflected onto the surface from the surroundings.

Beam Radiation

The ratio of the beam radiation flux falling on a tilted surface to that falling on a horizontal surface is called the *tilt factor* for beam radiation. It is denoted by the symbol r_b. For the case of a tilted surface facing south (i.e. $\gamma = 0°$),

$$\cos\theta = \sin\delta\sin(\phi-\beta) + \cos\delta\cos\omega\cos(\phi-\beta)$$

while for a horizontal surface

$$\cos\theta_z = \sin\phi\sin\delta + \cos\phi\cos\delta\cos\omega$$

Hence

$$r_b = \frac{\cos\theta}{\cos\theta_z} = \frac{\sin\delta\sin(\phi-\beta) + \cos\delta\cos\omega\cos(\phi-\beta)}{\sin\phi\sin\delta + \cos\phi\cos\delta\cos\omega} \tag{3.41}$$

Similarly, expressions for r_b can be derived for other situations in which the tilted surface is oriented in a different direction with $\gamma \neq 0°$.

Diffuse Radiation

The tilt factor r_d for diffuse radiation is the ratio of the diffuse radiation flux falling on the tilted surface to that falling on a horizontal surface. The value of this tilt factor depends upon the distribution of diffuse radiation over the sky and on the portion of the sky dome seen by the tilted surface. Assuming that the sky is an isotropic source of diffuse radiation, we have for a tilted surface with a slope β,

$$r_d = (1 + \cos\beta)/2 \tag{3.42}$$

since $(1 + \cos\beta)/2$ is the radiation shape factor for a tilted surface with respect to the sky.

In reality, the distribution of diffuse radiation over the sky is anisotropic. It has been established experimentally that a small portion of the sky near the sun disc has a larger intensity of diffuse radiation due to forward scattering of a part of the beam radiation. Similarly, the intensity of diffuse radiation near the horizon is higher because of multiple Rayleigh scattering by a larger air mass. Keeping these factors in mind, some sky models taking account of the anisotropic distribution have been developed. However, for most engineering calculations, it is usual to consider the distribution of the diffuse radiation to be isotropic.

Reflected Radiation

Since (1 + cos β)/2 is the radiation shape factor for a tilted surface with respect to the sky, it follows that (1 – cos β)/2 is the radiation shape factor for the surface with respect to the surrounding ground. Assuming that the reflection of the beam and diffuse radiations falling on the ground is diffuse and isotropic, and that the reflectivity is ρ, the tilt factor for reflected radiation is given by

$$r_r = \rho\,(1 - \cos\beta)/2 \tag{3.43}$$

Flux on Tilted Surface

The flux I_T falling on a tilted surface at any instant is thus given by

$$I_T = I_b r_b + I_d r_d + (I_b + I_d) r_r \tag{3.44}$$

where the values of r_b, r_d and r_r are as given in Eqs (3.41), (3.42) and (3.43). It should be noted that Eq. (3.41) is valid only for a tilted surface with $\gamma = 0°$, whereas Eqs (3.42) and (3.43) are valid for any tilted surface with a slope β. Dividing both sides of Eq. (3.44) by I_g, we obtain the ratio of the flux falling on a tilted surface at any instant to that on a horizontal surface.

$$\frac{I_T}{I_g} = \left(1 - \frac{I_d}{I_g}\right) r_b + \frac{I_d}{I_g} r_d + r_r \tag{3.45}$$

One difficulty associated with using Eq. (3.44) is that the value of the diffuse reflectivity ρ is not known for most situations. A value around 0.2, generally expected with surfaces of concrete or grass, can be used. Fortunately, the reflected radiation term does not very often contribute much to the total.

Equation (3.45) can be used for calculating the hourly radiation falling on a tilted surface if the value of ω is taken at the mid-point of the hour. It can also be used for calculating the monthly average hourly value ($\overline{I}_T$) if the calculations are done for the representative day of the month as specified in Section 3.7.1. Equation (3.45) is then used in the modified form

$$\frac{\overline{I}_T}{\overline{I}_g} = \left(1 - \frac{\overline{I}_d}{\overline{I}_g}\right) \overline{r}_b + \frac{\overline{I}_d}{\overline{I}_g} \overline{r}_d + \overline{r}_r \tag{3.46}$$

where $\overline{r}_b = r_b$ on the representative day,

$\overline{r}_d = r_d = (1 + \cos\beta)/2$,

$\overline{r}_r = r_r = \rho\,(1 - \cos\beta)/2$

The daily radiation falling on a tilted surface is also of interest in many applications. Liu and Jordan [37] have proposed that the ratio of the daily radiation falling on such a surface (H_T) to the daily global radiation on a horizontal surface

(H_g) is given by an equation having a form similar to Eq. (3.45). Thus

$$\frac{H_T}{H_g} = \left(1 - \frac{H_d}{H_g}\right) R_b + \frac{H_d}{H_g} R_d + R_r \tag{3.47}$$

For a south-facing surface ($\gamma = 0°$), Liu and Jordan show that

$$R_b = \frac{\omega_{st} \sin \delta \sin (\phi - \beta) + \cos \delta \sin \omega_{st} \cos (\phi - \beta)}{\omega_s \sin \phi \sin \delta + \cos \phi \cos \delta \sin \omega_s} \tag{3.48}$$

$$R_d = r_d = (1 + \cos \beta)/2 \tag{3.49}$$

$$R_r = r_r = \rho (1 - \cos \beta)/2 \tag{3.50}$$

In Eq. (3.48), ω_{st} and ω_s are the sunrise or sunset hour angles (expressed in radians) for the tilted surface and a horizontal surface respectively.

Equation (3.47) can also be used for calculating the monthly average daily radiation falling on a tilted surface if the values required are calculated for the representative day of the month. Equation (3.47) is then used in the modified form

$$\frac{\bar{H}_T}{\bar{H}_g} = \left(1 - \frac{\bar{H}_d}{\bar{H}_g}\right) \bar{R}_b + \frac{\bar{H}_d}{\bar{H}_g} \bar{R}_d + \bar{R}_r \tag{3.51}$$

where $\bar{R}_b = R_b$ on the representative day,

$\bar{R}_d = R_d = (1 + \cos \beta)/2$

$\bar{R}_r = R_r = \rho (1 - \cos \beta)/2$

Example 3.10

Calculate the monthly average hourly radiation falling on a flat-plate collector facing south ($\gamma = 0°$) with a slope of 15°, given the following data

Location : Chennai (13°00′ N)
Month : October
Time : 1100 – 1200 h (LAT)
$\bar{I}_g$: 2408 kJ/m^2-h
$\bar{I}_d$: 1073 kJ/m^2-h

Assume ground reflectivity to be 0.2.

Solution

The calculations will be done for the representative day in October, viz. Oct. 15.

On this day, $n = 288$

From Eq. (3.4), $\delta = -9.599°$

Substituting in Eq. (3.41),

$$r_b = \frac{\sin (-9.599°) \sin (13.00° - 15°) + \cos (-9.599°) \cos 7.5° \cos (13.00° - 15°)}{\sin 13.00° \sin (-9.599°) + \cos 13.00° \cos (-9.599°) \cos 7.5°}$$

$= 1.074$

From Eqs (3.42) and (3.43),

$$r_d = (1 + \cos 15°)/2 = 0.9830$$

$$r_r = 0.2\,(1 - \cos 15°)/2 = 0.0034$$

Putting $\overline{r}_b$ equal to r_b on October 15, $\overline{r}_d = r_d$ and $\overline{r}_r = r_r$, and substituting in Eq. (3.46) we have

$$\frac{\overline{I}_T}{\overline{I}_g} = \left(1 - \frac{1073}{2408}\right) \times 1.074 + \frac{1073}{2408} \times 0.9830 + 0.0034$$

$$= 1.0369$$

$$\overline{I}_T = 2497 \text{ kJ/m}^2\text{-h}$$

PROBLEMS

3.1 (a) Find the temperature of the sun's surface if it is assumed to be black and the value of the solar constant is 1367 W/m^2.

(b) Using Planck's law, calculate the spectral distribution for the temperature calculated in (a). Compare this distribution with measured values of the distribution given in Table 3.1.

3.2 Estimate the rate at which the sun emits energy. What fraction of this energy is intercepted by the earth and what is the amount intercepted?

3.3 Calculate the angle of incidence of beam radiation on a flat-plate collector for the following situation:

Location	:	Nagpur (21°06′ N, 79°03′ E)
Slope of collector	:	31°
Surface azimuth angle	:	15°
Date	:	December 1, 1979
Time	:	0900 h (IST)

3.4 Calculate and plot the variation of the angle of incidence of beam radiation on May 1st on a flat surface located in Mumbai (19°07′ N, 72°51′ E). The surface is tilted at an angle of 10° with the horizontal and is pointing due south. Would you prefer to plot the variation against IST or LAT?

3.5 Calculate the angle of incidence of beam radiation on a photovoltaic array located in Mumbai (19° 07′ N, 72° 51′ E). Given the following data:

Date	:	August 18, 2005
Time	:	1030 h (LAT)
Slope of PV array	:	28°
Surface azimuth angle	:	20°

3.6 Find the days of the year on which the sun is directly overhead at 1200 h (LAT) at Pune (18°32′ N).

3.7 An array of flat-plate collectors is facing due south and is inclined at an angle β to the horizontal. Reflecting mirrors of the same length L as the collectors are used to boost the output (Fig. 3.16). Using the criterion that the sun's rays striking the top of the mirrors (line *AA)* at 1200 h (LAT) should be reflected to the top of the collectors (line *BB*), derive an expression for the correct inclination ψ of the booster mirrors in terms of β, δ and ϕ.

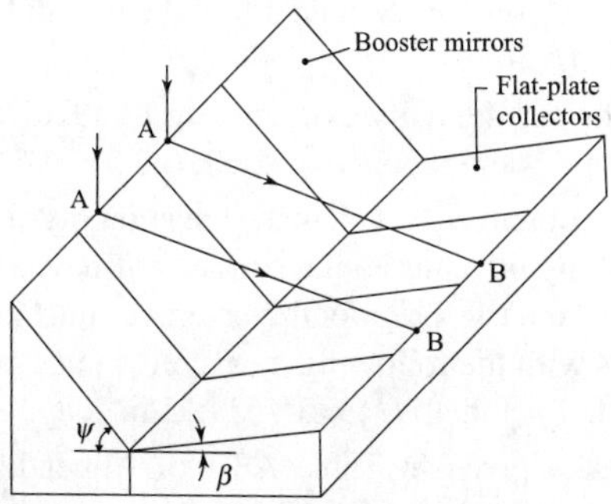

Fig. 3.16

3.8 Plot the variation of the day length on a horizontal surface through the year for the following locations:

Srinagar	:	(34°06′ N, 74°51′ E)
Kolkata	:	(22°39′ N, 88°27′ E)
Vishakhapatnam	:	(17°43′ N, 83°14′ E)
Thiruvananthapuram	:	(8°29′ N, 76°57′ E)

3.9 Calculate the day length on May 1 and December 1 for a south-facing surface tilted at an angle of 40° and located at New Delhi (28°35′ N, 77°12′ E).

3.10 Calculate the hour angle at sunrise and sunset on a plane surface tilted at an angle of 40°. Given $\phi = 28°$ N, $\delta = -21°$ and $\gamma = 48°$.

3.11 Calculate the length and direction of the shadow cast on the ground by a 1 m long vertical stick for the following situation.

Location	:	Mumbai (19°07′ N, 72°51′ E)
Date	:	February 13, 1980
Time	:	1000 h (IST)

3.12 A horizontal stick, 1 m long, is fixed at right angles to a vertical south-facing wall ($\gamma = 0°$). Calculate the length and direction of the shadow cast by the stick on the wall for the following situation.

Location	:	Jodhpur (26°18′ N, 73°01′ E)
Date	:	November 5
Time	:	0900 h (LAT)

3.13 A long vertical wall (4.2 m high) is built along the southern boundary of an open ground. A solar collector system is installed in the ground north of the wall. The collectors are tilted at an angle of 25° and face due south. Calculate the minimum distance that should be left on the ground between the wall and the collectors, so that the wall does not shade the collectors for two hours before as well as two hours after solar noon during the month of April. The latitude and longitude of the location are 27°50′ N and 85°32′ E.

3.14 The south wall of a building at a place (27°43′ N, 76°28′ E) has a rectangular projection (overhang) at its top edge. The length and height of the wall are 5 and

3.5 m, respectively. The overhang is constructed normal to the wall. Its length is equal to that of the wall and its depth is 0.55 m. Calculate the per-centage of the area of the wall that is shaded on April 10 at 1130 h (IST).

3.15 Test the validity of the recommendation that the value of H_o for July can be taken equal to the value of H_o on the seventeenth of the month by doing calculations for Shillong (25°34′ N, 91°56′ E).

3.16 Derive an expression for the daily extraterrestrial radiation which would fall on a surface having a slope β and facing due south (i.e., $\gamma = 0°$).

3.17 Use the prediction equations to calculate the monthly average daily global and diffuse radiation falling on a horizontal surface at Bhavnagar (21°45′ N, 72°11′ E) during the month of January. The monthly average sunshine hours is 9.8. Compare the predicted values with the following measured values given in the Handbook of Solar Radiation Data for India, $\overline{H}_g$ = 18 511 kJ/m^2-day, $\overline{H}_d$ = 4198 kJ/m^2-day.

3.18 Use the radiation data given in Table A3.1 of Appendix 3 for New Delhi and estimate the constants *a* and *b* of Eq. (3.20). Compare the calculated values with those given in Table 3.3.

3.19 Use the prediction equations to calculate the variation of the monthly average hourly global and diffuse radiation at Mumbai (19°07′ N, 72°51′ E) during May. Given that the monthly average sunshine hours is 9.3. Compare the predicted values with measured data given in Appendix 3, Tables A3.3 and A3.4.

3.20 Use the ASHRAE-2013 method to calculate the hourly global and diffuse radiation on March 10th at Chennai (13°00′ N, 80°11′ E). Assume that there is a clear sky and that the values of constants given in Table 3.7 are valid.

3.21 Calculate the variation of the tilt factor r_b with slope β varying from 0° to 90° for a surface facing south ($\gamma = 0°$), given the following data,

Location	:	Nagpur (21°06′ N, 79°03′ E)
Date	:	March 1
Time	:	1200 h (LAT) and 1700 h (LAT)

Comment on the nature of the variation obtained.

3.22 At a certain location (25°47′ N, 75°39′ E) in India, an unshaded pyranometer (kept on a horizontal surface) and a pyrheliometer record 695 and 721 W/m^2 respectively at 1000 h (IST) on September 15. A flat-plate collector is installed with a tilt of 13° and is oriented east of south at an angle of 35°. Calculate the flux incident on the collector. Given that the reflectivity of the surfaces surrounding the collector is 0.2.

What is the percentage reduction in the value of the flux incident on the collector due to the presence of the atmosphere?

3.23 A flat surface is located at a place in the southern hemisphere and is tilted towards the equator with a slope of 40°. The latitude and longitude of the place are 30°S and 73°E, respectively. Calculate the beam and total solar radiation incident on the tilted surface.

Given:	Time	:	1030 h (LAT)
	Date	:	February 10
	I_g	:	720 W/m^2
	I_d	:	144 W/m^2
	ρ	:	0.2

REFERENCES

1. Thekaekara, M.P. and Drummond, A.J. 1971. Standard values for the solar constant and its spectral components. *Nat. Phys. Sci.,* 229: 6.
2. Froelich, C. and Brusa, R.W. 1981. Solar radiation and its variation in time. *Solar Physics,* 74: 209.
3. Froelich, C. and Wehrli, C. 1981. Spectral distribution of solar irradiance from 25000 nm to 250 nm. *World Radiation Centre,* Davos, Switzerland.
4. Marty, C. and Philipona, R. 2000. The clear-sky index to separate clear-sky from cloudy-sky situations in climate research. *Geophysical Research Letters*, 27: 2649.
5. India Meteorological Department, Ministry of Earth Sciences, Government of India, New Delhi, India.
   (http://www.imd.gov.in)
6. Mani, A. 1981. *Handbook of Solar Radiation Data for India*, Allied Publishers, New Delhi, India.
7. Ministry of New and Renewable Energy, Government of India, New Delhi, India.
   (http://www.mnre.gov.in)
8. National Institute of Wind Energy, Chennai, India.
   (http://www.niwe.res.in)
9. World Radiation Data Centre, St. Petersburg, Russia,
   (http://wrdc-mgo.nrel.gov)
10. Mani, A. and Chacko, O. 1973. Solar radiation climate of India. *Solar Energy*, 14: 139.
11. Ohmura, A. 2009. Observed decadal variations in surface solar radiation and their causes. *Journal of Geophysical Research*, 114: D05.
12. Cooper, P.I. 1969. The absorption of solar radiation in solar stills. *Solar Energy*, 12: 3.
13. Iqbal, M. 1983. *An Introduction to Solar Radiation*, Academic Press, Canada.
14. Braun, J.E. and Mitchell, J.C. 1983. Solar geometry for fixed and tracking surfaces. *Solar Energy*, 31: 439.
15. Page, J.K. 1961. The estimation of monthly mean values of daily total short-wave radiation of vertical and inclined surfaces from sunshine records for latitudes 40°N–40°S. *Proc. UN Conf. New Sources of Energy*, 4: 378.
16. Lof, G.O.G., Duffie, J.A. and Smith, CO. 1966. World distribution of solar radiation. *Solar Energy*, 10: 27.
17. Modi, Vijay and Sukhatme, S.P. 1979. Estimation of daily total and diffuse insolation in India from weather data. *Solar Energy*, 22: 407.
18. Klein, S.A. 1977. Calculation of monthly average insolation on tilted surfaces. *Solar Energy*, 19: 325.
19. Gopinathan, K.K. 1988. A general formula for computing the coefficients of the correlation connecting global solar radiation to sunshine duration. *Solar Energy*, 41: 499.

20. Liu, B.Y.H. and Jordan, R.C. 1960. The interrelationship and characteristic distribution of direct, diffuse and total solar radiation. *Solar Energy*, 4:1.
21. Kreith, F. and Kreider, J.F. 1978. *Principles of Solar Engineering*, Chapter 2, McGraw-Hill Book Co, New York, USA.
22. Gopinathan, K.K. and Soler, A. 1995. Diffuse radiation models and monthly-average, daily diffuse data for a wide latitude range. *Energy*, 20: 657.
23. Garg, H.P. and Garg, S.N. 1985. Correlation of monthly-average daily global, diffuse and beam radiation with bright sunshine hours. *Energy Conversion and Management*, 25: 409.
24. Collares-Pereira, M. and Rabi, A. 1979. The average distribution of solar radiation correlations between diffuse and hemispherical and between daily and hourly insolation values. *Solar Energy*, 22: 155.
25. Gueymard, C. 1986. Mean daily averages of beam radiation received by tilted surfaces as affected by atmosphere. *Solar Energy*, 37: 261.
26. Gueymard, C. 2000. Prediction and performance assessment of mean hourly global radiation. *Solar Energy*, 68: 285.
27. Satyamurty, V.V. and Lahiri, P.K. 1992. Estimation of symmetric and asymmetric hourly global and diffuse radiation from daily values. *Solar Energy*, 48: 7.
28. ASHRAE-1972. *Handbook of Fundamentals*, American Society of Heating, Refrigerating and Air-conditioning Engineers, pp. 385-443.
29. Threlkeld, J.L. and Jordan, R.C. 1958. Direct solar radiation available on clear days. *ASHRAE Transactions*, 64: 45.
30. ASHRAE-2013. *Handbook of Fundamentals*, American Society of Heating, Refrigerating and Air-conditioning Engineers, pp. 14.9-14.10.
31. Ineichen, P. 2006. Comparison of eight clear sky broadband models against 16 independent data banks. *Solar Energy*, 80: 468.
32. Mueller, R.W., Dagestad, K.F., Ineichen, P., Schroedter-Homscheidt, M., Cros, S., Dumortier, D., Kuhlemann, R., Olseth, J.A., Piernavieja, G., Reise, C., Wald, L. and Heinemann, D. 2004. Rethinking satellite-based solar irradiance modeling: the SOLIS clear-sky model. *Remote Sensing of Environment*, 91: 160.
33. Sengupta, M. and Gotseff, F. 2013. Evaluation of clear sky models for satellite-based irradiance estimates. *Technical Report NREL/TP-5D00-60735*. National Renewable Energy Laboratory, Golden, Colorado, U.S.A.
34. Ineichen, P. 2008. A broad band simplified version of the Solis clear sky model. *Solar Energy*, 82: 758.
35. Bird, R.E. and Hulstrom, R.L., 1981. A simplified clear sky model for direct and diffuse insolation on horizontal surfaces. *Technical Report SERI/TR-642-761*. Solar Energy Research Institute (now NREL), Golden, Colorado, U.S.A. (http://www.nrel.gov/docs/legosti/old/761.pdf)
36. Gueymard, C. 2012. Clear sky irradiance predictions for solar resource mapping and large scale applications. *Solar Energy*, 86: 2145.
37. Liu, B.Y.H. and Jordan, R.C. 1961. Daily insolation on surfaces tilted towards the equator. *ASHRAE Transactions*, 67: 526.

Liquid Flat-Plate Collectors

CHAPTER 4

INTRODUCTION 4.1

The liquid flat-plate collector is the most widely used solar collector because it is simple in design, has no moving parts and requires little maintenance. A brief description of this collector has been given in Section 2.2, and its varied applications have been described in Section 2.3. The applications include water heating, space heating and cooling, and low-temperature cycles for power generation.

In this chapter, the detailed procedure for analyzing the performance of a flat-plate collector is given in Sections 4.3 – 4.9. The effect of various parameters on the performance is then described in Section 4.10. This is followed by a transient analysis (Section 4.12) and a description of standard procedures for thermal testing of collectors (Section 4.13). The chapter concludes with brief descriptions of collectors which have been developed as alternatives to the conventional collector (Section 4.14).

GENERAL 4.2

The basic parts that make up a conventional liquid flat-plate collector are (i) the absorber plate, (ii) the tubes fixed to the absorber plate through which the liquid to be heated flows, (iii) the transparent cover, and (iv) the collector box. The main advantage of a flat-plate collector is that it utilises both the beam and diffuse components of the solar radiation. In addition, because of its simple stationary design, it requires little maintenance. Its principal disadvantage is that because of the absence of optical concentration, the area from which heat is lost is large. As a result, the collection efficiency is generally low.

The liquid heated is generally water. However, sometimes mixtures of water and ethylene glycol are used if ambient temperatures below 0°C are likely to be encountered. The absorber plate is usually made from a thin metal sheet ranging in thickness from 0.2 to 0.7 mm, while the tubes, which are also of metal, range in diameter from 1 to 1.5 cm. They are soldered, brazed or pressure bonded to the bottom of the absorber plate with the pitch ranging from 5 to 12 cm. In some designs, the tubes are bonded to the top or are in-line and integral with the absorber plate. The metal most commonly used, both for the absorber plate and the tubes, is copper. The header pipes, which lead the liquid in and out of the collector and distribute it to the tubes, are made of the same metal as the tubes and have slightly larger diameters (2 to 2.5 cm).

The cover should be made of a material, which is highly transparent to incoming solar radiation and at the same time, opaque to long wavelength re-radiation emitted by the absorber plate. Glass with a low ferric oxide content satisfies these requirements. Toughened glass of 4 or 5 mm thickness is the most favoured material. This type of glass is able to withstand thermal shock as well as the impact of objects which may fall on the collector face. The usual practice is to have one cover with a spacing ranging from 1.5 to 3 cm between the cover and the absorber plate.

The bottom and sides are usually insulated by mineral wool, rock wool or glass wool with a covering of aluminium foil and has a thickness ranging from 2.5 to 8 cm. The whole assembly is contained within a box which is tilted at a suitable angle. The collector box is usually made of aluminium with an epoxy coating on the outside for protection.

The face areas of most commercially available collectors are around 2 m^2, with the length (along the sloping direction) being usually larger than the width.

More details of the components used in collectors are given in IS 12933 (Part 2)[1].

In the last few years, the use of plastic materials for the absorber plate, the tubes as well as the cover has increased. This is particularly true for applications involving lower temperatures up to 60° or 70°C. Initially plastics were not used because they degraded on exposure to sunlight. They also have low thermal conductivities and high coefficients of expansion as compared to metals. However, advances in polymer technology have resulted in the development of suitable plastic materials which can withstand long exposures to sunlight. Plastics have the advantages of being light in weight and easy to manufacture. They also cost less and require less energy input for their manufacture than metals like copper and aluminium. However, it has to be remembered that they generally originate from fossil fuels. As the volume of production of flat-plate collectors increases, the above considerations of energy input and raw material origin will become increasingly important.

In recent years, nano-fluids have been tried instead of water with a view towards improving the performance of flat-plate collectors. Nano-fluids are liquids or liquid mixtures with a small percentage of nano-sized particles (1–100 nm) suspended in them. The nano-fluids tested with flat-plate collectors are water or water/ethylene glycol mixtures having suspensions of nano-particles of titanium dioxide, copper, silicon dioxide, aluminium oxide, etc. Multiwall carbon nano-tubes (MWCNT) have also been tried. Nano-particles and nano-tubes have a tendency to agglomerate over a period of time. For this reason, investigators have also added small amounts of surfactants which help to keep the particles dispersed. Data obtained by some investigators have yielded improvements of 20 to 30 per cent in the collector efficiency. However, the data presented have generally lacked in consistency [2]. For this reason, liquid flat-plate collectors using nano-fluids are not yet available commercially.

The present rate of production of liquid flat-plate collectors in the world, as well as in India, is steadily increasing. About 8 900 000 m^2 of collector area has been installed in India up to 2015, the typical cost of a good quality collector being in the range of Rs 4000 to 4500 per m^2.

PERFORMANCE ANALYSIS 4.3

We will now take up for detailed consideration the performance analysis of a liquid flat-plate collector. The analysis will first be done for a steady state situation* in which the liquid is flowing through tubes bonded on the under-side of the absorber plate. Later on, the results for other types of flat-plate collectors will be given and transient effects will be considered.

An energy balance on the absorber plate yields the following equation for a steady state

$$q_u = A_p S - q_l \tag{4.1}$$

in which

q_u = useful heat gain, i.e. the rate of heat transfer to the working fluid,
S = incident solar flux absorbed in the absorber plate,
A_p = area of the absorber plate, and
q_l = rate at which heat is lost by convection and re-radiation from the top, and by conduction and convection from the bottom and sides.

From Section 3.8, the flux incident on the top cover of the collector is given by Eq. (3.44)

$$I_T = I_b r_b + I_d r_d + (I_b + I_d) r_r$$

Each of the terms in the above equation is multiplied by a term called the transmissivity-absorptivity product $(\tau\alpha)$ in order to determine the flux S absorbed in the absorber plate. Thus,

$$S = I_b r_b (\tau\alpha)_b + \{I_d r_d + (I_b + I_d) r_r\} (\tau\alpha)_d \tag{4.2}$$

in which,

* Strictly speaking, a steady state situation is never obtained. However, quasi-steady state situations (in which the values of the parameters of interest change very slowly) are often obtained over extended periods of time.

τ = transmissivity of the glass cover system, the ratio of the solar radiation coming through after reflection at the glass-air interfaces and absorption in the glass to the radiation incident on the glass cover system,

α = absorptivity of the absorber plate,

$(\tau\alpha)_b$ = transmissivity-absorptivity product for beam radiation falling on the collector (defined in Section 4.5), and

$(\tau\alpha)_d$ = transmissivity-absorptivity product for diffuse radiation falling on the collector.

The other terms have been defined in Chapter 3.

Thus, in order to evaluate q_u in Eq. (4.1), it is necessary to derive expressions for calculating the values of $(\tau\alpha)_b$, $(\tau\alpha)_d$ and q_l. This calculation will therefore be taken up in the sections which follow.

At this stage, it will be worthwhile to define two terms, the instantaneous collection efficiency and stagnation temperature. The *instantaneous collection efficiency* is given by

$$\eta_i = \frac{\text{Useful heat gain}}{\text{Radiation incident on the collector}} = \frac{q_u}{A_c I_T} \tag{4.3}$$

where A_c is the *collector gross area* (the area of the topmost cover including the frame). A_c is usually 15 to 20 per cent more than A_p.

If the liquid flow rate through the collector is stopped, there is no useful heat gain and the efficiency is zero. In this case, the absorber plate attains a temperature such that $A_p S = q_l$. This temperature is the highest that the absorber plate can attain and is sometimes referred to as the *stagnation temperature.* Knowledge of the stagnation temperature is useful as an indicator for comparing different collector designs. It also helps in choosing proper materials for construction of the collector.

It has been stated earlier (Section 3.7.3) that many solar processes occur at a relatively slow pace. As a result, the time base of an hour is often convenient. Thus Eq. (4.3) is also valid as an expression for calculating the hourly collection efficiency, if q_u is the useful heat gain in one hour (kJ/h) and I_T is the energy incident on the collector face in one hour (kJ/m^2-h).

4.4 TRANSMISSIVITY OF THE COVER SYSTEM

The transmissivity of the cover system of a collector can be obtained with adequate accuracy by considering reflection-refraction and absorption separately, and is given by the product form

$$\tau = \tau_r \tau_a \tag{4.4}$$

where τ_r = transmissivity obtained by considering only reflection and refraction, and

and τ_a = transmissivity obtained by considering only absorption.

4.4.1 Transmissivity Based on Reflection-Refraction

When a beam of light of intensity I_{bn} travelling through a transparent medium 1 strikes the interface separating it from another transparent medium 2, it is reflected and refracted (Fig. 4.1). The reflected beam has reduced intensity I_r and has a direction such that the angle of reflection is equal to the angle of incidence. On the other hand, the directions of the incident and refracted beams are related to each other by Snell's law which states that

$$\frac{\sin\theta_1}{\sin\theta_2} = \frac{n_2}{n_1} \tag{4.5}$$

where θ_1 = angle of incidence,
θ_2 = angle of refraction, and
n_1, n_2 = refractive indices of the two media.

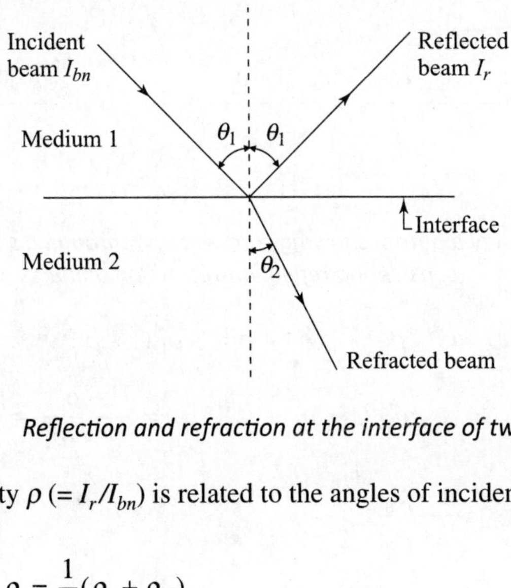

Fig. 4.1 *Reflection and refraction at the interface of two media*

The reflectivity ρ $(= I_r/I_{bn})$ is related to the angles of incidence and refraction by the equations

$$\rho = \frac{1}{2}(\rho_{\text{I}} + \rho_{\text{II}})$$

$$\rho_{\text{I}} = \frac{\sin^2(\theta_2 - \theta_1)}{\sin^2(\theta_2 + \theta_1)} \tag{4.6}$$

$$\rho_{\text{II}} = \frac{\tan^2(\theta_2 - \theta_1)}{\tan^2(\theta_2 + \theta_1)} \tag{4.7}$$

ρ_{I} and ρ_{II} being the reflectivities of the two components of polarisation.

For the special case of normal incidence ($\theta_1 = 0°$), it can be shown that

$$\rho = \rho_{\text{I}} = \rho_{\text{II}} = \left(\frac{n_1 - n_2}{n_1 + n_2}\right)^2 \tag{4.8}$$

The transmissivity τ_r is given by an expression similar to that for ρ. Thus

$$\tau_r = \frac{1}{2}(\tau_{r\text{I}} + \tau_{r\text{II}}) \tag{4.9}$$

where $\tau_{r\text{I}}$ and $\tau_{r\text{II}}$ are the transmissivities of the two components of polarisation.

Consider one of the components of polarisation of a beam incident on a single cover. Because of the fact that there are two interfaces, multiple reflections and refractions will occur as shown is Fig. 4.2.

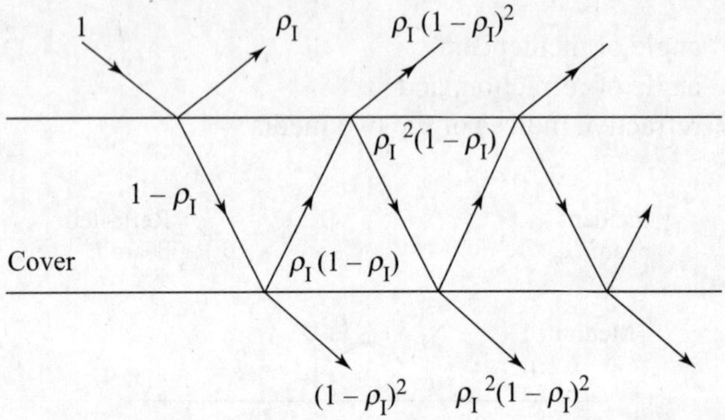

Fig. 4.2 *Ray diagram showing transmission through a single cover considering reflection-refraction alone*

Hence,

$$\tau_{r\text{I}} = (1 - \rho_\text{I})^2 + \rho_\text{I}^2(1 - \rho_\text{I})^2 + \rho_\text{I}^4(1 - \rho_\text{I})^2 + \ldots$$

$$= (1 - \rho_\text{I})^2(1 + \rho_\text{I}^2 + \rho_\text{I}^4 + \ldots) = \frac{(1 - \rho_\text{I})^2}{1 - \rho_\text{I}^2} = \frac{1 - \rho_\text{I}}{1 + \rho_\text{I}} \tag{4.10}$$

Similarly,

$$\tau_{r\text{II}} = \frac{1 - \rho_\text{II}}{1 + \rho_\text{II}} \tag{4.11}$$

These results can be readily extended to a system of M covers for which it can be shown that

$$\tau_{r\text{I}} = \frac{1 - \rho_\text{I}}{1 + (2M - 1)\rho_\text{I}} \tag{4.12}$$

and

$$\tau_{r\text{II}} = \frac{1 - \rho_\text{II}}{1 + (2M - 1)\rho_\text{II}} \tag{4.13}$$

4.4.2 Transmissivity Based on Absorption

The transmissivity based on absorption can be obtained by assuming that the attenuation due to absorption is proportional to the local intensity (Bouger's law). Consider a beam of intensity I_{bn} incident normally on a transparent cover of thickness δ_c and emerging with an intensity I_l (Fig. 4.3).

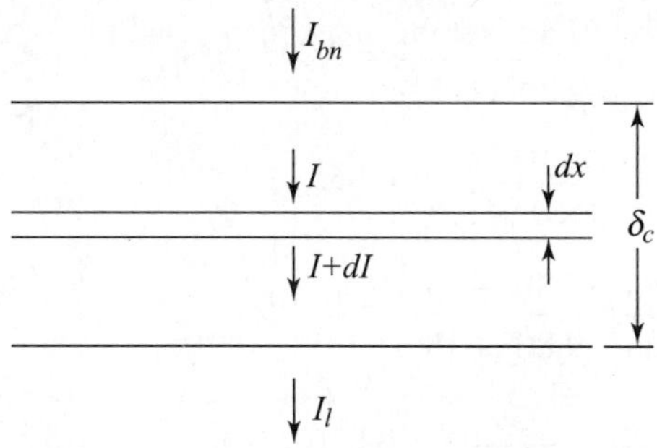

Fig. 4.3 *Absorption in a transparent cover*

From Bouger's law

$$dI = -KI\,dx$$

where K is a constant of proportionality and is called the *extinction coefficient.* It will be assumed to have a value independent of wavelength. Integrating over the length traversed by the beam, we have

$$\tau_a = \frac{I_l}{I_{bn}} = e^{-K\delta_c} \tag{4.14}$$

In case the beam is incident at an angle θ_1, the path traversed through the cover would be $(\delta_c/\cos\theta_2)$, where θ_2 is the angle of refraction. Then Eq. (4.14) gets modified to the form

$$\tau_a = e^{-K\delta_c \cos\theta_2} \tag{4.15}$$

The extinction coefficient K is a property of the cover material. Its value varies from about 4 to 25 m^{-1} for different qualities of glass. A low value is obviously desirable.

Equations (4.14) and (4.15) are derived for one cover. If there are M covers, the exponent in these equations would be multiplied by M.

Example 4.1

Plot the variation of τ_r, τ_a and τ with the angle of incidence for the following cover system:

Material	:	Glass
Number of covers	:	2
Thickness of each cover	:	4 mm
Refractive index of glass relative to air	:	1.52
Extinction coefficient of glass	:	15 m^{-1}

Solution

The calculation is given in detail for one angle of incidence, viz. $\theta_1 = 15°$

Hence, $\theta_2 = \sin^{-1}[(\sin 15°)/1.52] = 9.80°$

$$\rho_{\mathrm{I}} = \frac{\sin^2 (9.80° - 15°)}{\sin^2 (9.80° + 15°)} = 0.047$$

$$\tau_{r\mathrm{I}} = \frac{1 - 0.047}{1 + (3 \times 0.047)} = 0.835$$

$$\rho_{\mathrm{II}} = \frac{\tan^2 (9.80° - 15°)}{\tan^2 (9.80° + 15°)} = 0.039$$

$$\tau_{r\mathrm{II}} = \frac{1 - 0.039}{1 + (3 \times 0.039)} = 0.860$$

$$\tau_r = \frac{1}{2}(0.835 + 0.860) = 0.848$$

$$\tau_a = \exp\left[-(2 \times 15 \times 4 \times 10^{-3}/\cos 9.80°)\right] = 0.885$$

$$\tau = 0.848 \times 0.885 = 0.750$$

The transmissivities for other angles of incidence are obtained in a similar manner. Their variation with the angle of incidence is shown in Fig. 4.4. It will be seen that the values are essentially constant up to angles of incidence of 45°. Thereafter, the values drop rather sharply to zero as the angle of incidence increases to 90°.

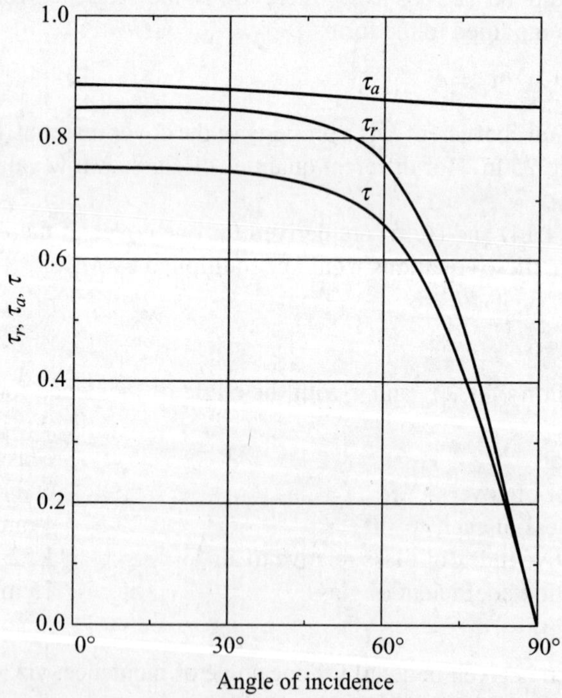

Fig. 4.4 *Example 4.1 - Variation of τ_r, τ_a and τ with angle of incidence*

4.4.3 Transmissivity for Diffuse Radiation

The preceding considerations apply only to beam radiation. Calculation of the transmissivity of a cover system when diffuse radiation is incident on it presents some difficulty, because the radiation comes from many directions. The usual practice is to assume that the diffuse radiation is equivalent to beam radiation coming at an angle of incidence of 60°. This angle is arrived at by considering the variation of τ as seen in Fig. 4.4 and by assuming that the amount of diffuse radiation coming from all directions is the same.

TRANSMISSIVITY–ABSORPTIVITY PRODUCT 4.5

The *transmissivity-absorptivity product* is defined as the ratio of the flux absorbed in the absorber plate to the flux incident on the cover system, and is denoted by the symbol $(\tau\alpha)$, an appropriate subscript (b or d) being added to indicate the type of incident radiation. An expression for the transmissivity-absorptivity product will now be derived.

Out of the fraction τ transmitted through the cover system, a part is absorbed and a part reflected diffusely. Out of the reflected part, a portion is transmitted through the cover system and a portion reflected back to the absorber plate. The process of absorption and reflection at the absorber plate surface (Fig. 4.5) goes on indefinitely, the quantities involved being successively smaller.

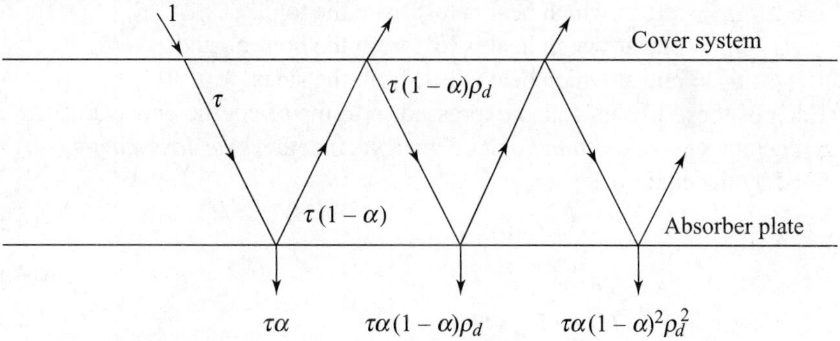

Fig. 4.5 *Absorption and reflection at the absorber plate*

Thus, the net fraction absorbed $(\tau\alpha)$

$$= \tau\alpha[1 + (1-\alpha)\rho_d + (1-\alpha)^2 \rho_d^2 + \ldots]$$

$$= \frac{\tau\alpha}{1-(1-\alpha)\rho_d} \tag{4.16}$$

The symbol ρ_d represents the diffuse reflectivity of the cover system. It can be found by determining the value $\tau_a(1 - \tau_r)$ for the cover system for an incidence angle of 60°.

From Example 4.1, it is seen that $\rho_d = 0.21$ for a two-glass cover system. Similarly, for a one-glass cover system, the value of ρ_d can be shown to be 0.15.

OVERALL LOSS COEFFICIENT AND HEAT TRANSFER CORRELATIONS 4.6

It is convenient from the point of view of analysis to express the heat lost from the collector in terms of an overall loss coefficient defined by the equation

$$q_l = U_l A_p(T_{pm} - T_a) \tag{4.17}$$

where U_l = overall loss coefficient,
A_p = area of the absorber plate,
T_{pm} = average temperature of the absorber plate, and
T_a = temperature of the surrounding air (assumed to be the same on all sides of the collector).

The heat lost from the collector is the sum of the heat lost from the top, the bottom and the sides. Thus,

$$q_l = q_t + q_b + q_s$$

where q_t = rate at which heat is lost from the top,
q_b = rate at which heat is lost from the bottom, and
q_s = rate at which heat is lost from the sides.

Each of these losses is also expressed in terms of coefficients called the *top loss coefficient*, the *bottom loss coefficient* and the *side loss coefficient* and defined by the equations

$$q_t = U_t A_p(T_{pm} - T_a) \tag{4.18}$$

$$q_b = U_b A_p(T_{pm} - T_a) \tag{4.19}$$

$$q_s = U_s A_p(T_{pm} - T_a) \tag{4.20}$$

It will be noted that the definition of each of the coefficients is based on the area A_p and the temperature difference $(T_{pm} - T_a)$. This is done for convenience and helps in giving the simple additive equation

$$U_l = U_t + U_b + U_s \tag{4.21}$$

The losses can also be pictured in terms of thermal resistances as shown in Fig. 4.6. The overall loss coefficient is an important parameter since it is a measure of all the losses. Typical values range from 2 to 10 W/m^2-K.

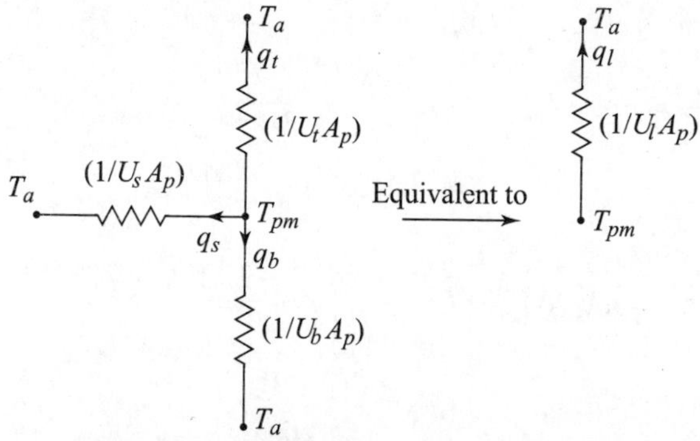

Fig. 4.6 *Thermal resistance network showing collector losses*

4.6.1 Top Loss Coefficient

The top loss coefficient U_t is evaluated by considering convection and re-radiation losses from the absorber plate in the upward direction. For purposes of calculation, it is assumed that the transparent covers and the absorber plate constitute a system of infinite parallel surfaces and that the flow of heat is one-dimensional and steady [3]. It is further assumed that the temperature drop across the thickness of the covers is negligible and that the interaction between the incoming solar radiation absorbed by the covers and the outgoing loss may be neglected. The outgoing re-radiation is of large wavelengths. For these wavelengths, the transparent cover will be assumed to be opaque. This is a very good assumption if the material is glass.

A schematic diagram for a two-cover system is shown in Fig. 4.7. In a steady state, the heat transferred by convection and radiation between (i) the absorber plate and the first cover, (ii) the first cover and the second cover, and (iii) the second cover and the surroundings must be equal.

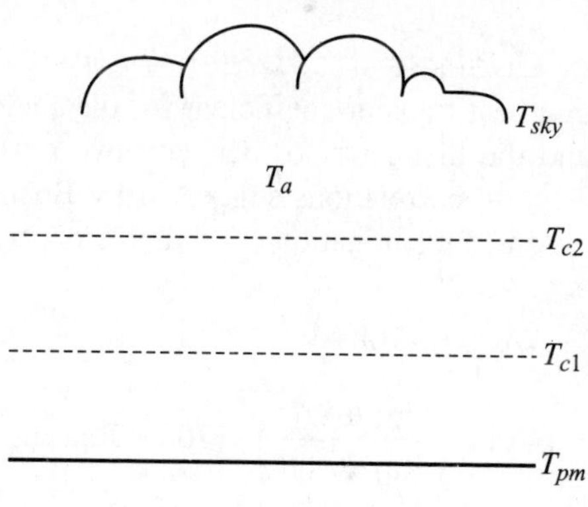

Fig. 4.7 *Calculation of the top loss coefficient*

Hence,

$$\frac{q_t}{A_p} = h_{p-c1}(T_{pm} - T_{c1}) + \frac{\sigma(T_{pm}^4 - T_{c1}^4)}{\left(\frac{1}{\varepsilon_p} + \frac{1}{\varepsilon_c} - 1\right)} \tag{4.22}$$

$$= h_{c1-c2}(T_{c1} - T_{c2}) + \frac{\sigma(T_{c1}^4 - T_{c2}^4)}{\left(\frac{1}{\varepsilon_c} + \frac{1}{\varepsilon_c} - 1\right)} \tag{4.23}$$

$$= h_w (T_{c2} - T_a) + \sigma \varepsilon_c (T_{c2}^4 - T_{sky}^4) \tag{4.24}$$

where

h_{p-c1} = convective heat transfer coefficient between the absorber plate and the first cover,

h_{c1-c2} = convective heat transfer coefficient between the first and second covers,

h_w = convective heat transfer coefficient between the topmost cover (in this case the second) and the surrounding air,

T_{c1}, T_{c2} = temperatures attained by the two covers,

T_{sky} = effective temperature of the sky with which the radiative exchange takes place,

ε_p = emissivity of the absorber plate for long wavelength radiation, and

ε_c = emissivity of the covers for long wavelcngth radiation.

Equations (4.22), (4.23) and (4.24) constitute a set of three non-linear equations which have to be solved for the unknowns q_t, T_{c1} and T_{c2}. However, before this can be done it will be necessary to have some correlations for calculating the convective heat transfer coefficients h_{p-c1}, h_{c1-c2} and h_w, and the sky temperature T_{sky}.

❑ *Heat Transfer Coefficient between Inclined Parallel Surfaces*

The natural convection heat transfer coefficient for the enclosed space between the absorber plate and the first cover or between two covers is calculated by using one of the following correlations suggested by Buchberg *et al.* [4]. The correlations are based on an examination of available experimental data and all previous correlations.

$$Nu_L = 1\ ;\ Ra_L \cos\beta < 1708$$

$$Nu_L = 1 + 1.446\left(1 - \frac{1708}{Ra_L \cos\beta}\right);\ 1708 < Ra_L \cos\beta < 5900$$

$$Nu_L = 0.229\,(Ra_L \cos\beta)^{0.252};\quad 5900 < Ra_L \cos\beta < 9.23\times10^4$$

$$Nu_L = 0.157\,(Ra_L \cos\beta)^{0.285};\quad 9.23\times10^4 < Ra_L \cos\beta < 10^6 \tag{4.25}$$

Nu_L and Ra_L are the Nusselt and Rayleigh numbers, respectively. The characteristic dimension L is the spacing between the surfaces, while properties are evaluated at the arithmetic mean of the surface temperatures.

❑ *Heat Transfer Coefficient at the Top Cover* The convective heat transfer coefficient (h_w) at the top cover is often referred to as the wind heat transfer coefficient. It has generally been calculated from the following empirical correlation suggested by McAdams [5],

$$h_w = 5.7 + 3.8\, V_\infty \tag{4.26}$$

in which h_w is in W/m^2-K and V_∞ is the wind speed in m/s. This correlation is based on the experiments performed by Jurges in 1924 for flow of air at room temperature parallel to a heated vertical plate 0.5 m square. Since the direction of the wind is, in general, not parallel to the top glass cover, it is obvious that Eq. (4.26) suffers from certain limitations.

Sparrow and Tien [6] have suggested the following dimensionless correlation

$$j = 0.86\,(\mathrm{Re}_L^*)^{-1/2} \tag{4.27}$$

where j = j-factor given by $(h_w / \rho C_p V_\infty)\,\mathrm{Pr}^{2/3}$

Re_L^* = Reynolds number ($V_\infty L^*/\nu$) based on the characteristic dimension $L^* = 4A_c/C_c$

A_c = collector gross area, and

C_c = circumference associated with the collector gross area.

Equation (4.27) was recommended on the basis of extensive experiments performed in a low turbulence wind tunnel on square and rectangular plates inclined at various angles of attack and yaw [7].

Test *et al.* [8] presented results of an experimental investigation to determine the value of h_w from the upper surface of a rectangular plate of size 1.220 m × 0.813 m inclined at an angle of 40°. The experiments were conducted in the natural environment at mean wind speeds ranging from 1.5 to 5.6 m/s. Based on their data, they suggested the following equation:

$$h_w = 8.55 + 2.56\, V_\infty \tag{4.28}$$

where h_w is in W/m^2-K and V_∞ is in m/s

The values predicted by Eq. (4.28) are typically about twice the values obtained from Eq. (4.27). Test *et al.* state that there is a considerable amount of turbulence in the wind in a natural environment. As a result, tests conducted in a low turbulence intensity wind tunnel (as was done by Sparrow and his co-workers) significantly underestimate the value of h_w. Results of other investigators (Kumar *et al.* [9]) substantiate the arguments of Test *et al.* It is, therefore, recommended that Eq. (4.28) be used for calculating the value of the wind heat transfer coefficient.

❑ *Sky Temperature* The effective temperature of the sky is usually calculated from the following simple empirical relation in which temperatures are expressed in Kelvin.

$$T_{sky} = T_a - 6 \tag{4.29}$$

4.6.2 Bottom Loss Coefficient

The bottom loss coefficient U_b is evaluated by considering conduction and convection losses from the absorber plate in the downward direction through the bottom of the collector. It will be assumed that the flow of heat is one-dimensional and steady (Fig. 4.8). In most cases, the thickness of insulation provided is such that the thermal resistance associated with conduction dominates. Thus, neglecting the convective resistances, we have

$$U_b = \frac{k_i}{\delta_b} \tag{4.30}$$

where k_i = thermal conductivity of the insulation,
δ_b = thickness of the insulation.

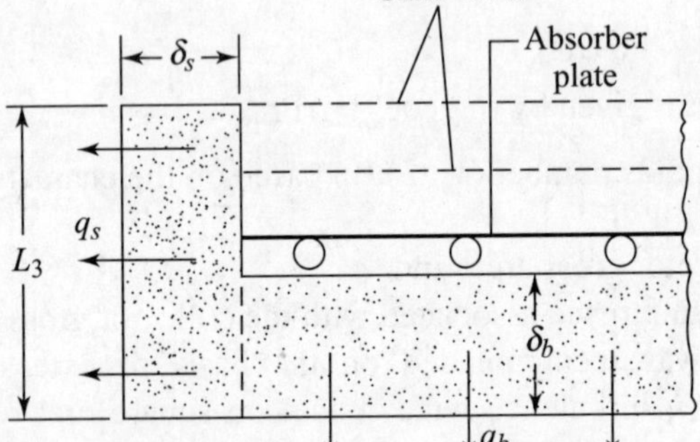

Fig. 4.8 *Bottom and side losses from a flat-plate collector*

4.6.3 Side Loss Coefficient

As in the case of the bottom loss coefficient, it will be assumed that the conduction resistance dominates and that the flow of heat is one-dimensional and steady. The one-dimensional approximation can be justified on the grounds that the side loss coefficient is always much smaller than the top loss coefficient.

If the dimensions of the absorber plate are $L_1 \times L_2$ and the height of the collector casing is L_3, then the area across which heat flows sideways is $2(L_1 + L_2)L_3$. The temperature drop across which the heat flow occurs varies from $(T_{pm} - T_a)$ at the absorber plate level to zero both at the top and bottom. Assuming, therefore, that the average temperature drop across the side insulation is $(T_{pm} - T_a)/2$ and that the thickness of this insulation is δ_s, we have

$$q_s = 2L_3(L_1 + L_2)k_i \frac{(T_{pm} - T_a)}{2\delta_s} \tag{4.31}$$

Thus, from Eq. (4.20),

$$U_s = \frac{(L_1 + L_2)L_3 k_i}{L_1 L_2 \delta_s} \tag{4.32}$$

Example 4.2

Calculate the overall loss coefficient for a flat-plate collector with two glass covers. Given the following data:

Size of the absorber plate ($L_1 \times L_2$)	:	1.90 m × 0.90 m
Spacing between plate and first glass cover (L)	:	4 cm
Spacing between first and second glass cover (L)	:	4 cm
Plate emissivity (ε_p)	:	0.92
Glass cover emissivity (ε_c)	:	0.88
Collector tilt (β)	:	20°
Mean plate temperature (T_{pm})	:	70°C
Ambient air temperature (T_a)	:	24°C
Wind speed (V_∞)	:	2.5 m/s
Back insulation thickness (δ_b)	:	8 cm
Side insulation thickness (δ_s)	:	4 cm
Thermal conductivity of insulation (k_i)	:	0.05 W/m-K

Solution

From Eq. (4.29),

$$T_{sky} = 297.2 - 6 = 291.2 \text{ K}$$

Substituting this value and the given values of T_{pm}, T_a, ε_p and ε_c in Eqs (4.22) to (4.24), we have

$$\frac{q_t}{A_p} = h_{p-c1}(343.2 - T_{c1}) + \frac{5.67 \times 10^{-8}(343.2^4 - T_{c1}^4)}{\left(\frac{1}{0.92} + \frac{1}{0.88} - 1\right)}$$

$$= h_{p-c1}(343.2 - T_{c1}) + 4.6350 \times 10^{-8}(138.736 \times 10^8 - T_{c1}^4) \tag{4.33}$$

$$\frac{q_t}{A_p} = h_{c1-c2}(T_{c1} - T_{c2}) + \frac{5.67 \times 10^{-8}(T_{c1}^4 - T_{c2}^4)}{\left(\frac{1}{0.88} + \frac{1}{0.88} - 1\right)}$$

$$= h_{c1-c2}(T_{c1} - T_{c2}) + 4.455 \times 10^{-8}(T_{c1}^4 - T_{c2}^4) \tag{4.34}$$

and

$$\frac{q_t}{A_p} = h_w(T_{c2} - 297.2) + 5.67 \times 10^{-8} \times 0.88(T_{c2}^4 - 291.2^4)$$

$$= h_w(T_{c2} - 297.2) + 4.990 \times 10^{-8}(T_{c2}^4 - 71.9061 \times 10^8) \tag{4.35}$$

Equations (4.33) to (4.35) have to be solved for the unknowns (q_t/A_p), T_{c1} and T_{c2}. For this, the values of h_{p-c1}, h_{c1-c2} and h_w are needed. Since these values depend upon T_{c1} and T_{c2}, a trial-and-error method becomes necessary.

Assume $\quad T_{c1} = 325$ K, $\quad T_{c2} = 305$ K

We use the correlations (4.25) to calculate h_{p-c1} and h_{c1-c2}, and correlation (4.28) to calculate h_w.

Calculation of h_{p-c1}:

$$\text{Mean temperature of air between plate and first cover} = \frac{343.2 + 325}{2} = 334.1 \text{ K}$$

At this temperature,†

$$k = 0.0291 \text{ W/m-K}$$

$$\nu = 19.06 \times 10^{-6} \text{ m}^2\text{/s}$$

$$\text{Pr} = 0.696$$

$$\text{Ra}_L \cos\beta = 9.81 \times \frac{1}{334.1} \frac{(343.2 - 325) \times 0.04^3}{19.06^2 \times 10^{-12}} \times 0.696 \cos 20°$$

$$= 61\,544$$

Therefore,

$$\text{Nu}_L = 0.229\,(61\,544)^{0.252} = 3.6873$$

$$h_{p-c1} = \frac{3.6873 \times 0.0291}{0.04} = 2.683 \text{ W/m}^2\text{-K}$$

Calculation of h_{c1-c2}:

$$\text{Mean temperature of air between first and second cover} = \frac{325 + 305}{2} = 315 \text{ K}$$

Proceeding in a similar manner, we obtain

$$h_{c1-c2} = 2.803 \text{ W/m}^2\text{-K}$$

Calculation of h_w:

$$h_w = 8.55 + 2.56 \times 2.5 = 14.95 \text{ W/m}^2\text{-K}$$

After substituting these values into Eqs (4.33) – (4.35), we obtain the values of T_{c1} and T_{c2} which satisfy the equations. The calculation is shown in tabular fashion:

T_{c1} (K)	T_{c2} (K)	(q_t/A_p) from		
		Eq. (4.33)	Eq. (4.34)	Eq. (4.35)
325	305	174.8	167.6	189.6
325	304	174.8	175.4	169.0
325	304.5	174.8	171.5	179.3
325.1	304.3	173.9	174.0	175.0

† Properties of air and water are given in Appendix 4.

The values of (q_t/A_p) in the last line of the table are reasonably close to each other. Therefore the average value of 174.3 W/m^2 is acceptable. Since the values of $T_{c1} = 325.1$ K and $T_{c2} = 304.3$ K are close to the original assumptions of 325 K and 305 K, it will not be necessary to repeat the calculations for h_{p-c1} and h_{c1-c2} with these new values.

Therefore,

$$U_t = \frac{174.3}{(343.2 - 297.2)} = 3.79 \text{ W/m}^2\text{-K}$$

Using Eq. (4.30), bottom loss coefficient

$$U_b = \frac{0.05}{0.08} = 0.63 \text{ W/m}^2\text{-K}$$

From Eq. (4.32), side loss coefficient

$$U_s = \frac{(0.90 + 1.90) \times 0.16 \times 0.05}{0.90 \times 1.90 \times 0.04} = 0.33 \text{ W/m}^2\text{-K}$$

Therefore, overall loss coefficient

$$U_l = 3.79 + 0.63 + 0.33 = 4.75 \text{ W/m}^2\text{-K}$$

4.6.4 Empirical Equation for Top Loss Coefficient

From Example 4.2, it is clear that a tedious iterative calculation is required for obtaining the value of the top loss coefficient. Based on calculations for a large number of cases covering the entire range of conditions normally expected for flat-plate collectors, Klein [10] has developed the following convenient empirical equation for calculating the top loss coefficient.

$$U_t = \left[\frac{M}{\left(\dfrac{C}{T_{pm}}\right)\left(\dfrac{T_{pm} - T_a}{M + f}\right)^{0.33}} + \frac{1}{h_w} \right]^{-1} + \left[\frac{\sigma(T_{pm}^2 + T_a^2)(T_{pm} + T_a)}{\dfrac{1}{\varepsilon_p + 0.05M(1 - \varepsilon_p)} + \dfrac{(2M + f - 1)}{\varepsilon_c} - M} \right] \tag{4.36}$$

where $f = (1 - 0.04\, h_w + 0.0005\, h_w^2)(1 + 0.091\, M)$

$C = 365.9(1 - 0.00883\, \beta + 0.0001298\, \beta^2)$

$M =$ number of glass covers

While using Eq. (4.36), T_{pm} and T_a are expressed in K, h_w in W/m^2-K, σ in W/m^2-K^4, and β in degrees. The value of U_t is obtained in W/m^2-K. The range of conditions over which Eq. (4.36) has been developed are as follows:

$$320 \leq T_{pm} \leq 420 \text{ K}$$
$$260 \leq T_a \leq 310 \text{ K}$$
$$0.1 \leq \varepsilon_p \leq 0.95$$
$$0 \leq V_\infty \leq 10 \text{ m/s}$$
$$1 \leq M \leq 3$$
$$0 \leq \beta \leq 90°$$

Although not explicitly stated, it appears that the values of ε_c and L have been taken to be constant and equal to 0.88 and 2.54 cm, respectively. In developing the correlation, Klein has assumed that the value of h_w could be obtained from Eq. (4.26).

A number of empirical correlations having the same form as Eq. (4.36) have also been suggested. Garg and Datta [11] have made a detailed study of these correlations as well as Klein's correlation and have concluded that the correlation suggested by Malhotra *et al.*[12] agrees best with the exact iterative solution. Malhotra *et al.* have given the following correlation.

$$U_t = \left[\frac{M}{\left(\dfrac{C}{T_{pm}}\right)\left(\dfrac{T_{pm} - T_a}{M + f}\right)^{0.252}} + \frac{1}{h_w} \right]^{-1} + \left[\frac{\sigma(T_{pm}^2 + T_a^2)(T_{pm} + T_a)}{\dfrac{1}{\varepsilon_p + 0.0425\, M(1 - \varepsilon_p)} + \dfrac{2M + f - 1}{\varepsilon_c} - M} \right] \qquad (4.37)$$

where $f = \left(\dfrac{9}{h_w} - \dfrac{30}{h_w^2}\right)\left(\dfrac{T_a}{316.9}\right)(1 + 0.091M)$

$C = 204.429 (\cos \beta)^{0.252} / L^{0.24}$, and

$L =$ spacing (m).

COLLECTOR EFFICIENCY FACTOR 4.7

In Section 4.6, procedures for calculating the overall loss coefficient were described. The heat lost from the collector can thus be calculated, if the average plate temperature is known. However, this temperature is generally not known. It will, therefore, be necessary to consider the flow of heat in the absorber plate and across the fluid tubes to the fluid so that the values of T_{pm} can be related to the value of the inlet fluid temperature, which is a known quantity.

In order to simplify the problem, the approach adopted will be to conduct a number of one-dimensional analyses. First, the one-dimensional flow of heat in the absorber plate in a direction at right angles to the direction of fluid flow will be considered. This will be followed by a consideration of the heat flow from the plate to the fluid across the tube wall. Finally, the one-dimensional flow of fluid inside the tubes will be analysed.

Consider a collector having an absorber plate of length L_1 and width L_2. Assume that there are N fluid tubes and that the pitch of the tubes is $W (= L_2/N)$. Let D_i and D_o be the inside and outside diameters of the tubes.

Consider a section of the absorber plate with two adjacent fluid tubes. The temperature in the plate (T_p) will vary in the x-direction in the manner as shown in Fig. 4.9. It will be assumed that the same distribution exists between any two tubes. Above the fluid tubes, the temperature will be constant, while in between the tubes, the temperature will pass through a maximum. Taking a slice dy along the flow direction and neglecting heat conduction in the plate in that direction, we can write an energy balance for an element $dx \times dy$ of the plate.

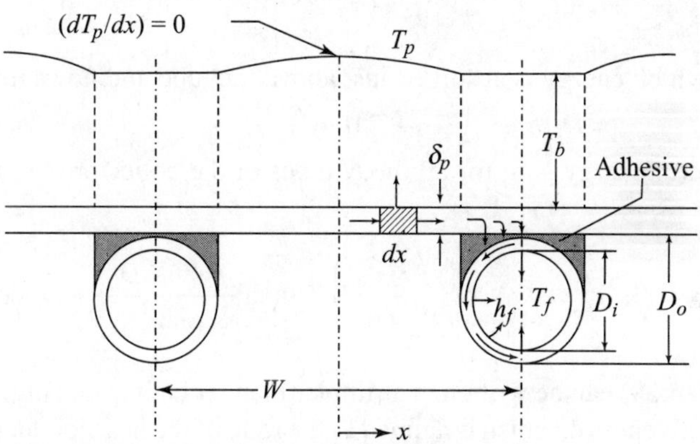

Fig. 4.9 *Temperature distribution in an absorber plate in a direction at right angles to the fluid flow*

(Net heat conducted into element) + (Incident energy absorbed)
= Heat lost from element

$$k_p \delta_p \frac{d^2T_p}{dx^2} dx\, dy + S\, dx\, dy = U_l\, dx\, dy(T_p - T_a)$$

Simplifying,
$$\frac{d^2T_p}{dx^2} = \frac{U_l}{k_p \delta_p}\left[T_p - T_a - \frac{S}{U_l}\right] \tag{4.38}$$

Equation (4.38) is solved using the boundary conditions

$$x = 0\,, \quad \frac{dT_p}{dx} = 0 \quad \text{and} \quad x = \left(\frac{W - D_o}{2}\right), \quad T_p = T_{po}$$

This yields the solution

$$\frac{T_p - \left(T_a + \dfrac{S}{U_l}\right)}{T_{po} - \left(T_a + \dfrac{S}{U_l}\right)} = \frac{\cosh mx}{\cosh\left[\dfrac{m(W - D_o)}{2}\right]}$$

where $m = (U_l/k_p\delta_p)^{1/2}$.

The temperature distribution obtained is similar to that for a long rectangular fin. The rate at which energy is conducted through the plate to one fluid tube from both sides

$$= -\,2k_p\,\delta_p\left(\frac{dT_p}{dx}\right)_{x=(W-D_o)/2} dy$$

$$= 2\left(\frac{k_p\delta_p}{U_l}\right)^{1/2}[S - U_l(T_{po} - T_a)]\tanh\left(\frac{m(W - D_o)}{2}\right)dy$$

The rate at which energy is absorbed just above the tube and flows in

$$= D_o[S - U_l(T_{po} - T_a)]\;dy$$

Thus the useful energy gain for all the N tubes of the collector over a length dy is given by

$$dq_u = N[S - U_l\,(T_{po} - T_a)] \times \left[2\left(\frac{k_p\,\delta_p}{U_l}\right)^{1/2}\tanh\frac{m(W - D_o)}{2} + D_o\right]dy \qquad (4.39)$$

Equation (4.39) can be written in a simpler manner by introducing the concept of plate effectiveness ϕ, which is defined as the ratio of the heat conducted through the plate to the fluid tube, to the heat which would have been conducted if the thermal conductivity of the plate material was infinite. It is easily shown that

$$\phi = \frac{\tanh\,[m(W - D_o)/2]}{[m(W - D_o)/2]}$$

Thus, $$\frac{1}{N}\left(\frac{dq_u}{dy}\right) = [S - U_l\,(T_{po} - T_a)]\;[\phi(W - D_o) + D_o] \qquad (4.40)$$

We next consider the flow of heat from the plate to the fluid. The three thermal resistances in the path are due to the adhesive used for attaching the tubes to the absorber plate, the tube wall and the heat transfer coefficient at the inner surface of the tube. Assuming the thermal resistance of the tube wall to be negligible,

$$\frac{1}{N}\left(\frac{dq_u}{dy}\right) = \frac{(T_{po} - T_f)}{\left(\dfrac{\delta_a}{k_a D_o} + \dfrac{1}{\pi D_i h_f}\right)} \tag{4.41}$$

where δ_a = average thickness of the adhesive,
k_a = thermal conductivity of the adhesive material,
T_f = local fluid temperature, and
h_f = heat transfer coefficient on the inside surface of the tube.

Combining Eqs (4.40) and (4.41) so as to eliminate the intermediate temperature T_{po}, we have

$$\frac{1}{N}\left(\frac{dq_u}{dy}\right) = \frac{[S - U_l(T_f - T_a)]}{U_l\left[\dfrac{1}{U_l[(W - D_o)\phi + D_o]} + \dfrac{\delta_a}{k_a D_o} + \dfrac{1}{\pi D_i h_f}\right]} \tag{4.42}$$

Let us define a term called the *collector efficiency factor* (F') as follows:

$$F' = \frac{1}{WU_l\left[\dfrac{1}{U_l[(W - D_o)\phi + D_o]} + \dfrac{\delta_a}{k_a D_o} + \dfrac{1}{\pi D_i h_f}\right]} \tag{4.43}$$

Substituting this definition into Eq. (4.42), we get

$$\frac{1}{N}\left(\frac{dq_u}{dy}\right) = WF'\,[S - U_l(T_f - T_a)] \tag{4.44}$$

where F' represents the ratio of the actual useful gain rate per tube per unit length to the gain which would occur if the collector absorber plate were at the temperature T_f.

4.8 COLLECTOR HEAT-REMOVAL FACTOR

The final one-dimensional analysis will be performed along the direction of fluid flow with the objective of determining the variation of fluid temperature. This analysis will help in linking the useful heat gain rate with the fluid inlet temperature.

Consider as a control volume, an elementary length *dy* of one tube (Fig. 4.10). Applying the first law of thermodynamics,

Rate of change of enthalpy of the fluid flowing through the control volume

= Rate of heat transfer to fluid inside the control volume

Thus, $$\left(\frac{\dot{m}}{N}\right) C_p\, dT_f = \frac{1}{N} dq_u = WF'\,[S - U_l(T_f - T_a)]\, dy$$

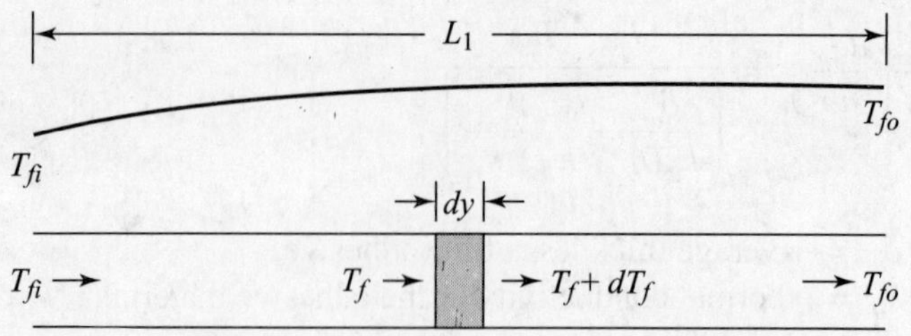

Fig. 4.10 *Variation of fluid temperature in flow direction*

$$\frac{dT_f}{dy} = \frac{WF'U_l}{(\dot{m}/N)C_p}\left[\left(\frac{S}{U_l} + T_a\right) - T_f\right] \tag{4.45}$$

Integrating and using the inlet condition $y = 0$, $T_f = T_{fi}$, we obtain the temperature distribution

$$\frac{\left(\dfrac{S}{U_l} + T_a\right) - T_f}{\left(\dfrac{S}{U_l} + T_a\right) - T_{fi}} = \exp\left\{\frac{-L_2F'U_ly}{\dot{m}C_p}\right\} \tag{4.46}$$

The fluid outlet temperature T_{fo} is obtained by substituting $T_f = T_{fo}$ and $y = L_1$ in Eq. (4.46). Thus,

$$\frac{\left(\dfrac{S}{U_l} + T_a\right) - T_{fo}}{\left(\dfrac{S}{U_l} + T_a\right) - T_{fi}} = \exp\left\{\frac{F'U_lL_1L_2}{\dot{m}C_p}\right\} \tag{4.47}$$

Subtracting both sides of Eq. (4.47) from unity,

$$\frac{(T_{fo} - T_{fi})}{\left(\dfrac{S}{U_l} + T_a\right) - T_{fi}} = 1 - \exp\left\{-\frac{F'U_lA_p}{\dot{m}C_p}\right\} \tag{4.48}$$

Thus, the useful heat gain rate for the collector

$$q_u = \dot{m}C_p(T_{fo} - T_{fi})$$

$$= \frac{\dot{m}C_p}{U_l}[S - U_l(T_{fi} - T_a)]\left[1 - \exp\left\{-\frac{F'U_lA_p}{\dot{m}C_p}\right\}\right]$$

$$q_u = F_RA_p[S - U_l(T_{fi} - T_a)] \tag{4.49}$$

where,
$$F_R = \frac{\dot{m}C_p}{U_lA_p}\left[1 - \exp\left\{-\frac{F'U_lA_p}{\dot{m}C_p}\right\}\right] \tag{4.50}$$

The term F_R is called the *collector heat-removal factor.* It is an important design parameter since it is a measure of the thermal resistance encountered by the absorbed solar radiation in reaching the collector fluid. From Eq. (4.49), it can be seen that F_R represents the ratio of the actual useful heat gain rate to the gain which would occur if the collector absorber plate was at the temperature T_{fi} everywhere. As such its value ranges between 0 and 1.

Equation (4.49) is a very convenient expression for calculating the useful energy gain because the inlet fluid temperature is usually a known quantity. It is often referred to as the *Hottel-Whillier-Bliss* (HWB) equation.

A NUMERICAL EXAMPLE 4.9

We will illustrate the use of the equations derived in the preceding sections through a detailed numerical example.

Example 4.3

A flat-plate collector is made up of a copper absorber plate, copper tubes fixed on the under side and one glass cover. The following data is given:

Length of collector	: 2.08 m
Width of collector	: 1.07 m
Length of absorber plate	: 2.00 m
Width of absorber plate	: 0.98 m
Plate to cover spacing	: 2.5 cm
Thermal conductivity of plate material	: 350 W/m-K
Plate thickness	: 0.15 mm
Plate absorptivity for solar radiation	: 0.94
Plate emissivity for re-radiation	: 0.14
Outer diameter of tube	: 13.7 mm
Inner diameter of tube	: 12.5 mm
Tube centre-to-centre distance	: 11.3 cm
Glass cover emissivity/absorptivity	: 0.88
Extinction coefficient of glass	: 19.0 m^{-1}
Thickness of glass cover	: 4 mm
Refractive index of glass relative to air	: 1.526
Location of collector	: Pune (18°32′ N, 73°51′ E)
Date	: May 15
Time	: 12 noon (IST)
Collector tilt	: latitude angle
Surface azimuth angle	: 0°
I_b	: 725 W/m^2
I_d (uniformly distributed over the sky)	: 230 W/m^2
Adhesive resistance	: Negligible

Fluid to tube heat transfer coefficient	:	205 W/m^2-K
Water flow rate	:	70 kg/h
Water inlet temperature	:	60°C
Ambient temperature	:	25°C
Wind speed	:	3.1 m/s
Back insulation thickness	:	5 cm
Insulation thermal conductivity	:	0.04 W/m-K
Reflectivity of the surrounding surfaces	:	0.2

Assume that the side loss coefficient is 10 per cent of the bottom loss coefficient.

Calculate,

1. The angle of incidence of beam radiation on the collector,
2. The total solar flux incident on the collector,
3. $(\tau\alpha)_b$ and $(\tau\alpha)_d$
4. The incident flux absorbed by the absorber plate,
5. The collector heat-removal factor and overall loss coefficient,
6. The water outlet temperature, and
7. The instantaneous efficiency.

Solution

1. Angle of Incidence of Beam Radiation

On May 15, $n = 135$. From Eq. (3.4),

$$\delta = 23.45 \sin\left[\frac{360}{365}(284 + 135)\right] = 18.79°$$

From Fig. (3.15),

$$\begin{aligned} \text{LAT} &= 12\text{ h} - 4(82.50 - 73.85)\text{ minutes} + (+2.6\text{ minutes}) \\ &= 12\text{ h} - 32.0\text{ minutes} \\ &= 11\text{ h }28\text{ minutes} \end{aligned}$$

$$\omega = 32 \times 15/60 = 8.0°$$

Therefore, substituting into Eq. (3.9), we have

$$\cos\theta = \sin 18.79° \sin 0° + \cos 18.79° \cos 8.0° \cos 0° = 0.9375$$

$$\theta = 20.36°$$

2. Solar Flux Incident on Collector

From Eqs (3.41)–(3.43),

$$r_b = \frac{0.9375}{\sin 18.79° \sin 18.53° + \cos 18.79° \cos 18.53° \cos 8.0°}$$

$$= 0.9458$$

$$r_d = \frac{1 + \cos 18.53°}{2} = 0.9741$$

$$r_r = 0.2\left(\frac{1-\cos 18.53°}{2}\right) = 0.0052$$

Therefore, from Eq. (3.44),

$$I_T = (725 \times 0.9458) + (230 \times 0.9741) + (955 \times 0.0052)$$
$$= 914.7 \text{ W/m}^2$$

3. $(\tau\alpha)_b$ and $(\tau\alpha)_d$

Angle of incidence = 20.36°
Therefore, angle of refraction = $\sin^{-1}$ (sin 20.36°/1.526) = 13.18°
From Eqs (4.6), (4.7), (4.10), (4.11) and (4.9),

$$\rho_{\text{I}} = \frac{\sin^2 (20.36° - 13.18°)}{\sin^2 (20.36° + 13.18°)} = 0.0512$$

$$\rho_{\text{II}} = \frac{\tan^2 (20.36° - 13.18°)}{\tan^2 (20.36° + 13.18°)} = 0.0361$$

$$\tau_{r\text{I}} = \frac{1-0.0512}{1+0.0512} = 0.903$$

$$\tau_{r\text{II}} = \frac{1-0.0361}{1+0.0361} = 0.930$$

$$\tau_r = (0.903 + 0.930)/2 = 0.9165$$

From Eq. (4.15),

$$\tau_a = \exp[-(19.0 \times 0.004/\cos 13.18°)] = 0.925$$

Therefore, from Eq. (4.4),

$$\tau = 0.9165 \times 0.925 = 0.848$$

For diffuse radiation, the angle of incidence is taken to be 60°. Therefore, angle of refraction

$$= \sin^{-1} (\sin 60°/1.526) = 34.58°$$

Following the same procedure as for beam radiation, it can be shown that

$$\rho_{\text{I}} = 0.1854 \qquad \rho_{\text{II}} = 0.0014$$
$$\tau_{r\text{I}} = 0.6872 \qquad \tau_{r\text{II}} = 0.9972$$
$$\tau_r = (0.6872 + 0.9972)/2 = 0.8422$$
$$\tau_a = \exp[-(19.0 \times 0.004/\cos 34.58°)] = 0.912$$
$$\tau = 0.8422 \times 0.912 = 0.768$$

Thus, $\rho_d = 0.912 - 0.768 = 0.144$

Therefore, substituting into Eq. (4.16),

$$(\tau\alpha)_b = \frac{0.848 \times 0.94}{1-(1-0.94) \times 0.144} = 0.804$$

$$(\tau\alpha)_d = \frac{0.768 \times 0.94}{1-(1-0.94)\times 0.144} = 0.728$$

4. Incident Flux Absorbed by Absorber Plate

The amount of incident flux absorbed by the absorber plate is given by Eq. (4.2).

$$S = (725 \times 0.9458 \times 0.804) + (230 \times 0.9741 \times 0.728) + (955 \times 0.0052 \times 0.728)$$

$$= 718.0 \text{ W/m}^2$$

5. Collector Heat-Removal Factor and Overall Loss Coefficient

An iterative procedure will be required since both F_R and U_l cannot be directly determined and the value of one is dependent on the other.

First Iteration: Assume $U_l = 4.0$ W/m^2-K. This is a reasonable assumption for a collector with one glass cover and a selective absorber surface.

$$m = \left(\frac{4.0}{350 \times 0.15 \times 10^{-3}}\right)^{1/2} = 8.73 \text{ m}^{-1}$$

$$\frac{m(W-D_o)}{2} = \frac{8.73(0.113-0.0137)}{2} = 0.433$$

Effectiveness $\phi = \tanh 0.433/0.433 = 0.942$

From Eq. (4.43), the collector efficiency factor

$$F' = \frac{1}{0.113 \times 4.0\left[\dfrac{1}{4.0\{(0.113-0.0137)\times 0.942+0.0137\}} + \dfrac{1}{205\times\pi\times 0.0125}\right]}$$

$$= 0.901$$

$$\frac{\dot{m}C_p}{A_p} = \frac{70}{2.0 \times 0.98} \times \frac{4.18 \times 1000}{3600} = 41.47 \text{ W/m}^2\text{-K}$$

Therefore, from Eq. (4.50), the collector heat-removal factor

$$F_R = \frac{41.47}{4.0}\left[1 - \exp\left\{-\frac{4.0 \times 0.901}{41.47}\right\}\right] = 0.863$$

and from Eq. (4.49), the useful heat gain

$$q_u = F_R A_p [S - U_l (T_{fi} - T_a)]$$

$$= 0.863 \times 2.0 \times 0.98\,[718.0 - 4.0\,(60 - 25)] = 977.7 \text{ W}$$

Therefore $\quad q_l = 718.0 \times 2.0 \times 0.98 - 977.7 = 429.6$ W

From Eq. (4.17),

$$q_l = 4.0 \times 2.0 \times 0.98 \times (T_{pm} - 25)$$

Thus, $T_{pm} = 54.80 + 25 = 79.80°\text{C} = 353.0$ K

We will calculate the value of the top loss coefficient using the value of $T_{pm} = 353.0$ K. Following the same procedure as in Example 4.2, we have

$$T_{sky} = 298.2 - 6 = 292.2 \text{ K}$$

$$\frac{q_t}{A_p} = h_{p-c}(353.0 - T_c) + \frac{5.67 \times 10^{-8}(353.0^4 - T_c^4)}{\left(\frac{1}{0.14} + \frac{1}{0.88} - 1\right)}$$

$$= h_{p-c}(353.0 - T_c) + 7.78929 \times 10^{-9}(155.274 \times 10^8 - T_c^4) \quad (4.51a)$$

$$\frac{q_t}{A_p} = h_w(T_c - 298.2) + 5.67 \times 10^{-8} \times 0.88\,(T_c^4 - 292.2^4)$$

$$= h_w(T_c - 298.2) + 4.990 \times 10^{-8}(T_c^4 - 72.899 \times 10^8) \quad (4.51b)$$

Assume $T_c = 307$ K.

Calculation of h_{p-c}

Mean temperature = (353.0 + 307.0)/2 = 330.0 K = 56.8°C

$$k = 0.0288 \text{ W/m-K}$$

$$\nu = 18.66 \times 10^{-6} \text{ m}^2\text{/s}$$

$$\text{Pr} = 0.697$$

$$\text{Ra}_L \cos\beta = 9.81 \times \frac{1}{330.0} \times \frac{(353.0 - 307.0) \times 0.025^3}{18.66^2 \times 10^{-12}} \times 0.697 \times \cos 18.53°$$

$$= 40\,553$$

$$\text{Nu}_L = 0.229 \times (40\,553)^{0.252} = 3.319$$

$$h_{p-c} = 3.319 \times 0.0288/0.025 = 3.823 \text{ W/m}^2\text{-K}$$

Calculation of h_w

$$h_w = 8.55 + 2.56 \times 3.1 = 16.486 \text{ W/m}^2\text{-K}$$

A trial-and-error method involving small adjustments in the value of T_c will now be required such that the values of (q_t/A_p) calculated from Eqs (4.51a) and (4.51b) agree with each other. This is shown in the following table.

T_c(K)	q_t/A_p (W/m²)	
	Eq. (4.51a)	Eq. (4.51b)
307	175.9 + 51.8 = 227.7	145.1 + 79.5 = 224.6
307.1	175.5 + 51.7 = 227.2	146.7 + 80.1 = 226.8

$$(q_t/A_p)_{avg} = 227.0 \text{ W/m}^2$$

$$U_t = \frac{227.0}{(353.0 - 298.2)} = 4.142 \text{ W/m}^2\text{-K}$$

Now from Eq. (4.30)

$$U_b = 0.04/0.05 = 0.8 \text{ W/m}^2\text{-K}$$

Therefore,

$$U_s = 0.08 \text{ W/m}^2\text{-K}$$

$$U_l = 4.142 + 0.8 + 0.08 = 5.022 \text{ W/m}^2\text{-K}$$

Since this value of U_l is considerably different from the assumed value, a second iteration will be necessary.

Second Iteration:

$$m = \left(\frac{5.022}{350 \times 0.15 \times 10^{-3}}\right)^{1/2} = 9.78 \text{ m}^{-1}$$

$$\frac{m(W - D_o)}{2} = \frac{9.78(0.113 - 0.0137)}{2} = 0.4856$$

$$\text{Effectiveness } \phi = \tanh 0.4856/0.4856 = 0.9282$$

$$F' = \frac{1}{0.113 \times 5.022\left[\frac{1}{5.022\{(0.113 - 0.0137) \times 0.9282 + 0.0137\}} + \frac{1}{205 \times \pi \times 0.0125}\right]}$$

$$= 0.879$$

$$F_R = \frac{41.47}{5.022}\left[1 - \exp\left\{-\frac{5.022 \times 0.879}{41.47}\right\}\right] = 0.834$$

$$q_u = 0.834 \times 2.0 \times 0.98\,[718.0 - 5.022\,(60 - 25)] = 886.4 \text{ W}$$

$$q_l = 718.0 \times 2.0 \times 0.98 - 886.4 = 520.9 \text{ W}$$

$$= 5.022 \times 2.0 \times 0.98 \times (T_{pm} - 25)$$

$$T_{pm} = 77.92°\text{C} = 351.1 \text{ K}$$

This value of T_{pm} is different from the value of 353.0 K obtained in the first iteration. It will therefore be necessary to find out the top loss coefficient using this value of T_{pm}.

$$\frac{q_t}{A_p} = h_{p-c}\,(351.1 - T_c) + \frac{5.67 \times 10^{-8}\,(351.1^4 - T_c^4)}{\left(\frac{1}{0.14} + \frac{1}{0.88} - 1\right)}$$

$$= h_{p-c}\,(351.1 - T_c) + 7.78929 \times 10^{-9}(151.957 \times 10^8 - T_c^4) \quad (4.52)$$

Equation (4.51b) remains unchanged.

Assume $T_c = 307$ K.

Calculation of h_{p-c}

Mean temperature = (351.1 + 307.0)/2 = 329.1 K = 55.9°C

$$k = 0.0287 \text{ W/m-K}$$

$$\nu = 18.56 \times 10^{-6} \text{ m}^2\text{/s}$$

$$\text{Pr} = 0.697$$

$$\text{Ra}_L \cos\beta = 9.81 \times \frac{1}{329.1} \times \frac{(351.1 - 307.0) \times 0.025^3}{18.56^2 \times 10^{-12}} \times 0.697 \times \cos 18.53°$$

$$= 39\,406$$

$$\text{Nu}_L = 0.229 \times (39\,406)^{0.252} = 3.295$$

$$h_{p-c} = 3.295 \times 0.0287/0.025 = 3.783 \text{ W/m}^2\text{-K}$$

The value of h_w remains unchanged.

A trial-and-error method involving small adjustments in the value of T_c will now be repeated such that the values of (q_t/A_p) calculated from Eqs (4.51b) and (4.52) agree with each other. This is shown in the following table:

T_c(K)	q_t/A_p (W/m^2)	
	Eq. (4.52)	Eq. (4.51b)
307	166.8 + 49.2 = 216.0	145.1 + 79.5 = 224.6
306.7	168.0 + 49.4 = 217.5	140.1 + 77.8 = 217.9

$$(q_t/A_p)_{avg} = 217.7 \text{ W/m}^2$$

$$U_t = \frac{217.7}{(351.1 - 298.2)} = 4.115 \text{ W/m}^2\text{-K}$$

Therefore,

$$U_l = 4.115 + 0.8 + 0.08 = 4.995 \text{ W/m}^2\text{-K}$$

This value of U_l is not very different from the value of 5.022 W/m^2-K obtained after the first iteration. With U_l = 4.995 W/m^2-K, the values of ϕ, F', F_R, q_u, q_l and T_{pm} respectively are 0.9285, 0.8794, 0.8344, 888.3 W, 519.0 W and 351.2 K. These values are very close to those obtained with 5.022 W/m^2-K. Hence no more iterations are necessary.

6. Water Outlet Temperature

The water outlet temperature is obtained from the heat balance equation. Substituting,

$$70 \times 4.18\,(T_{fo} - 60) = 888.3 \times 3600/1000$$

Hence,
$$T_{fo} = 70.93°\text{C} = 344.1 \text{ K}$$

7. Instantaneous Efficiency

Using Eq. (4.3), the instantaneous efficiency is given by

$$\eta_i = \frac{888.3}{914.7 \times 2.08 \times 1.07} = 0.436, \text{ i.e. } 43.6\ \%$$

Considering the fact that the water inlet temperature is only 60°C, the efficiency of the given collector is rather low. This is so because the glass cover used is not of very good quality and has a low transmissivity. A collector fitted with a better glass cover having a lower extinction coefficient would, under comparable conditions, yield a higher efficiency (see Section 4.10.6).

Performance Over a Day

It is of interest to study the performance of a collector over a whole day. This is done for the same collector by using radiation data measured over a whole day. For the sake of simplicity, the water flow rate, water inlet temperature, ambient temperature and wind speed are all assumed to be constant at the values given earlier.

The radiation data used and the results obtained are given in Table 4.1. It is seen that the values of the useful heat gain and the efficiency (Fig. 4.11) increase sharply from 0900 to 1100 h, touch a peak around noon and then drop sharply after 1500 h. The variation obtained is typical for a flat-plate collector and indicates the strong dependence of these factors on the radiation incident on the collector. It is also seen that the value of the top loss coefficient does not vary much.

The average efficiency over the whole period, during which useful energy is collected, can be approximately calculated if it is assumed that the values of instantaneous efficiency and solar radiation are valid for half an hour on either side of the instant considered. Making this approximation, the efficiency averaged over 10 hours from 0830 to 1630 h works out to be 39.6 per cent.

Table 4.1 Performance of a flat-plate collector over a whole day

IST(h)	0900	1000	1100	1200	1300	1400	1500	1600
I_b(W/m^2)	390	547	665	725	715	615	476	337
I_d(W/m^2)	192	210	230	230	233	239	221	185
I_T(W/m^2)	536.3	712.7	852.9	914.7	908.2	814.7	657.9	482.4
T_{pm}(K)	340.8	345.8	349.5	351.2	351.0	348.5	344.3	339.5
U_t(W/m^2-K)	3.99	4.05	4.10	4.12	4.12	4.09	4.03	3.96
q_u(W)	377.1	619.2	805.6	888.3	879.2	755.4	548.8	313.0
T_{fo}(K)	337.8	340.8	343.1	344.1	344.0	342.5	340.0	337.1
η_i(%)	31.6	39.0	42.4	43.6	43.5	41.7	37.5	29.1

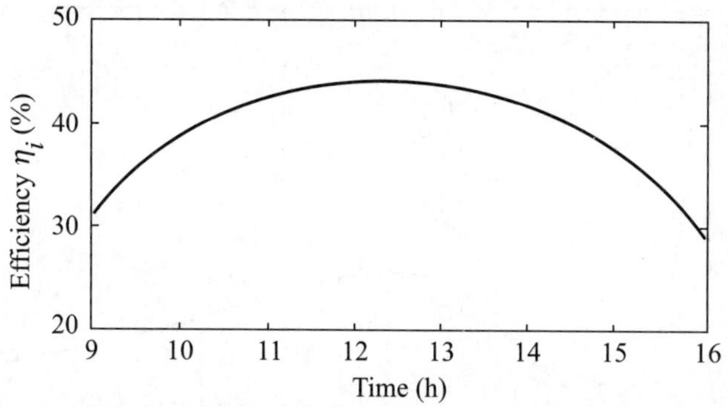

Fig. 4.11 *Variation of instantaneous efficiency of a collector over a day*

EFFECTS OF VARIOUS PARAMETERS ON PERFORMANCE 4.10

It is evident from the preceding sections and from Example 4.3 that a large number of parameters influence the performance of a liquid flat-plate collector. These parameters could be classified as design parameters, operational parameters, meteorological parameters and environmental parameters. In this section, the effects of some of these will be considered. The parameters discussed are the selectivity of the absorber surface, the number of glass covers, the spacing between the covers, the tilt of the collector, the fluid inlet temperature, the transmissivity of the glass, and dust settlement on the top glass cover.

4.10.1 Selective Surfaces

Absorber plate surfaces which exhibit the characteristics of a high value of absorptivity for incoming solar radiation and a low value of emissivity for outgoing re-radiation are called *selective surfaces.* Such surfaces are desirable because they maximize the absorption of solar energy and minimize the emission of the radiative loss. Obviously they would yield higher collector efficiencies than are obtained when the absorptivity and emissivity are equal.

The possibility of having selective absorber plate surfaces for flat-plate collectors was suggested first by Tabor [13] and later by Gier and Dunkle [14]. The basis for the suggestion can be understood if one compares the spectral distribution of extraterrestrial solar radiation with the black body radiation from a source at 350 K (which corresponds approximately to the temperature of the absorber plate). These are shown in Fig. 4.12 (a) and it is immediately obvious that there is almost no overlap between the two. Unlike solar radiation, which lies almost exclusively in the wavelength region up to 4 μm, the radiation coming off from the absorber plate is of large wavelengths with a maximum at 8.3 μm. It follows therefore, that if a surface that has a high absorptivity for wavelengths less than 4 μm and a low emissivity for wavelengths greater that 4 μm can be prepared, it would have the characteristics desirable for an absorber plate surface to act in

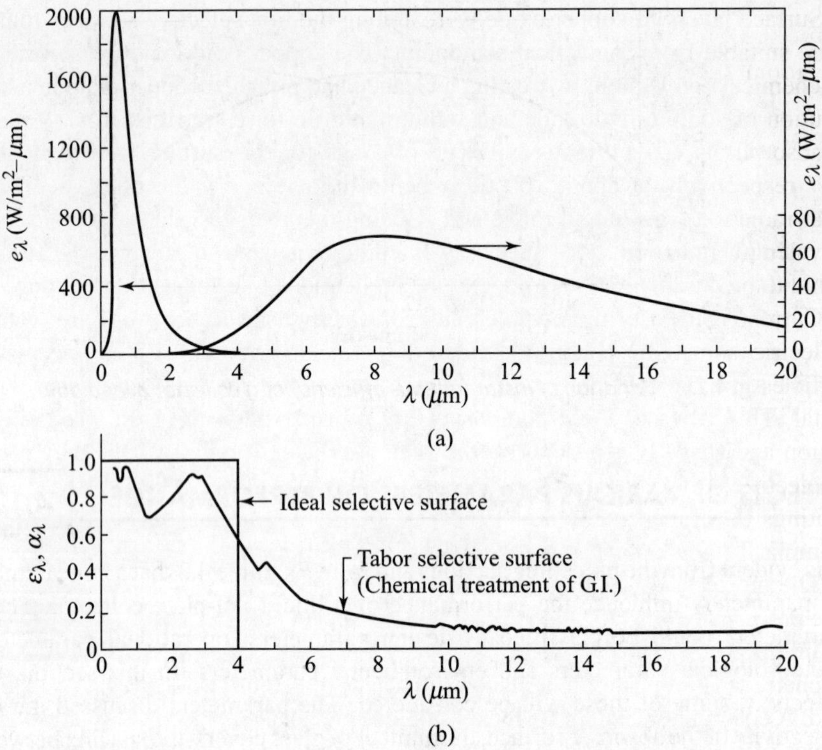

Fig. 4.12 *(a) Spectral distribution of extraterrestrial solar radiation and black body radiation from a source at 350 K (b) Monochromatic emissivity/absorptivity variation desired for an ideal selective surface and variation obtained for a selective surface by Tabor [13]*

a selective fashion. The characteristics desired for an ideal selective surface ($\alpha_\lambda = \varepsilon_\lambda = 1$ for $\lambda < 4$ μm and $\alpha_\lambda = \varepsilon_\lambda = 0$ for $\lambda > 4$ μm) are shown in Fig. 4.12 (b). For comparison, the variation obtained for one of the earliest surfaces synthesised by Tabor is also shown.

The development of selective coatings on various metal substrates has been the subject of intensive work for many years. As a result, a number of surfaces having characteristics approaching those of an ideal surface have been synthesised and a few have been commercialized. In most of these surfaces, the selectivity is achieved by having a polished and cleaned metal base and depositing on it a thin layer which is transparent to large wavelengths, but highly absorbing for small wavelength solar radiation. The layer is less than 1 μm in thickness and is deposited by a variety of methods. These include electroplating, chemical conversion, anodic oxidation, rf-magnetron sputtering and reactive DC magnetron sputtering. Some of the successful developments in this field will now be described.

Surface layers of copper oxide were among the first selective surfaces found to be suitable from a practical standpoint. The copper oxide layer was formed by chemical conversion, by treating a cleaned and polished copper plate in a hot solution of sodium hydroxide and sodium chlorite for a specified time. Values of absorptivity (α) and emissivity (ε_p) obtained for this surface were 0.89 and 0.17 respectively at about 100°C, α being the average value of α_λ over the solar radiation wavelength range and ε_p being the average value of ε_λ for large wavelength radiation. This surface was commercialized in Australia [15] and found to be durable for low temperature applications like solar water heating.

Currently, most of the commercialized selective surface coatings are metal-dielectric composite coatings, known as *cermets*. They consist of fine metal particles in a dielectric or ceramic matrix, or a porous oxide impregnated with metal. Thin films of these composites are transparent in the high wavelength region and strongly absorbing in the solar wavelength region. Thus they form a selective surface when deposited on a highly reflective metal surface. The coatings developed include nickel-black, black chrome and nickel-pigmented alumina. They are suitable for flat-plate collector applications up to 100°C.

The nickel-black coating was initially developed in Israel 40 years ago [16]. The process involved the cleaning of a metal sheet and subsequent electroplating by immersion of the sheet as the cathode in an aqueous electrolytic bath of nickel sulphate, zinc sulphate, ammonium sulphate, ammonium thiocyanate and citric acid. A coating of nickel, zinc, nickel sulphide and zinc sulphide was obtained and values of $\alpha = 0.81$ and $\varepsilon_p = 0.16$ to 0.18 were reported. Subsequently, different investigators [17-19] have modified the electro-deposition bath and the process to produce coatings with significantly better properties. Values of α ranging from 0.88 to 0.96 and of ε_p ranging from 0.03 to 0.10 have been obtained. These values are for a surface temperature of 100°C. Mason and Brendel [20] reported a higher value of $\alpha = 0.97$ on a nickel-black coating commercially available as Maxorb. This value was obtained by incorporating micro-roughness on the coating.

Recently Klochko *et al.* [21] have reported an investigation on producing zinc oxide-nickel cermet coatings. Zinc oxide arrays with vertically aligned ZnO nanorods were first laid on a fluorine based tin oxide substrate through a series of pulse electro-deposition steps. This was followed by filling the gaps between the nanorods with nickel nanoparticles again using pulse electro-deposition techniques. Values of α for different samples tested ranged from 0.83 to 0.92, while values of ε_p ranged from 0.14 to 0.28. According to the authors, optimization of various experimental parameters should lead to an improvement in the values of α and ε_p. They also feel that these coatings could be cheap and affordable for use in solar flat-plate collectors.

One of the most successful selective surfaces developed so far is 'black chrome'. It consists of a Cr particle/Cr_2O_3 composite electroplated on a copper, nickel-plated copper or stainless steel base. The commercial product is available from MTI in USA, BATEC in Denmark and Energie Solaire in Switzerland.

Values of α ranging from 0.95 to 0.97 and ε_p from 0.09 to 0.15 have been reported. In India, 'black chrome' coatings on copper are made by a number of manufacturers. Most of them use the Nalsun technology developed by NAL Bengaluru ($\alpha = 0.96$, $\varepsilon_p = 0.12$). The Nalsun technology uses a black chromium bath, the electroplating being done at room temperature. This helps to save on energy. In addition, the process does not require an undercoat of nickel on the copper sheet. In addition to its excellent properties, 'black chrome' has been found to be very durable. No significant degradation in performance has been observed even after its exposure to high humidity and high temperature as prescribed in an accelerated life testing procedure [22]. Because of environmental concerns with electro-deposition processes, a considerable amount of research has been carried out to develop physical vapor deposition (PVD) processes which are more environmental friendly. Selective coatings have been made by evaporation and sputtering methods. A few of these coatings have been commercialized. For example, a CrN/Cr_2O_3 coating with $\alpha = 0.95$ and $\varepsilon_p = 0.07$ at 100°C has been developed. The cermet was deposited on copper substrates using a DC magnetron sputtering technique. It has been shown to be durable and is being commercially produced under the name Sunselect by Alanod Solar in Germany [23].

Andersson *et al.* [24] have reported the development of a durable metal-dielectric composite coating formed by anodic oxidation. This cermet coating consists of grains of nickel embedded in a 0.7 μm thick porous layer of Al_2O_3. Development of stable nickel-alumina selective surface coatings on aluminium substrates have been reported by many investigators [25, 26]. These have absorptivity values in the range of 0.85 to 0.97 and emissivity values in the range of 0.08 to 0.21. The surfaces are commercially produced by Techno Term Energi in Sweden and Showa in Japan.

It is worth mentioning that the low-temperature solar hot-water collectors using evacuated tubes (described in Section 4.14) have a double-layer cermet coating of stainless steel-aluminium nitride. These are produced by reactive DC magnetron sputtering under the trade name of Turbo Sun in China. The α and ε_p values for this coating are reported to be 0.95 and 0.05 respectively.

Finally, we note that some spectrally selective paint coatings have also been commercially developed for low temperature applications (Solec, USA). These coatings are not as effective as electro-deposited coatings. The absorptivity value ranges from 0.88 to 0.94, while the emissivity value on carefully cleaned metal substrates ranges from 0.28 to 0.49.

An excellent review of solar selective surface coatings is given by Kennedy [27] and by Selvakumar and Barshilia [28].

❑ *Effect on Collector Performance* The effect of a selective surface on the performance of a collector can be best illustrated by taking specific situations. The collector of Section 4.9 is considered again and its performance without a selective surface ($\alpha = \varepsilon_p = 0.94$) and with a different commercially available selective surface ($\alpha = 0.95$, $\varepsilon_p = 0.085$) is calculated. The calculations are carried out in a manner similar to that adopted earlier and the results obtained are indicated in Table 4.2.

Table 4.2 Effect of a selective surface on performance of collector

	Selective surface $\alpha = 0.94, \varepsilon_p = 0.14$ (Sec. 4. 9)	Non-selective surface $\alpha = \varepsilon_p = 0.94$	Selective surface $\alpha = 0.95, \varepsilon_p = 0.085$
T_{pm}(K)	351.2	346.2	351.9
U_t (W/m^2-K)	4.12	7.26	3.83
q_u (W)	888.3	642.0	924.1
T_{fo}(K)	344.1	341.1	344.6
η_i (%)	43.6	31.5	45.4

It is seen from Table 4.2 that with a non-selective absorber plate, the top loss coefficient is 7.26 W/m^2-K and the efficiency is 31.5 per cent. The top loss coefficient increases by 3.14 W/m^2-K, while the efficiency decreases by 12.1 per cent. Thus, significant differences are observed. With the other selective surface, in which the value of α is marginally higher and that of ε_p is less, it is observed that the value of U_t decreases and the efficiency increases a little compared to the first selective surface.

4.10.2 Number of Covers

We will study the effect of the number of covers on performance of a collector by again considering the collector of Section 4.9. Table 4.3 shows the results obtained by using two or three covers. For comparison, the results obtained earlicr with one cover are also presented. Table 4.4 presents the same results as in Table 4.3 with the difference that the absorber surface is now non-selective.

Table 4.3 Effect of number of covers on performance of collector (selective surface, α = 0.94, ε_p= 0.14)

	Number of covers		
	1	2	3
$(\tau\alpha)_b$	0.8041	0.6892	0.5932
$(\tau\alpha)_d$	0.7284	0.6008	0.5114
U_t(W/m^2-K)	4.12	2.68	1.99
η_i (%)	43.6	41.0	36.6

The results in Table 4.3 show that the highest value of efficiency is obtained with one cover if the absorber plate surface is selective. With the addition of more covers, the efficiency goes on decreasing. On the other hand, when the absorber surface is non-selective, the efficiency increases as the number of covers is increased from one to two. Thereafter, the efficiency goes on decreasing with

Table 4.4 Effect of number of covers on performance of collector (non-selective surface, $\alpha = \varepsilon_p = 0.94$)

	Number of covers		
	1	2	3
$(\tau\alpha)_b$	0.8041	0.6892	0.5932
$(\tau\alpha)_d$	0.7284	0.6008	0.5114
U_t(W/m^2-K)	7.26	4.04	2.75
η_i (%)	31.5	35.3	33.4

the addition of more covers. Thus it is optimum to use only one cover in the present case if the absorber plate surface is selective and two covers if the surface is non-selective. This result is found to be generally true for many collectors.

The above results regarding the optimum number of covers can be explained as follows:

As the number of covers increases, the values of both $(\tau\alpha)_b$ and $(\tau\alpha)_d$ decrease. Thus, the flux S absorbed in the absorber plate decreases. The addition of more covers also causes the value of U_t and hence the heat loss, to decrease. However, the amount of decrease is not the same in both cases. For this reason, the useful heat gain (and consequently the efficiency) goes through a maximum value with a certain number of covers.

4.10.3 Spacing

The proper spacing to be kept between the absorber plate and the first cover, or between two covers has been the subject of considerable discussion. From the point of view of the heat loss from the top, it is evident that the spacing must be such that the values of the convective heat transfer coefficients are minimized. It is, therefore, useful to examine the behaviour of the correlating equations (4.25). This is done in Fig. 4.13 in which the variation of the heat transfer coefficient with spacing is drawn. Curves for two temperature differences are plotted with the mean air temperature and the tilt being kept fixed.

It is seen that for a particular temperature difference, the value of h first decreases and reaches a minimum. This variation corresponds to the conduction regime in which $\text{Nu}_L = 1$, with the minimum occurring at a spacing corresponding to $\text{Ra}_L \cos \beta = 1708$. Thereafter, as the spacing increases the value of h increases first sharply and then gradually. It goes through a maximum and then gradually decreases. With large values of spacing, the value drops below the previous minimum value.

It will be noted that the spacings at which the minimum and maximum values occur vary with the temperature difference. They also vary with the tilt. Since collectors are designed to operate at different locations with varying tilts and under varying service conditions, an optimum value of spacing is difficult to specify. It appears best to use a sufficiently large spacing away from the local minimum and maximum. Spacings from 4 to 8 cm have been suggested from this point of view by Buchberg *et al.* [4].

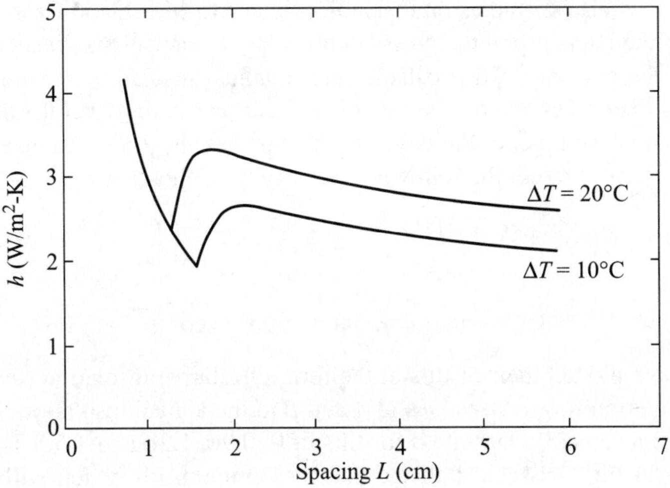

Fig. 4.13 *Plot of Eq. (4.25) showing variation of heat transfer coefficient with spacing.* $T_{mean} = 70°C$, $\beta = 20°$

Buchberg *et al.* have also studied the effect of spacing on the performance of a collector. They have done calculations for a single and double cover collector with a non-selective absorber plate and a single cover collector with a selective absorber plate. With the same operating conditions, two spacings have been tried out and the efficiency calculated. It has been shown that by using the larger spacing, around 5 cm, collector area requirements can be reduced by 2 to 8 per cent, the higher reduction being obtained with the collector having the selective absorber plate [4].

❑ *Effect of Shading* The main problem associated with the use of larger spacings is that shading of the absorber plate by the side walls of the collector casing increases. Some shading always occurs in every collector and needs to be corrected for. The shading is particularly important in the early morning and late evening hours. It is estimated that for most designs using spacings of 2 to 3 cm between the covers, shading reduces the radiation absorbed by about 3 per cent. Accordingly, it is recommended that the absorbed flux S be calculated in the usual manner but with a multiplying factor of 0.97. With larger spacings of 5 cm, it would be necessary to use a smaller multiplying factor around 0.95. Thus the gain obtained by using a larger spacing is offset by the reduction in S. For this reason, spacings of 2 or 3 cm are generally used by all manufacturers. It may be noted that in some collectors the inside of the side walls is lined with a reflecting surface in order to alleviate the effects of shading.

4.10.4 Collector Tilt

Flat-plate collectors are normally fixed in one position and do not track the sun. The question of the amount of tilt one should give to them is therefore of considerable importance. The basis for arriving at an optimum tilt will now be discussed.

One of the earliest studies on the subject is due to Morse and Czarnecki [29] who simplified the problem by assuming that extraterrestrial insolation was falling on the collector. They calculated the annual insolation per unit area by integrating the expression for the flux on a tilted surface first over the day length and then summing up over the days of the year. Taking $\gamma = 0°$, so that the daily insolation is maximized, the following expression is obtained:

$$\text{Annual insolation} = \sum_{n=1}^{365} I_{sc} \int_{-\omega_s}^{+\omega_s} \left(1 + 0.033 \cos \frac{360n}{365}\right)$$

$$\times (\sin \delta \sin \overline{\phi - \beta} + \cos \delta \cos \omega \cos \overline{\phi - \beta})\, d\omega$$

They have plotted their results in the form of relative insolation (the ratio of annual insolation for given values of ϕ and β to the annual insolation for $\phi = 0°$ and $\beta = 0°$) against the latitude ϕ for tilts of 0, 0.9ϕ, 1.2ϕ and 1.5ϕ. The results are shown in Fig. 4.14 and indicate that the optimum tilt which will result in maximizing the annual insolation is given by $\beta = 0.9\ \phi$. It is also seen that for latitudes up to 30°, small deviations of a degree or two from the optimum tilt will not cause much change in the relative insolation.

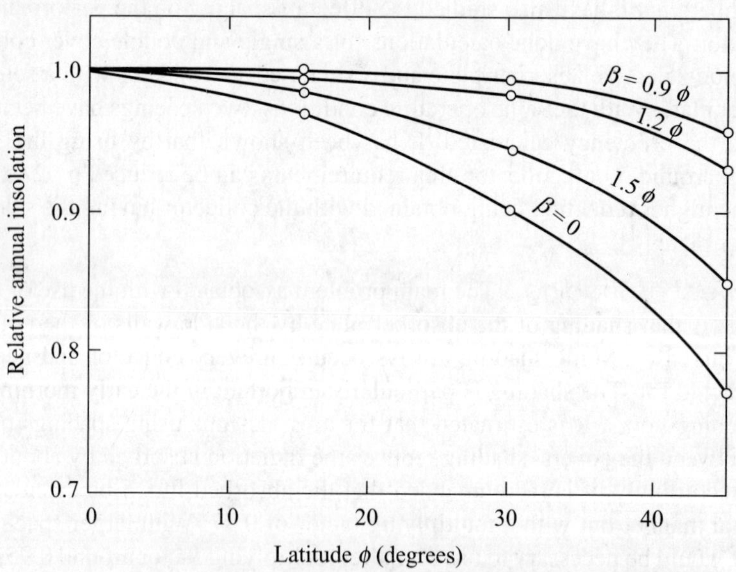

Fig. 4.14 *Variation of relative annual insolation with latitude for collectors tilted at various values of β and with $\gamma = 0°$*

The effect of having surface azimuth angles other than zero, i.e. collectors which do not slope precisely towards the south in the northern hemisphere (or the north in the southern hemisphere) has also been considered by Morse and Czarnecki. In this case also, it has been seen from their results that surface azimuth angles up to ±5° result in very small reductions (less than 1 per cent) in the relative insolation.

The above analysis for optimum tilt can be criticised on the grounds that it considers only beam radiation falling on the collector and assumes it to have the extraterrestrial value. Other analyses which consider the measured radiation data for specific locations have therefore been performed. One such simple analysis is due to Kern and Harris [30]. They have considered a collector oriented with a zero azimuth angle and having a tilt β, and used the monthly averages of the daily beam and diffuse radiation measured at a location. The geometry of the situation existing at solar noon has been used for obtaining the appropriate tilt factor for beam radiation, while for the diffuse part the tilt factor is taken to be unity. Thus, the monthly average value of the energy falling on the collector plane per unit area is determined approximately to be

$$\bar{H}_b \frac{\sin(\bar{\alpha}_a + \beta)}{\sin \bar{\alpha}_a} + \bar{H}_d$$

where $\bar{\alpha}_a$ is the monthly average altitude angle at noon.

The total flux falling on the collector over the year obtained by summing over all the months is found to be

$$\sum_{i=1}^{12} \bar{H}_{bi} \frac{\sin(\bar{\alpha}_{ai} + \beta)}{\sin \bar{\alpha}_{ai}} + \bar{H}_{di}$$

For this annual flux to be maximum, it can be shown by differentiation that the optimum tilt β is given by the expression

$$\beta_{\text{opt}} = \tan^{-1}\left[\left\{\sum_{i=1}^{12} \bar{H}_{bi} \tan|\phi - \delta_i|\right\} \Big/ \left\{\sum_{i=1}^{12} \bar{H}_{bi}\right\}\right] \tag{4.53}$$

Calculations by Kern and Harris for four locations in South Africa show that the value of the optimum tilt is approximately equal to the latitude of the location and that the energy collected is not very sensitive to small deviations from the best angle. Thus the conclusion reached is similar to that of Morse and Czarnecki.

Equation (4.53) has been derived for the situation $\gamma = 0°$ and $\gamma_s = 0°$ existing at solar noon for all months of the year. We shall call this situation (1). Situation (1) would generally occur at a location having a latitude greater than 23.45°. Apart from situation (1), it is possible to have the following situations for some or all the months of the year.

Situation	γ(deg)	γ_s(deg)
(2)	0	180
(3)	180	0
(4)	180	180

For situations (2) and (3), we would have to substitute $(\bar{\alpha}_{ai} - \beta)$ for $(\bar{\alpha}_{ai} + \beta)$ in the expression for the monthly average value of flux falling on the collector plane while for situation (4), the expression is the same as for situation (1). Thus the expression for the annual flux would have to be suitably modified and the optimum tilt is now given by

$$\beta_{opt} = \tan^{-1}\left[\left\{\sum_{i=1}^{12} \pm \overline{H}_{bi} \tan|\phi - \delta_i|\right\}\Big/\left\{\sum_{i=1}^{12} \overline{H}_{bi}\right\}\right]$$

The plus sign before $\overline{H}_{bi}$ is valid for the months in which situations (1) and (4) exist, while the negative sign is valid for the months in which situations (2) and (3) exist.

Example 4.4

A flat-plate collector array is to be installed in New Delhi (28.58°N). Calculate the optimum tilt for the following three situations: (i) Insolation falling on the array over the whole year is to be maximized. (ii) Insolation for the months of April, May and June is to be maximized. (iii) Insolation for the months of December, January and February is to be maximized. The following radiation data are given:

Month	J	F	M	A	M	J	J	A	S	O	N	D
$\overline{H}_g$ (kWh /m^2-day)	3.897	5.001	6.138	6.935	7.287	6.544	5.334	5.053	5.602	5.355	4.523	3.843
$\overline{H}_d$ (kWh/ m^2-day)	1.240	1.474	1.861	2.473	2.922	3.540	3.135	2.727	2.152	1.465	1.141	1.117

Solution

At New Delhi, $\gamma_s = 0°$ at solar noon throughout the year. We will use Eq. (4.53) under the assumption[‡] that situation (1) exists, i.e. the collector array is facing due south with $\gamma = 0°$. For part (i), the summation in Eq. (4.53) will be done over all the 12 months, while for parts (ii) and (iii), the summation will be done only for the months specified. The calculations are shown in Table 4.5, with the declination angles being calculated for the middle days of the month. It is seen that the values obtained for the optimum tilt are in approximate agreement with the recommendations given earlier.

(i) $$\beta_{opt} = \tan^{-1}\left[\frac{24.345}{40.355}\right] = 31.1°$$

(ii) $$\beta_{opt} = \tan^{-1}\left[\frac{1.517 + 0.734 + 0.277}{4.462 + 4.365 + 3.004}\right] = 12.1°$$

‡If this assumption is incorrect, one would obtain a negative value of β_{opt}. It would then be necessary to reverse the assumption, take $\gamma = 180°$ and recalculate the value of β_{opt}.

$$\text{(iii)} \quad \beta_{opt} = \tan^{-1}\left[\frac{3.483 + 3.237 + 3.198}{2.726 + 2.747 + 3.527}\right] = 47.8°$$

Table 4.5 Calculations for optimum tilt for a collector array in New Delhi

	$\bar{H}_b$ (kWh/m²)	n	δ (deg)	$(\phi - \delta)$ (deg)	$\bar{H}_b \tan(\phi - \delta)$ (kWh/m²)
J	2.747	16	–21.10	49.68	3.237
F	3.527	45	–13.62	42.20	3.198
M	4.277	75	–2.42	31.00	2.570
A	4.462	105	9.41	19.17	1.517
M	4.365	136	19.03	9.55	0.734
J	3.004	166	23.31	5.27	0.277
J	2.199	197	21.35	7.23	0.279
A	2.326	228	13.45	15.13	0.629
S	3.450	258	2.22	26.36	1.710
O	3.890	289	–8.97	37.55	2.990
N	3.382	319	–19.15	47.73	3.721
D	2.726	350	–23.37	51.95	3.483
Σ	40.355				24.345

A large number of papers have been written on the subject of optimum tilt. Many of them use different models or more accurate procedures for calculating the radiation falling on a tilted surface. Yakup and Malik [31] use Eq. (3.51) for calculating monthly averages of the daily radiation falling on a tilted surface from measured radiation data. Despotovic and Nedic [32] have found out the yearly, biannual, seasonal, monthly, fortnightly and daily optimum tilt angles from the point of view of receiving the maximum solar radiation on a collector surface. Stanciu and Stanciu [33] have reported the monthly and yearly optimum tilt angles at different geographical locations for receiving the maximum solar radiation on a collector surface. From these investigations and others reported in the literature, the conclusion which can be drawn is that for maximizing the annual solar radiation availability on a collector surface or a PV panel, it is best to have the tilt equal to the latitude.

It is to be noted, however, that the optimum tilt would be different if the nature of the energy demand is different. For example for an application like space heating, the demand may be high in the winter months of December, January and February. On the other hand, if solar energy were to be used for running an absorption refrigeration plant, the duty would be highest in months like April, May and June having high ambient temperatures. In such cases, it would obviously be desirable to use a tilt greater than the latitude for a winter application and the reverse for a summer application. The usual practice is to recommend values of (ϕ + 10°) or (ϕ + 15°) for the former and (ϕ – 10°) or (ϕ – 15°) for the latter.

4.10.5 Fluid Inlet Temperature

The fluid inlet temperature is an operational parameter which strongly influences the performance of a flat-plate collector. The effect is best illustrated by again doing calculations for the case of the collector of Section 4.9. Results are obtained with fluid inlet temperature varying from 30° to 90°C, while the values of the other parameters are held constant (Table 4.6).

Table 4.6 Effect of fluid inlet temperature on performance of collector

T_{fi}(°C)	30	40	50	60	70	80	90
T_{pm} (K)	326.5	334.8	343.0	351.2	359.3	367.5	375.5
U_t(W/m^2-K)	3.78	3.90	4.02	4.12	4.22	4.31	4.39
q_u(W)	1149.0	1065.0	977.4	888.3	796.6	703.4	608.9
T_{fo}(K)	317.3	326.3	335.2	344.1	353.0	361.9	370.7
η_i (%)	56.4	52.3	48.0	43.6	39.1	34.6	29.9

It is seen from Table 4.6 that the efficiency of the collector decreases more or less linearly with increasing values of T_{fi} (see also Fig. 4.15), the value falling from 56.4 to 29.9 per cent as T_{fi} increases from 30° to 90°C. This decrease is because of the higher temperature level at which the collector as a whole operates when the fluid inlet temperature increases. Because of this, the top loss coefficient as well as the temperature difference with the surroundings increases, the heat lost increases and the useful heat gain decreases.

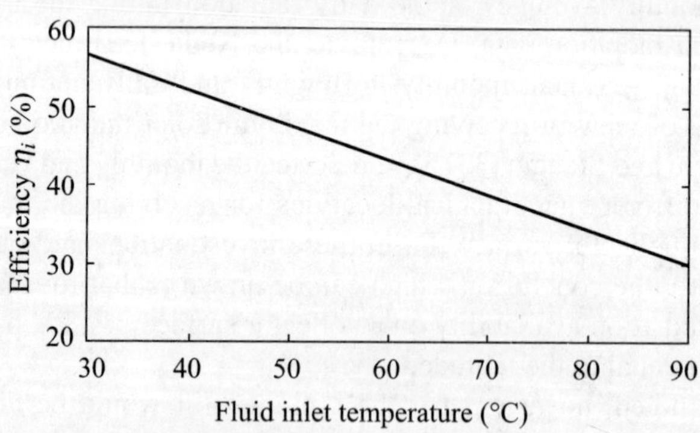

Fig. 4.15 *Variation of efficiency of a collector with fluid inlet temperature*

4.10.6 Cover Transmissivity

The transmissivity of the cover affects the performance of a collector significantly. The higher the transmissivity (that is, lower the extinction coefficient of the cover material), the better is the performance of the collector. The effect is demonstrated by repeating the calculations for the collector of Section 4.9 with a lower extinction coefficient, the values of all the other parameters being the same (Table 4.7). It is seen that the efficiency increases from 43.6 per cent to 47.4 per cent as the extinction coefficient decreases from 19.0 m^{-1} to 4 m^{-1}. The values of $(\tau\alpha)_b$ and $(\tau\alpha)_d$ increase from 0.8041 and 0.7284 to 0.8554 and 0.7836 respectively. The calculations have also been done for a better selective surface, in which case the efficiency increases further from 47.4 per cent to 49.2 per cent.

Table 4.7 Effect of cover transmissivity on performance of collector (extinction coefficient = 4 m^{-1})

	Selective surface	
	$\alpha = 0.94, \varepsilon_p = 0.14$	$\alpha = 0.95, \varepsilon_p = 0.085$
$(\tau\alpha)_b$	0.8554	0.8631
$(\tau\alpha)_d$	0.7836	0.7907
T_{pm}(K)	352.8	353.5
U_t(W/m^2-K)	4.14	3.86
T_{fo}(K)	345.1	345.5
q_u(W)	964.7	1001.6
η_i (%)	47.4	49.2

4.10.7 Dust on the Top Cover

The preceding calculations of the flux transmitted through the covers of the collector have been done under the assumption that the top cover is clean and has no dust accumulated on it. This assumption is acceptable only if the cover is continuously cleaned. However, in any practical situation, this is not possible. Cleaning is generally done once every few days. For this reason, it is recommended that the incident flux be multiplied by a correction factor which accounts for the reduction in intensity because of the accumulation of dust. The correction factor is the ratio of the normal transmissivity of a dust-laden cover to the normal transmissivity of a clean cover.

There is, however, considerable difficulty in assigning a value to the correction factor in a specific situation because of its dependence on a number of parameters. The value depends obviously on the location of the collector and the time of the year. It also depends upon the material of the cover (glass or plastic), the tilt of the collector and the frequency of cleaning.

The results of some studies are as follows.

For collectors inclined at 30° in Boston, USA, Hottel and Woertz [3] found a reduction of less than 1 per cent in the transmitted radiation and accordingly recommended a correction factor of 0.99. However this recommendation seems to be on the higher side. Garg [34] conducted studies at Roorkee on glass covers inclined at various angles. The experiments were performed during May and June, which are relatively dusty months. For a collector inclined at 45°, Garg obtained a correction factor of 0.92 for a cleaning frequency of 20 days.

Hegazy [35] measured the dust accumulation on glass plates with different tilt angles ranging from 0° to 90° in Minia, Egypt (28°N, 30.5°E). The location was surrounded by agricultural fields but was not far from the eastern and western deserts of Egypt. Data were collected over a period of one year so that different weather conditions prevailing in the Minia region were considered. The measurements show that for tilt angles from 20° to 40°, a correction factor of about 0.95 was obtained after 4 days of exposure and that a value of about 0.92 was obtained after 7 days. Based on his data, Hegazy recommends that in a moderately dusty place, the transparent covers of a collector system should be definitely cleaned at least once a week as part of the maintenance routine. He also recommends that the covers be cleaned immediately after a dust storm.

For most parts of India, the dust accumulation would be less than the values measured in Minia by Hegazy. Thus if a cleaning frequency of one week is adopted, the correction factor would be greater than 0.92. It is recommended that for India a value ranging from 0.92 to 0.96 be adopted. The actual value would depend upon the location in India and the time of the year. Finally, it should be noted that this recommendation is also applicable for photovoltaic systems.

Elminir *et al.* [36] have conducted a systematic study of the reduction in transmissivity of glass plates because of deposition of dust. The experiments were conducted on a sandy lime stone plateau outside the town of Helwan, Egypt (29° 52′ N, 31° 20′ E). Glass samples were oriented at different tilts (0 to 90°) and at different surface azimuth angles. The deposition of dust on these samples (in g/m^2) and their transmissivities were measured at monthly intervals over a period of seven months. As expected, maximum deposition occurred on horizontal surfaces and minimum deposition on vertical surfaces. The value of the surface azimuth angle had a relatively weak effect on the change in transmissivity. The important result obtained was that there was a good correlation between the dust deposition density and the reduction in transmissivity expressed as a percentage of transmissivity of a clean glass. It was seen that reduction in transmissivity increases as the dust deposition density increases, but with a decreasing rate. The reduction seems to reach an upper limit beyond a certain value of the dust deposition density. Elminir *et al.* suggest that the correlation equation developed along with its constants could be used in other regions with similar climates. They extended their study to PV modules and obtained data on the decrease in output of these modules due to dust deposition if they were left unattended.

More recently, Jing *et al.* [37] have studied the effect of dust on the thermal performance of solar collectors in Lanzhou, China. Collectors were exposed to ambient conditions and left without cleaning for three months. These collectors were then tested for their performance in comparison to clean collectors placed beside them. All the tests were conducted with the collectors having a tilt of 36° and facing south. Jing *et al.* defined a term called the 'dust shading coefficient' as the ratio of the value of $F_R(\tau\alpha)_{av}$ for a dust-laden collector to the value of $F_R(\tau\alpha)_{av}$ for an identical but clean collector. They obtained an average daily value of 0.867 for the dust shading coefficient. The dust shading coefficient takes into account changes in F_R, τ and α. It needs to be distinguished from the correction factor mentioned earlier which accounts only for changes in the normal transmissivity.

Sarver *et al.* [38] have conducted a review of the literature concerned with the impact of dust on the performances of solar energy devices. One of the important conclusions reached by them is that the effect of dust is not a major concern in many parts of USA and Europe where low-soiling conditions exist and/or periodic precipitation cleans the surfaces naturally. However this is not the case in most of India and in many other parts of the world. Thus mitigation techniques, which help in restoring the affected surfaces to as near their original condition as possible, have to be adopted. The strategy most often adopted is to remove the dust by washing, followed by wiping. This is done at regular intervals. Water is the most effective medium. Demineralized water is the best, but adds to the cost.

Mani and Pillai [39] have recommended cleaning and mitigation approaches based on classification of various climatic zone groups and the dust-concern level. The regions in India where solar collectors and PV modules are likely to be used would have latitudes from about 15 to 25°N and climates which can be described as either dry and tropical or wet-dry and tropical in Mani and Pillai's classification. The dry and tropical regions (parts of Rajasthan, north Gujarat, etc.) have dusty desert environments, low annual rainfall (less than 25 cm) and low humidity, and large variation in temperature through the year from 5 to 45°C. For such regions, Mani and Pillai recommend cleaning at least once a week. In addition, immediate cleaning is suggested subsequent to a dust storm. The wet-dry and tropical regions (parts of Madhya Pradesh, Maharashtra, Andhra Pradesh, Karnataka, etc.) have less dusty environments, annual rainfalls ranging from 30 to 100 cm, and temperatures varying from 10 to 40°C through the year. Solar devices are used in these regions where the land is arid and not suitable for agricultural purposes. For such regions, Mani and Pillai recommend weekly cleaning. More frequent or daily cleaning is suggested only during certain periods (April/May) when there is more dust accumulation.

ANALYSIS OF COLLECTORS SIMILAR TO THE CONVENTIONAL COLLECTOR 4.11

So far we have considered only the conventional liquid flat-plate collector having the geometry of a flat absorber plate with tubes bonded below. We now consider two other types, which are fairly similar to the one analysed thus far. In one of these, the tubes are bonded on the top side of the absorber plate, while in the other, the tubes are in line (Fig. 4.16). It can be shown for both these types that the method of analysis is identical to the case considered. The only difference is that different expressions are obtained for the collector efficiency factor, F'.

In the case of tubes bonded above,

$$F' = \frac{1}{WU_l\left[\dfrac{1}{\left\{\dfrac{1}{U_l(W-D_o)\phi}+\dfrac{\delta_a}{k_a D_o}\right\}^{-1}+U_l D_o}+\dfrac{1}{\pi D_i h_f}\right]} \quad (4.54)$$

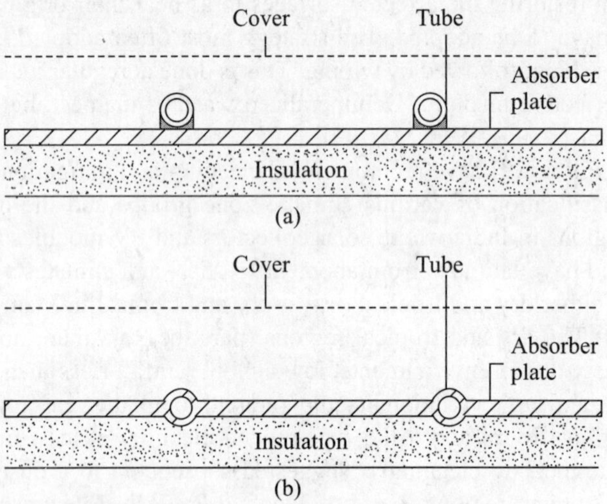

Fig. 4.16 *Liquid flat-plate collectors (a) Tubes bonded above absorber plate (b) Tubes in-line with absorber plate*

In the case of tubes in line with the absorber plate,

$$F' = \frac{1}{WU_l\left[\dfrac{1}{U_l\{(W-D_o)\phi+D_o\}}+\dfrac{1}{\pi D_i h_f}\right]} \quad (4.55)$$

In the second case, there is no term involving the bond (adhesive) resistance because the tubes are usually fabricated integral with the absorber plate.

TRANSIENT ANALYSIS 4.12

The analysis done so far has assumed that a quasi-steady state exists. Although this is a valid assumption for most situations, there are also some situations which require a transient analysis.

We will now consider one type of transient problem. On any day, a collector takes a few hours in the morning to reach its operating temperature, after which it may operate under quasi-steady conditions. Thus, a transient analysis is needed for these hours (from say 7 to 10 a.m.) to consider the heating of the collector from its overnight low temperature to its operating temperature. Following Duffie and Beckman [40], we will do an approximate lumped-parameter analysis for a single-cover collector for this situation. We assume that the absorber plate and tubes, the bottom and side insulation, and the water in the tubes have a heat capacity $(mC)_p$ and are at the same temperature T_{pm}, while the cover has a heat capacity $(mC)_c$ and is at a temperature T_c. Since there is no useful heat gain during this period, we obtain the following energy balance equations for the absorber plate and the cover

$$\frac{(mC)_p}{A_p}\frac{dT_{pm}}{dt} = S - \left[h_{p-c}(T_{pm} - T_c) + \frac{\sigma(T_{pm}^4 - T_c^4)}{\left(\dfrac{1}{\varepsilon_p} + \dfrac{1}{\varepsilon_c} - 1\right)}\right] - (U_b + U_s)(T_{pm} - T_a) \tag{4.56}$$

$$\frac{(mC)_c}{A_p}\frac{dT_c}{dt} = \left[h_{p-c}(T_{pm} - T_c) + \frac{\sigma(T_{pm}^4 - T_c^4)}{\left(\dfrac{1}{\varepsilon_p} + \dfrac{1}{\varepsilon_c} - 1\right)}\right] - [h_w(T_c - T_a) + \sigma\varepsilon_c(T_c^4 - T_{sky}^4)] \tag{4.57}$$

We combine the convective and radiative losses from the top by defining overall coefficients U_{t1} and U_{t2} as follows

$$U_{t1}(T_{pm} - T_c) = h_{p-c}(T_{pm} - T_c) + \frac{\sigma(T_{pm}^4 - T_c^4)}{\left(\dfrac{1}{\varepsilon_p} + \dfrac{1}{\varepsilon_c} - 1\right)}$$

and $$U_{t2}(T_c - T_a) = h_w(T_c - T_a) + \sigma\varepsilon_c(T_c^4 - T_{sky}^4)$$

Thus, Eqs (4.56) and (4.57) become

$$\frac{(mC)_p}{A_p}\frac{dT_{pm}}{dt} = S - U_{t1}(T_{pm} - T_c) - (U_b + U_s)(T_{pm} - T_a) \tag{4.58}$$

$$\frac{(mC)_c}{A_p}\frac{dT_c}{dt} = U_{t1}(T_{pm} - T_c) - U_{t2}(T_c - T_a) \tag{4.59}$$

Now in the steady state

$$U_{t2}(T_c - T_a) = U_t(T_{pm} - T_a) \tag{4.60}$$

We assume that this relation holds during the transient state also and that T_a is a constant. Differentiating Eq. (4.60), we have

$$U_{t2}\frac{dT_c}{dt} = U_t\frac{dT_{pm}}{dt} \tag{4.61}$$

Substituting for U_{t2} and $\frac{dT_c}{dt}$ in Eq. (4.59) and then adding Eqs (4.61) and (4.59), we have

$$\frac{1}{A_p}\left[(mC)_p + (mC)_c\frac{U_t}{U_{t2}}\right]\frac{dT_{pm}}{dt} = S - U_l(T_{pm} - T_a) \tag{4.62}$$

The quantity within the square brackets represents an effective heat capacity of the collector and will be denoted by the symbol $(mC)_e$. The differential equation (4.62) can be integrated under the assumption that the values of S and T_a are constants. This is valid if the time interval of the integration is kept reasonably small. Using the initial condition

$$t = 0,\ T_{pm} = T_{pi}$$

we get

$$\frac{S - U_l(T_{pm} - T_a)}{S - U_l(T_{pi} - T_a)} = \exp\left[-\frac{A_p U_l t}{(mC)_e}\right] \tag{4.63}$$

Equation (4.63) can be used either to find the time taken by the collector to reach the mean plate temperature corresponding to its prescribed fluid inlet temperature or to determine the mean plate temperature after a certain amount of time has elapsed.

Example 4.5

A water-heating flat-plate collector operates at a mean plate temperature of 70°C. Determine the time by which it will reach this temperature if there is no useful heat gain till that time. The following data is given:

Length of absorber plate	:	1.6 m
Width of absorber plate	:	1 m
Number of covers	:	1
Overall loss coefficient	:	7.2 W/m^2-K
Top loss coefficient	:	6.4 W/m^2-K
U_{t2}	:	22.8 W/m^2-K
(mC) for plate and tubes	:	4.5 kJ/K
(mC) for insulation	:	2.0 kJ/K
Volume of water in collector	:	2.1 litres
(mC) for glass cover	:	16 kJ/K
Mean plate temperature at 6 a.m.	:	14°C

Time (a.m.)	Flux incident on collector I_T (W/m^2)	Average transmissivity-absorptivity product $(\tau\alpha)_{av}$	Ambient temperature T_a (°C)
6–7	30	0.25	14.0
7–8	175	0.50	15.0
8–9	380	0.70	16.0
9–10	555	0.86	18.0
10–11	700	0.89	20.5

Solution

Starting with the initial condition at 6 a.m., we apply Eq. (4.63) for intervals of one hour or less.

$$(mC)_e = 4.5 + 2.0 + (2.1 \times 4.18) + \left(16 \times \frac{6.4}{22.8}\right) = 19.79 \text{ kJ/K}$$

From 6 to 7 a.m., the absorbed flux

$$S = 30 \times 0.25 = 7.5 \text{ W/m}^2$$

Applying Eq. (4.63),

$$\frac{7.5 - 7.2(T_{pm} - 14)}{7.5 - 7.2(14 - 14)} = \exp\left[-\frac{1.6 \times 1 \times 7.2 \times 1}{19.79} \times \frac{3600}{1000}\right] = 0.123$$

Therefore, T_{pm} at 7 a.m. = 14.91°C

Similarly, from 7 to 8 a.m., S = 87.5 W/m^2 and T_{pm} at 8 a.m. = 25.65°C.

From 8 to 9 a.m., S = 266 W/m^2 and T_{pm} at 9 a.m. = 49.59°C.

Since the temperature at 10 a.m. will cross the operating value, we shall find out the time after 9 a.m. which will be required to reach 70°C. Assuming that the value of S = 477.3 W/m^2 from 9 to 10 a.m. holds for part of the hour also, we get from Eq. (4.63),

$$\frac{477.3 - 7.2(70 - 18)}{477.3 - 7.2(49.59 - 18)} = \exp\left[-\frac{1.6 \times 1 \times 7.2 \times t}{19.79} \times \frac{3600}{1000}\right]$$

Solving, we get

$$t = 0.42 \text{ h} = 25 \text{ minutes}$$

Thus the collector will attain a mean plate temperature of 70°C at 9.25 a.m.

TESTING PROCEDURES 4.13

In this section, we will describe certain standard procedures for the thermal testing of collectors. Standardized testing and rating procedures provide an equitable basis for comparing the efficiency of different types of collectors and an essential basis for their selection for a given application as well as their design improvements. The procedure described here, ASHRAE Standard 93, is followed in many countries [41] and has been adopted as an Indian Standard IS 12933 (part 5) [42].

A schematic diagram showing the essential features of the test set-up is shown in Fig. 4.17. It is a closed loop consisting of the flat-plate collector under test, a liquid pump, a heat exchanger with a cooling coil, and a storage tank with an electric immersion heater. A bypass is provided around the pump so that the mass flow rate can be adjusted to the prescribed value. The purpose of the heat exchanger with a cooling coil is to remove heat. Thus the combination of the heat exchanger and the storage tank with an electric heater provides a means for adjusting and controlling the inlet fluid temperature to the collector to a desired value.

The standard specifies that the collector shall be tested under clear sky conditions in order to determine its efficiency characteristics. On any given day, data is recorded under steady state conditions for fixed values of $\dot{m}$ and T_{fi}. For each set of fixed values, it is recommended that an equal number of tests be conducted symmetrically before and after solar noon. Thus, for example, if data of four tests are recorded, these could be at 1100, 1130, 1230 and 1300 h (LAT). In this way, any bias because of transient effects is eliminated. If such data is recorded for four inlet temperatures on different days, then a total of 16 data sets are obtained. This is the minimum number recommended.

The principal measurements made in each data set are the fluid flow rate ($\dot{m}$), the fluid inlet and outlet temperatures of the collector (T_{fi} and T_{fo}), the solar radiation incident on the collector plane (I_T), the ambient temperature (T_a), and the wind speed (V_∞). The efficiency is calculated from the equation

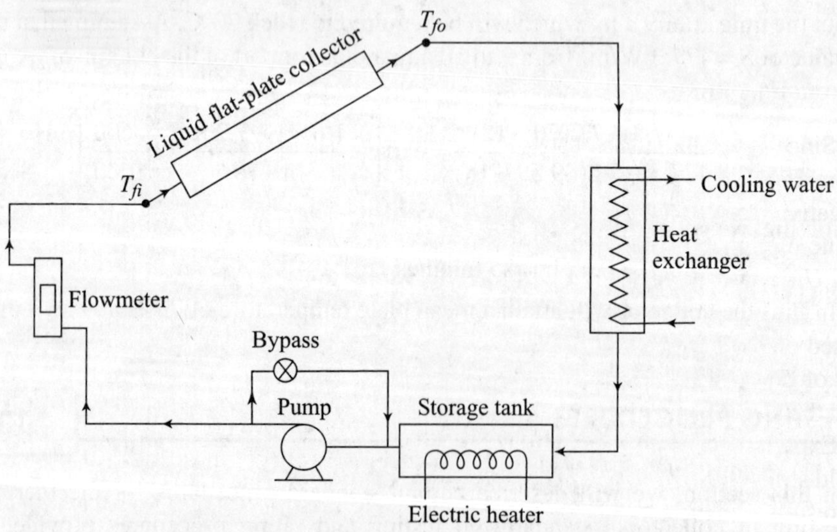

Fig. 4.17 *Schematic diagram of a closed loop set-up for testing liquid flat-plate collectors*

$$\eta_i = \frac{q_u}{A_c I_T} = \frac{\dot{m}C_p(T_{fo} - T_{fi})}{A_c I_T} \tag{4.64}$$

As stated earlier, readings are recorded under steady state conditions. A collector is considered to be operating under steady state conditions if the deviation of the experimental parameters is less than the following specified limits over a 15 minute period:

Global radiation incident on collector plane	$\pm$ 50 W/m^2
Ambient temperature	$\pm$ 1°C
Fluid flow rate	$\pm$ 1 %
Fluid inlet temperature	$\pm$ 0.1°C
Temperature rise across collector	$\pm$ 0.1°C

In addition, it is specified that the value of I_T should be greater than 700 W/m^2, the wind speed should be between 2 and 5 m/s, and the fluid flow rate should be set at approximately 0.02 kg per second per square metre of collector gross area. It is to be noted that the above procedure can also be carried out in an open test loop and that it is also applicable for indoor testing with a solar simulator.

The efficiency values calculated from Eq. (4.64) are plotted against the parameter $(T_{fi} - T_a)/I_T$. The reason for doing this is apparent if one considers the Hottel-Whillier-Bliss equation. Dividing both sides of Eq. (4.49) by $A_c I_T$, we have

$$\eta_i = F_R \left(\frac{A_p}{A_c}\right)\left[\frac{S}{I_T} - U_l \frac{(T_{fi} - T_a)}{I_T}\right] \tag{4.65}$$

We put $\qquad S = I_T(\tau\alpha)_{av} \qquad (4.66)$

where $(\tau\alpha)_{av}$ is an average transmissivity-absorptivity product for both beam and diffuse radiation. Hence

$$\eta_i = F_R \frac{A_p}{A_c}\left[(\tau\alpha)_{av} - U_l \frac{(T_{fi} - T_a)}{I_T}\right] \tag{4.67}$$

Since the values of F_R, $(\tau\alpha)_{av}$ and U_l are essentially constant, it is seen from Eq. (4.67) that if η_i is plotted against $(T_{fi} - T_a)/I_T$ a straight line with a negative slope would be obtained. The intercept on the *y*-axis would give the value of $[F_R (\tau\alpha)_{av} A_p/A_c]$, while the slope of the line would give the value of $[F_R U_l A_p/A_c]$.

In Eq. (4.64), the value of η_i is based on the collector gross area. It could be based on the absorber plate area also. In that case, the term (A_p/A_c) would drop out of Eqs (4.65) and (4.67). The intercept on the *y*-axis would then be $F_R(\tau\alpha)_{av}$ and the slope of the line would be $F_R U_l$.

Experimental values of η_i plotted against the parameter $(T_{fi} - T_a)/I_T$ generally yield straight lines. However, the scatter of the data is always large. A typical set of results obtained by testing a commercially available, conventional collector is shown in Fig. 4.18. A straight line fitted to the data by the method of least squares yields the following equation.

$$\eta_i = 0.692 - 4.024\,(T_{fi} - T_a)/I_T \tag{4.68}$$

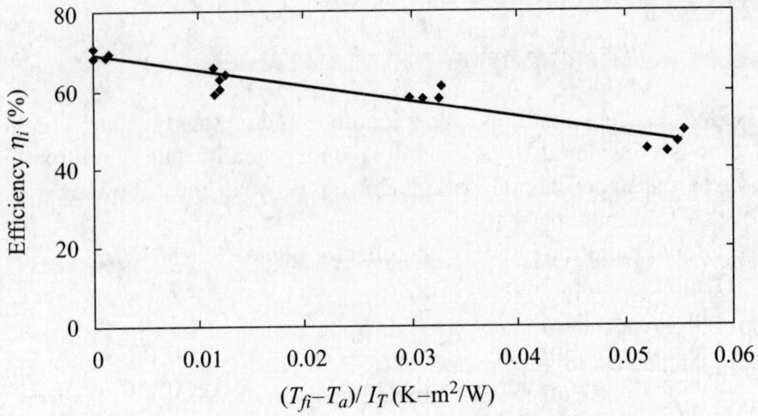

Fig. 4.18 *Efficiency curve for a commercial flat-plate collector of the conventional type (single cover, selective copper absorber plate. $A_c = 2.270\ m^2$; $\dot{m} = 0.0456$ kg/s)*

For the given collector, $A_p/A_c = 0.909$

Thus $\quad F_R(\tau\alpha)_{av} = 0.692/0.909 = 0.761$

and $\quad F_R U_l = 4.024/0.909 = 4.427$ W/m²-K

It is to be noted that for liquid flat-plate collectors, changes in mass flow rate do not appreciably affect the performance because of the relatively high value of the liquid side heat transfer coefficient h_f. For this reason, although the efficiency curve of a collector is determined for a particular value of mass flow rate, it can also be used for predicting the behaviour of the collector for other flow rates which differ a little from the value used during testing.

The standard procedure described for determining the collector efficiency requires that the tests be conducted on a clear day, close to solar noon, so that the value of I_T would be greater than 700 W/m². As a result the beam radiation component is dominant and the angle of incidence of the beam radiation is small (less than 15°). Thus the $(\tau\alpha)_{av}$ term in the parameter $[F_R(\tau\alpha)_{av}A_p/A_c]$ is effectively the transmissivity-absorptivity product for normal incidence beam radiation. In order to characterise collector performance early and late in the day, when the angle of incidence of beam radiation is high, a term called the *incidence angle modifier* ($K_{\tau\alpha}$) is defined. It is the ratio of the $(\tau\alpha)$ product at any angle of incidence to the $(\tau\alpha)$ product at normal incidence. It is found that the variation of $K_{\tau\alpha}$ with θ can be approximated by Eq. (4.69).

$$K_{\tau\alpha} = 1 - b_0\left(\frac{1}{\cos\theta} - 1\right) \tag{4.69}$$

where b_0 is a constant which has to be determined by conducting additional tests under clear skies. If $K_{\tau\alpha}$ is incorporated into Eq. (4.67), then the efficiency η_i can be calculated for different angles of incidence. Thus we have

$$\eta_i = F_R \frac{A_p}{A_c}\left[K_{\tau\alpha}(\tau\alpha)_{av} - \frac{U_l (T_{fi} - T_a)}{I_T} \right] \tag{4.70}$$

One more parameter of interest in the testing of collectors is the *time constant,* which is a measure of the heat capacity of a collector. The time constant is defined as the time required for the exit fluid temperature T_{fo} to change by a prescribed amount when the collector is subjected to a step change in the incident solar radiation or in the inlet fluid temperature. For more details regarding the determination of the incidence angle modifier and the time constant, the reader is referred to ASHRAE Standard 93-86 and Indian Standard 12933 (Part 5).

In comparison to the testing procedure just described, the practice followed in Europe is to use European standard EN 12975-2 [43]. This standard uses an expression for efficiency based on the collector efficiency factor (F') and the temperature difference between the mean of the liquid inlet and outlet temperature and the ambient temperature

$$\eta_i = F' \frac{A_p}{A_c}(\tau\alpha)_{av} - F' \frac{A_p U_l (\overline{T}_f - T_a)}{I_T} \tag{4.71}$$

Equation (4.71) is obtained by integrating Eq. (4.44) over the length L_1 of the absorber plate and by defining the mean fluid temperature by the expression

$$\overline{T}_f = \frac{1}{L_1}\int_0^{L_1} T_f \, dy$$

In practical situations, $\overline{T}_f$ can be taken as the arithmetic mean of T_{fi} and T_{fo}. The first term on the right hand side of Eq. (4.71) is multiplied by $K_{\tau\alpha}$ if its applicability is to be extended to higher angles of incidence.

Tests performed on collectors show that the plot of η_i against $(\overline{T}_f - T_a)/I_T$ generally yield a graph which is slightly non-linear. An extra term involving $(\overline{T}_f - T_a)$ is therefore introduced to account for the non-linearity and an equation of the form

$$\eta_i = a_0 - a_1 \frac{(\overline{T}_f - T_a)}{I_T} - a_2 \frac{(\overline{T}_f - T_a)^2}{I_T} \tag{4.72}$$

is fitted to the data obtained from tests conducted on the collector. Multilinear regression analysis is used for obtaining the values of a_0, a_1 and a_2. EN 12975-2 recommends that the second order fit be used only if the value of a_2 is positive. Keeping in mind the linear equation (4.71) derived earlier, it is clear that the constants a_0 and a_1 obtained from the regression analysis will be equal to $F'(A_p/A_c)(\tau\alpha)$ and $F'(A_p/A_c)U_l$ respectively when $a_2 = 0$

Kalogirou [44] has shown that once the values of $[F_R(\tau\alpha)_{av}A_p/A_c]$ and $[F_R U_l A_p/A_c]$ or $[F'(\tau\alpha)_{av} A_p/A_c]$ and $[F' U_l A_p/A_c]$ are known for a large number of flat-plate collectors, it is possible to use the Artificial Neural Network (ANN) technique to predict the values of these parameters for new collectors belonging to the same broad category. The training and testing of the artificial neural network was done with the help of data available on 110 flat-plate collectors. Apart from the performance parameters, the data fed were the volume flow rate and the physical characteristics of the collectors-type of collector, dimensions, absorber material and type, absorber coating, thermal insulation, glazing, etc. After the training and validation, the network developed was tested on 8 new collectors. The physical characteristics of these collectors were fed in and the performance parameters $[F'(\tau\alpha)_{av}A_p/A_c]$ and $[F'U_lA_p/A_c]$ predicted. The predicted values were found to agree very closely with the actual experimental values for these collectors.

In contrast to the steady-state methods described thus far, transient methods for thermal performance testing of flat-plate collectors have also been developed. The motivation behind the development of these methods is the need to circumvent the difficulties associated with satisfying the strict limits on weather and operating conditions specified in the steady-state method. Because of these limits, it takes a long time to test a collector and in some regions, which experience rapid weather fluctuations, it is often not possible to conduct steady-state outdoor tests. A number of transient methods have been developed over the years by many investigators. Nayak and Amer [45] have presented a critical evaluation of nine such methods. Based on these methods, experimental tests can be carried out relatively quickly. They can also be conducted throughout the year even when weather conditions may be changing rapidly and there are cloudy skies. Keeping in mind the advantages of transient methods, European Standard EN 12975-2 also prescribes a quasi-dynamic test (QDT) for evaluating flat-plate collectors. The QDT is performed outdoors and has the advantage that it can be run in conditions with variable radiation and ambient temperature. Data can therefore be collected on many more days than is possible with the steady state method.

Finally, it is worth noting that an international standard ISO 9806: 2013 [46] has been adopted recently. It incorporates the test methods suggested in EN 12975-2 for both steady state testing as well as quasi-dynamic testing. The release of ISO 9806:2013 has led to a global certification scheme for solar thermal collectors.

Example 4.6

Based on the observations recorded during the testing of a liquid flat-plate collector, the following data are available:

$(\bar{T}_f - T_a)/I_T$ (K-m^2/W)	$(\bar{T}_f - T_a)^2/I_T$ (K^2-m^2/W)	Efficiency
0.0011	0.097	0.631
0.019	0.289	0.588
0.029	0.673	0.529
0.041	1.345	0.451

(*Contd.*)

0.049	1.921	0.394
0.060	2.880	0.310
0.071	4.033	0.220

The same collector is installed to heat water and the arithmetic mean temperature of water across the collector is found to be 45.0°C at some instant of time. If the corresponding ambient temperature and the total solar radiation incident on the plane of the collector are 30.0°C and 800 W/m^2 respectively, calculate the instantaneous efficiency of the collector and the inlet and outlet temperature of the fluid.

Given: Flow rate = 1.5 kg/minute, C_p = 4.18 kJ/kg-°C and A_c = 2.0 m^2

Solution

We carry out a multilinear regression analysis of the data using Eq. (4.72) and standard software (e.g. Microsoft Excel). The correlation coefficients are found to be

$$a_0 = 0.685$$
$$a_1 = 4.549 \text{ W/m}^2\text{-K}$$
$$a_2 = 0.035 \text{ W/m}^2\text{-K}^2$$

Substituting the given values of $\overline{T}_f$, T_a and I_T, we get

$$\frac{\overline{T}_f - T_a}{I_T} = (45 - 30)/800 = 0.01875 \text{ (m}^2\text{– K)/W}$$

$$\frac{(\overline{T}_f - T_a)^2}{I_T} = (45 - 30)^2/800 = 0.28125 \text{ (m}^2\text{– K}^2\text{)/W}$$

Therefore, from Eq. (4.72),

$$\eta_i = 0.685 - 4.549 \times 0.01875 - 0.035 \times 0.28125$$
$$= 0.590, \text{ i.e. } 59.0 \text{ per cent}$$

Efficiency can also be written as

$$\eta_i = \frac{\dot{m} C_p (T_{fo} - T_{fi})}{A_c\, I_T}$$

Thus

$$0.590 = \frac{1.5 \times 4.18 \times 10^3 \times (T_{fo} - T_{fi})}{60 \times 2.0 \times 800}$$

$$= 0.06531\ (T_{fo} - T_{fi})$$

$$(T_{fo} - T_{fi}) = \frac{0.590}{0.06531} = 9.034°\text{C}$$

$$T_{fi} = 45.0 - 9.034/2 = 40.48°\text{C}$$

$$T_{fo} = 45.0 + 9.034/2 = 49.52°\text{C}$$

Note: Suppose the value of the inlet temperature (T_{fi}) is given instead of $\bar{T}_f$ and the value of the outlet temperature (T_{fo}) is to be determined. One can still solve the above type of problem, but by using an iterative technique. Assume a value of $\bar{T}_f$ (about 2 to 5°C more than T_{fi}) and calculate $(\bar{T}_f - T_a)/I_T$ and $(\bar{T}_f - T_a)^2/I_T$. Obtain η_i by substituting these values in Eq. (4.72). Then use Eq. (4.64) to calculate $(T_{fo} - T_{fi})$ and hence $\bar{T}_f$. Use the new value of $\bar{T}_f$ and repeat the calculation until $\bar{T}_f$ (assumed) matches with $\bar{T}_f$ (calculated).

ALTERNATIVES TO THE CONVENTIONAL COLLECTOR 4.14

A number of novel designs of solar collectors have been suggested and developed over the years as alternatives to the conventional liquid flat-plate collector. Some of these have been commercialized. The objectives in developing these designs have been varied. They include a desire to improve the efficiency, to reduce cost, to increase the operating temperature, or to reduce the weight of the collector. Some of these designs will now be described.

4.14.1 Evacuated Tube Collectors

One way of improving the performance of a liquid flat-plate collector is to reduce or suppress the heat lost by convection from the top. This is done by having a vacuum above the absorber plate. As a consequence, it becomes essential to use a glass tube as the cover because only a tubular surface is able to withstand the stresses introduced by the pressure difference.

A number of evacuated tube collector (ETC) designs have been developed. One design consists of a number of long cylindrical flat-plate collector modules side-by-side. Each module (Fig. 4.19a) is an evacuated, cylindrical glass tube containing a rectangular metal absorber plate. The absorber plate has a selective surface coating and a heat pipe is attached to it. A glass-to-metal seal is provided between the heat pipe and the end cover of the glass tube. The length of the heat pipe inside the evacuated glass tube constitutes the evaporator section in which heat is absorbed and the fluid inside the heat pipe evaporates. The evaporated fluid rises to the condenser section where it condenses. The heat of condensation is conducted to the fluid flowing in the collector header pipe through an aluminium block clamped on the heat pipe and the header pipe.

The design described in Fig. 4.19(a) requires a glass-to-metal seal whose integrity is difficult to maintain. The need for such a seal is eliminated by developing designs using all-glass double-walled evacuated tubes. Here each module consists of a long Dewar type evacuated tube with the outer surface of the inner wall being the absorber surface. This surface is selectively coated. The absorbed heat is conducted inwards through the inner glass tube wall and is removed by a number of ways. Collectors differ mainly in the way the heat is extracted from the inside. Figure 4.19(b) shows a design in which a heat pipe

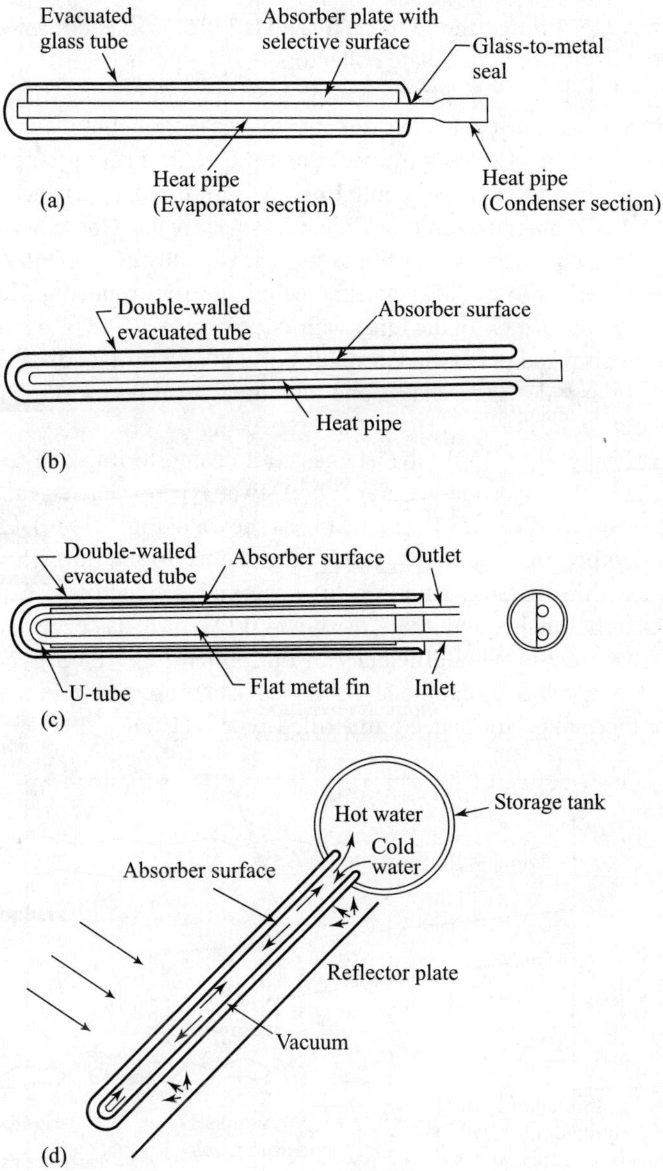

Fig. 4.19 *Various designs of ETC modules (a) Flat-plate type with heat pipe (b) Double-walled evacuated tube with heat pipe (c) Double-walled evacuted tube with U-tube heat removal system (d) Water-in-glass type*

inside the evacuated tube transports the heat. Aluminium spacers [not shown in Fig. 4.19(b)] are used to provide the thermal contact between the inner glass wall and the heat pipe. Figure 4.19(c) shows another design in which a U-tube fixed to a flat metal fin is inserted in the evacuated tube. The metal fin makes good thermal contact with the inner glass wall and helps the transfer of heat to the liquid flowing through the U-tube. Evacuated tube collectors are expensive compared to conventional flat-plate collectors.

Figure 4.19(d) shows a design used for domestic water heating. In this design, the evacuated tubes are connected directly to a storage tank. Thus the inside of the evacuated tube is filled with water and the water heating takes place by thermosyphon circulation. This configuration is referred to as the 'water-in-glass' design and is manufactured in China. It is widely used there besides being exported to other countries [47]. The design is reasonably cost effective because of the development of low cost sputtering techniques for producing the selective surface on the outer surface of the inner wall. Morrison *et al.* [48, 49] have carried out detailed testing and performance studies of this design. The disadvantage of the 'water-in-glass' design is that if one tube breaks, all the working fluid may get drained out.

Recently, Liang *et al.* [50] have suggested a design to improve the thermal contact between the inner glass wall of the Dewar type evacuated tube and the water to be heated. Instead of a flat metal fin, as shown in Fig. 4.19 (c), Liang *et al.* filled the space between the inner glass wall and the U-tube through which the water flows with a material having a high thermal conductivity. Compressed graphite with a thermal conductivity of about 100 W/m-K has been used for the purpose. Tests show that the efficiency of this 'filled-type' ETC is better than the values obtained with a 'flat metal fin' ETC. Data presented in Fig. 4.20 show the superior thermal performance of the filled-type ETC.

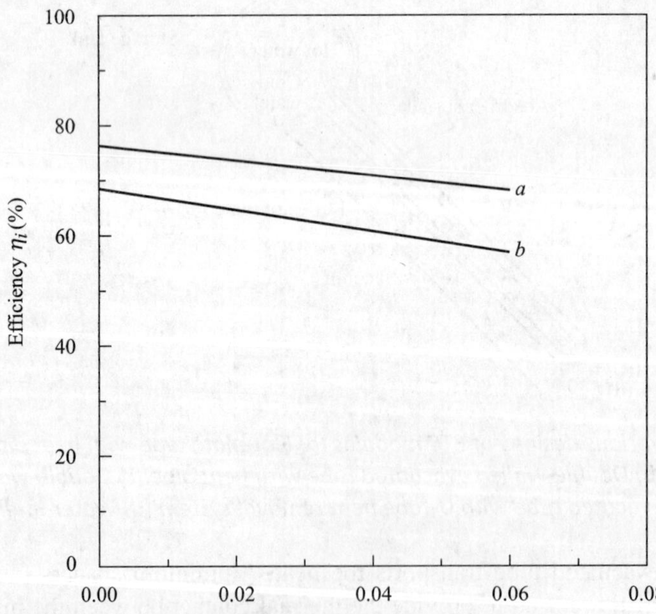

Fig. 4.20 *Efficiency of (a) 'filled-type' ETC and (b) 'flat metal fin' ETC. (Adapted from Liang et al. [50])*

Two types of layouts are used for the modules in evacuated tube collectors. In one type, the modules are stacked side-by-side with no gap between them, while in another, a spacing of a few centimetres is kept between the modules and a back reflector is used. The back reflector may be a plane white surface acting as a diffuse reflector or a curved surface acting as a specular reflector[§] (Fig. 4.21). Thus the radiation falling on the absorber surface in each module consists of the beam and diffuse radiation falling directly as well as the radiation reflected from the back reflector.

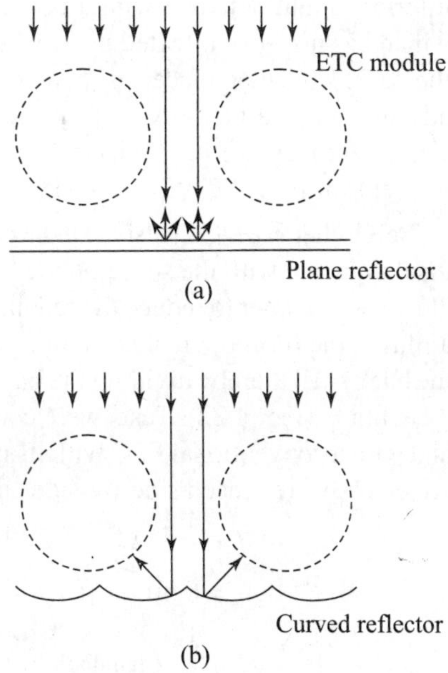

Fig. 4.21 *Reflectors used in evacuated tube collectors (a) Plane diffuse reflector (b) Curved specular reflector*

Because of the suppression of convection, the overall loss coefficient of an ETC is low. Consequently its efficiency is significantly higher than that of a conventional collector at high inlet fluid temperatures. This is readily seen if one compares the efficiency values in Fig. 4.20 with those for a conventional flat-plate collector presented in Fig. 4.18. For example, for $[(T_{fi} - T_a)/I_T] = 0.05$, the efficiency of the 'flat metal fin' ETC is 58 per cent and that of the 'filled-type' ETC is 70 per cent. The corresponding value from Fig. 4.18 is 50 per cent.

4.14.2 The Flat-Plate Collector (a) With no mineral wool insulation on the back side (b) With an Al film insulation

Conventional flat-plate collectors have mineral wool insulation about 4 to 6 cm thick below the absorber plate to reduce heat losses from the bottom. When new, this insulation results in a bottom loss coefficient U_b of 0.8 – 1.0 W/m^2-K. In practice, it is seen that over a period of time the value of U_b increases because the insulation absorbs moisture. Beikircher *et al.* [51] have suggested that mineral wool insulation should not be used and that instead the 4-6 cm space below the absorber plate be reduced to about 2 cm. They suggest that the inclined space so created would lose heat by natural convection and shown by calculations that the equivalent value of U_b would be in the range of 1.5 to 2 W/m^2-K. Beikercher *et al.* conducted tests on such a collector and showed that the extrapolated value of efficiency for $[(\overline{T}_f - T_a)/I_T] = 0$, was 83.5 per cent. With a wind speed of 1.6 m/s, the η decreased to 44 per cent for $[(\overline{T}_f - T_a)/I_T] = 0.06$. This value of η is slightly

[§] In some commercial ETCs, the curved surface is a truncated CPC reflector (see Sec. 6.6).

inferior to that obtained with a conventional FPC with mineral wool insulation which has not been affected by moisture. Experiments were also conducted with the same collector placed in contact with the flat roof top of a house so that advantage could be derived from the insulation below the roof of the house. A significant increase in efficiency was obtained with η increasing from 44 to 53 per cent for $[(\bar{T}_f - T_a)/I_T] = 0.06$.

Beikircher *et al.* [51] also tested a conventional FPC without any mineral wool insulation and with the space on the back side of the absorber place not reduced. However, in order to reduce the radiative losses from the back side, they stretched a film in the form of an aluminum foil having a low emissivity insulation across the back side, thereby dividing the back space into two compartments, each about 2 cm thick. (Fig. 4.22). Tests were conducted simultaneously on such a collector and with a conventional FPC with its mineral wool not affected by humidity. The collectors were placed side-by-side and yielded more or less identical results:

$$\eta = 0.753 \text{ for } [(\bar{T}_f - T_a)/I_T] = 0. \tag{4.73}$$

$$\eta = 0.50 \text{ for } [(\bar{T}_f - T_a)/I_T] = 0.06 \tag{4.74}$$

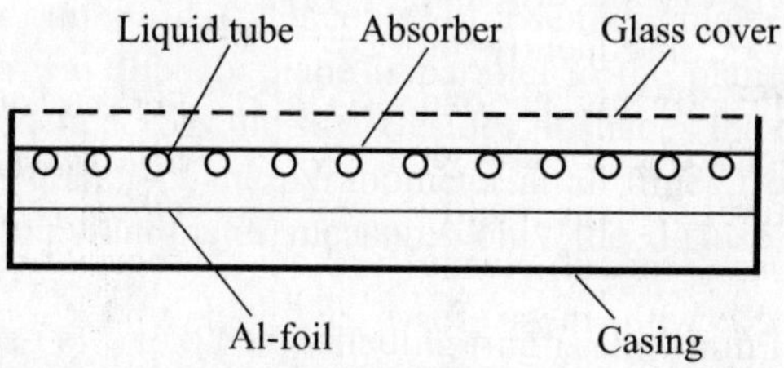

Fig. 4.22 *Flat-plate collector with no mineral wool insulation but with Al-foil insulation to reduce radiative losses from the back side (Adapted from Beikircher et al. [51])*

The overall conclusion reached was that both the designs suggested had an important role to play. The conventional FPC without any insulation but with a reduced space of about 2 cm could economically deliver domestic hot water from 45–60°C if placed in contact with the roof top of the house. On the other hand, the second design with a film insulation was able to match the conventional FPC with mineral wool with the same efficiency.

4.14.3 The Polymer Solar Collector

Kudish *et al.* [52,53] report data on collectors (1 m long, 0.5 m wide) having a double- walled polymer absorber plate with channels (Fig. 4.23). A selective coating was deposited on the upper surface of the plate, while the water to be heated flowed through the channels between the two walls. A 4-mm thick glass glazing was used over the absorber surface. Performance tests in Israel yielded average daily efficiencies in the range of 50–60 per cent and maximum fluid outlet temperatures in excess of 60°C.

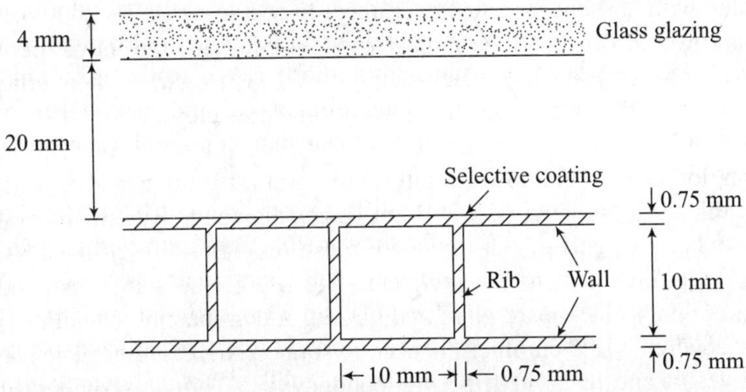

Fig. 4.23 *Collector with double-walled polymer absorber plate*

A disadvantage of using polymer materials in the absorber plate is their low thermal conductivity as compared to the values for metal absorber plates. As a result, one has to consider the thermal resistance offered by the walls of the plate. Tsilingiris [54] has idealised the absorber plate to be a parallel plate channel and assumed one-dimensional heat flow perpendicular to the plate. He has considered the thermal resistance of the upper and lower walls of the absorber plate and assumed that in comparison the thermal resistance to convective heat transfer from the walls to the fluid is negligible. His analysis shows that the Hottel-Whillier-Bliss Eq. (4.49) is still valid with the collector heat-removal factor F_R defined as in Eq. (4.50). However, the overall loss coefficient U_l and the collector efficiency factor F' are now redefined as follows

$$U_l = U_t + U_b \left\{ \frac{1 + (U_t / H)}{1 + (U_b / H)} \right\} \tag{4.75}$$

$$F' = \frac{1}{1 + (U_t / H)} \tag{4.76}$$

where $H = k_p / \delta_p$

4.14.4 The Concrete Collector

A large amount of metal (copper, aluminium, galvanized iron, etc.) is used in conventional flat-plate collector systems. But studies on the energy inputs required for the production of different materials indicate that metals need a large amount of fossil fuel energy for their production. Furthermore, individual collector modules are connected to form a large array to meet the required demand. Thus, the solar system forms a separate entity, which has its own individual cost and adds dead loads on a building structure. In the long run, it would seem desirable that solar collectors be made an integral part of building elements, like the roof and wall panels. Thus, the separate solar system investment would be partially merged into the building construction investment. With these facts in mind, concrete solar collectors have been developed and tested at IIT Bombay [55]. Such collectors integrated with building structures are likely to cost less than Rs 2000 per square metre and would obviously be more cost effective than conventional collectors for providing domestic hot water at 50° to 60°C. However, their efficiencies would be lower than the values for conventional collectors.

Figure 4.24 shows a cross-section of the concrete collector. The absorber (1.31 m long and 0.68 m wide) is made from a thin concrete slab about 3.5 cm thick with a network of 1.7 cm (inner diameter) PVC tubes embedded inside. A layer of galvanized iron wire mesh on either side of the PVC tubes provides reinforcement to the concrete. The top of the slab is painted black with an ordinary blackboard paint and glazed. A commercially available slab of cellular concrete (12.5 cm thick) is used to support the absorber plate. This slab is lightweight and has a low thermal conductivity. Thus, it provides adequate insulation on the back side of the collector. The PVC network is arranged in such a manner that it provides a parallel flow through the individual tubes. The metal mesh reinforcement helps to increase the ability of the concrete to withstand handling stresses. A concrete collector is thermally massive; hence, the concept of instantaneous efficiency does not carry any significant meaning with respect to its thermal performance. On the other hand, the performance is best judged on the basis of the daily efficiency (η), which is defined as the ratio of the useful energy obtained during the day to the total radiation incident on the collector area during the day.

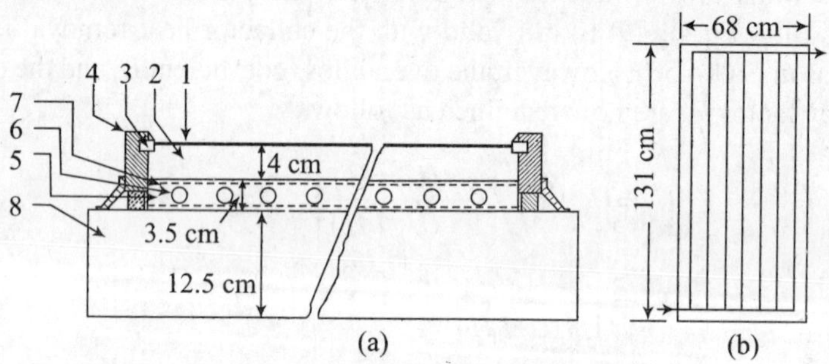

Fig. 4.24 *(a) Cross-section of concrete collector (1. Glass cover; 2. air gap; 3. rubber gasket; 4. wooden frame; 5. concrete slab; 6. PVC tube; 7. wire mesh; 8. cellular concrete insulation) (b) Schematic of PVC tube network*

Extensive tests have been carried out to find out the effect of tube spacings, flow rates and fluid inlet temperatures on the thermal performance of the collector. Of the collectors tested, it has been found that the collector with a pitch of 6 cm, operating at a flow rate of 1.2 litre per minute gives the best thermal performance.

Typical results of a whole day's testing at a fixed fluid inlet temperature are shown in Fig. 4.25. The figure shows the variation of the useful energy gain and solar flux incident on the collector plane. It is observed that the useful energy of the collector follows the same pattern of variation as that of the solar radiation. However, unlike a conventional collector, the useful energy from the concrete collector is available only after some time has elapsed in the morning. This is because the concrete collector has a large thermal mass and requires more time to heat up to the operating conditions. For the same reason, the maximum of the useful energy occurs after the maximum of solar flux, the phase lag being about 40 to 45 minutes. Furthermore, the collector continues to deliver useful energy for some time even after sunset.

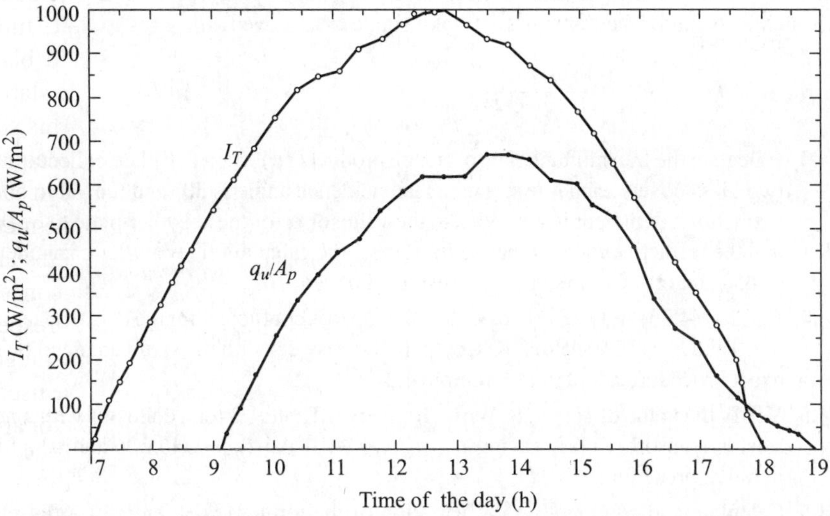

Fig. 4.25 *Variation of solar radiation incident on the collector surface and useful heat gain from the concrete collector (pitch = 6 cm) for a flow rate of 1.2 lpm and average fluid inlet temperature of 35°C*

Experimental values of the daily efficiency of the collector are plotted against the parameter $(T_{fi} - \bar{T}_a)/H_T$, where T_{fi} is the fluid inlet temperature and $\bar{T}_a$ is the average ambient temperature for the day. It is seen that the daily efficiency varies linearly with $(T_{fi} - \bar{T}_a)/H_T$ and decreases as $(T_{fi} - \bar{T}_a)/H_T$ increases. This behaviour is similar to the dependence of the instantaneous efficiency on the parameter $(T_{fi} - T_a)/I_T$ in the case of a conventional collector. The daily efficiency of a concrete collector can be expressed by an equation of the form

$$\eta = A - B[(T_{fi} - \bar{T}_a)/H_T] \tag{4.77}$$

where A and B are constants. For the collector having a tube pitch of 6 cm and a flow rate of 1.2 lpm, $A = 0.527$ (dimensionless) and $B = 0.239$ MJ/m^2-day-°C.

It may be mentioned that an experimental study on solar concrete collectors of a slightly different design has been reported by Chaurasia [56] in Jodhpur. His collectors consisted of a network of aluminium tubes partly embedded in a concrete slab with a coating of blackboard paint on top. No glazing was used. Although the absence of a glazing makes such collectors less efficient than the collectors tested by Nayak *et al.*, Chaurasia showed that they can provide moderately hot water to meet various requirements during the day time even in the winter season in Jodhpur.

PROBLEMS

4.1 Prove Eq. (4.8).

4.2 Consider absorption and reflection simultaneously and derive the following expression for the transmissivity of a single glass cover

$$\tau = \frac{\{(1-\rho)^2\, e^{-K\delta_c/\cos\theta_2}\}}{\{1-\rho^2\, e^{-2K\delta_c/\cos\theta_2}\}}$$

4.3 Calculate the transmittance-absorptance product $(\tau\alpha)_b$ for a flat-plate collector with two glass covers each 4 mm thick. The incidence angle is 30° and the value of the extinction coefficient is 4 m^{-1}. Take the value of α for the absorber plate to be 0.86 and the refractive index to be 1.526. Repeat the calculation for angles of incidence of 45° and 60°. Comment on the results obtained.

4.4 Use Eqs (4.36) and (4.37) to calculate the top loss coefficient for the data of Example 4.2. Take h_w = 14.95 W/m^2-K. Compare the answers with the value obtained by the exact procedure adopted in Example 4.2.

4.5 Verify the value of U_t = 7.26 W/m^2-K given in Table 4.2 for a collector with a non-selective absorber plate having $\alpha = \varepsilon_p$ = 0.94. Take T_{pm} = 346.2 K and adopt the iterative procedure.

4.6 Calculate and comment on the variation of the top loss coefficient of a flat-plate collector as the absorber plate emissivity varies from 0 to 1, given T_{pm} = 80°C, T_a = 20°C, M = 1, β = 25°, L = 3 cm, h_w = 10 W/m^2-K, ε_c = 0.88. Use Eq. (4.37).

4.7 A liquid flat-plate collector for heating water is made of a non-selective aluminium absorber plate with integral in-line tubes as shown in Fig. 4.16(b).

The following data is given:

Length of collector	1.5 m
Width of collector	0.83 m
Length of absorber plate	1.4 m
Width of absorber plate	0.755 m
Thickness of absorber plate	2 mm
Thermal conductivity of aluminium	200 W/m-K
Plate absorptivity/emissivity	0.95
Number of glass covers	2
Spacing between absorber plate and first glass cover	12 mm
Spacing between first and second glass cover	12 mm
Tube centre-to-centre distance	3.4 cm
Tube outer diameter	1.1 cm
Tube inner diameter	0.9 cm
Glass cover emissivity/absorptivity	0.88
$K\delta_c$	0.037 per plate
Index of refraction	1.526
Location of collector	Mumbai (19.11°N)
Date	February 12, 1980
Time	1400 h (LAT)

Collector tilt	Latitude angle
Surface azimuth angle	0°
I_b	645 W/m^2
I_d	142 W/m^2 (assume uniformly distributed)
Water flow rate	33 litres/h
Water side heat transfer coefficient	325 W/m^2-K
Water inlet temperature	70°C
Ambient temperature	36°C
Wind speed	1.25 m/s
Back insulation thickness	9 cm
Side insulation thickness	5 cm
Insulation thermal conductivity	0.05 W/m-K

Calculate the instantaneous efficiency and the exit temperature of the water.

4.8 (a) Calculate the variation in efficiency of a liquid flat-plate collector with the inlet temperature. Given:

Geometry of collector	Absorber plate with tubes bonded below
Length of collector	2.15 m
Width of collector	2.15 m
Width of plate	2 m
Length of plate	2 m
Plate thickness	1.2 mm
Plate material	Aluminium (k = 210 W/m-K)
Tube material	Aluminium
Tube outer diameter	15.9 mm
Tube inner diameter	12.7 mm
Tube centre-to-centre distance	20 cm
Bond resistance	Negligible
Water inlet temperature	20, 50, 80°C
Ambient temperature	20°C
Number of glass covers	2
$(\tau\alpha)_{av}$	0.75
I_T	1000 W/m^2
Overall loss coefficient	4 W/m^2-K
Fluid to tube heat transfer coefficient	300 W/m^2-K
Water flow rate	90 kg/h

(b) Calculate the efficiency again if the following changes are made in the data:

Number of glass-covers	1
$(\tau\alpha)_{av}$	0.89
U_l	7.5 W/m^2-K

Comment on the variations obtained in cases (a) and (b) and compare them with each other.

(c) Calculate the collector stagnation temperatures for cases (a) and (b).

4.9 Starting with the Hottel-Whillier-Bliss equation (4.49), derive an explicit expression for useful heat gain rate (q_u) of a liquid flat-plate collector in terms of the temperature difference ($T_{fo} - T_a$) and parameters F_R, A_p, U_l, $\dot{m}$, C_p and S. Derive a similar expression in which the temperature difference ($T_{fo} - T_a$) is replaced by $[\{(T_{fo} + T_{fi})/2\} - T_a]$.

4.10 Derive Eq. (4.53).

4.11 The following radiation measurements are available for Thiruvananthapuram (8.48°N, 76.95°E):

Month	J	F	M	A	M	J	J	A	S	O	N	D
$\bar{H}_g$ (kWh/m^2-day)	5.932	6.356	6.682	6.166	5.490	5.250	5.027	5.559	5.942	5.232	4.954	5.131
$\bar{H}_d$ (kWh/m^2-day)	1.770	1.914	2.196	2.617	2.827	2.974	3.133	3.135	2.765	2.557	2.264	2.030

Calculate the optimum tilt of a flat-plate collector for the three situations studied in Example 4.4.

4.12 Calculate the optimum tilt angle for a flat-plate collector array located at Mumbai (19.12°N) if (i) insolation falling over the whole year is to be maximized, (ii) insolation falling for all months excepting July and August is to be maximized. Use the data given in Table A3.3 and A3.4. Comment on the results obtained.

4.13 Develop a more accurate formulation than the procedure indicated in Section 4.10.4 for the optimum tilt problem by using Eq. (3.51).

4.14 A liquid flat-plate collector with tubes in-line with the absorber plate has an overall heat loss coefficient of 5.8 W/m^2-K. The inner and outer diameters of the tubes are 14 mm and 18 mm respectively and the tube centre-to-centre distance is 12 cm. The fluid-to-tube heat transfer coefficient is 205 W/m^2-K. Calculate the collector efficiency factor for the following conditions:

(i) Absorber material: Copper (0.2 mm thick)
(ii) Absorber material: Galvanized iron (1.3 mm thick)
(iii) Absorber material: Aluminium (0.5 mm thick)

Given: The thermal conductivities of copper, galvanized iron and aluminium are 350, 35 and 180 W/m-K, respectively.

4.15 Extend the transient analysis of Section 4.12 to the situation of a flat-plate collector having two glass covers and show that the final result given by Eq. (4.63) still holds true with $(mC)_e$ given by

$$(mC)_e = (mC)_p + (mC)_{c1}\, U_t \left\{\frac{1}{U_{t2}} + \frac{1}{U_{t3}}\right\} + (mC)_{c2}\,\frac{U_t}{U_{t3}}$$

In the above expression, the subscripts $c1$ and $c2$ refer to the inner and outer glass covers respectively, U_{t2} is the overall convective-radiative coefficient based on the temperature difference $(T_{c1} - T_{c2})$, and U_{t3} is the overall coefficient based on the temperature difference $(T_{c2} - T_a)$.

4.16 A water heating flat-plate collector is fitted with two glass covers and a non-selective absorber plate of dimensions 1 m × 2 m. The collector is tested by the standard procedure described in Section 4.13 and the following data is obtained:

T_{fi} (°C)	T_{fo} (°C)	T_a (°C)	I_T (W/m^2)
84.95	93.98	23.0	885
79.97	89.83	22.4	879
75.63	85.51	22.1	862

(Contd.)

69.54	79.24	21.6	841
68.19	77.36	21.3	827
50.05	61.19	20.4	819
43.37	54.06	20.4	792
38.17	49.79	19.3	770
33.92	45.44	19.0	761

Given: $(\tau\alpha)_{av} = 0.74$; $\dot{m} = 1.10$ kg/minute; $C_p = 4.18$ kJ/kg-°C; $A_c/A_p = 1.2$.

(a) Calculate the values of η_i and plot these against the parameter $(T_{fi} - T_a)/I_T$. Draw a best fit straight line and determine the values of U_l and F_R.

(b) How does the value of F_R change if the value of $\dot{m}$ is increased to 1.5 kg/minute? Assume that the value of F' does not change significantly because of the increase in $\dot{m}$.

4.17 A liquid flat-plate collector has the following characteristic parameters:

$$F_R(\tau\alpha)_{av} = 0.73, F_R U_L = 3.50 \text{ W/m}^2\text{-K}$$

In an experiment on such a collector, the following observations are recorded:

Collector water flow rate 0.02 kg/s-m^2 of absorber area
Collector outlet temperature 72°C
Ambient temperature 29°C

An unshaded pyranometer kept on a horizontal surface and a pyrheliometer record 780 W/m^2 and 650 W/m^2 respectively. Calculate the collector inlet temperature.

Given:

Location 22°39′ N, 88°27′ E
Date September 13
Time 1100 h (IST)
Slope of the collector 20°
Surface azimuth angle of the collector 45°
Reflectivity of the surrounding surface 0.2

4.18 The temperature rise across a flat-plate collector is measured to be 7.4°C under the following conditions:

Solar flux in the collector plane 970 W/m^2
Ambient temperature 25°C
Collector fluid flow rate 144 kg/h
Collector overall loss coefficient 4.5 W/m^2-K
Collector absorber area 2 m^2
$(\tau\alpha)_{av}$ 0.80

If the inlet fluid temperature is decreased by 10°C, the temperature rise across the collector is 7.9°C. All other data remains the same. Calculate the collector efficiency factor.

4.19 A liquid flat-plate collector has the following characteristic parameters:

$$F_R(\tau\alpha)_{av} = 0.68, F_R U_l = 6.1 \text{ W/m}^2\text{-K}$$

It is operating under the following conditions:

Solar flux incident on the collector plane 900 W/m^2
Water flow rate 0.015 kg/s-m^2 of absorber area
Ambient temperature 20°C
Inlet water temperature 40°C

(i) Calculate the mean absorber plate temperature of the collector if the collector efficiency factor is 0.90.

(ii) If the circulating pump fails, what is the maximum temperature attained by the plate?

4.20 The following data are recorded during the testing of a liquid flat-plate collector:

I_T (W/m^2)	T_a (°C)	T_{fi} (°C)	T_{fo} (°C)
705	23.1	30.1	35.4
753	25.2	29.9	36.0
820	26.1	30.5	37.3
931	28.3	30.4	38.1
701	23.5	40.1	44.9
765	24.4	39.9	45.5
822	25.6	40.3	46.6
941	27.7	40.2	47.7
710	23.2	51.0	54.5
759	24.1	50.5	55.0
829	27.2	50.3	56.1
920	28.6	50.1	57.0
703	23.9	61.5	64.0
745	24.9	61.0	64.6
840	26.1	61.1	65.7
951	27.9	60.9	66.8

The flow rate is kept constant at 0.02 kg/(s-m^2). Carry out a multilinear regression analysis and find the collector correlation coefficients in Eq. (4.72).

4.21 On the basis of tests conducted on a liquid flat-plate collector, the following correlation coefficients are obtained:

$a_0 = 0.734$, $a_1 = 4.751$ W/(m^2-K) and $a_2 = 0.032$ W/(m^2-K^2)

The collector is used to heat water with an inlet temperature of 35°C and a flow rate of 2.3 kg/minute. If the total solar radiation incident on the plane of the collector is 825 W/m^2 and the ambient temperature is 27°C, calculate the outlet temperature of the water.

Given: Gross collector area = 2.1 m^2; specific heat of water = 4.174 kJ/kg-K

4.22 Consider the liquid flat-plate collector described in Section 4.14.3 having a double-walled absorber plate made from a polymeric material having a low thermal conductivity. Perform a one-dimensional analysis in which the thermal resistance of the walls is considered and the thermal resistance to convective heat transfer from the walls to the fluid is assumed to be negligible by comparison. Show that the overall loss coefficient and the collector efficiency factor for this collector are given by Eqs (4.75) and (4.76).

REFERENCES

1. IS 12933 (Part 2):2003. *Solar flat-plate collector – specification, part 2 – components*, Bureau of Indian Standards, New Delhi, India.
2. Hordy, N., Rabilloud, D., Meunier, J. and Coulombe, S. 2014. High temperature and long-term stability of carbon nanotube nanofluids for direct absorption solar thermal collectors. *Solar Energy*, 105: 82.
3. Hottel, H.C. and Woertz, B.B. 1942. Performance of flat-plate solar-heat collectors. *Trans. ASME*, 64: 91.
4. Buchberg, H., Catton, I. and Edwards, D.K. 1976. Natural convection in enclosed spaces – a review of application to solar energy collection. *Journal of Heat Transfer, Trans. ASME*, 98: 182.
5. McAdams, W.A. 1954. *Heat Transmission,* 3rd ed. McGraw-Hill Book Co., New York, p.249.
6. Sparrow, E.M. and Tien, K.K. 1977. Forced convection heat transfer at an inclined and yawed square plate – application to solar collectors. *Journal of Heat Transfer, Trans. ASME*, 99: 507.
7. Sparrow, E.M., Ramsey, J.W. and Mass, E.A. 1979. Effect of finite width on heat transfer and fluid flow about an inclined rectangular plate. *Journal of Heat Transfer, Trans. ASME*, 101: 199.
8. Test, F.L., Lessman, R.C.L. and Johary, A. 1981. Heat transfer during wind flow over rectangular bodies in the natural environment. *Journal of Heat Transfer, Trans. ASME*, 103:262.
9. Kumar, S., Chourasia, B.K. and Mullick, S.C. 2005. Wind heat transfer coefficient in flat plate solar collectors. *SESI Journal*, 15: 30.
10. Klein, S.A. 1975. Calculation of flat-plate collector loss coefficients. *Solar Energy*, 17: 79.
11. Garg, H.P. and Datta, G. 1984. The top loss calculation for flat-plate solar collectors. *Solar Energy*, 32: 141.
12. Malhotra, A., Garg, H.P. and Palit, A. 1981. Heat loss calculation of flat-plate solar collectors. *Journal of Thermal Energy*, 2: 2.
13. Tabor, H. 1956. Selective radiation. *Bulletin Research Council of Israel,* 5A: 119.
14. Gier, J.T. and Dunkle, R.V. 1958. Selective spectral characteristics as an important factor in the efficiency of solar collectors. *Trans. Conf on the Use of Solar Energy*, 2, Part 1:41.
15. Close, D.J. 1962. Flat plate solar absorbers: the production and testing of a selective surface for copper absorber plates. Report E.D. 7, C.S.I.R.O., Melbourne, Australia.
16. Tabor, H., Harris, J., Weinberger, H. and Doron, B. 1964. Further studies on selective black coatings. *Proc. UN. Conference on New Sources of Energy*, 4: 618.
17. Bogaerts, W.F. and Lampert, CM. 1983. Materials for photothermal solar conversion. *Journal of Mat. Sci.*, 18: 2847.
18. Koltun, M. Gukhman, G. and Gavrilina, A. 1994. Stable selective coating black nickel for solar collector surfaces. *Solar Energy Materials and Solar Cells*, 33: 41.

19. Wackelgard, E. 1998. Characterization of black nickel solar absorber coatings electroplated in a nickel chlorine aqueous solution. *Solar Energy Materials and Solar Cells*, 56: 35.
20. Mason, J.J. and Brendel, T.A. 1982. Maxorb - a new selective surface on nickel. *Proc. International Society for Optical Engineering*, 324: 139.
21. Klochko, N.P., Klepikova, K.S., Tyukhov, I.I., Myagchenko, Y.O., Melnychuk, E.E., Kopach, V.R., Khrypunov, G.S., Lyubov, V.M., Kopach, A.V., Starikov, V.V. and Kirichenko, M.V. 2015. Zinc oxide-nickel cermet selective coatings obtained by sequential electrodeposition. *Solar Energy*, 117: 1.
22. Brunold, S., Frei, U., Carlsson, B., Moller, K., and Kohl, M. 2000. Accelerated life testing of solar absorber coatings: testing procedure and results. *Solar Energy*, 68: 313.
23. Sunselect coatings (http://alanod-solar.com; accessed on February 03, 2016)
24. Andersson, A., Hunderi, O. and Granqvist, C.G. 1980. Nickel pigmented anodic aluminium oxide for selective absorption of solar energy. *Journal of App. Physics*, 51: 754.
25. Kadirgan, F., Wackelgard, E. and Sohmen, M. 1999. Electrochemical characterization of Al_2O_3 - Ni thin film selective surface on aluminium. *Turk. J Chem.*, 23: 381.
26. Salmi, J., Bonino, J.P. and Bes, R.S. 2000. Nickel pigmented anodized aluminium as solar selective absorbers. *Journal of Mat. Sci.*, 35: 1347.
27. Kennedy, C.E. 2002. Review of mid- to high-temperature solar selective absorber materials. Technical Report NREL/TP-520-31267, National Renewable Energy Laboratory, Colorado.
28. Selvakumar, N. and Barshilia, H.C. 2012. Review of physical vapor deposited (PVD) spectrally selective coatings for mid- and high-temperature solar thermal applications. *Solar Energy Materials and Solar Cells*, 98:1
29. Morse, R.N. and Czarnecki, J.T. 1958. Flat-plate solar absorbers: the effect on incident radiation of inclination and orientation. *Report E.E. 6*, Mechanical Engineering Division, C.S.I.R.O., Melbourne, Australia.
30. Kern, J. and Harris, I. 1975. On the optimum tilt of a solar collector. *Solar Energy*, 17: 97.
31. Yakup, M.A.H.M. and Malik, A.Q. 2001. Optmum tilt angle and orientation for solar collector in Brunei Darussalam. *Renewable Energy*, 24: 223.
32. Despotovic, M. and Nedic, V. 2015. Comparison of optimum tilt angles of solar collectors determined at yearly, seasonal and monthly levels. *Energy Conversion and Management*, 97: 121.
33. Stanciu, C. and Stanciu, D. 2014. Optimum tilt angle for flat-plate collectors all over the world – a declination dependence formula and comparison of three solar radiation models. *Energy Conversion and Management*, 81: 133.
34. Garg, H.P. 1974. Effect of dirt on transparent covers in flat-plate solar energy collectors. *Solar Energy*, 15: 299.
35. Hegazy, A.A. 2001. Effect of dust accumulation on solar transmittance through glass covers of plate-type collectors. *Renewable Energy*, 22: 525.
36. Elminir, H.K., Ghitas, A.E., Hamid, R.H., El-Hussainy, F., Beheary, M.M. and Abdel-Moneim, K.M. 2006. Effect of dust on the transparent cover of solar collectors. *Energy Conversion and Management*, 47: 3192.
37. Jing, Z., Zhiping, W., Kezhen, W. and Jianbo, L. 2015. Dust effect on thermal performance of flat-plate solar collector. *Journal of Solar Energy Engineering, Trans. ASME*, 137: 014502-1.

38. Sarver, T., Al-Qaraghuli, A. and Kazmerski, L.L. 2013. A comprehensive review of the impact of dust on the use of solar energy: history, investigations, results, literature, and mitigation approaches. *Renewable and Sustainable Energy Reviews*, 22: 698.

39. Mani, M. and Pillai, R. 2010. Impact of dust on solar photovoltaic (PV) performance: research status, challenges and recommendations. *Renewable and Sustainable Energy Reviews*, 14: 3124.

40. Duffie, J.A. and Beckman, W.A. 1974. *Solar Energy Thermal Processes*, John Wiley, New York.

41. ASHRAE Standard 93-86: 1986. *Method of testing to determine thermal performance of solar collectors*. American Society of Heating, Refrigerating and Air-conditioning Engineers, Atlanta, USA.

42. IS 12933 (Part 5):2003. *Solar flat-plate collector – specification, part 5 – test methods*. Bureau of Indian Standard, New Delhi, India.

43. BS EN 12975 – 2. 2006. *Thermal solar systems and components – solar collectors–part 2: test methods*. European Standard, Brussels, Belgium

44. Kalogirou, S.A. 2006. Prediction of flat-plate collector performance parameters using artificial neural networks. *Solar Energy*, 80: 248.

45. Nayak, J.K. and Amer, E.H. 2000. Experimental and theoretical evaluation of dynamic test procedures for solar flat-plate collectors. *Solar Energy*, 69: 377.

46. ISO 9806: 2013. *Solar thermal collectors – test methods*. International Organization for Standardization, ISO central Secretariat, Geneva, Switzerland.

47. Yin, Z., Zuqing, X. and Jian, Z. 1997. The evacuated absorber tube industry in China. *Proc. ISES 1997 Solar World Congress*, Taejon, Korea, 2:105.

48. Morrison, G.L., Budihardeja, I. and Behnia, M. 2004. Water-in- glass evacuated tube solar water heaters. *Solar Energy*, 76: 135.

49. Morrison, G.L., Budihardeja, I. and Behnia, M. 2005. Measurement and simulation of flow rate in a water-in-glass evacuated tube solar water heater. *Solar Energy*, 78: 257.

50. Liang, R., Ma, L., Zhang, J. and Zhao, D. 2011. Theoretical and experimental investigation of the filled-type evacuated tube solar collector with U-tube. *Solar Energy*, 85: 1735.

51. Beikircher, T., Berger, V., Osgyan, P., Reub, M. and Streib, G. 2014. Low- e confined air chambers in solar flat-plate collectors as an economic new type of rear side insulation avoiding moisture problems. *Solar Energy,* 105:280.

52. Kudish, A.I., Evseev, E.G., Rommel, M., Kohl, M., Walter, G. and Leukefeld, T. 1999. Research and development of solar collectors fabricated from polymeric material. *Proc. ISES Solar World Congress*, Jerusalem, p. 1126.

53. Kudish, A.I., Evseev, E.G., Walter, G. and Leukefeld, T. 2002. Simulation study of a solar collector with a selectively coated polymeric double-walled absorber plate. *Energy Conversion and Management*, 43: 651.

54. Tsilingiris, P.T. 1999. Towards making solar water heating technology feasible – the polymer solar collector approach. *Energy Conversion and Management*, 40: 1237.

55. Nayak, J.K., Sukhatme, S.P., Limaye, R.G. and Bopshetty, S.V. 1989. Performance studies on solar concrete collectors. *Solar Energy*, 42: 45.

56. Chaurasia, P.B.L. 2000. Solar water heaters based on concrete collectors. *Energy*, 25: 703.

Solar Air Heaters

CHAPTER 5

INTRODUCTION 5.1

This chapter describes and analyzes various types of solar air heaters. The principal applications in which solar air heaters are used include drying for agricultural and industrial purposes, and space heating. Indeed, they are the logical choice for these applications, compared to liquid flat-plate collectors, because they eliminate the need to transfer heat from one fluid to another.

A brief description of the conventional solar air heater is given in Section 5.2. This is followed by a performance analysis of the heater (Section 5.3). Variations on the conventional type are described in Section 5.4.1 and methods for analysing them are given. Section 5.4.2 deals with some novel designs–the matrix air heater, the plastic air heater, the transpired air heater and the evacuated tube solar air heater. The chapter concludes (Section 5.5) with a description of standard procedures for thermal testing of solar air heaters.

GENERAL 5.2

A conventional solar air heater generally consists of an absorber plate with a parallel plate below forming a small passage through which the air to be heated flows. As in the case of the liquid flat-plate collector, a transparent cover system is provided above the absorber plate, while a sheet metal container filled with insulation is provided on the bottom and sides. The arrangement is sketched in Fig. 5.1 (a). Two other arrangements, which are not so common, are also shown in Fig. 5.1. In the arrangement shown in Fig. 5.1 (b), the air to be heated flows between the cover and the absorber plate itself instead of through a separate passage, while in Fig. 5.1 (c), the air flows between the cover and the absorber plate, as well as through the passage below the absorber plate.

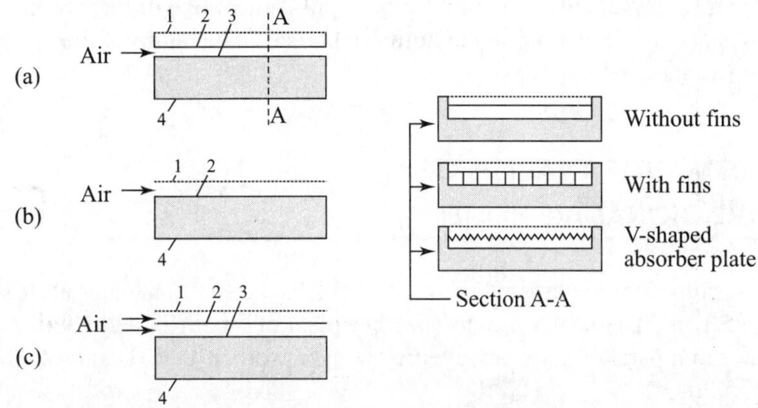

Fig. 5.1 *Various types of solar air heaters (1. Cover; 2. Absorber plate; 3. Bottom plate; 4. Insulation)*

Like a liquid flat-plate collector, a solar air heater is simple in design and requires little maintenance. In addition, since the fluid does not freeze, the solar air heater has the advantage of not requiring any special attention at temperatures below 0°C. Corrosion and leakage problems are also less severe. However, the value of the heat transfer coefficient between the absorber plate and the air is low and this results in a lower efficiency. For this reason, the surfaces are sometimes roughened or longitudinal fins are provided in the air-flow passage. Another variation is to use a V-shaped or corrugated absorber plate [see Section A-A, Fig. 5.1 (a)].

A further disadvantage associated with the use of a solar air heater is that large volumes of fluid have to be handled. As a result, the electrical power required to blow the air through a system can be significant if the pressure drop is not kept within prescribed limits.

The face areas of solar air heaters range from 1 to 2 m^2. Materials of construction and sizes are similar to those used with liquid flat-plate collectors (see Section 4.2). Thus, the absorber plate is a metal sheet about 1 mm in thickness, usually made of GI or steel. Glass of thickness 4 to 5 mm is the most commonly used cover material. However, plastics are being used in increasing numbers. Mineral wool or glass wool of thickness 5 to 8 cm is used for the bottom and side insulation. The whole assembly is contained in a sheet metal box and inclined at a suitable angle.

Compared to liquid flat-plate collectors, the pace of commercialisation for the production of solar air heaters has been slow all over the world. This is true in India as well, where they have been used primarily in systems for forced convection drying of various kinds of agricultural products. The reason for the slow pace is the fact that the drying of agricultural products is a seasonal activity, requiring energy for only a few months. As a result, the drying systems remain idle for a large part of the year and the economics in terms of the payback period is poor. It seems essential to pursue other applications like drying for industrial purposes and space heating in the northern parts of the country if the market for solar air heaters is to increase.

PERFORMANCE ANALYSIS OF A CONVENTIONAL AIR HEATER 5.3

We now consider the performance analysis of the conventional air heater shown in Fig. 5.1 (a). The heater has an absorber plate of length L_1 and width L_2. The air flows in a parallel plate passage below the absorber plate. Details are shown in Fig. 5.2.

The analysis is due to Whillier [1] and proceeds along lines identical to those adopted for the liquid flat-plate collector (Chapter 4) for the calculation of $(\tau\alpha)_b$, $(\tau\alpha)_d$, U_t and U_b. Considering a slice of width L_2 and thickness dx at a distance x from the inlet, we write down energy balances for the absorber plate, the plate below it, and the air flowing in between. We assume that (i) the bulk mean temperature of the air changes from T_f to $(T_f + dT_f)$ as it flows through the distance dx, (ii) the air mass flow rate is $\dot{m}$, (iii) the mean temperatures of the absorber plate and the plate below are T_{pm} and T_{bm}, respectively and their variation may be neglected, and (iv) side losses can be neglected. The following equations are obtained:

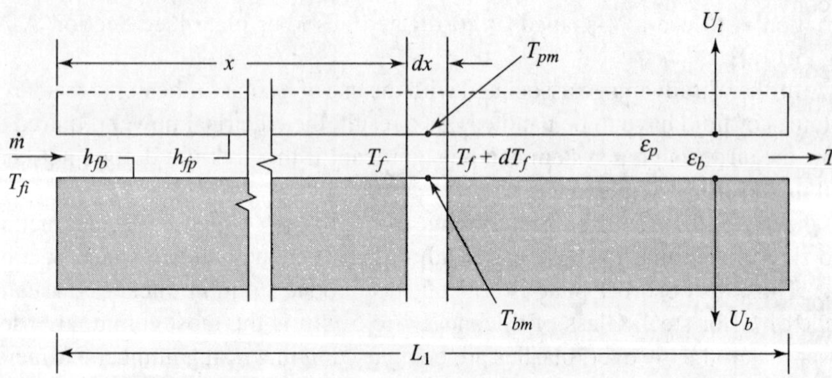

Fig. 5.2 *Analysis of conventional solar air heater*

❑ *For Absorber Plate*

$$SL_2\,dx = U_t L_2\,dx\,(T_{pm} - T_a) + h_{fp}\,L_2\,dx\,(T_{pm} - T_f) + \frac{\sigma\,L_2\,dx}{[(1/\varepsilon_p) + (1/\varepsilon_b) - 1]}(T_{pm}^4 - T_{bm}^4) \tag{5.1}$$

- For Bottom Plate

$$\frac{\sigma L_2\, dx}{[(1/\varepsilon_p)+(1/\varepsilon_b)-1]}(T_{pm}^4 - T_{bm}^4)$$

$$= h_{fb} L_2\, dx\,(T_{bm} - T_f) + U_b L_2\, dx\,(T_{bm} - T_a) \tag{5.2}$$

- For Air Stream

$$\dot{m}C_p\, dT_f = h_{fp}\, L_2\, dx\,(T_{pm} - T_f) + h_{fb}\, L_2\, dx\,(T_{bm} - T_f) \tag{5.3}$$

In the above equations,

S = flux absorbed in the absorber plate,
U_t = top loss coefficient based on the temperature difference $(T_{pm} - T_a)$,
U_b = bottom loss coefficient based on the temperature difference $(T_{bm} - T_a)$,
h_{fp} = convective heat transfer coefficient between the absorber plate and the air stream,
h_{fb} = convective heat transfer coefficient between the bottom plate and the air stream,
ε_p = emissivity of the absorber plate surface, and
ε_b = emissivity of the bottom plate surface.

While writing the energy balance Eqs (5.1) – (5.3), we distinguish between h_{fp}, the convective heat transfer coefficient for the absorber plate and the air, and h_{fb}, the convective heat transfer coefficient for the bottom plate and the air. However this distinction is usually not made in the correlations available for calculating the convective heat transfer coefficients. Thus while using these correlations, we will have to assume $h_{fp} = h_{fb}$.

We introduce an equivalent radiative heat transfer coefficient h_r defined in this case by

$$h_r(T_{pm} - T_{bm}) = \frac{\sigma}{[1/\varepsilon_p)+(1/\varepsilon_b)-1]}(T_{pm}^4 - T_{bm}^4) \tag{5.4}$$

For small values of the temperature difference $(T_{pm} - T_{bm})$, it is readily shown that the expression $(T_{pm}^4 - T_{bm}^4)$ can be approximated by the expression $4T_{av}^3(T_{pm} - T_{bm})$ where $T_{av} = (T_{pm} + T_{bm})/2$. Then

$$h_r = \frac{4\sigma T_{av}^3}{[(1/\varepsilon_p)+(1/\varepsilon_b)-1]} \tag{5.4a}$$

Further, it is assumed for purposes of simplification that the bottom loss coefficient U_b is much smaller in magnitude than the top loss coefficient U_t. Consequently the bottom loss term can be deleted from Eq. (5.2) and clubbed with the top loss term in Eq. (5.1). Equations (5.1) – (5.3) thus reduce to

$$S = U_l(T_{pm} - T_a) + h_{fp}(T_{pm} - T_f) + h_r(T_{pm} - T_{bm}) \tag{5.5}$$

$$h_r(T_{pm} - T_{bm}) = h_{fb}(T_{bm} - T_f) \tag{5.6}$$

$$\frac{\dot{m}C_p}{L_2}\frac{dT_f}{dx} = h_{fp}(T_{pm} - T_f) + h_{fb}(T_{bm} - T_f) \tag{5.7}$$

From Eq. (5.6) we get

$$T_{bm} = \frac{h_r T_{pm} + h_{fb} T_f}{h_r + h_{fb}} \tag{5.8}$$

Substituting this expression into Eq. (5.5), we have

$$T_{pm} = \frac{S + U_l T_a + h_e T_f}{U_l + h_e} \tag{5.9}$$

where
$$h_e = \left[h_{fp} + \frac{h_r h_{fb}}{h_r + h_{fb}}\right] \tag{5.10}$$

is the effective heat transfer coefficient between the absorber plate and the air stream.

Hence, $(T_{pm} - T_a) = \dfrac{S + h_e(T_f - T_a)}{U_l + h_e}$

From Eqs (5.5) – (5.7), we have

$$\frac{\dot{m}C_p}{L_2}\frac{dT_f}{dx} = S - U_l(T_{pm} - T_a) \tag{5.11}$$

Substituting the expression for $(T_{pm} - T_a)$ into Eq. (5.11) we get the differential equation

$$\frac{\dot{m}C_p}{L_2}\frac{dT_f}{dx} = \frac{1}{\left(1 + \dfrac{U_l}{h_e}\right)}\{S - U_l(T_f - T_a)\} \tag{5.12}$$

In an analogous manner to the liquid flat-plate collector, we now define a *collector efficiency factor F′* given by

$$F' = \left(1 + \frac{U_l}{h_e}\right)^{-1} \tag{5.13}$$

Equation (5.12) thus becomes
$$\frac{\dot{m}C_p}{L_2}\frac{dT_f}{dx} = F'\{S - U_l(T_f - T_a)\} \tag{5.14}$$

Equation (5.14) is identical to Eq. (4.45) derived earlier. The solution now proceeds along identical lines and we obtain the fluid temperature distribution

$$\frac{\left(\frac{S}{U_l}+T_a\right)-T_f}{\left(\frac{S}{U_l}+T_a\right)-T_{fi}} = \exp\left[-\frac{L_2F'U_lx}{\dot{m}C_p}\right] \tag{5.15}$$

Similarly, the useful heat gain rate for the collector is given by

$$q_u = F_RA_p[S - U_l(T_{fi} - T_a)] \tag{5.16}$$

where F_R = collector heat-removal factor

$$= \frac{\dot{m}C_p}{U_lA_p}\left[1-\exp\left\{-\frac{F'U_lA_p}{\dot{m}C_p}\right\}\right] \tag{5.17}$$

and A_p = area of the absorber plate = L_1L_2.

It is worth noting that if the simplifying assumption of deleting U_b from Eq. (5.2) and clubbing it with Eq. (5.1) had not been made, we would obtain the following differential equation instead of Eq. (5.14)

$$\frac{\dot{m}C_p}{L_2}\frac{dT_f}{dx} = F'\{S-U_l''(T_f-T_a)\} \tag{5.18}$$

where U_l'' is an equivalent overall loss coefficient.
F' and U_l'' are defined as follows:

$$F' = \left(1+\frac{U_l'}{h_e}\right)^{-1} \tag{5.19}$$

$$U_l'' = U_l' + \frac{1}{F'}\frac{U_bh_{fb}}{(h_r+h_{fb}+U_b)} \tag{5.20}$$

where

$$U_l' = U_t + \frac{h_rU_b}{(h_r+h_{fb}+U_b)} \tag{5.21}$$

and

$$h_e = h_{fp} + \frac{h_rh_{fb}}{(h_r+h_{fb}+U_b)} \tag{5.22}$$

The useful heat gain rate for the collector is then given by

$$q_u = F_RA_p[S - U_l''(T_{fi} - T_a)] \tag{5.23}$$

where F_R = collector heat-removal factor

$$= \frac{\dot{m}C_p}{U_l''A_p}\left[1-\exp\left\{\frac{-F'U_l''A_p}{\dot{m}C_p}\right\}\right] \tag{5.24}$$

Heat Transfer and Pressure Drop in a Parallel Plate Duct

In order to calculate the performance based on Eq. (5.16) or Eq. (5.23), we need to know the values of the convective heat transfer coefficient to the heated air. The situation corresponds to one of turbulent flow with one of the long sides heated and the other insulated. It can be considered to be fully developed if the length-to-equivalent diameter ratio exceeds a value of about 30. The following two correlations are then appropriate if the surfaces are smooth,

$$\text{Nu} = 0.0158\,\text{Re}^{0.8} \tag{5.25}$$

$$\text{Nu} = \frac{0.01344\,\text{Re}^{0.75}}{1-1.586\,\text{Re}^{-0.125}} \tag{5.26}$$

Equation (5.25) is based on the data of Kays [2], while Eq. (5.26) has been suggested by Malik and Buelow [3]. In the above equations, the characteristic dimension is the equivalent diameter d_e given by

$$d_e = \frac{4 \times \text{Cross-sectional area of duct}}{\text{Wetted perimeter}}$$

Properties are evaluated at the arithmetic mean of the fluid inlet and outlet temperature, and the values of h_{fp} and h_{fb} are taken to be equal. Nusselt numbers calculated from Eqs (5.25) and (5.26) agree within 10 per cent for Reynolds numbers ranging from 10 000 to 20 000. These values are normally obtained in solar air heater applications.

The dimensionless pressure drop in the duct can be calculated from the well-known Blasius equation which is valid for smooth surfaces.

$$f = 0.079\,\text{Re}^{-0.25} \tag{5.27}$$

where f is the friction factor and the characteristic dimension is again the equivalent diameter d_e.

Example 5.1

Calculate the performance of a conventional solar air heater with the following data:

Length of collector	: 2.1 m
Width of collector	: 1.1 m
Length of absorber plate (L_1)	: 2 m
Width of absorber plate (L_2)	: 1 m
Spacing between absorber plate and bottom plate (L)	: 1.5 cm
Air flow rate ($\dot{m}$)	: 200 kg/h
Air inlet temperature (T_{fi})	: 50°C
Ambient temperature (T_a)	: 20°C
Solar flux incident on the collector face (I_T)	: 950 W/m^2
$(\tau\alpha)_{av}$	: 0.85
Top loss coefficient (U_t)	: 6.2 W/m^2-K

Bottom loss coefficient (U_b) : 0.8 W/m^2-K

$\varepsilon_p = \varepsilon_b$: 0.95

Solution

Neglect heat loss from the sides.

We will first calculate the value of the convective heat transfer coefficients h_{fp} and h_{fb}. Assuming that the mean fluid temperature, T_{fm} is 55°C, the properties of air are

$$\rho = 1.077 \text{ kg/m}^3$$
$$C_p = 1.005 \text{ kJ/kg-K}$$
$$\mu = 19.85 \times 10^{-6} \text{ N-s/m}^2$$
$$k = 0.0287 \text{ W/m-K}$$

$$\text{Equivalent diameter} = \frac{4 \times 1 \times 0.015}{2(1 + 0.015)} = 0.0296 \text{ m}$$

$$\text{Therefore, } \frac{L_1}{d_e} = \frac{2}{0.0296} = 68$$

$$\text{Average air velocity} = \frac{200}{3600 \times 1.077 \times 1 \times 0.015} = 3.439 \text{ m/s}$$

$$\text{Re} = \frac{\rho V d_e}{\mu} = \frac{1.077 \times 3.439 \times 0.0296}{19.85 \times 10^{-6}} = 5515$$

Hence, the flow is turbulent and fully developed. Assuming the surfaces to be smooth and using Eq. (5.25) we have

$$\text{Nu} = 0.0158 \times 5515^{0.8} = 15.56$$

Therefore,
$$h_{fp} = h_{fb} = 15.56 \times \frac{0.0287}{0.0296} = 15.08 \text{ W/m}^2\text{-K}$$

From Eq. (5.4), the radiative heat transfer coefficient is given by

$$h_r = \frac{\sigma}{\left(\frac{1}{\varepsilon_p} + \frac{1}{\varepsilon_b} - 1\right)} (T_{pm} + T_{bm})(T_{pm}^2 + T_{bm}^2)$$

For purposes of calculating h_r, we assume that the mean temperature of the absorber plate and the bottom plate can be taken to be equal to the mean fluid temperature T_{fm}. This is a good assumption since the absolute value of temperature has to be used in calculating h_r and the values of $(T_{pm} - T_{fm})$ and $(T_{bm} - T_{fm})$ are likely to be small, of the order of 10 K. Thus

$$h_r = \frac{5.67 \times 10^{-8} \times 4 \times (273.2 + 55)^3}{\left(\frac{1}{0.95} + \frac{1}{0.95} - 1\right)} = 7.24 \text{ W/m}^2\text{-K}$$

From Eq. (5.10), $h_e = 15.08 + \dfrac{7.24 \times 15.08}{7.24 + 15.08} = 19.97$ W/m^2-K.

From Eq. (5.13), collector efficiency factor $F' = \left(1 + \dfrac{6.2 + 0.8}{19.97}\right)^{-1} = 0.740$

$$\frac{\dot{m}C_p}{U_l A_p} = \frac{200}{3600} \times \frac{1.005 \times 1000}{7 \times 1 \times 2} = 3.988$$

Hence from Eq. (5.17),
Collector heat-removal factor $F_R = 3.988\ (1 - e^{-0.740/3.988}) = 0.675$
Useful heat gain rate $q_u = 0.675 \times 1 \times 2[950 \times 0.85 - 7(50 - 20)] = 806.6$ W

Instantaneous efficiency $\eta_i = \dfrac{806.6}{950 \times 1.1 \times 2.1} = 0.368$

The air outlet temperature is obtained from the energy balance equation

$$\frac{200}{3600} \times 1.005 \times (T_{fo} - 50) = \frac{806.6}{1000}$$

Hence $T_{fo} - 50 = 14.45$°C and $T_{fo} = 64.45$°C

The pressure drop across the collector is calculated from Eq. (5.27). We have

$$f = 0.079 \times 5515^{-0.25} = 0.009167$$

Hence pressure drop $= \dfrac{4 f \rho L_1 V^2}{2 d_e}$

$$= \frac{4 \times 0.009167 \times 1.077 \times 2 \times 3.439^2}{2 \times 0.0296}$$

$$= 15.78 \text{ N/m}^2 = 1.61 \text{ mm of water}$$

Finally, it may be noted that if we use the more correct performance equations, viz. Eqs (5.18) to (5.24), we would obtain the following slightly different results,

$$F' = 0.754; \quad F_R = 0.686; \quad q_u = 813.9 \text{ W} \quad \text{and} \quad \eta_i = 0.371$$

OTHER TYPES OF AIR HEATERS 5.4

5.4.1 Variations on the Conventional Type

We now describe solar air heaters which are variations on the conventional type. Some of them are shown in Fig. 5.1.

Flow Between the Cover and Absorber Plate

Consider first the air heater of Fig. 5.1 (b) in which the air flows in a parallel plate passage between the cover and the absorber plate. The method of analysis of this type is similar to that adopted for the conventional type. Making the same assumptions as in Section 5.3, and writing energy balances for the absorber plate, the cover plate and the air flowing in between, we get

- *For Absorber Plate*

$$S = h_{fp} (T_{pm} - T_f) + h_r (T_{pm} - T_c) + U_b (T_{pm} - T_a) \tag{5.28}$$

- *For Cover*

$$h_r (T_{pm} - T_c) = U_t (T_c - T_a) + h_{fc} (T_c - T_f) \tag{5.29}$$

- *For Air Stream*

$$\frac{\dot{m} C_p}{L_2} \frac{dT_f}{dx} = h_{fp} (T_{pm} - T_f) + h_{fc} (T_c - T_f) \tag{5.30}$$

where
$$h_r = \frac{\sigma}{\left(\dfrac{1}{\varepsilon_p} + \dfrac{1}{\varepsilon_c} - 1\right)} \frac{(T_{pm}^4 - T_c^4)}{(T_{pm} - T_c)} \tag{5.31}$$

In the above equations, values of the convective heat transfer coefficients h_{fc} and h_{fp} would be calculated from either Eqs (5.25) or (5.26). It is also to be noted that the top loss coefficient U_t is based on the temperature difference $(T_c - T_a)$, while the bottom loss coefficient U_b is based on the temperature difference $(T_{pm} - T_a)$. Solving Eqs (5.28)–(5.30), one again obtains Eqs (5.23) and (5.24) for the useful heat gain rate and the collector heat removal factor. However, the equivalent overall loss coefficient and the collector efficiency factor are now given by the following expressions:

$$U_l'' = \frac{(U_t + U_b)(h_{fc}\, h_{fp} + h_{fc}\, h_r + h_{fp}\, h_r) + U_t U_b (h_{fc} + h_{fp})}{h_{fc}\, h_r + h_{fp}\, U_t + h_{fp} h_r + h_{fc}\, h_{fp}} \tag{5.32}$$

$$F' = \frac{h_{fc} h_r + h_{fp}\, U_t + h_{fp}\, h_r + h_{fc}\, h_{fp}}{(U_t + h_r + h_{fc})(U_b + h_r + h_{fp}) - h_r^2} \tag{5.33}$$

Solar air heaters of this type are simpler in construction than the conventional design. However, they yield lower efficiencies because of higher losses from the top.

Conventional Air Heater with Continuous Longitudinal Fins

The addition of continuous longitudinal fins to the bottom side of the absorber plate improves the heat transfer. This is desirable because it increases the efficiency. We now analyse such a heater in which fins of height L_f and thickness δ_f are spaced at a distance W centre-to-centre apart (Fig. 5.3). The distance between the absorber plate and the bottom plate is L. Consequently the clearance between the fins and the bottom plate is $(L - L_f)$. Considering a slice of width W and thickness dx at a distance x from the inlet, we again write down energy balances for the absorber plate, the bottom plate, and the air flowing in between. The assumptions made earlier in Section 5.3 are again made. We get

$$S\,W\,dx = U_t W\,dx\,(T_{pm} - T_a) + h_{fp}\,W\,dx\,(T_{pm} - T_f) + 2L_f\,dx\,\phi_f\,h_{ff}\,(T_{pm} - T_f) + h_r\,W\,dx\,(T_{pm} - T_{bm}) \quad (5.34)$$

$$h_r W\,dx\,(T_{pm} - T_{bm}) = h_{fb} W\,dx\,(T_{bm} - T_f) + U_b W\,dx\,(T_{bm} - T_a) \quad (5.35)$$

$$(W/L_2)\,\dot{m}C_p\,dT_f = h_{fp}\,W\,dx\,(T_{pm} - T_f) + 2L_f\,dx\,\phi_f\,h_{ff}(T_{pm} - T_f) + h_{fb}\,W\,dx\,(T_{bm} - T_f) \quad (5.36)$$

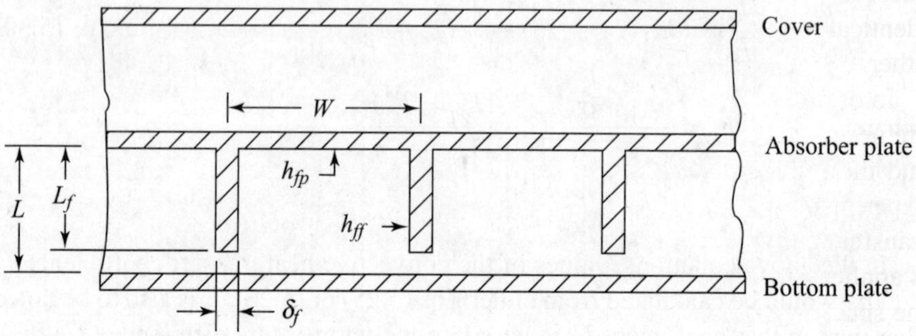

Fig. 5.3 *Conventional air heater with fins*

It will be seen that additional terms are introduced in Eqs (5.34) and (5.36) to take account of heat transfer from the fin surfaces. In these equations,

ϕ_f = fin effectiveness,*
h_{ff} = convective heat transfer coefficient between the fin surface and the air stream, and
h_r = equivalent radiative heat transfer coefficient†

As in Section 5.3, we delete the bottom loss term from Eq. (5.35) and club it with the top loss term in Eq. (5.34). Equations (5.34) to (5.36) then simplify to

$$S = U_l(T_{pm} - T_a) + h_{fp}\left(1 + \frac{2L_f\,\phi_f\,h_{ff}}{W\,h_{fp}}\right)(T_{pm} - T_f) + h_r\,(T_{pm} - T_{bm}) \quad (5.37)$$

$$h_r(T_{pm} - T_{bm}) = h_{fb}\,(T_{bm} - T_f) \quad (5.38)$$

$$\frac{\dot{m}C_p}{L_2}\frac{dT_f}{dx} = h_{fp}\left(1 + \frac{2L_f\phi_f\,h_{ff}}{Wh_{fp}}\right)(T_{pm} - T_f) + h_{fb}\,(T_{bm} - T_f) \quad (5.39)$$

*The fin effectiveness is given by $\phi_f = (\tanh mL_f)/mL_f$ where $m = (2h_{ff}/k_f\,\delta_f)^{1/2}$

†The procedure for calculating h_r is not described here.

Equations (5.37) to (5.39) become the same as Eqs (5.5) to (5.7) if $h_{fp}\left(1+\frac{2L_f\,\phi_f\,h_{ff}}{W\,h_{fp}}\right)$ is replaced by h_{fp}. Thus, the solutions given in Eqs (5.16) and (5.17) for the useful heat gain rate and the collector heat removal factor are still valid. The expression for the collector efficiency factor F' is also given by Eq. (5.13) with the effective heat transfer coefficient h_e now being given by

$$h_e = h_{fp}\left(1+\frac{2L_f\,\phi_f\,h_{ff}}{Wh_{fp}}\right)+\frac{h_r\,h_{fb}}{h_r+h_{fb}} \tag{5.40}$$

Whillier has also derived an expression for the collector efficiency factor of a finned air heater by neglecting the radiative heat transfer but by considering an absorber plate effectiveness ϕ_p. The two expressions for F' can be shown to be identical if h_r is made equal to zero in one and ϕ_p is made equal to unity in the other.

In order to calculate the performance of a finned solar air heater, correlations are needed for determining the values of the convective heat transfer coefficient and the pressure drop. Thombre and Sukhatme [4] have conducted extensive experiments for this situation and suggested appropriate correlations. For the heat transfer coefficient, their data show that the well-known Dittus-Boelter equation is applicable if the clearance-to-spacing ratio $(L-L_f)/(W-\delta_f)$ is less than one or the spacing-to-fin height ratio $(W-\delta_f)/L_f$ is greater than one. These constraints are usually satisfied in practice. The Nusselt number is then given by

$$\text{Nu} = 0.023\,\text{Re}^{0.8}\,\text{Pr}^{0.4} \tag{5.41}$$

where the characteristic dimension used in the definitions of Nu and Re is the equivalent diameter d_e given by

$$d_e = \frac{4\times\text{Cross-sectional area of a fin channel}}{\text{Wetted perimeter of a fin channel}} = \frac{4(WL-\delta_f\,L_f)}{2(W+L_f)} \tag{5.42}$$

Properties are evaluated at the arithmetic mean of the air inlet and outlet temperature, and the values of h_{fp}, h_{ff} and h_{fb} are taken to be equal.

The following correlation is developed for calculating the pressure drop

$$f = M\,\text{Re}^{-m} \tag{5.43}$$

where

$$\begin{aligned} M &= 0.040[2.058-\{(L-L_f)/L_f\}^{0.313}] && \text{for} \quad \text{Re} < 15\,000 \\ &= 0.033[1.394-\{(L-L_f)/L_f\}^{0.408}] && \text{for} \quad \text{Re} > 15\,000 \\ m &= 0.075\,[3.40-\{(L-L_f)/L_f\}^{0.711}] && \text{for} \quad \text{Re} < 15\,000 \\ &= 0.138[1.435-\{(L-L_f)/L_f\}^{0.773}] && \text{for} \quad \text{Re} > 15\,000 \end{aligned}$$

f is the friction factor and the characteristic dimension in Re is again the equivalent diameter d_e.

Example 5.2

Consider again the solar air heater of Example 5.1 with continuous longitudinal fins fixed to the bottom side of the absorber plate. Assume the following additional data:

Centre-to-centre distance between fins	: 2.5 cm
Fin height	: 1.3 cm
Fin thickness	: 0.3 cm

Take the value of the radiative heat transfer coefficient to be the same as in Example 5.1. Calculate

1. air outlet temperature,
2. instantaneous efficiency, and
3. pressure drop.

Solution

We will first use Eq. (5.41) to calculate the values of the convective heat transfer coefficients. Assuming a mean fluid temperature of 60°C for the purpose of evaluating the properties of air, we have

$$\rho = 1.060 \text{ kg/m}^3$$

$$C_p = 1.005 \text{ kJ/kg-K}$$

$$k = 0.0290 \text{ W/m-K}$$

$$\nu = 18.97 \times 10^{-6} \text{ m}^2\text{/s}$$

$$\text{Pr} = 0.696$$

$$\text{Equivalent diameter } d_e = \frac{4(2.5 \times 1.5 - 0.3 \times 1.3)}{2(2.5 + 1.3)} = \frac{4 \times 3.36}{7.6} = 1.768 \text{ cm}$$

$$\text{Average air velocity} = \frac{200}{3600 \times 1.060 \times 40 \times 3.36 \times 10^{-4}} = 3.900 \text{ m/s}$$

$$\text{Re} = \frac{Vd_e}{\nu} = \frac{3.900 \times 1.768 \times 10^{-2}}{18.97 \times 10^{-6}} = 3635$$

Substituting into Eq. (5.41),

$$\text{Nu} = 0.023 \times 3635^{0.8} \times 0.696^{0.4} = 14.035$$

Thus, $$h_{fp} = h_{ff} = h_{fb} = 14.035 \times \frac{0.0290}{1.768 \times 10^{-2}} = 23.02 \text{ W/m}^2\text{-K.}$$

Next we calculate the fin effectiveness assuming that the fins are made of steel ($k = 50$ W/m-K).

$$mL_f = \left(\frac{2 \times 23.02}{50 \times 0.3 \times 10^{-2}}\right)^{1/2} \times 1.3 \times 10^{-2} = 0.2277$$

Therefore $\phi_f = \dfrac{\tanh mL_f}{mL_f} = \dfrac{0.2238}{0.2277} = 0.9830$

From Eq. (5.40), the effective heat transfer coefficient h_e

$$= 23.02\left(1+\frac{2\times1.3\times10^{-2}\times0.9830}{2.5\times10^{-2}}\right)+\left(\frac{7.24\times23.02}{7.24+23.02}\right)$$

$$= 52.05\ \text{W/m}^2\text{-K}$$

From this point in the calculation, the equations for a conventional solar air heater are applicable. From Eqs (5.13) and (5.17),

$$F' = \left(1+\frac{7.0}{52.05}\right)^{-1} = 0.8815$$

$$\frac{\dot{m}C_p}{U_l A_p} = \frac{200}{3600}\times\frac{1.005\times1000}{7.0\times1\times2} = 3.988$$

$$F_R = 3.988(1 - e^{-0.8815/3.988}) = 0.7909$$

$$q_u = 0.7909\times1\times2[950\times0.85 - 7.0(50-20)] = 945.1\ \text{W}$$

The air outlet temperature is obtained from the equation

$$\frac{200}{3600}\times1.005\times(T_{fo}-50) = \frac{945.1}{1000}$$

Hence $(T_{fo} - 50) = 16.93°\text{C}$ and $T_{fo} = 66.93°\text{C}$

Instantaneous efficiency $\eta_i = \dfrac{945.1}{950\times1.1\times2.1} = 0.431$

The pressure drop across the air heater is calculated from Eq. (5.43). We have

$$M = 0.040\left[2.058-\left(\frac{0.2}{1.3}\right)^{0.313}\right] = 0.06006$$

$$m = 0.075\left[3.40-\left(\frac{0.2}{1.3}\right)^{0.711}\right] = 0.2352$$

Therefore $f = 0.06006\times3635^{-0.2352} = 0.008734$

Pressure drop $= \dfrac{4\times0.008734\times1.060\times2\times3.900^2}{2\times0.01768} = 31.85\ \text{N/m}^2$

Comparing these results with those of Example 5.1, we see that the efficiency of the air heater has increased significantly from 36.8 per cent to 43.1 per cent, a gain of 6.3 per cent in absolute terms. However, the pressure drop has also increased by a factor of 2 from 15.78 N/m^2 to 31.85 N/m^2.

Some experimental studies [5] on finned solar air heaters are available. These also indicate a substantial improvement in efficiency. However, as seen in Example 5.2, the addition of fins introduces an extra pressure drop. Thus, there is an optimum spacing below which it does not pay to increase the number of fins because of the increased pressure drop.

Instead of fins, many investigators have enhanced the heat transfer by roughening the under side of the absorber plates. One way of doing this is to fix rows or ribs (1 or 2 mm in height) either across the air flow or at an angle to it. The rib may be continuous or broken. They serve as turbulence promoters and increase the heat transfer coefficient [6].

Two-pass Solar Air Heater

Satcunanathan and Deonarine [7] have suggested the use of a two-pass solar air heater in order to reduce the losses from the top. They constructed a unit in which the air was first passed between the covers of a two-glass cover heater and then under the absorber plate (Fig. 5.4 (a)). When operated as an open system with inlet air at ambient temperature, it was found that the outer glass cover temperature was lowered by 2° to 5°C and that it operated nearer the ambient temperature. As a result, the losses were reduced and the efficiency of the collector was measured to be 10 to 15 per cent higher than of a conventional heater.

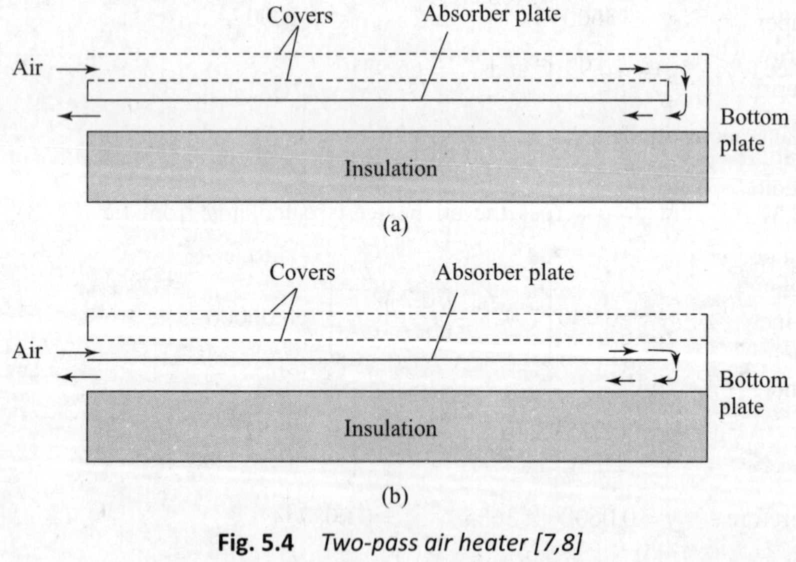

Fig. 5.4 *Two-pass air heater [7,8]*

Subsequently Wijeysundera *et al.* [8] have studied the two-pass concept in greater detail both analytically and experimentally. Two two-pass flow arrangements were considered. One arrangement was the same as the one studied by Satcunanathan and Deonarine (Fig. 5.4 (a)), while in the other, the inlet air flowed first above the absorber plate and then under it (Fig. 5.4 (b)). For open systems, with inlet air at ambient temperature, both the two-pass arrangements gave an efficiency about 10–15 per cent more than the conventional single pass arrangement over a wide range of operating conditions. However, for closed air recirculating systems, the two-pass arrangements yielded a better performance only up to a certain value of the difference between the air inlet temperature to the collector and the ambient temperature. With the arrangement shown in Fig. 5.4 (a), the two-pass design was found to be better than the single pass design up to an inlet air temperature difference of 20°C, while with the arrangement shown in Fig. 5.4 (b), the two-pass design was better up to an inlet air temperature difference of 50°C.

Karim and Hawlader [5] have reported the performance of single cover air heaters operated in the single and two-pass mode. Three types of absorber plates – flat, flat with fins and V-shaped – were tested [Section A-A, Fig. 5.1 (a)]. The performance of all collectors was examined over a wide range of operating conditions. It was found that two-pass operation improved the efficiency of all the three types of absorber plates tested and that the V-shaped absorber plate collector was the most efficient of the three types.

Ho *et al.* [9] have shown that if a part of the hot air leaving a two-pass air heater is mixed with the inlet air externally and recycled, the collector efficiency improves. By recycling part of the hot air, the temperature of the inlet air increases. Thus the temperature rise across the collector is likely to decrease. However, the enhancement of heat transfer to the air, due to increase in its velocity by recycling, compensates for the decrease of temperature rise. Consequently the efficiency improves. Ho *et al.* have validated their theoretical predictions by carrying out experiments on a solar air heater having two covers operated in the two-pass mode (Fig. 5.5). Theoretical calculations show that for a flow rate of 38.52 kg/h, solar radiation of 830 W/m^2, and air inlet as well as ambient temperature of 30°C, the collector efficiency increases from 0.404 to 0.518 for a recycle ratio of 0.5 and from 0.438 to 0.577 for a recycle ratio of 1. Ho *et al.* have also shown that the extra power consumption due to recycling is insignificant. Further, the collector efficiency can be enhanced by decreasing the width of the upper flow channel and increasing that of the lower one, for recycle ratios greater than 0.5. However, if the recycle ratio is less than 0.5, the width of the upper flow channel needs to be increased compared to that of the lower one.

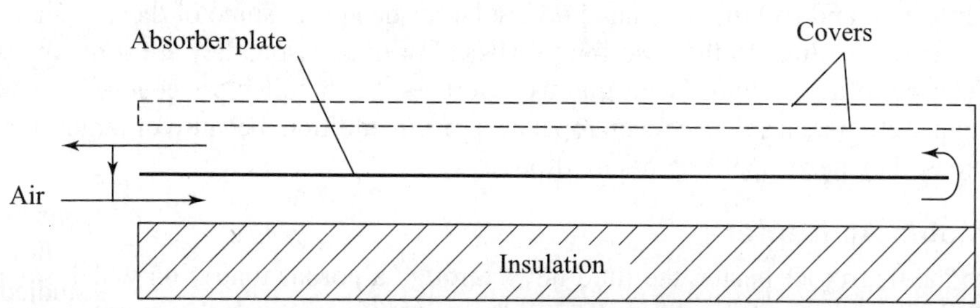

Fig. 5.5 *Two-pass air heater with recycle [9]*

Promising results have been reported by Sopian *et al.* [10] for a solar air heater operated in the two-pass mode and having a porous medium (steel wool) in the second channel (Fig. 5.6) Experiments were carried out in an indoor facility, where solar radiation as well as the width of the flow channels (both upper and lower) were varied. Sopian *et al.* concluded that the typical thermal efficiency of the two-pass air heater with a porous medium in the second channel is about 60 to 70 per cent and the value obtained is 10 to 20 per cent higher than that for the collector without a porous medium. Othman *et al.* [11] have reported field tests on six such solar collectors, arranged in an array of 2 banks each having 3 collectors in series. An outlet temperature of about 90°C was obtained at an average solar radiation intensity between 900 to 1000 W/m^2 for a flow rate of 0.0995 kg/s.

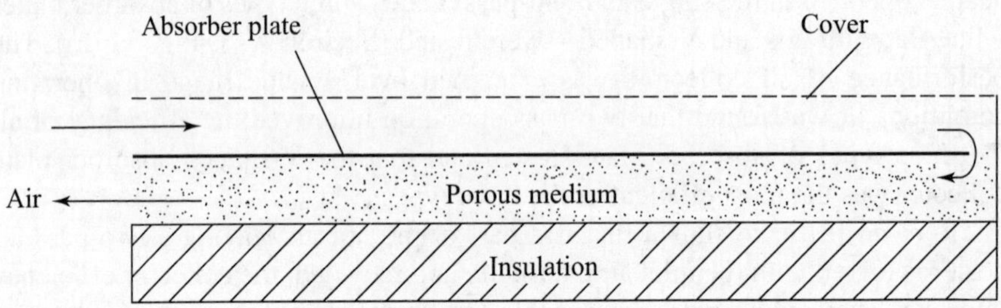

Fig. 5.6 *Two-pass air heater with porous medium [10]*

Dhiman and Singh [12] have analyzed the performance of a two-pass solar air heater with recycle (as in Fig. 5.5) in which a porous medium is placed either in the upper or lower channel. Thus they have tried to combine the advantages of recycling with the performance improvement obtained in the presence of a porous medium. The results obtained show that the use of a porous medium with recycle yields higher values of efficiency and that it is more beneficial to pack the porous medium in the upper channel.

5.4.2 Some Novel Designs

A number of novel designs of solar air heaters have also been suggested from time to time by many investigators. Some of these will now be briefly described. They are (i) the matrix air heater, (ii) the plastic air heater, (iii) the transpired air heater, and (iv) the evacuated tube solar air heater. In some of these designs, the air flows through the absorbing surface. For this reason, they are some times referred to as collectors with porous absorbers. Such collectors generally yield higher efficiencies than conventional designs. In addition, because of larger flow areas, they have smaller pressure drops.

Matrix Air Heater

In the matrix air heater, the fluid flows through a porous matrix on which solar radiation is directly incident. The radiation thus penetrates the matrix and is gradually absorbed. This is quite unlike a conventional non-porous absorber surface in which radiation is only absorbed at the surface. The inlet air is introduced at the top and is heated as it flows down through the matrix (Fig. 5.7). This flow direction is preferred to the other possible arrangement in which the air flows up through the matrix in the reverse manner. The reason for this is that in the first arrangement both the glass cover and the top surface of the matrix are in contact with the incoming air and are thus at the lowest possible temperatures. As a result, top losses are reduced. Matrix collectors provide large heat transfer area to volume ratios. Also, the heat transfer coefficients are relatively high due to increased turbulence of air flowing through the matrix. Consequently, the collector efficiencies are high.

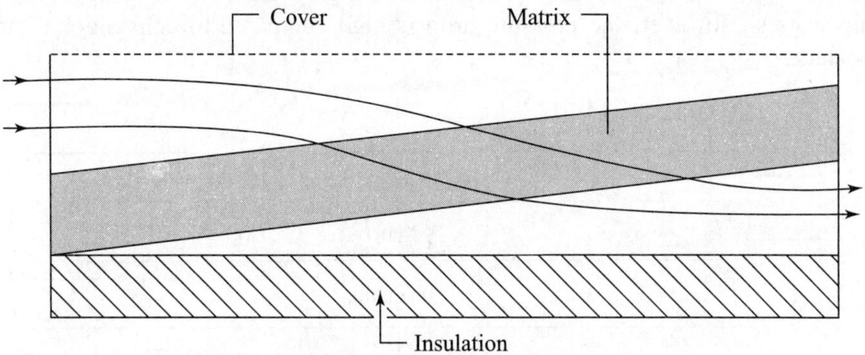

Fig. 5.7 *Matrix air heater*

The matrices used have been made by stacking wire screen meshes or slit-and-expanded metallic foils. Inexpensive materials like glass beads, crushed glass, wool and Raschig rings made of hard plastic have also been used. Hamid and Beckman [13] have studied the absorption characteristics and temperature profiles in matrices made by stacking copper wire screens. Heat transfer and pressure drop characteristics of such matrices are available in Kays and London [14]. Characteristics of a number of slit-and-expanded aluminium foil matrices have been obtained by Chiou *et al.* [15]. Based on their studies, they find a matrix depth of about 4 cm to be adequate. With an inlet air temperature of 21°C, it is estimated that an efficiency of 75 per cent would be obtained with the matrix air heater as opposed to a value of 58 per cent obtained with a conventional solar air heater. The pressure drop in the matrix is also very small. It is estimated to be only (1/200) times that in the passage of a conventional heater. However, it is to be noted that the above estimates require a uniform flow though the matrix. This is rather difficult to achieve.

Kolb *et al.* [16] have tested a matrix air heater with black oxidized or black galvanized fine-meshed copper wire-screens. They have used a matrix with spectrally selective radiative properties for reducing emissivity since layered fibrous materials have large values of emissivities due to multiple reflections [17].

Kolb *et al.* have reported that for a flow rate of 0.027 kg/m^2-s, an inlet temperature of 37°C and a channel gap of 20 mm (same for both upper and lower channels), the performance improves by 20 per cent compared to a two cover conventional air heater and 40 per cent compared to a single cover one. The pressure drop is found to be about 40 per cent less than that in a conventional air heater.

Mohamad [18] has reported an analysis of a novel type of solar air heater, which is operated in the two-pass mode and has a matrix in the second channel (Fig. 5.8). The collector has two glass covers. Air flows in the upper channel (between the two glass covers) and into the second channel (between the second glass cover and the back plate). Mohamad has reported that the thermal efficiency of this type of air heater is much higher than that of conventional air heaters. However, the pressure drop is high compared to conventional air heaters.

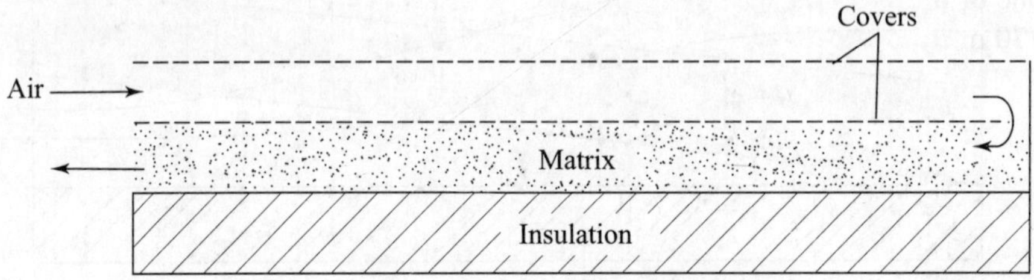

Fig. 5.8 *Two-pass air heater with matrix [18]*

Experimental investigations on an air heater with a wire mesh screen matrix have been reported by Varshney and Saini [19]. They have varied geometrical parameters like wire diameter, pitch and number of layers of the matrix (Table 5.1). Based on the experimental results, generalized correlations for heat transfer coefficient and friction factor were developed. Mittal and Varshney [20] have carried out theoretical investigations on air heaters with matrices having the geometrical parameters given in Table 5.1. They have found that the air heater with matrix no.6 yields the highest efficiency in the flow rate range 0.005 to 0.05 kg/s.

Table 5.1 Geometrical parameters of wire mesh screen matrices [20]

Matrix No.	Diameter of wire (mm)	Pitch (mm)	Number of layers	Bed porosity	Bed depth (mm)
1	0.360	2.72	14	0.958	25
2	0.450	2.08	10	0.939	25
3	0.590	2.23	10	0.902	25
4	0.795	3.19	9	0.887	25
5	0.795	3.19	7	0.905	25
6	0.795	3.19	5	0.937	25

Plastic Air Heater

A number of designs of solar air heaters fabricated from plastics have been reported. These are generally suitable for handling large volumes of air, which have to be heated only by a few degrees. A few of these designs will be described.

Bansal *et al.* [21] have developed all-plastic solar air heaters fabricated from flexible plastic sheets. These heaters attain their shape because of the pressure of the air blown through them. The absorber is a porous black textile of polyester. It is attached at the edges to two transparent sheets of polyvinyl chloride and is covered by them from both sides (Fig. 5.9). Ring holes are provided at the edges for fixing the stretched heaters on a stand. Three collectors were fabricated and tested, two having an area of about 10 m^2 and the third having an area of about 20 m^2. All the collectors were about 1 m wide. Different back insulations were tried, best results being obtained with 6 cm thick polyethylene. With this insulation, one of the 10 m^2 heaters gave an efficiency of 67.9 per cent with a flow rate of 770 m^3/h of ambient air and incident radiation of 759 W/m^2.

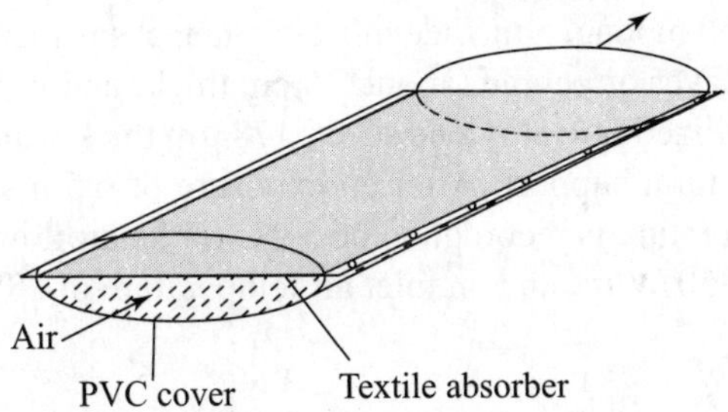

Fig. 5.9 *All-plastic air heater [21]*

Ahmad [22] has reported data on a simple, cylindrical plastic air heater made by wrapping low cost polyethylene sheeting around a flexible helical metal wire. Such a heater is suitable for agricultural drying operations requiring warm air. Each collector was 5 m long and 0.36 m in diameter. A black film was fixed along the lower circumference of the inner side of the polyethylene cover along the whole length. The collector was divided into two halves, each 2.5 m long. The first half had only polyethylene sheet, while the second half had another layer of plastic wrapping film with air bubbles to reduce the convection heat loss to the surroundings. Figure 5.10 shows a section of the assembly connected to a green house. Ahmad tested the collector under natural air flow conditions and placed it at an inclined angle with the opening facing the wind. He reported that a temperature difference of around 10°C was measured across the collector for a solar radiation of 981 W/m^2 and an ambient temperature of 23.5°C.

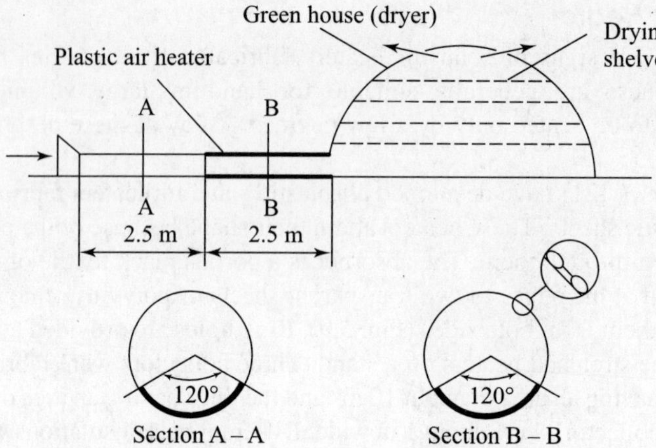

Fig. 5.10 *Solar drying system – Plastic air heater connected to green house [22]*

Flores-Irigollen *et al.* [23] have reported a theoretical and experimental investigation on a 50 m long, 5 m wide inflatable-tunnel solar air heater (Fig. 5.11). The absorber is a layer of pebble (about 5.5 cm thick) and is painted black. The cover is a UV stabilized polyethylene sheet, 178 μm thick which is held in place by a metallic structural support. After an exposure of 6.5 h since the morning, the outlet air temperature is recorded to be 53°C for an air flow rate of 4.36 kg/s, solar radiation of 850 W/m^2 and an inlet air temperature of 38°C.

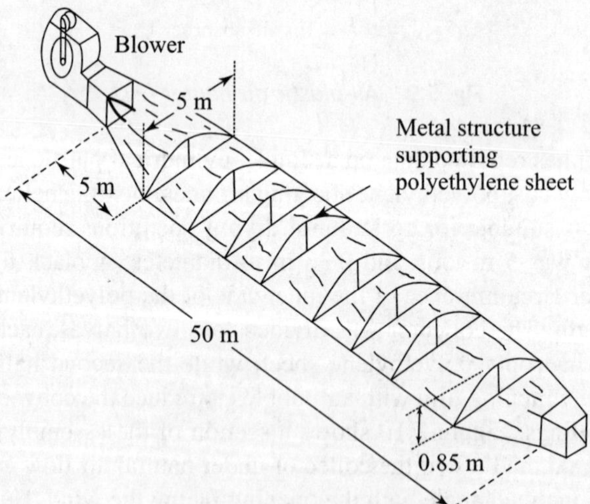

Fig. 5.11 *Inflatable-tunnel plastic solar air heater [23]*

Unglazed Transpired Solar Air Heater

Unglazed transpired collectors (UTC) have become popular in cold countries for space heating applications. They can also be used for crop drying. In these air heaters, a perforated dark-coloured metallic plate is exposed to solar radiation. Ambient air is drawn through the perforations in a once-through manner by a fan and gets heated as it flows through a plenum behind the plate (Fig. 5.12). Typically the porosity is about 0.5 per cent, the mass flow rate of the air passing through the collector is between 0.01 and 0.05 kg/s-m^2 and the plenum depth is in the range 5 to 30 cm.

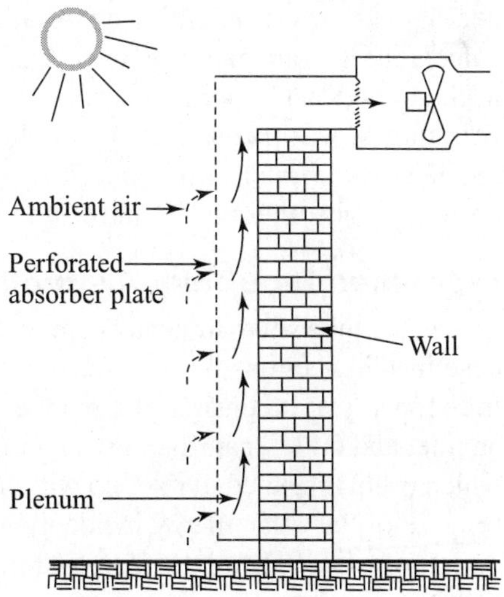

Fig. 5.12 *Unglazed transpired collector installed on a vertical wall*

For space heating purposes, the UTC is usually oriented vertically and fastened in front of a south-facing wall (in the northern hemisphere). The hot air moves up the plenum between the absorber plate and the wall of the building and then is distributed in the living space through appropriate ducts. Some UTC component designs and systems have been patented [24–26].

Gawlik *et al.* [27] have investigated the effect of using materials with low-conductivity for the perforated plates in UTCs. They have demonstrated that by using styrene or polyethylene, there is no significant drop in performance. On the other hand, there is benefit in the form of cost savings and reduced losses due to corrosion. Further, collectors can be made of flexible sheets, which can be rolled up and stored when not required.

Kutscher *et al.* [28] and Decker *et al.* [29] have carried out extensive studies on the heat-exchange effectiveness of a UTC and derived appropriate relations for isothermal and non-isothermal plates. Plate porosity, plate thickness, hole shapes and diameters play a significant role in the performance of a UTC. The flow distribution has been investigated by Gunnewiek *et al.* [30] in detail. For a particular UTC (height = 4.5 m, plenum depth = 0.15 m, plate hydraulic impedance = 2250 $\sqrt{m/s}$, net solar radiation absorbed = 650 W/m^2 and ratio of plate radiation heat loss coefficient to its heat exchanger effectiveness = 5.0 W/m^2-K), they have shown that the volumetric flow rate per unit area which is lowest at the top of the plenum should be greater than 0.0125 m/s to avoid reverse flow in the top part of the plate. Thus the fan should be sized such that the flow should be more than this limiting value. Further, the UTC should be designed such that the flow should be non-uniform with higher flow at the bottom compared to the top.

Recently, Gholampour and Ameri [31] have developed a simple model for analyzing the performance of a UTC. Energy balance equations have been written for the perforated absorber plate, the plenum chamber and the wall, and available correlations have been used for calculating the convective heat transfer coefficients on various surfaces. Available correlations have also been used for calculating the air pressure drop. The model developed was first tested for its accuracy against experimental data on performance of UTCs available in the literature. It was then used to study the effect of important parameters like hole diameter, hole pitch, suction velocity, plenum depth, amplitude and pitch of corrugated absorber plates, and wind velocity.

Evacuated Tube Solar Air Heater

Evacuated tube solar air heaters are the counterparts of evacuated tube collectors described in Section 4.14.1. A schematic view of a single pass air heater is shown in Fig. 5.13(a). It consists of a number of evacuated tube modules connected to an inlet and outlet manifold. Fresh air enters at the bottom in the inlet manifold, which is not insulated. It rises up within the evacuated tube modules and heats up because of the solar energy incident on the modules. The hot air then enters the outlet manifold. The outlet manifold is insulated and is connected to the suction side of a blower.

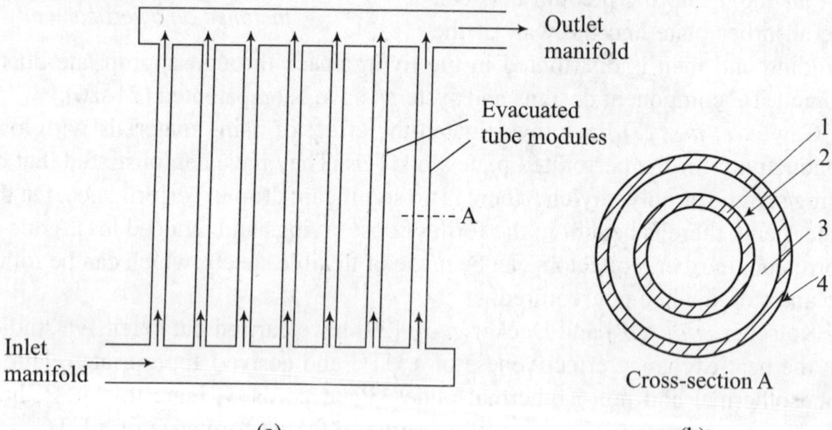

Fig. 5.13 *(a) Schematic of an evacuated tube solar air heater. (b) Cross section of a module. 1. Inner tube with air flowing inside. May be of metal or glass. 2. Selective surface on outside of inner tube. 3. Evacuated annular space. 4. Outer tube of glass.*

The cross-section of a module is shown in Fig. 5.13(b). It consists of two co-axial tubes. The inner tube may be of metal or glass and has a selective surface on the outside. The outer tube is made of glass, and the annular space between the two tubes is evacuated.

Paradis *et al* [32] have analyzed the performance of the solar air heater described above. The effect of parameters like mass flow rate, solar radiation, ambient temperature and wind speed on performance indicators like outlet air temperature, efficiency and pressure drop has been studied.

TESTING PROCEDURES 5.5

The standard procedures suggested for testing solar air heaters are similar in most respects to those described in Section 4.13 for testing liquid flat-plate collectors. A schematic diagram showing the essential features of the test set-up is shown in Fig. 5.14. It is a closed loop consisting of the solar air heater to be tested, a blower and an apparatus for reconditioning the air which ensures that the air enters the air heater at the desired temperature T_{fi}. Provision is made for measuring the same quantities specified earlier. Some precautions are, however, necessary. Since the fluid is air, it has to be ensured that it is well mixed at the exit from the air heater before its temperature is measured. The mixing is achieved

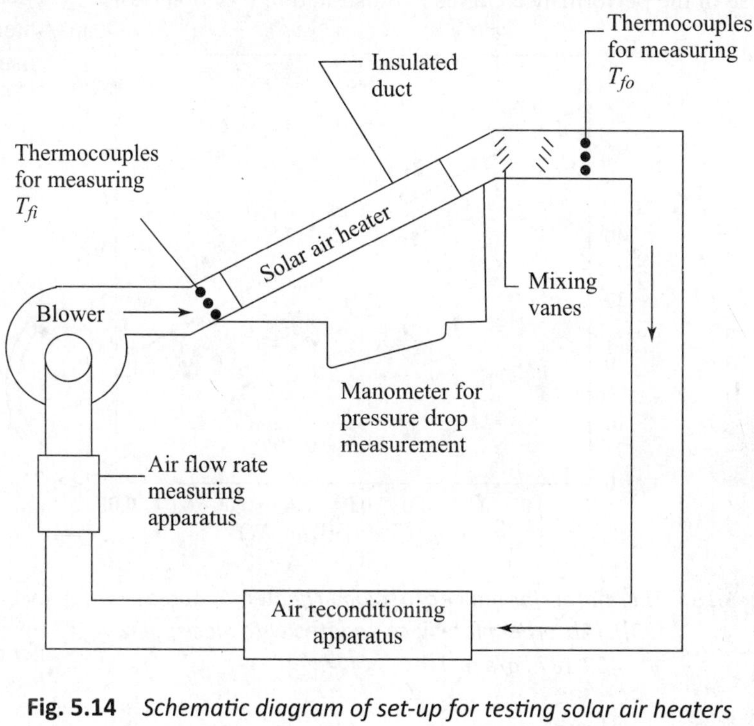

Fig. 5.14 *Schematic diagram of set-up for testing solar air heaters*

with the help of vanes. As an additional precaution, the temperature both at the inlet and exit of the air heater is measured at a number of locations across the duct cross-section. The closed loop is modified into an open loop if ambient air is used as the inlet air for the solar air heater. The modification essentially consists in removing the air reconditioning apparatus shown in Fig. 5.14 keeping the inlet of the solar air heater open to the atmosphere and connecting the outlet to the suction side of the blower.

Measurements are made under the conditions specified earlier and the results are also presented in the same manner. A typical set of results given by Parker *et al.* [33] is shown in Fig. 5.15. It will be seen that the scatter of the data is large. Currently, the international standard ISO 9806:2013 forms the basis for the testing of solar air heaters.

It has been mentioned in Section 4.13 that for conventional liquid flat-plate collectors, changes in the value of $\dot{m}$ do not appreciably affect the performance because of high values of the liquid side heat transfer coefficient h_f. A single test curve is, therefore, generally adequate for predicting the behaviour of such collectors. In the case of solar air heaters, however, changes in the values of $\dot{m}$ appreciably affect the performance because the value of the air side heat transfer coefficient (h_{fp}) is relatively low. For this reason, in order to obtain complete information on a solar air heater, it becomes necessary to conduct tests over a range of mass flow rates with each flow rate yielding its own efficiency curve. The use of the performance curves is illustrated in Example 5.3.

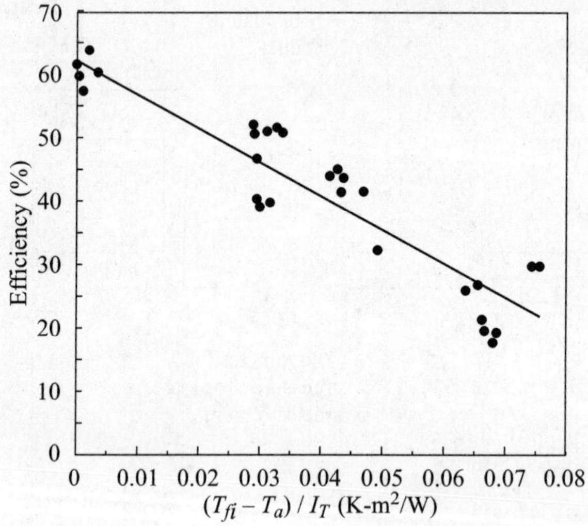

Fig. 5.15 *Thermal performance of a solar air heater (Adapted from Parker et al. [33]). (Absorber plate: v-corrugated with black paint, $A_c = 2\ m^2$ and $\dot{m} = 0.0187\ kg/s\text{-}m^2$)*

Example 5.3

The efficiency curves shown in Fig. 5.16 are obtained for a solar air heater ($L_1 = 1.2$ m, $L_2 = 0.9$ m) which is tested over a range of flow rates varying from 25 to 200 kg/h. Find the efficiency which would be obtained and the corresponding mass flow rate if the air heater is used under the following conditions:

Air inlet temperature	: 55°C
Air outlet temperature	: 75°C
Ambient temperature	: 27°C
Solar flux incident on collector face	: 950 W/m^2

Solution

For the given conditions, the x-axis parameter

$$(T_{fi} - T_a)/I_T = \frac{(55-27)}{0.950} = 29.5 \text{ K-m}^2\text{/W}$$

A trial-and-error procedure will be necessary in order to find the required values of η_i and $\dot{m}$.

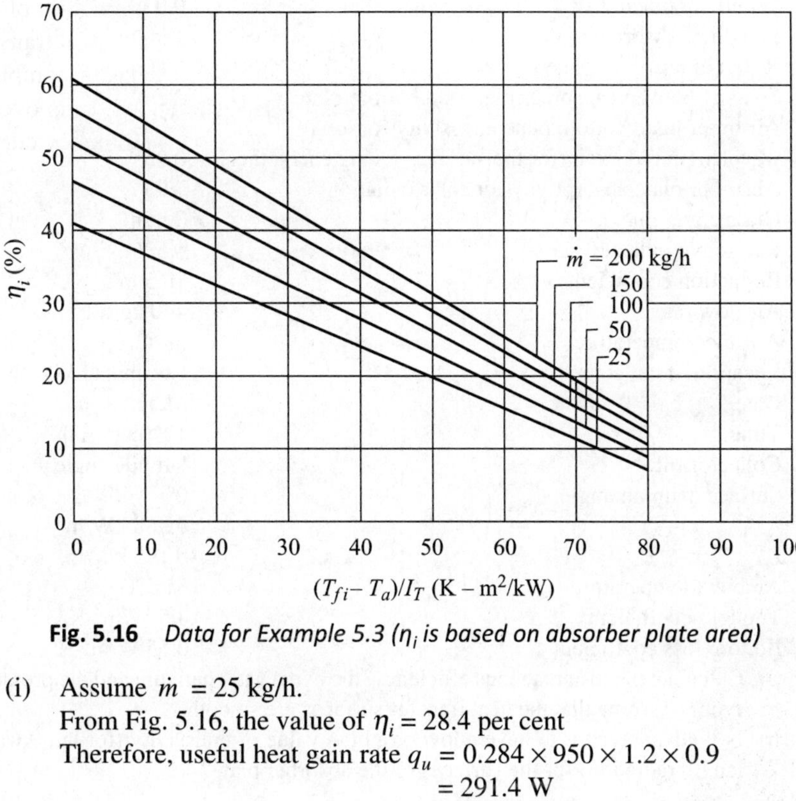

Fig. 5.16 *Data for Example 5.3 (η_i is based on absorber plate area)*

(i) Assume $\dot{m}$ = 25 kg/h.
From Fig. 5.16, the value of η_i = 28.4 per cent
Therefore, useful heat gain rate $q_u = 0.284 \times 950 \times 1.2 \times 0.9$
$= 291.4$ W

$$\dot{m} = \frac{q_u}{C_p(T_{fo} - T_{fi})} = \frac{291.4 \times 3600}{1.007 \times (75-55) \times 1000} = 52.1 \text{ kg/h}$$

(ii) Since the value of $\dot{m}$ calculated from the useful heat gain rate does not match the assumed value, we assume $\dot{m}$ = 60 kg/h. This yields

η_i = 33.6 per cent $\quad q_u = 344.7$ W and $\dot{m}$ = 61.6 kg/h

(iii) Assume $\dot{m}$ = 61.6 kg/h.
This yields η_i = 33.7 per cent

q_u = 345.8 W and $\dot{m}$ = 61.8 kg/h

We accept these values as the solution to the problem.

PROBLEMS

5.1 The following data is given for a conventional solar air heater with one glass cover:

Length of collector	2.0 m
Width of collector	0.9 m
Length of absorber plate	1.90 m
Width of absorber plate	0.80 m
Spacing between absorber plate and bottom plate	2 cm
Absorber plate/bottom plate emissivity for long wavelength radiation (for the surfaces facing each other)	0.93
Absorber plate absorptivity for solar radiation	0.93
Glass cover thickness	3 mm
Refractive index of glass	1.526
Extinction coefficient	10.5 m^{-1}
Air flow rate	440 kg/h
Air inlet temperature	48°C
Location of collector	Chennai (13.00° N)
Date	March 3
Time	1230 h(LAT)
Collector tilt	latitude angle
Surface azimuth angle	0°
I_g	0.968 kW/m^2
I_d	0.195 kW/m^2
Ambient temperature	31.5°C
Top loss coefficient	4.5 W/m^2-K
Bottom loss coefficient	0.55 W/m^2-K

(i) Calculate the instantaneous efficiency, the exit air temperature and the pressure drop. (Assume that the heat transfer surfaces are smooth.)

(ii) Is it advantageous to have a low or a high value of emissivity (for large wave length radiation) for the surfaces of the absorber plate?

5.2 How do the results obtained in Problem 5.1 change if the heat transfer surfaces in the air duct are roughened? Assume that because of the roughening, the convective heat transfer coefficient increases by 40 per cent, while the friction factor is doubled.

5.3 How do the results obtained in Problem 5.1 change if longitudinal steel fins (k = 50 W/m-K) are fixed to the bottom side of the absorber plate? Take W = 2 cm, L_f = 1.8 cm, and δ_f = 1 mm. Assume for the sake of simplicity that the value of the radiative heat transfer coefficient does not change because of the presence of the fins.

5.4 Repeat the calculations of Example 5.1 assuming that the upper surface of the absorber plate has a selective coating and that the value of U_t = 4.0 W/m^2-K. Compare the values of q_u and η_i with those obtained in Example 5.1.

5.5 Repeat the calculations of Example 5.2 assuming that the upper surface of the absorber plate has a selective coating and that the value of U_t = 4.0 W/m^2-K.

Compare the values of air outlet temperature, instantaneous efficiency and pressure drop with those obtained in Example 5.2.

5.6 Calculate the performance of a solar air heater of the type shown in Fig. 5.1 (b). Use the data given in Example 5.1 in the text with the following changes/additions:

Spacing between absorber plate and cover	:	1.5 cm
Top loss coefficient	:	12.7 W/m^2-K
ε_c	:	0.88

Data regarding the spacing between the absorber plate and bottom plate, and the emissivity of the bottom plate may be deleted as it is not applicable in this case.

5.7 Derive the governing energy balance equations for the solar air heater shown in Fig. 5.1 (c) in which the air flows between the cover and the absorber plate as well as between the absorber plate and the bottom plate.

5.8 A solar air heater is used for heating ambient air in a particular application. The characteristic parameters of the air heater are $F_R(\tau\alpha)_{av} = 0.63$, $F_R U_l = 6.2$ W/m^2-K. If the solar flux incident on the plane of the collector is 705 W/m^2, calculate the useful heat gain rate.

5.9 The value of $F_R U_l$ for a solar air heater is 6.8 W/m^2-K. In an experiment, the blower is turned off and the plate temperature is measured to be 105°C. The blower is then started with the inlet air at 29°C. What is the useful heat gain rate from the collector if the solar radiation, the ambient temperature and the characteristic parameters of the air heater do not change?

5.10 The following observations were recorded in experiments on a single pass solar air heater (Fig. 5.1a) and on a double pass solar air heater with a porous medium (Fig. 5.6):

Inlet air temperature	26°C
Outlet air temperature	40.0°C (single pass air heater) 47.0°C (double pass with porous medium)
Air flow rate	0.033 kg/s
Collector length	1.8 m
Collector width	0.7 m
Location	Jodhpur (26°18′ N, 73° 01′ E)
Day	March 16
Time	1030 h (IST)
Tilt of the collector	Latitude of the place
Azimuth of the collector	0°
Global solar radiation on horizontal surface	640 W/m^2
Diffuse solar radiation on horizontal surface	160 W/m^2
Reflectivity of the surrounding surface	0.2

Calculate the values of η_i for the above two cases.

When the porous medium is removed, the outlet air temperature in the double pass heater is recorded to be 44.0°C. Calculate the value of η_i for this situation also. Comment on the values obtained.

5.11 The temperature rise (ΔT) of air through a vertical south-facing unglazed transpired collector (UTC) is found to satisfy the following empirical relation:

$$\Delta T = 0.03 I_T + 3.0$$

for an air flow rate of 36 m^3/h-m^2 of UTC. I_T is the total solar radiation incident on UTC in W/m^2; ΔT is in °C.

Assuming this relation to be valid, calculate the efficiency of a vertical south-facing UTC for the following data:

Location	New Delhi (28°35′ N, 77°12′ E)
Date	December 10
Time	1100 h (1ST)
Air flow rate	36 m^3/h-m^2 of UTC
Global solar radiation on horizontal surface	543 W/m^2
Diffuse solar radiation on horizontal surface	144 W/m^2
Reflectivity of the surrounding surface	0.2

REFERENCES

1. Whillier, A. 1963. Black-painted solar air heaters of conventional design. *Solar Energy,* 8: 31.
2. Kays, W.M. 1966. *Convective Heat and Mass Transfer,* McGraw-Hill Book Co., New York.
3. Malik, M.A.S. and Buelow, F.H. 1975. Hydrodynamic and heat transfer characteristics of a heated air duct. *Heliotechnique and Development,* 2:3.
4. Thombre, S.B. and Sukhatme, S.P. 1995. Turbulent flow heat transfer and friction factor characteristics of shrouded fin arrays with uninterrupted fins. *Experimental Thermal and Fluid Science,* 10: 388.
5. Karim, M.A. and Hawlader, M.N.A. 2006. Performance investigation of flat-plate, v-corrugated and finned air collectors. *Energy,* 31: 452.
6. Maithani, R. and Saini, J.S. 2016. Heat transfer and friction factor correlations for a solar air heater duct roughened artificially with v-ribs with symmetrical gaps. *Experimental Thermal and Fluid Science*, 70: 220.
7. Satcunanathan, S. and Deonarine, S. 1973. A two-pass solar air heater. *Solar Energy,* 15: 41.
8. Wijeysundera, N.E, Ah, L.L. and Tjioe, L.E. 1982. Thermal performance study of two-pass solar air heaters. *Solar Energy,* 28: 363.
9. Ho, C.D., Yeh, H.M. and Wang, R.C. 2005. Heat-transfer enhancement in double-pass flat-plate solar air heaters with recycle. *Energy,* 30: 2796.
10. Sopian, K., Supranto, Daud, W.R.W., Othman, M.Y. and Yatim, B. 1999. Thermal performance of the double-pass solar collector with and without porous media. *Renewable Energy,* 18: 557.
11. Othman, M.Y.H., Sopian, K., Yatim, B. and Daud, W.R.W. 2006. Development of advanced solar assisted drying systems. *Renewable Energy,* 31: 703.
12. Dhiman, P. and Singh, S. 2015. Thermal and thermo-hydraulic performance investigation of double-pass packed bed solar air heaters under external recycle. *International Journal of Sustainable Energy*, DOI: 10.1080/14786451.2015.1017496.
13. Hamid, Y.H. and Beckman, W.A. 1971. Performance of air-cooled radiatively heated screen matrices. *Journal of Engg. for Power, Trans. ASME,* 93: 221.

14. Kays, W.M. and London, A.L. 1964. *Compact Heat Exchangers,* McGraw-Hill Book Co., New York.

15. Chiou, J.P., El-Wakil, M.M. and Duffie, J.A. 1965. A slit-and-expanded aluminium-foil matrix solar collector. *Solar Energy,* 9: 73.

16. Kolb, A., Winter, E.R.F. and Viskanta, R. 1999. Experimental studies on a solar air collector with metal matrix absorber. *Solar Energy,* 65: 91.

17. Golombok, M. and Shirvill, L.C. 1988. Emissivity of layered fibrous materails. *Applied Optics,* 27: 3921.

18. Mohamad, A.A. 1997. High efficiency solar air heater. *Solar Energy,* 60: 71.

19. Varshney, L. and Saini, J.S. 1998. Heat transfer and friction factor correlations for rectangular solar air heater duct packed with wire mesh screen matrices. *Solar Energy,* 62: 255.

20. Mittal, M.K. and Varshney, L. 2006. Optimal thermohydraulic performance of a wire mesh packed solar air heater. *Solar Energy,* 80: 1112.

21. Bansal, N.K., Uhlemann, R. and Boettcher, A. 1982. Plastic solar air heaters of a novel design–testing and performance. *Report No. 26, Kernforschungs anlage Julich,* Germany.

22. Ahmad, N.T. 2001. Agricultural solar air collector made from low-cost plastic packing film. *Renewable Energy,* 23: 663

23. Flores-Irigollen, A., Fernandez, J.L., Rubio-Cerda, E. and Poujol, F.T. 2004. Heat transfer dynamics in an inflatable-tunnel solar air heater. *Renewable Energy,* 29: 1367.

24. Wienecke, F. 1981. Solardach Absorber, *Patent No. 29 29219,* Germany.

25. Hollick, J.C. and Peter, W. 1990. Method and apparatus for preheating ventilation air for building, *Patent No. 4934338,* USA.

26. Christensen, C.B., Kutscher, C.F. and Gawlik, K.M. 1997. Unglazed transpired solar collector having a low-thermal conductance absorber, *Patent No. 5692491,* USA.

27. Gawlik, K., Christensen, C. and Kutscher, C. 2005. A numerical and experimental investigation of low-conductivity unglazed, transpired solar air heaters. *Journal of Solar Energy Engineering, Trans. ASME,* 127: 153.

28. Kutscher, C., Christensen, C. and Barker, G. 1991. Unglazed transpired solar collectors: an analytical model and test results. *Proc. of ISES Solar World Congress,* Elsevier Science, Vol. 2, part I: 1245.

29. Decker, G.W.E.V., Hollands, K.G.T. and Brunger, A.P. 2001. Heat-exchange relations for unglazed transpired solar collectors with circular holes on a square or triangular pitch. *Solar Energy,* 71:33.

30. Gunnewiek, L.H., Brundrett, E. and Hollands, K.G.T. 1996. Flow distribution in unglazed transpired plate solar air heaters of large area. *Solar Energy,* 58: 227.

31. Gholampour, M. and Ameri, M. 2014. Design considerations of unglazed transpired collector: energetic and exergetic studies. *Journal of Solar Energy Engg., Trans. ASME.* 136: 031004-1.

32. Paradis, P., Rousse, D.R., Halle, S., Lamarche, L. and Quesada, G. 2015. Thermal modeling of evacuated tube solar air collectors. *Solar Energy*, 115: 708.

33. Parker, B.F., Lindley, M.R., Colliver, D.G. and Murphy, W.E. 1993. Thermal performance of three solar air heaters. *Solar Energy*, 51: 467.

Concentrating Collectors

CHAPTER 6

INTRODUCTION 6.1

In Chapters 4 and 5, we have considered flat-plate collectors for heating liquids and gases to temperatures up to and around 100°C. We now take up the description and analysis of some types of concentrating collectors. These are needed when higher temperatures are required. Typical thermal applications requiring the use of concentrators are medium or high temperature energy conversion cycles and numerous systems for supplying industrial process heat at intermediate temperatures from 100° to 400°C or at high temperatures above 400°C.

Brief descriptions of a few concentrating collectors have been given in Section 2.2. We begin this chapter by mentioning briefly the characteristics associated with concentrating collectors (Section 6.2). After this, various terms are defined and typical collector geometries described. Flat-plate collectors with reflectors are considered in Section 6.3, and the cylindrical parabolic collector in Section 6.4. The tracking modes adopted with it are listed and compared, and a performance analysis of the collector is given. The linear Fresnel reflecting collector is considered in detail in Section 6.5. This is followed by an analysis of the compound parabolic collector in Section 6.6. The chapter concludes with descriptions of the paraboloid dish collector in Section 6.7 and the central receiver collector in Section 6.8.

DESCRIPTION, CLASSIFICATION AND TYPES 6.2

6.2.1 General Characteristics

Concentration of solar radiation is achieved using a reflecting arrangement of mirrors or a refracting arrangement of lenses. The optical system directs the solar radiation on to an absorber of smaller area which is usually surrounded by a transparent cover. Because of the optical system, certain losses (in addition to those which occur while the radiation is transmitted through the cover) are introduced. These include reflection or absorption losses in the mirrors or lenses, and losses due to geometrical imperfections in the optical system. The combined effect of all such losses is indicated through the introduction of a term called *optical efficiency.* The introduction of more optical losses is compensated for by the fact that the flux incident on the absorber surface is concentrated on a smaller area. As a result, the thermal loss terms do not dominate to the same extent as in a flat-plate collector and the collection efficiency is usually higher.

It has been noted earlier that some of the attractive features of a flat-plate collector are simplicity of design and ease of maintenance. The same cannot be said of a concentrating collector. Because of the presence of an optical system, a concentrating collector usually has to follow or 'track' the sun so that the beam radiation is directed on to the absorber surface. The method of tracking adopted and the precision with which it has to be done varies considerably. In collectors giving a low degree of concentration, it is often adequate to make one or two adjustments of the collector orientation every day. These can be made manually. On the other hand, with collectors giving a high degree of concentration, it is necessary to make continuous adjustments of the collector orientation. The need for some form of tracking introduces a certain amount of complexity in the design. Maintenance requirements are also increased. All these factors add to the cost. An added disadvantage is the fact that much of the diffuse radiation is lost because it does not get focussed.

In the last few years, significant advances have been made in the development of concentrating collectors and a number of types have been commercialised abroad.

6.2.2 Definitions

In order to be consistent in the use of terms, we will use the phrase 'concentrating collector' to denote the whole system. The term 'concentrator' will be used only for the optical subsystem which directs the solar radiation on to the absorber, while the term 'receiver' will normally be used to denote the subsystem consisting of the absorber, its cover and other accessories.

We will now define four terms: aperture, area concentration ratio, intercept factor and acceptance angle.

The *aperture* (W) is the plane opening of the concentrator through which the solar radiation passes. For a cylindrical or linear concentrator, it is characterised by the width, while for a surface of revolution, it is characterised by the diameter of the opening.

The *area concentration ratio* (C) is the ratio of the effective area of the aperture to the surface area of the absorber. Values of the concentration ratio vary from unity (which is the limiting case for a flat-plate collector) to a few thousand for a paraboloid dish. This quantity is also referred to as the *geometric concentration ratio* or simply *concentration ratio.*

The *intercept factor* (γ_{int}) is the fraction of the radiation, which is reflected or refracted from the concentrator and is incident on the absorber. The value of the intercept factor is generally close to unity.

The *acceptance angle* ($2\theta_a$) is the angle over which beam radiation may deviate from the normal to the aperture plane and yet reach the absorber. Collectors with large acceptance angles require only occasional adjustments, while collectors with small acceptance angles have to be adjusted continuously.

6.2.3 Methods of Classification

Concentrating collectors are of various types and can be classified on the basis of different criteria. They may be classified as reflecting type utilizing mirrors or of the refracting type utilizing Fresnel lenses. The reflecting surfaces used may be parabolic, spherical or flat. They may be continuous or segmented. Classification is also possible from the point of view of the formation of the image, the concentrator being either imaging or non-imaging. Further, the imaging concentrator may focus on a line or at a point.

The concentration ratio is also used as a measure for classifying concentrating collectors. Since this ratio approximately determines the operating temperature, this method of classification is equivalent to classifying the collector by its operating temperature range.

A final possibility is to describe concentrating collectors by the type of tracking adopted. Depending upon the acceptance angle, the tracking may be intermittent (one adjustment daily or every few days) or continuous. Further, the tracking may be required about one axis or two axes.

6.2.4 Types of Concentrating Collectors

A number of concentrating collector geometries are shown in Fig. 6.1.

The first type shown in Fig. 6.1(a) is a flat-plate-collector with adjustable mirrors at the edges to reflect radiation on to the absorber plate. It is simple in design, has a concentration ratio a little above unity and is useful for giving temperatures about 20° or 30°C higher than those obtained with a flat-plate collector alone. It is discussed further in Section 6.3.

A *compound parabolic concentrating collector* (CPC) is shown in Fig. 6.1 (b). The concentrator consists of curved segments which are parts of two parabolas. Like the first type, this collector is also non-imaging. The concentration ratio is moderate and generally ranges from 3 to 10. The main advantage of the compound parabolic collector is that it has a high acceptance angle and consequently requires only occasional tracking. In addition, its concentration ratio is equal to the maximum value possible for a given acceptance angle. The CPC is considered in Section 6.6.

The next type of collector [Fig. 6.1 (c)] is a *cylindrical parabolic collector* in which the image is formed on the focal axis of the parabola. Many commercial versions of this type are now available. For this reason, it is described and analysed in detail in Section 6.4.

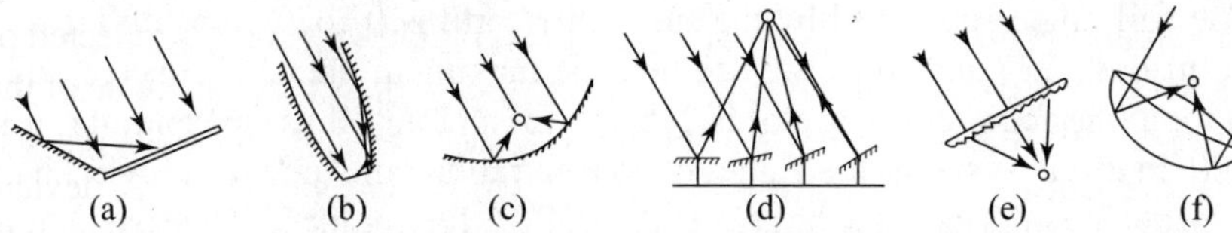

Fig. 6.1 *Types of concentrating collectors: (a) Flat-plate collector with plane reflectors, (b) Compound parabolic collector, (c) Cylindrical parabolic collector, (d) LFR collector, (e) Fresnel lens concentrating collector, (f) Paraboloid dish collector*

Like the cylindrical parabolic collector, the linear Fresnel reflector collector is essentially a line-focussing collector. The concentrator is made up of a number of long mirror strips held side-by-side in a horizontal plane, while the receiver's absorber tube is held parallel to and above the mirror strips. The mirror strips can rotate about a horizontal axis and thereby track the sun. The LFR collector has been commercialized extensively in the last ten to fifteen years. It is discussed in detail in Section 6.5.

Concentration is also achieved by using lenses. The most commonly used device is the *Fresnel lens* shown in Fig. 6.1 (e). The one shown in the figure is a thin sheet, flat on one side and with fine longitudinal grooves on the other. The angles of these grooves are such that radiation is brought to a line focus. The lens is usually made of extruded acrylic plastic sheets. Line focussing collectors like the ones shown in Fig. 6.1 (c), (d), (e) usually have concentration ratios between 10 and 80 and yield temperatures between 150° to 400°C.

In order to achieve higher concentration ratios and temperatures, it becomes necessary to have point focussing rather than line focussing. The point focussing *paraboloid dish collector* has been mentioned earlier in Chapter 2. A sketch is shown in Fig. 6.1(f). Such collectors can have concentration ratios ranging from 100 to a few thousand and have yielded temperatures up to 2000°C. However, from the point of view of the mechanical design, there are limitations to the size of the concentrator and hence, the amount of energy which can be collected by one dish. Commercial versions have been built with dish diameters up to 17 m. In order to collect larger amounts of energy at one point, the *central receiver concept* (see Fig. 2. 24) has been adopted. In this case, beam radiation is reflected from a number of independently controlled mirrors called heliostats to a central receiver located at the top of a tower.

Rabl [1] has shown that for a given acceptance angle ($2\theta_a$), the maximum possible concentration ratio of a two-dimensional (line-focus) concentrator is

$$C_{m,2D} = \frac{1}{\sin \theta_a} \tag{6.1}$$

For a three-dimensional (point-focus) concentrator, he has shown that

$$C_{m,3D} = \frac{1}{\sin^2 \theta_a} \tag{6.2}$$

The half-angle subtended by the sun at the earth is 0.267°. Substituting this value in Eqs (6.1) and (6.2), we see that the maximum value of concentration ratio for a line-focus concentrator is 215 and for a point-focus concentrator, it is 46 000. In actual systems, the values of concentration ratio are much lower since the acceptance angle needs to be greater than 0.267° for a number of reasons. These include tracking errors, imperfections in the reflecting- or refracting-component of the concentrator, mechanical misalignment between the concentrator and the receiver, etc.

6.2.5 Thermal Analysis of Concentrating Collectors

We will now discuss the thermal analysis of a concentrating collector. Like a flat-plate collector, an energy balance on the absorber yields the following equation under steady-state conditions

$$q_u = A_a S - q_l \tag{6.3}$$

where q_u = rate of useful heat gain,
A_a = effective area of the aperture of the concentrator,
S = solar beam radiation per unit effective aperture area absorbed in the absorber, and
q_l = rate of heat loss from the absorber.

Equation (6.3) is written under the assumption that the contribution of the diffuse component of solar radiation is negligible.

Similar to Eq. (4.17), q_l can be written in terms of an overall loss coefficient defined by the equation

$$q_l = U_l A_p (T_{pm} - T_a) \tag{6.4}$$

where U_l = overall loss coefficient,
A_p = area of the absorber surface,
T_{pm} = average temperature of the absorber surface, and
T_a = temperature of the surrounding air.

We combine Eqs (6.3) and (6.4) to obtain

$$q_u = A_a \left[S - \frac{U_l}{C}(T_{pm} - T_a) \right] \tag{6.5}$$

where $C = (A_a / A_p)$ is the concentration ratio.

6.3 FLAT-PLATE COLLECTOR WITH PLANE REFLECTORS

A flat-plate collector with plane reflectors is a simple non-imaging concentrating collector and represents an effective means of getting slightly higher temperatures than are obtainable with a flat-plate collector alone. With a single collector, it is possible to use four reflectors all around. On the other hand, with an array of flat-plate collectors, it is possible to have only two arrays of reflectors, one of which faces north and the other south (Fig. 6.2). The reflectors used may reflect the radiation in a specular or diffuse manner. The concentration ratios obtained are low and normally range from one to four. Operating temperatures up to 130° to 140°C can be attained. An advantage associated with this type of concentrating collector is that the diffuse component of the incoming solar radiation is not entirely wasted.

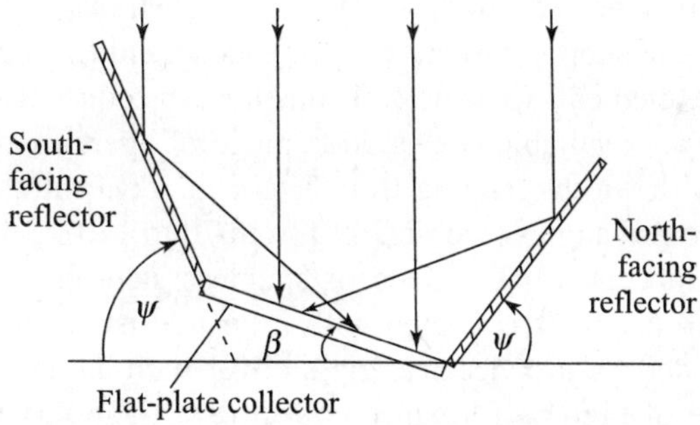

Fig. 6.2 *Flate-plate collector with reflectors*

With an array of flat-plate collectors, the usual practice is to use an array of north-facing reflectors only, since these are more convenient to handle and adjust than south-facing reflectors. The inclination of the reflectors is usually adjusted once every few days. For the case of a north-facing specular reflector array whose dimensions are equal to those of the flat-plate collector array, it can be shown (Chapter 3, Problem 3.7) that the inclination ψ of the reflectors should be

$$\psi = (\pi - \beta - 2\phi + 2\delta)/3 \tag{6.6}$$

where β = slope of the collectors. Equation (6.6) is derived under the assumption that the reflector mirrors are adjusted in such a way that the sun's rays striking the top edge of the mirrors at 1200 h (LAT) are reflected to the top edge of the collectors. If we assume the location to be Mumbai and take the collector slope to be equal to the latitude, the value of ψ is found to vary from 25° to 57° as the declination varies from its minimum value of –23.45° to its maximum value of +23.45°.

At times other than noon, only a certain fraction of the radiation falling on the reflectors will be reflected usefully on to the collectors with the remaining falling on the sides of the collector. McDaniels *et al.* [2], and Grassie and Sheridan [3] have given results for determining this fraction for the two cases of specular and diffuse reflection. It has also been shown that specular reflectors are more effective in augmenting the radiation than diffuse reflectors.

CYLINDRICAL PARABOLIC COLLECTOR 6.4

6.4.1 Description

The cylindrical parabolic collector is also referred to as a parabolic trough or a linear parabolic collector (Plate 6.1). The basic elements making up a conventional collector are (i) the absorber tube located at the focal axis through which the liquid to be heated flows, (ii) the concentric transparent cover, (iii) the reflector, and (iv) the support structure. Elements (i) and (ii) together constitute the receiver, while elements (iii) and (iv) constitute the concentrator.

The collectors are available over a wide range of aperture areas from about 1 to 60 m^2 and with widths ranging from 1 to 6 m. Concentration ratios range from 10 to 80, and rim angles (see Fig. 6.4) from 70 to 120°.

The absorber tube is usually made of stainless steel or copper and has a diameter of 2.5 to 5 cm. It is coated with a heat resistant black paint and is generally surrounded by a concentric glass cover with an annular gap of 1 or 2 cm. In the case of high-performance collectors, the absorber tube is coated with a selective surface and the space between the tube and the glass cover is evacuated. In some small collectors, the concentric cover is replaced by a glass or plastic sheet covering the whole aperture area of the collector. Such an arrangement helps in protecting the reflecting surface from the weather.

The liquid heated in the collector depends upon the temperature required. Usually organic heat transfer liquids (referred to as thermic fluids) are used. Because of their low thermal conductivities, these liquids yield low heat transfer coefficients. Augmentative devices in the form of twisted tapes or central plugs (which create annular passages) are, therefore, used to increase the value of the heat transfer coefficient.

The reflecting surface is generally curved back silvered glass. It is fixed on a light-weight structure usually made of aluminium sections. The proper design of this supporting structure and of the system for its movement is important, since it influences the shape and orientation of the reflecting surface. Some of the factors to be considered in designing the structure are that it should not distort significantly due to its own weight and that it should be able to withstand wind loads.

Compared to flat-plate collectors, there are very few manufacturers of concentrating collectors all over the world. The volume of production is also low. In India, many experimental collectors have been built and tested. However, commercial manufacture has not yet begun.

6.4.2 Orientation and Tracking Modes

A cylindrical parabolic collector is oriented with its focal axis pointed either in the east-west (E-W) or the north-south (N-S) direction. In the east-west orientation, the focal axis is horizontal, while in the north-south orientation, the focal axis may be horizontal or inclined. The various tracking modes, which can be adopted, are as follows:

Plate 6.1 *Cylindrical parabolic collector. Length 3.6 m, Aperture 1.25 m, Concentration ratio 9.6, Reflector material: Electropolished anodized aluminium sheet. Location: Indian Institute of Technology Bombay.*

❑ *Mode I* The focal axis is E-W and horizontal. The collector is rotated about a horizontal E-W axis and adjusted once every day so that the solar beam is normal to the collector aperture plane at solar noon on that day.

In this mode, the aperture plane is an imaginary surface with either $\gamma = 0°$ or $\gamma = 180°$. The case of $\gamma = 0°$ occurs when $(\phi - \delta) > 0$, while the case of $\gamma = 180°$ occurs when $(\phi - \delta) < 0$. In order to find the slope β of the aperture plane, we substitute the condition at solar noon, viz. $\omega = 0°$, $\theta = 0°$ in Eq. (3.5). This yields,

$$\beta = (\phi - \delta) \quad \text{for } \gamma = 0° \tag{6.7a}$$

and

$$\beta = (\delta - \phi) \quad \text{for } \gamma = 180° \tag{6.7b}$$

The angle of incidence of the beam radiation on the aperture plane throughout the day is obtained by putting Eqs (6.7a) and (6.7b) in Eq. (3.5). For both cases, $\gamma = 0°$ and $\gamma = 180°$, we obtain the same relation

$$\cos\theta = \sin^2\delta + \cos^2\delta\cos\omega \tag{6.8}$$

❑ *Mode II* The focal axis is E-W and horizontal. The collector is rotated about a horizontal E-W axis and adjusted continuously so that the solar beam makes the minimum angle of incidence with the aperture plane at all times.

In this mode also, the aperture plane is an imaginary surface with either $\gamma = 0°$ or $\gamma = 180°$. Equation (3.5) is applicable with $\gamma = 0°$ or $180°$. In order to find the condition to be satisfied for θ to be a minimum, we differentiate the right hand side of the resulting equation with respect to β and equate it to zero. Thus, we get

$$\tan(\phi - \beta) = [\tan\delta/\cos\omega] \quad \text{for } \gamma = 0° \tag{6.9a}$$

and

$$\tan(\phi + \beta) = [\tan\delta/\cos\omega] \quad \text{for } \gamma = 180° \tag{6.9b}$$

Equations (6.9a) and (6.9b) can be used for finding the slope of the aperture plane. Equation (6.9a) corresponding to $\gamma = 0°$ is used if the magnitude of the solar azimuth angle γ_s is less than 90°, while Eq. (6.9b) corresponding to $\gamma = 180°$ is used if the magnitude of the solar azimuth angle is greater than 90°.

The expression for the corresponding minimum angle of incidence is obtained by substituting Eqs (6.9a) and (6.9b) in the appropriate version of Eq. (3.5). For both cases, we obtain

$$\cos\theta = (1 - \cos^2\delta\sin^2\omega)^{1/2} \tag{6.10}$$

❑ *Mode III* The focal axis is N-S and horizontal. The collector is rotated about a horizontal N-S axis and adjusted continuously so that the solar beam makes the minimum angle of incidence with the aperture plane at all times.

In this mode, the surface azimuth angle $\gamma = +90°$ before noon, and $-90°$ after … before noon, Eq. (3.5) becomes

$$\ldots \sin\delta + \cos\phi\cos\delta\cos\omega)\cos\beta + \cos\delta\sin\omega\sin\beta \tag{6.11}$$

… the condition to be satisfied for θ to be a minimum, we differentiate … side of Eq. (6.11) with respect to β and equate it to zero. Thus, we

$$\beta = \tan^{-1}\left[\frac{\cos\delta\sin\omega}{\sin\phi\sin\delta + \cos\phi\cos\delta\cos\omega}\right] \tag{6.12}$$

Equation (6.12) is used for finding the slope of the aperture plane at any time before noon. The expression for the corresponding minimum angle of incidence is obtained by substituting Eq. (6.12) in Eq. (6.11), giving

$$\cos\theta = [(\sin\phi\sin\delta + \cos\phi\cos\delta\cos\omega)^2 + \cos^2\delta\sin^2\omega]^{1/2} \tag{6.13}$$

After noon, i.e. with $\gamma = -90°$, we would obtain

$$\beta = \tan^{-1}\left[\frac{-\cos\delta\sin\omega}{\sin\phi\sin\delta + \cos\phi\cos\delta\cos\omega}\right] \tag{6.14}$$

The expression for $\cos\theta$ remains the same.

❑ *Mode IV* The focal axis is N-S and inclined at a fixed angle equal to the latitude. Thus, it is parallel to the earth's axis. This orientation is sometimes referred to as a *polar mount*. The collector is rotated about an axis parallel to the earth's axis at an angular velocity equal and opposite to the earth's rate of rotation (15° per hour). In the northern hemisphere, it is adjusted such that at solar noon the aperture plane is an inclined surface facing due south. Thus, putting $\beta = \phi$ and $\omega = 0$ in Eq. (3.8), we get

$$\theta = \delta \tag{6.15}$$

For the southern hemisphere, we get the same expression for θ by putting $\beta = -\phi$ and $\omega = 0$ in Eq. (3.10).

This is also seen from Fig. 6.3 in which the circle represents the longitude through the location of the collector. At all other times, since the collector is rotated at speed equal to the earth's rate of rotation and about an axis parallel to the earth's axis, it follows that Eq. (6.15) is still valid.

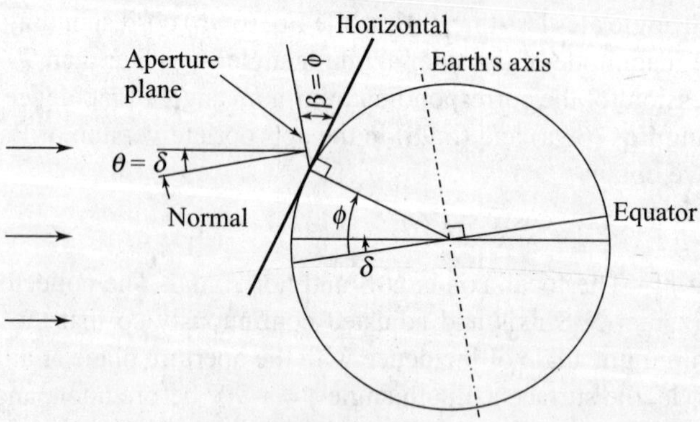

Fig. 6.3 *Tracking mode IV for a cylindrical parabolic collector. Angle of incidence θ is equal to the declination δ*

❑ *Mode V* The focal axis is N-S and inclined. The collector is rota continuously (but not at a constant angular velocity) about an axis para the focal axis, as well as about a horizontal axis perpendicular to this axis, and adjusted so that the solar beam is normally incident on the aperture plane at all times. In this situation, obviously cos $\theta = 1$. It is easy to show that at solar noon,

$$\beta = |\phi - \delta| \tag{6.16}$$

It is of interest to compare the amounts of beam radiation which would be incident on a collector's aperture plane over a day if one adopted the various tracking modes. This comparison is made through a numerical example.

Example 6.1

A cylindrical parabolic collector is used in New Delhi (28.58° N, 77.20° E). Compare the beam radiation which would fall on one square metre of the aperture plane of this collector from 0600 to 1800 h (LAT) on June 10 for the five tracking modes just described. The following values of I_b are available.

Time (h)	I_b (W/m^2)	Time (h)	I_b (W/m^2)
0630	110	1230	523
0730	240	1330	495
0830	333	1430	445
0930	424	1530	322
1030	495	1630	220
1130	550	1730	118

Solution

On June 10, $n = 161$ and $\delta = 23.012°$.

We show a sample calculation for one time, viz. 1030 h. From Eq. (3.7),

$$\cos\theta_z = \sin 28.58° \sin 23.012° + \cos 28.58° \cos 23.012° \cos 22.5° = 0.9338$$

For tracking mode I, from Eq. (6.8),

$$\cos\theta = \sin^2 23.012° + \cos^2 23.012° \cos 22.5° = 0.9355$$

Therefore, tilt factor r_b for the aperture plane $= \dfrac{\cos\theta}{\cos\theta_z} = \dfrac{0.9355}{0.9338} = 1.002$

Beam flux incident normally on aperture plane

$$= I_b r_b = 495 \times 1.002 = 495.9 \text{ W/m}^2$$

Similarly for the other tracking modes we obtain the following results:

	Tracking mode			
	II	III	IV	V
$\cos\theta$	0.9359	0.9980	0.9204	1
r_b	1.0023	1.0688	0.9857	1.0709
$I_b r_b$ (W/m^2)	496.1	529.0	487.9	530.1

The values of $I_b r_b$ obtained over the day are given in Table 6.1. The total incident energy from 0600 to 1800 h is also obtained by adding the values of $I_b r_b$. The assumption made is that the instantaneous value is also the average value for a one-hour period. It is seen that the maximum total is obtained with mode V. This is as expected since mode V involves two-axis tracking and gives normal incidence. The other modes involve continuous one-axis tracking or one daily adjustment and yield lesser totals. The results obtained are obviously dependent on the latitude of the location, the day of the year and the input radiation data.

Table 6.1 Comparison of tracking modes - Example 6.1

LAT	$I_b r_b$ (W/m^2)				
(h)	Mode I	Mode II	Mode III	Mode IV	Mode V
0630	99.1	153.8	360.4	346.1	376.1
0730	230.7	254.5	476.1	445.1	483.6
0830	327.9	335.0	489.0	451.3	490.4
0930	422.3	424.0	511.9	471.2	511.9
1030	495.9	496.1	529.0	487.9	530.1
1130	552.4	552.4	554.0	512.2	556.4
1230	525.3	525.3	526.9	487.0	529.1
1330	495.9	496.1	529.0	487.9	530.1
1430	443.2	445.0	537.2	494.5	537.2
1530	317.0	323.9	472.8	436.4	474.2
1630	211.4	233.2	436.5	408.0	443.3
1730	106.3	165.0	386.6	371.3	403.4
Total (kWh/m^2)	4.227	4.404	5.809	5.399	5.866

In practice, modes II, III and IV are the most common. Mode V is not preferred because of the complexity of providing motion about two axes, while mode I is not used because it does not yield an image on the focal axis. From a design standpoint, modes II and III are the simpler of the three modes II, III and IV. Since the focal axis is horizontal, it is easy to connect collectors in series (the outlet of one collector being the inlet of the next) and to provide common shaft arrangements for tracking. Land space is also used more effectively in these modes.

6.4.3 Performance Analysis

We now consider the performance analysis of a cylindrical parabolic concentrating collector whose concentrator has an aperture W, length L and rim angle ϕ_r (Fig. 6.4). The absorber tube has an inner diameter D_i and an outer diameter

D_o, and it has a concentric glass cover of inner diameter D_{ci} and outer diameter D_{co} around it. The fluid being heated in the collector has a mass flow rate $\dot{m}$, a specific heat C_p, an inlet temperature T_{fi} and an outlet temperature T_{fo}.

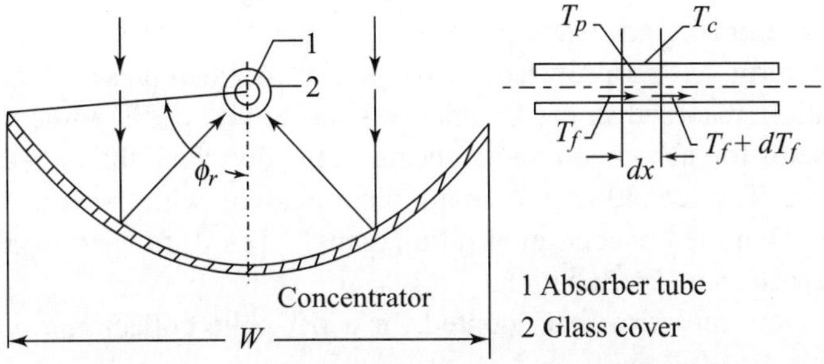

Fig. 6.4 *Cylindrical parabolic collector in cross-section*

The collector is operated in any one of the modes described in Section 6.4.2 and the beam radiation normally incident on its aperture is $I_b r_b$, whose value can be calculated from the equations derived in Section 6.4.2. In some of the tracking modes, the sun's rays are incident at an angle and will, therefore, come to a focus a little beyond the length of the concentrator. We assume that the absorber tube is long enough to intercept this image. In practice this would mean that the absorber tube might be a little longer (say about 10 per cent) than the concentrator and that the flux falling on the tube would not be uniform along the length. For the purposes of analysis, however, we will not take into account this extra tube length, and we will assume that the radiation flux is the same all along the length. We will also make the assumption that the temperature drops across the absorber tube and the glass cover are negligible.

The *concentration ratio* of the collector is given by*

$$C = \frac{\text{Effective aperture area}}{\text{Absorber tube area}} = \frac{(W - D_o)L}{\pi D_o L} = \frac{(W - D_o)}{\pi D_o} \tag{6.17}$$

The analysis which follows is in many respects similar to the analysis of a liquid flat-plate collector given in Sections 4.7 and 4.8. An energy balance on an elementary slice *dx* of the absorber tube, at a distance *x* from the inlet, yields the following equation for a steady state:

$$dq_u = [I_b r_b (W - D_o)\rho \gamma_{int} (\tau\alpha)_b + I_b r_b D_o (\tau\alpha)_b - U_l \pi D_o (T_p - T_a)]dx \tag{6.18}$$

in which

dq_u = useful heat gain rate for a length *dx*,

ρ = specular reflectivity of the concentrator surface,

γ_{int} = intercept factor, the fraction of the reflected radiation intercepted by the absorber tube,

---

*There are small but significant differences in the application of the definition of concentration ratio. Some authors take $(W - D_{co})L$ as the effective aperture area in the numerator, while others use the whole aperture area, viz. WL.

$(\tau\alpha)_b$ = average value of the transmissivity-absorptivity product for beam radiation,

U_l = overall loss coefficient,

T_p = local temperature of absorber tube, and

T_a = ambient temperature.

The first term on the right hand side in Eq. (6.18) represents the incident beam radiation absorbed in the absorber tube after reflection, while the second term represents the absorbed incident beam radiation which falls directly on the absorber tube. The second term is small in comparison with the first, but cannot be ignored when the concentration ratio is small. The third term represents the loss by convection and reradiation.†

In a manner similar to that adopted for a flat-plate collector, we define an absorbed flux S as follows,

$$S = I_b r_b \rho \gamma_{int} (\tau\alpha)_b + I_b r_b (\tau\alpha)_b \left(\frac{D_o}{W - D_o} \right) \tag{6.19}$$

Equation (6.18) thus becomes,

$$dq_u = \left[S - \frac{U_l}{C}(T_p - T_a) \right] (W - D_o)dx \tag{6.20}$$

The useful heat gain rate dq_u can also be written as

$$dq_u = h_f \pi D_i (T_p - T_f) dx \tag{6.21}$$

$$= \dot{m}\, C_p dT_f \tag{6.22}$$

where h_f = heat transfer coefficient on the inside surface of the tube,
and T_f = local fluid temperature.

We combine Eqs (6.20) and (6.21) in such a manner as to eliminate the absorber tube temperature T_p, and obtain

$$dq_u = F' \left[S - \frac{U_l}{C}(T_f - T_a) \right] (W - D_o)dx \tag{6.23}$$

where F' is the *collector efficiency factor* defined by

$$F' = \frac{1}{U_l \left[\dfrac{1}{U_l} + \dfrac{D_o}{D_i h_f} \right]} \tag{6.24}$$

Again, combining Eqs (6.22) and (6.23), we obtain the differential equation

$$\frac{dT_f}{dx} = \frac{F' \pi D_o U_l}{\dot{m} C_p} \left[\frac{CS}{U_l} - (T_f - T_a) \right] \tag{6.25}$$

Integrating and using the inlet condition at $x = 0$, $T_f = T_{fi}$ we have the temperature distribution

---

†Heat losses due to conduction at the ends through the supports for the receiver are generally significant and are also sometimes accounted for in the value of U_l.

$$\frac{\left(\dfrac{CS}{U_l}+T_a\right)-T_f}{\left(\dfrac{CS}{U_l}+T_a\right)-T_{fi}}=\exp\left\{-\frac{F'\pi D_o U_l x}{\dot{m}C_p}\right\} \tag{6.26}$$

The fluid outlet temperature is obtained by putting $T_f = T_{fo}$ and $x = L$ in Eq. (6.26). Making this substitution and subtracting both sides of the resulting equation from unity, we have

$$\frac{(T_{fo}-T_{fi})}{\dfrac{CS}{U_l}+T_a-T_{fi}}=1-\exp\left\{-\frac{F'\pi D_o U_l L}{\dot{m}C_p}\right\} \tag{6.27}$$

Thus, the useful heat gain rate

$$\begin{aligned} q_u &= \dot{m}\, C_p(T_{fo}-T_{fi}) \\ &= \dot{m}C_p\left[\frac{CS}{U_l}+T_a-T_{fi}\right]\left[1-\exp\left\{-\frac{F'\pi D_o U_l L}{\dot{m}C_p}\right\}\right] \\ &= F_R(W-D_o)L\left[S-\frac{U_l}{C}(T_{fi}-T_a)\right] \end{aligned} \tag{6.28}$$

where F_R is the *heat removal factor* defined by

$$F_R=\frac{\dot{m}C_p}{\pi D_o L U_l}\left[1-\exp\left\{-\frac{F'\pi D_o U_l L}{\dot{m}C_p}\right\}\right] \tag{6.29}$$

Equation (6.28) is the equivalent of the Hottel-Whillier-Bliss Eq. (4.49) for a flat-plate collector.

The *instantaneous collection efficiency* η_i is given by

$$\eta_i=\frac{q_u}{(I_b r_b+I_d r_d)WL} \tag{6.30}$$

if ground-reflected radiation is neglected. The instantaneous efficiency can also be calculated on the basis of beam radiation alone, in which case

$$\eta_{ib}=\frac{q_u}{I_b r_b WL} \tag{6.31}$$

6.4.4 Overall Loss Coefficient and Heat Transfer Correlations

In this section, we will give the procedure for calculating the *overall loss coefficient* U_l and the correlations required for calculating the individual heat transfer coefficients. The calculation of the overall loss coefficient based on convection and reradiation losses alone proceeds in a manner similar to that adopted in Section 4.6.1 for the top loss coefficient. We consider the absorber tube and the glass cover around it to constitute a system of long, concentric tubes. Making the same assumptions, we have

$$\frac{q_l}{L} = h_{p-c}(T_{pm} - T_c)\pi D_o + \frac{\sigma \pi D_o (T_{pm}^4 - T_c^4)}{\left\{\dfrac{1}{\varepsilon_p} + \dfrac{D_o}{D_{ci}}\left(\dfrac{1}{\varepsilon_c} - 1\right)\right\}} \tag{6.32}$$

$$= h_w(T_c - T_a)\pi D_{co} + \sigma \pi D_{co} \varepsilon_c (T_c^4 - T_{sky}^4) \tag{6.33}$$

where $\dfrac{q_l}{L}$ = heat loss rate per unit length,

h_{p-c} = convective heat transfer coefficient between the absorber tube and the glass cover,

T_{pm} = average temperature of the absorber tube, and

T_c = temperature attained by the cover.

All other symbols have been defined earlier. Equations (6.32) and (6.33) are a set of two non-linear equations which have to be solved for the unknowns (q_l/L) and T_c after substituting the values of h_{p-c} and h_w.

Heat Transfer Coefficient between the Absorber Tube and the Cover

The natural convection heat transfer coefficient h_{p-c} for the enclosed annular space between a horizontal absorber tube and a concentric cover is calculated using a correlation due to Raithby and Hollands [4]

$$\frac{k_{eff}}{k} = 0.317\,(\text{Ra}^*)^{1/4} \tag{6.34}$$

where k_{eff} = effective thermal conductivity defined as the thermal conductivity that the motionless air in the gap must have to transmit the same amount of heat as the moving air, and

Ra* = modified Rayleigh number related to the usual Rayleigh number by the following equation

$$(\text{Ra}^*)^{1/4} = \frac{\ln(D_{ci}/D_o)}{b^{3/4}\left(\dfrac{1}{D_o^{3/5}} + \dfrac{1}{D_{ci}^{3/5}}\right)^{5/4}}\text{Ra}^{1/4} \tag{6.35}$$

The characteristic dimension used for the calculation of the Rayleigh number is the radial gap $b = (D_{ci} - D_o)/2$. Properties are evaluated at the mean temperature $(T_{pm} + T_c)/2$. It is to be noted that the effective thermal conductivity k_{eff} cannot be less than the thermal conductivity k. Hence (k_{eff}/k) is put equal to unity if the use of Eq. (6.34) yields a value less than unity.

The relationship between the heat transfer coefficient h_{p-c} and the effective thermal conductivity can be found by equating expressions for the heat exchange rate per unit length. We have

$$\frac{2\pi k_{eff}}{\ln(D_{ci}/D_o)}(T_{pm} - T_c) = h_{p-c}\pi D_o (T_{pm} - T_c)$$

Thus $$h_{p-c} = \frac{2k_{eff}}{D_o \ln(D_{ci}/D_o)} \tag{6.36}$$

The limitations on using Eq. (6.34) are that Ra* should be less than 10^7, and b should be less than $0.3D_o$.

Heat Transfer Coefficient on the Outside Surface of the Cover

The convective heat transfer coefficient h_w on the outside surface of the cover (sometimes called the wind heat transfer coefficient) can be calculated using the well-known correlation based on the data of Hilpert [5] who conducted experiments on air flowing at right angles across cylinders of various diameters at low levels of free stream turbulence. Hilpert's data can be correlated by the equation

$$\text{Nu} = C_1 \text{Re}^n \tag{6.37}$$

where C_1 and n are constants having the following values:

For $40 < \text{Re} < 4000$, $C_1 = 0.615$, $n = 0.466$

For $4000 < \text{Re} < 40\,000$, $C_1 = 0.174$, $n = 0.618$

For $40\,000 < \text{Re} < 400\,000$, $C_1 = 0.0239$, $n = 0.805$

Churchill and Bernstein [6] have made a comprehensive analysis of the data available for cross flow across a cylinder and developed the following correlation

$$\text{Nu} = 0.3 + \frac{0.62\,\text{Re}^{1/2}\text{Pr}^{1/3}}{\left[1 + (0.4/\text{Pr})^{2/3}\right]^{1/4}}\left[1 + \left(\frac{\text{Re}}{282\,000}\right)^{5/8}\right]^{4/5} \tag{6.38}$$

Equation (6.38) is valid for all values of Re up to 10^7. For the range $20\,000 < \text{Re} < 400\,000$, Churchill and Bernstein recommended that the last term $[1 + (\text{Re}/282\,000)^{5/8}]^{4/5}$ be modified to $[1 + (\text{Re}/282\,000)^{1/2}]$.

D_{co} is the characteristic dimension to be used in Eqs (6.37) and (6.38). Properties are to be evaluated at the mean temperature $(T_c + T_a)/2$.

Equations (6.37) and (6.38) have been obtained for cross flow and at low levels of turbulence intensity. In practice, the flow may not be at right angles and the turbulence intensity in the wind may not be insignificant. As a result, there is an uncertainty in the value of h_w predicted by Eqs (6.37) and (6.38). Fortunately, this uncertainty does not affect the value of the overall loss coefficient significantly.

Heat Transfer Coefficient on the Inside Surface of the Absorber Tube

The convective heat transfer coefficient h_f on the inside surface of the absorber tube can be calculated under the assumption that the flow is fully developed. This assumption is justified because the length-to-diameter ratio (L/D_i) is large, usually greater than 20. For a Reynolds number less than 2000, the flow is laminar and the heat transfer coefficient may be calculated from the equation

$$\text{Nu} = 3.66 \tag{6.39}$$

On the other hand, for a Reynolds number greater than 2000, the flow is turbulent and the heat transfer coefficient may be calculated from the well-known Dittus-Boelter equation

$$\text{Nu} = 0.023\ \text{Re}^{0.8}\ \text{Pr}^{0.4} \tag{6.40}$$

The characteristic dimension used for calculating Nu and Re in Eqs (6.39) and (6.40) is D_i. Properties are evaluated at the mean temperature $(T_{fi} + T_{fo})/2$. It should be noted that Eqs (6.39) and (6.40) are also valid for calculating the value of h_f for a liquid flat-plate collector.

In most situations, the mass flow rate $\dot{m}$ is small and the flow is laminar. Equation (6.39) is therefore used. As a consequence, the value of h_f is sometimes so small as to adversely affect the value of F_R. This is particularly true when the liquid used is a heat transfer oil. These oils have high boiling points, but are characterized by low thermal conductivities and high Prandtl numbers. In such cases, it is desirable to use some kind of augmentative technique to increase the heat transfer coefficient. One of the simplest techniques is to use a twisted tape of width D_i inserted all along the inside of the absorber tube. Hong and Bergles [7] have suggested the following correlation for this case:

$$\text{Nu} = 5.172\ [1 + 0.005484\ \{\text{Pr}(\text{Re}/X)^{1.78}\}^{0.7}]^{0.5} \tag{6.41}$$

where X = tape twist ratio = H/D_i and H = length over which the tape is twisted through 180°.

The characteristic dimension used for calculating Nu and Re is D_i. Calculations with Eq. (6.41) show that the use of a twisted tape results in very significant increases in the heat transfer coefficient when the Prandtl number is high. At the same time, the pressure drop does not increase in the same proportion. The pressure drop may be calculated from the following correlations based on the work of Date and Singham [8]

$$\begin{aligned} f\,\text{Re} &= 38.4\ (\text{Re}/X)^{0.05} && \text{for } 6.7 \le (\text{Re}/X) \le 100 \\ &= C_2\ (\text{Re}/X)^{0.3} && \text{for } (\text{Re}/X) > 100 \end{aligned} \tag{6.42}$$

where f = friction factor, and

$$C_2 = 8.8201X - 2.1193\ X^2 + 0.2108\ X^3 - 0.0069\ X^4$$

Example 6.2

Calculate the overall loss coefficient U_l for the receiver of a cylindrical parabolic concentrating collector system. The receiver consists of a selectively-coated absorber tube with one glass cover around it. The following data is given:

Absorber tube, inner diameter (D_i)	:	7.5 cm
outer diameter (D_o)	:	8.1 cm
Glass cover, inner diameter (D_{ci})	:	14.4 cm
outer diameter (D_{co})	:	15.0 cm
Emissivity of absorber tube surface (ε_p)	:	0.15
Emissivity of glass (ε_c)	:	0.88
Mean temperature of absorber tube (T_{pm})	:	170°C

Ambient temperature (T_a) : 25°C
Wind speed (V_∞) : 4 m/s

Solution

Substituting the given data into Eqs (6.32) and (6.33), we get

$$\frac{q_l}{L} = h_{p-c}\,(443.2 - T_c)\pi \times 0.081 + \frac{5.67\times10^{-8} \times \pi \times 0.081(443.2^4 - T_c^4)}{\left\{\dfrac{1}{0.15} + \dfrac{0.081}{0.144}\left(\dfrac{1}{0.88} - 1\right)\right\}}$$

$$= 0.2545\,h_{p-c}\,(443.2 - T_c) + 0.2140 \times 10^{-8}(385.8 \times 10^8 - T_c^{\,4}) \qquad (6.43)$$

$$\frac{q_l}{L} = h_w(T_c - 298.2)\pi \times 0.15 + 5.67 \times 10^{-8} \times \pi \times 0.15 \times 0.88\,(T_c^{\,4} - 292.2^4)$$

$$= 0.4712\,h_w\,(T_c - 298.2) + 2.3513 \times 10^{-8}\,(T_c^{\,4} - 72.90 \times 10^8) \qquad (6.44)$$

Equations (6.43) and (6.44) have to be solved for the unknowns (q_l/L) and T_c. For this, the values h_{p-c} and h_w are needed. Since these values depend upon T_c, a trial-and-error method is necessary. Assume T_c = 310 K.

Calculation of h_{p-c}

Mean temperature of air between tube and cover

$$= \frac{443.2 + 310}{2} = 376.6 \text{ K} = 103.4°\text{C}$$

At this temperature,

$$k = 0.0323 \text{ W/m-K}$$

$$\nu = 23.52 \times 10^{-6} \text{ m}^2\text{/s}$$

$$\text{Pr} = 0.688$$

$$\text{Ra} = 9.81 \times \frac{1}{376.6} \times \frac{(443.2 - 310) \times 0.0315^3}{23.52^2 \times 10^{-12}} \times 0.688$$

$$= 134\,877$$

From Eq. (6.34),

$$\frac{k_{eff}}{k} = 0.317 \times \frac{\ln(0.144/0.081)}{0.0315^{3/4}\left(\dfrac{1}{0.081^{3/5}} + \dfrac{1}{0.144^{3/5}}\right)^{5/4}} \times (134\,877)^{1/4}$$

$$= 3.6349$$

and from Eq. (6.36),

$$h_{p-c} = \frac{2 \times 3.6349 \times 0.0323}{0.081 \ln(0.144/0.081)} = 5.036 \text{ W/m}^2\text{-K}$$

Calculation of h_w

Mean temperature of air between the cover and ambient

$$= \frac{310 + 298.2}{2} = 304.1 \text{ K} = 30.9°\text{C}$$

At this temperature,

$$k = 0.0268 \text{ W/m-K}$$
$$\nu = 16.09 \times 10^{-6} \text{ m}^2\text{/s}$$

Assuming that the wind velocity is at right angles to the axis of the collector and using Eq. (6.37), we have,

$$\text{Re} = \frac{4 \times 0.15}{16.09 \times 10^{-6}} = 37\,300$$

$$\text{Nu} = 0.174(37\,300)^{0.618} = 116.4$$

$$h_w = 116.4 \times \frac{0.0268}{0.15} = 20.79 \text{ W/m}^2\text{-K}$$

Substituting the values of h_{p-c} and h_w in Eqs (6.43) and (6.44) we obtain the values of T_c and (q_l/L) by trial-and-error.

T_c (K)	(q_l/L) from	
	Eq. (6.43)	Eq. (6.44)
310	233.5	161.3
315	225.9	224.7
315.1	225.6	226.0

The values of (q_l/L) in the last line are reasonably close to each other. The average value of 225.8 W/m will therefore be accepted. Since the values of h_{p-c} and h_w will not change much if the initial guess of T_c = 310 K is changed to 315.1 K, it will not be necessary to repeat the calculation.

Hence, $$U_l = \frac{225.8}{\pi \times 0.081 \times (170 - 25)} = 6.12 \text{ W/m}^2\text{-K.}$$

Empirical Equation for the Overall Loss Coefficient

Based on calculations for a large number of cases covering a broad range of conditions encountered with cylindrical parabolic collectors, Mullick and Nanda [9] have developed a semi-empirical equation for directly calculating the overall loss coefficient. This equation eliminates the need for an iterative calculation.

$$\frac{1}{U_l} = \frac{1}{C_3(T_{pm} - T_c)^{0.25} + \left[\sigma(T_{pm}^2 + T_c^2)(T_{pm} + T_c)\Big/\left\{\frac{1}{\varepsilon_p} + \frac{D_o}{D_{ci}}\left(\frac{1}{\varepsilon_c} - 1\right)\right\}\right]} + \left(\frac{D_o}{D_{co}}\right)\left(\frac{1}{h_w + \sigma\varepsilon_c(T_c^2 + T_a^2)(T_c + T_a)}\right) \tag{6.45}$$

The constant C_3 has been obtained from the correlation of Raithby and Hollands and is given by the expression

$$C_3 = \frac{17.74}{(T_{pm} + T_c)^{0.4} D_o (D_o^{-0.75} + D_{ci}^{-0.75})} \quad (6.46)$$

The cover temperature T_c is given by

$$\left(\frac{T_c - T_a}{T_{pm} + T_a}\right) = 0.04075\left(\frac{D_o}{D_{co}}\right)^{0.4} h_w^{-0.67}\left[2 - 3\varepsilon_p + \frac{(6 + 9\varepsilon_p)T_{pm}}{100}\right] \quad (6.47)$$

if $333 < T_{pm} < 513$ K, and by

$$\left(\frac{T_c - T_a}{T_{pm} - T_a}\right) = 0.163\left(\frac{D_o}{D_{co}}\right)^{0.4} h_w^{-0.67}\left[2 - 3\varepsilon_p + \frac{(1 + 3\varepsilon_p)T_{pm}}{100}\right] \quad (6.48)$$

if $513 < T_{pm} < 623$ K.

While using Eqs (6.45) – (6.48), T_{pm}, T_c and T_a are expressed in K, D_o, D_{ci} and D_{co} in m, σ in W/m^2–K^4, and h_w in W/m^2-K. The value of U_l is obtained in W/m^2-K. Equation (6.45) has been developed for the following range:

$$\begin{aligned} 0.1 &\le \varepsilon_p \le 0.95 \\ 0.0125 &\le D_o \le 0.15 \text{ m} \\ 15 &\le h_w \le 60 \text{ W/m}^2\text{-K} \\ 273 &\le T_a \le 313 \text{ K.} \end{aligned}$$

Equations (6.47) and (6.48) estimate the glass cover temperature to within ± 10°C. This estimate is good enough to obtain U_l from Eq. (6.45) to an accuracy of ± 1 per cent for $333 < T_{pm} < 513$ K, and to an accuracy of ± 2 per cent for $513 < T_{pm} < 623$ K.

Correlation for Heat Loss Rate

For an evacuated tube-type of receiver, NREL [10] have developed the following empirical correlation for the heat loss rate per unit length:

$$q_l/L = 0.141\,T_{pm} + 6.48 \times 10^{-9}\,T_{pm}^4 \quad (6.49)$$

where (q_l/L) is in W/m and T_{pm} is in °C.

Equation (6.49) has been obtained for the following geometry and conditions:

Inner diameter of the absorber tube	= 0.066 m
Outer diameter of the absorber tube	= 0.070 m
Glass cover outer diameter	= 0.125 m
Vacuum	= 10^{-3} mbar

It is valid for $100°\text{C} < T_{pm} < 500°\text{C}$.

6.4.5 A Numerical Example

We will now illustrate the procedure for calculating the performance of a cylindrical parabolic collector through a detailed numerical example.

Example 6.3

A cylindrical parabolic collector located in Mumbai, operating in tracking mode II, is used for heating a thermic fluid. The concentrator has an aperture of 1.25 m and a length of 3.657 m, while the absorber tube (3.81 cm inner and 4.135 cm outer diameter) has a concentric glass cover (5.60 cm inner and 6.30 cm outer diameter) around it. A twisted tape with a tape twist ratio of 4 is used inside the absorber tube. Values of other design parameters of the collector are as follows:

Specular reflectivity of concentrator surface	:	0.85
Glass cover transmissivity for solar radiation	:	0.85
Glass cover emissivity/absorptivity	:	0.88
Absorber tube emissivity/absorptivity	:	0.95
Intercept factor	:	0.95

Values of the operational and meteorological parameters are as follows:

Date	:	April 15
Time	:	1230 h (LAT)
I_b	:	705 W/m^2
I_g	:	949 W/m^2
Ambient temperature	:	31.9°C
Wind speed	:	5.3 m/s
Mass flow rate of thermic fluid	:	0.0986 kg/s
Inlet temperature	:	150°C

Calculate,

1. The slope of the aperture plane and the angle of incidence on the aperture plane,
2. The absorbed flux S,
3. The convective heat transfer coefficient on the inside surface of the absorber tube,
4. The collector heat-removal factor and overall loss coefficient,
5. The exit temperature of the thermic fluid,
6. The instantaneous efficiency, and
7. The pressure drop.

Solution

1. Slope of the Aperture Plane and Angle of Incidence

In tracking mode II, the slope of the aperture plane and angle of incidence are given by Eqs (6.9) and (6.10). On April 15, $n = 105$,

$$\delta = 23.45 \sin\left[\frac{360}{365}(284 + 105)\right] = 9.415°$$

Therefore, substituting $\delta = 9.415°$, $\omega = -7.5°$ and $\phi = 19.12°$ in Eq. (6.9) we have

$$\beta = 9.625°$$

and $$\cos\theta = [1 - \cos^2 9.415° \sin^2 (-7.5°)]^{1/2} = 0.9917$$

$$\theta = 7.398°$$

2. Absorbed Flux S

From Eq. (3.41),

$$r_b = \frac{0.9917}{\sin 19.12° \sin 9.415° + \cos 19.12° \cos 9.415° \cos(-7.5°)} = 1.0143$$

Therefore, substituting in Eq. (6.19) and taking $(\tau\alpha)_b = \tau\alpha$, we have

$$S = 705 \times 1.0143 \left[0.85 \times 0.95 \times 0.85 \times 0.95 + \frac{0.85 \times 0.95 \times 0.04135}{(1.25 - 0.04135)} \right]$$

$$= 486.03 \text{ W/m}^2$$

3. Convective Heat Transfer Coefficient h_f

We will use Eq. (6.41) for calculating h_f. The properties of the fluid are given in Fig. 6.5. Keeping in mind that the rise of temperature of the fluid will only be a few degrees in this case, properties will be taken at a mean fluid temperature of 152°C.

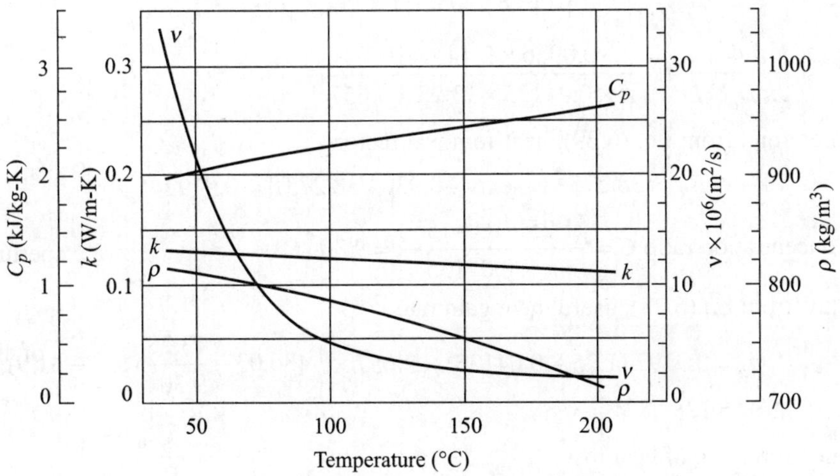

Fig. 6.5 *Example 6.3 – Properties of thermic fluid*

Thus, $\rho = 750.3 \text{ kg/m}^3$, $C_p = 2.449 \text{ kJ/kg-K}$,

$\nu = 2.42 \times 10^{-6} \text{ m}^2\text{/s}$, $k = 0.119 \text{ W/m-K}$

$$\text{Average velocity } V = \frac{\dot{m}}{\frac{\pi}{4} D_i^2 \rho} = \frac{0.0986}{\frac{\pi}{4} \times 0.0381^2 \times 750.3} = 0.1153 \text{ m/s}$$

$$\text{Reynolds number Re} = \frac{VD_i}{\nu} = \frac{0.1153 \times 0.0381}{2.42 \times 10^{-6}} = 1815$$

$$\text{Prandtl number Pr} = \frac{C_p \nu \rho}{k} = \frac{2.449 \times 2.42 \times 10^{-6} \times 750.3 \times 1000}{0.119} = 37.37$$

$$\text{Nusselt number Nu} = 5.172\left[1+0.005484\left\{37.37\left(\frac{1815}{4}\right)^{1.78}\right\}^{0.7}\right]^{0.5} = 61.70$$

$$\text{Therefore, } h_f = 61.70 \times \frac{0.119}{0.0381} = 192.7 \text{ W/m}^2\text{-K}$$

4. Collector Heat-removal Factor and Overall Loss Coefficient

As in the case of the flat-plate collector, an iterative procedure will be required since the values of F_R and U_l cannot be directly determined and the value of one depends on the other.

Assume U_l = 13.28 W/m^2-K[‡]. From Eq. (6.24), the collector efficiency factor

$$F' = \frac{1}{13.28\left[\dfrac{1}{13.28}+\dfrac{0.04135}{0.0381\times 192.7}\right]} = 0.9304$$

$$\frac{\dot{m}C_p}{\pi D_o U_l L} = \frac{0.0986\times 2.449\times 10^3}{\pi\times 0.04135\times 13.28\times 3.657} = 38.275$$

Therefore, from Eq. (6.29), heat-removal factor

$$F_R = 38.275\,[1-\exp(-0.9304/38.275)] = 0.9192$$

$$\text{Concentration ratio } C = \frac{(1.25-0.04135)}{\pi\times 0.04135} = 9.304$$

Thus from Eq.(6.28), useful heat gain rate

$$q_u = 0.9192\,(1.25-0.04135)\times 3.657\times\left[486.03-\frac{13.28}{9.304}(150-31.9)\right]$$
$$= 1289.8 \text{ W}$$

Therefore, rate of heat loss

$$= (W-D_o)LS - q_u$$
$$= (1.25-0.04135)\times 3.657\times 486.03 - 1289.8$$
$$= 858.46 \text{ W} = \pi D_o L U_l (T_{pm}-T_a)$$

$$\text{Hence}\qquad (T_{pm}-T_a) = \frac{858.46}{\pi\times 0.04135\times 3.657\times 13.28} = 136.07°\text{C}$$

$$T_{pm} = 167.97°\text{C} = 441.13 \text{ K}$$

We will now calculate the value of U_l corresponding to this value of T_{pm} and show that it is equal to the assumed value. The procedure of Example 6.2 will be followed.

[‡]In this case, the correct value of U_l has been determined by doing the iterative procedure on a computer. We therefore illustrate the procedure with the correct values of U_l = 13.28 W/m^2-K and T_c = 60.23°C.

Assume $T_c = 60.23°C = 333.39$ K

From Eqs (6.34) and (6.36), we get

$$h_{p-c} = 5.113 \text{ W/m}^2\text{-K}$$

From Eq. (6.37), $h_w = 34.119$ W/m²-K

Substituting these values of T_c, h_{p-c} and h_w in Eqs (6.32) and (6.33), we have

$$\frac{q_l}{L} = 5.113\,(441.13 - 333.39)\pi \times 0.04135 + \frac{5.67 \times 10^{-8} \times \pi \times 0.04135\,(441.13^4 - 333.39^4)}{\left\{\frac{1}{0.95} + \frac{0.04135}{0.0560}\left(\frac{1}{0.88} - 1\right)\right\}}$$

$$= 234.5 \text{ W/m}$$

and
$$\frac{q_l}{L} = 34.119\,(333.39 - 305.06)\pi \times 0.063 + 5.67 \times 10^{-8} \times 0.063 \times 0.88\,(333.39^4 - 299.06^4)$$

$$= 234.5 \text{ W/m}$$

The two values of (q_l/L) match with each other. The corresponding value of U_l is given by

$$U_l = \frac{234.5}{\pi \times 0.04135 \times (441.13 - 305.06)} = 13.27 \text{ W/m}^2\text{-K}$$

which also matches the original guess.

5. Exit Temperature

Equating the heat gained by the fluid to the useful heat gain rate, we get

$$0.0986 \times 2.449 \times (T_{fo} - 150) = \frac{1289.8}{1000}$$

Hence $T_{fo} = 155.34°C$

6. Instantaneous Efficiency

Using Eq. (6.30),

$$\eta_i = \frac{1289.8}{(705 \times 1.0143 + 244 \times 0.9930)\,1.25 \times 3.657} = 0.295$$

Using Eq. (6.31)

$$\eta_{ib} = \frac{1289.8}{705 \times 1.0143 \times 1.25 \times 3.657} = 0.395$$

7. Pressure Drop

We use Eq. (6.42) to get

$$\frac{\text{Re}}{X} = \frac{1815}{4} = 453.75$$

$$C_2 = 13.0964$$

Hence $f\,\text{Re} = 13.0964 \times (453.75)^{0.3} = 82.069$

$$f = 0.0452$$

$$\Delta p = \frac{4 f \rho L V^2}{2 D_i} = \frac{4 \times 0.0452 \times 750.3 \times 3.657 \times 0.1153^2}{2 \times 0.0381}$$

$$= 86.6 \text{ N/m}^2 = 0.88 \text{ cm of water}$$

This is a reasonable value.

The values of η_i and η_{ib} obtained in this example are low and the difference between them is significant. The low values are partly due to the fact that the inlet fluid temperature is rather high for the given concentration ratio, while the difference between the two values is due to the high diffuse component in the global radiation.

In order to obtain a break-up of the losses occurring in the collector, we distinguish between optical losses and thermal losses. Optical losses are those which occur in the path of the incident solar radiation before it is absorbed at the surface of the absorber tube, while thermal losses are due to convection and reradiation from the absorber tube and conduction through the ends. On this basis, we define an *optical efficiency* η_o as the fraction of the solar radiation incident on the aperture of the collector which is absorbed at the surface of the absorber tube. Thus,

$$\eta_o = \frac{I_b r_b \rho \gamma_{int} (\tau\alpha)_b (W - D_o) L + I_b r_b (\tau\alpha)_b D_o L}{I_b r_b W L}$$

$$= \rho \gamma_{int} (\tau\alpha)_b \frac{(W - D_o)}{W} + (\tau\alpha)_b \frac{D_o}{W} = \frac{S}{I_b r_b} \frac{(W - D_o)}{W} \quad (6.50)$$

Substituting the numerical values of this example, we get

$$\eta_o = \frac{486.03}{705 \times 1.0143} \times \frac{(1.25 - 0.04135)}{1.25} = 0.657$$

6.4.6 Parametric Study of Collector Performance[11]

As in the case of the flat-plate collector, Section 4.10, we now study the effect of some important parameters on the performance of a cylindrical parabolic collector. This has been done by writing a general computer programme which executes the procedure of Example 6.3.

Performance Over a Day with Different Tracking Modes

Using input data for solar radiation, ambient temperature and wind speed, the performance of the collector of Example 6.3 is first studied over two whole days, April 15 and December 15. The calculations are done for all the tracking modes described earlier in Section 6.4.2[§]. The input data and the results are presented in Tables 6.2 and 6.3. Average values of efficiency over a day are given in the last column.

[§] It is assumed that the heat transfer correlations (6.34) and (6.35) for calculating the value of h_{p-c} are valid for all the five tracking modes. Strictly speaking, this is not correct, but is necessary in view of the fact that correlations for an inclined annulus are not available.

Table 6.2 Performance of a cylindrical parabolic collector over a whole day in various tracking modes (Location: Mumbai, Date: April 15)

LAT(h)		0730	0830	0930	1030	1130	1230	1330	1430	1530	1630	Average value
I_b(W/m^2)		145	292	465	593	684	705	665	562	385	205	
I_d(W/m^2)		145	201	235	252	252	244	225	200	175	130	
T_a(°C)		27.8	29.6	31.0	31.8	32.0	31.9	31.8	31.5	31.0	30.2	
V_∞(m/s)		1.7	2.3	3.1	4.1	4.9	5.3	5.4	5.4	5.3	4.9	
Mode I:	q_u(W)		195.9	661.4	1000	1239	1289	1177	888.8	398.4		
	η_i(%)		8.7	20.6	25.7	28.7	29.5	28.7	25.4	15.6		22.1
	η_{ib}(%)		14.7	30.9	36.5	39.1	39.5	38.2	34.3	22.7		31.7
Mode II:	q_u(W)		200.0	662.5	1001	1240	1289	1183	890.1	403.8		
	η_i(%)		8.9	20.6	25.7	28.7	29.5	28.7	25.4	15.8		22.1
	η_{ib}(%)		15.0	30.9	36.5	39.1	39.5	38.2	34.4	22.9		31.7
Mode III:	q_u(W)	378.5	680.8	971.5	1111	1229	1280	1300	1260	1030	710.3	
	η_i(%)	18.1	23.5	26.7	27.5	28.6	29.3	30.4	31.1	29.5	26.2	27.7
	η_{ib}(%)	23.4	31.7	36.5	37.9	39.0	39.4	39.5	39.1	36.4	31.1	36.4
Mode IV:	q_u(W)	366.0	664.8	958.4	1106	1229	1280	1294	1245	1010	692.8	
	η_i(%)	16.3	22.0	26.0	27.3	28.8	29.5	30.2	30.5	28.8	24.5	27.1
	η_{ib}(%)	23.0	31.4	36.2	37.8	39.0	39.4	39.5	38.9	36.1	30.7	36.2
Mode V:	q_u(W)	379.0	682.0	979.7	1129	1255	1304	1321	1270	1032	710.9	
	η_i(%)	18.1	23.5	26.8	27.8	28.9	29.7	30.6	31.2	29.6	26.2	27.9
	η_{ib}(%)	23.5	31.7	36.6	38.1	39.3	39.6	39.7	39.2	36.4	31.1	36.5

Table 6.3 Performance of a cylindrical parabolic collector over a whole day in various tracking modes (Location: Mumbai, Date: December 15)

LAT(h)		0730	0830	0930	1030	1130	1230	1330	1430	1530	1630	Average value
I_b(W/m^2)		81	250	404	518	580	580	519	405	255	94	
I_d(W/m^2)		61	92	113	129	137	139	133	119	98	61	
T_a(°C)		23.6	25.3	27.4	29.0	30.0	30.4	30.3	29.8	28.8	27.4	
V_∞(m/s)		2.3	2.8	2.9	2.9	3.0	3.1	3.4	3.6	3.4	3.1	
Mode I:	q_u(W)		514.5	998.8	1341	1522	1521	1337	997.0	539.1		
	η_i(%)		22.4	31.4	35.1	36.6	36.5	34.8	31.0	22.9		29.8
	η_{ib}(%)		26.8	36.6	40.7	42.2	42.2	40.5	36.4	27.5		35.0
Mode II:	q_u(W)		539.8	1005	1342	1522	1521	1338	1004	564.9	44.5	
	η_i(%)		23.4	31.6	35.2	36.6	36.5	34.9	31.3	23.9	3.4	30.1
	η_{ib}(%)		27.5	36.7	40.7	42.2	42.2	40.5	36.5	28.2	4.0	35.1
Mode III:	q_u(W)	328.6	786.2	928.8	971.2	981.5	980.8	967.9	927.3	815.4	475.4	
	η_i(%)	18.5	29.1	30.3	30.0	29.7	29.6	29.7	30.0	29.4	23.4	28.6
	η_{ib}(%)	20.6	33.0	35.5	36.3	36.5	36.5	36.1	35.4	33.6	25.7	33.7
Mode IV:	q_u(W)	378.7	935.3	1178	1304	1362	1362	1301	1177	967.2	533.6	
	η_i(%)	19.4	30.7	33.3	34.2	34.5	34.4	33.9	32.9	31.0	24.0	31.7
	η_{ib}(%)	22.5	35.5	38.8	40.3	40.9	40.9	40.1	38.7	36.0	27.4	37.0
Mode V:	q_u(W)	467.6	1074	1339	1474	1537	1537	1471	1336	1109	636.4	
	η_i(%)	23.3	33.9	36.0	36.6	36.8	36.7	36.3	35.6	34.1	27.7	34.6
	η_{ib}(%)	25.5	37.4	40.5	41.8	42.3	42.3	41.6	40.3	37.9	30.0	39.0

Plots showing the variation of the angle of incidence and the instantaneous efficiency with time on one day, viz. December 15, are given in Figs 6.6 and 6.7. It will be noted from Fig. 6.7 that the general pattern of efficiency variation over a day is the same for all modes. The value of efficiency first

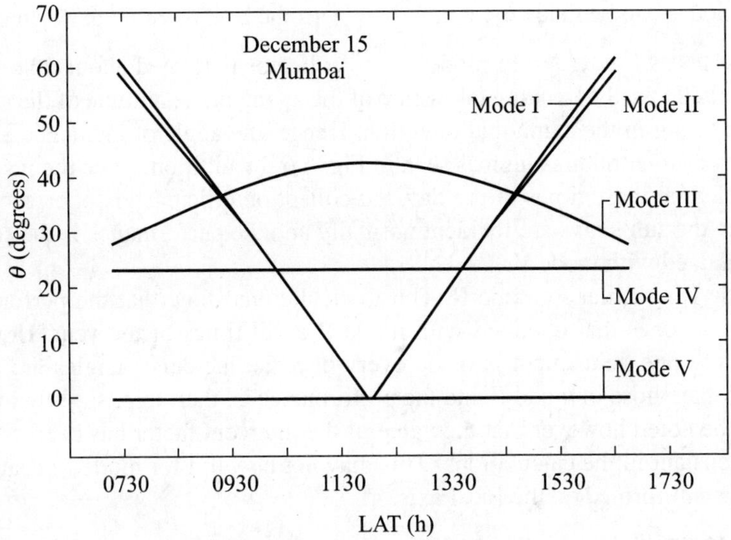

Fig. 6.6 *Variation of angle of incidence in different tracking modes – Data of Example 6.3*

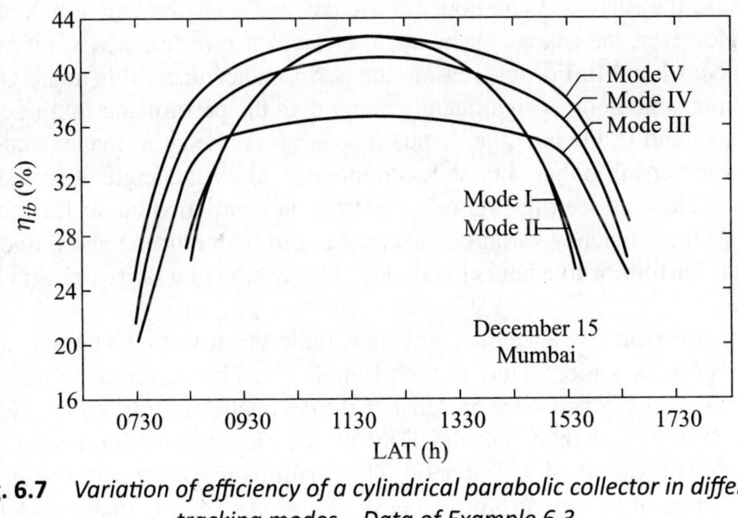

Fig. 6.7 *Variation of efficiency of a cylindrical parabolic collector in different tracking modes – Data of Example 6.3*

increases, reaches a peak value around noon and then decreases. This is because the efficiency is strongly influenced by the incident beam radiation and therefore follows its variation.

On any day and at any time, the performance of the collector under mode V has to be the best. Hence an assessment of the collector's performance under the other modes can be made by comparison with the best possible performance.

❑ *Modes I and II* In mode II, the collector is rotated about a horizontal E-W axis. Thus, the tracking takes care of the swing in the altitude of the sun, but not the swing in the azimuthal direction. Hence, the angle of incidence at times away from solar noon is high as seen in Fig. 6.6. In addition, since the insolation is also less at these times of the day, the collection efficiency is poor. However, at noon, the sun's rays are incident normally and the performance is the same as that obtained with mode V.

Mode I is similar to mode II. Thus, calculations show that the performance closely matches that obtained with mode II at all times of the year. However, since only one adjustment is made every day, the incidence angles are a little greater than those in mode II and the performance is, therefore, slightly inferior. It is to be noted however that the value of the intercept factor has been assumed to be constant in the calculations. This may not be valid for mode I because the image is not formed on the focal axis.

❑ *Mode III* In this mode, the collector is rotated about a horizontal N-S axis. Thus, the tracking takes care of the azimuthal swing of the sun, but not the swing in the altitude. On both the days considered, the angle of incidence is small in the early and late hours of the day and goes through a maximum at noon. However, the change in the angle of incidence over a day is not as large as in modes I and II. For this reason, the performance under this mode at times away from solar noon is significantly better than the performance at these times in modes I and II. On the other hand, at solar noon, the performance under this mode is not equal to that obtained with mode V, since the angle of incidence is not zero. Thus, as seen in Fig. 6.7, the efficiency variation curve for mode III intersects the efficiency variation curves obtained with modes I and II and tends to be flat for four or five hours every day. This results in a fairly uniform useful heat gain.

The performance of the collector is now quite sensitive to the latitude as well as the day of operation. Thus in the example considered, since the latitude of Mumbai is 19.12°N and since on April 15 the declination is only + 9.4°, the angle of incidence through the whole day is small. It varies from a minimum of 1.8° at 0730 h to a maximum of 9.7° at noon. The performance of the collector in mode III at all times of the day on April 15 is, therefore, very near to that obtained with mode V. On the other hand, on December 15 the declination is –23.3°, and the performance deviates significantly from that obtained in mode V.

❑ *Mode IV* In this mode, the collector is rotated about a N-S axis parallel to the earth's axis of rotation and the angle of incidence is always equal to the declination angle. As a result, the performance of the collector is independent of the latitude. Since the declination varies only between the limits of –23.45° and +23.45°, the performance in this mode is always close to that obtained in mode V. On the two equinox days of March 21 and September 21, when the declination is zero, the performance is the same as in mode V.

Effect of Inlet Temperature

As the fluid inlet temperature increases, the temperature of the absorber tube surface also increases. As a result, losses due to re-radiation and convection to the surroundings increase, resulting in a decrease in efficiency. This is apparent from Eq. (6.28) which gives the useful heat gain rate for the collector. In order to illustrate this effect, calculations are done for the case of the parabolic collector of Example 6.3 operating under the same conditions with only inlet temperature varying from 120°C to 180°C. The results are plotted in Fig. 6.8. It is seen that the value of η_{ib} decreases significantly with T_{fi}, the decrease being slightly non-linear. The non-linearity is due to the fact that the value of the overall loss coefficient increases slightly as T_{fi} increases. The value of the optical efficiency (which does not change with inlet temperature) is also shown in Fig. 6.8. The difference between the values of η_o and η_{ib} is a measure of the losses due to reradiation and convection.

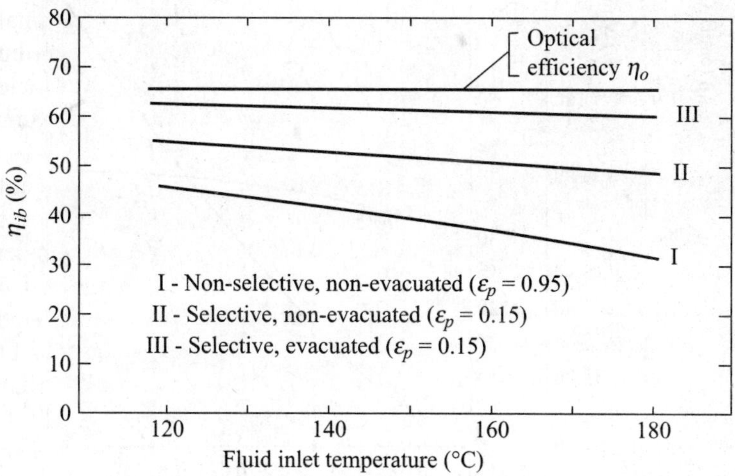

Fig. 6.8 *Variation of efficiency with fluid inlet temperature for three receiver designs – Data of Example 6.3*

Effect of Mass Flow Rate

An increase in the mass flow rate of the thermic fluid increases the value of the inside heat transfer coefficient h_f. Due to this, the collector efficiency factor and the collector heat-removal factor increase and the efficiency increases. This effect is illustrated in Fig. 6.9, in which the mass flow rate is varied from 0.0329 kg/s to 0.1315 kg/s. It is seen that the slope of the efficiency curve goes on decreasing with increasing values of $\dot{m}$ and that the value of η_{ib} tends to some asymptotic value. At the same time, the pressure drop increases, thereby increasing the demand for pumping power. Fortunately this increase is not so rapid because of the high Prandtl number of the fluid and the presence of a twisted tape. Thus, an optimum value of $\dot{m}$ would be one for which the asymptotic value of η_{ib} has almost been attained without an unduly high pressure drop. In the present case, this optimum value would seem to lie around 0.12 kg/s.

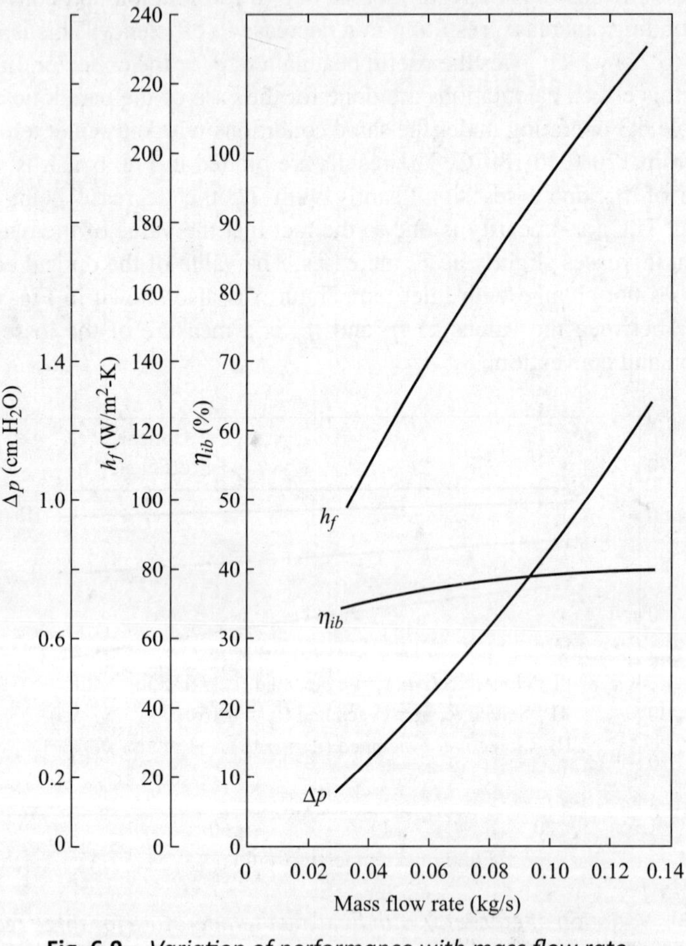

Fig. 6.9 *Variation of performance with mass flow rate–Data of Example 6.3*

Effect of Selectivity of Absorber Tube Surface and Evacuation of Annulus

From Fig. 6.8 it is evident that losses due to reradiation and convection are very high. Any means to reduce or suppress these losses would obviously help in significantly improving the collection efficiency. Losses due to reradiation are reduced by the use of a selective surface, while losses due to convection are reduced by having a vacuum in the annular space between the absorber tube and the glass cover. The effect of introducing these measures is also shown in Fig. 6.8 in which the variation of η_{ib} with T_{fi} is shown for (i) a selective surface with $\varepsilon_p = 0.15$ and a non-evacuated annulus, and (ii) a selective surface with $\varepsilon_p = 0.15$ and an evacuated annulus. It will be seen that there is a dramatic improvement in efficiency and that the rate of decrease of η_{ib} with T_{fi} is also reduced. The increase in η_{ib} is of course due to the decrease in the value of U_l. For example, at T_{fi} = 120°C, the value of U_l decreases from 12.28 W/m^2-K to 6.34 W/m^2-K and 1.65 W/m^2-K respectively for the above two cases.

Effect of Concentration Ratio

The effect of increasing the concentration ratio by decreasing the size of the absorber tube is shown in Fig. 6.10. It is seen that the efficiency increases.

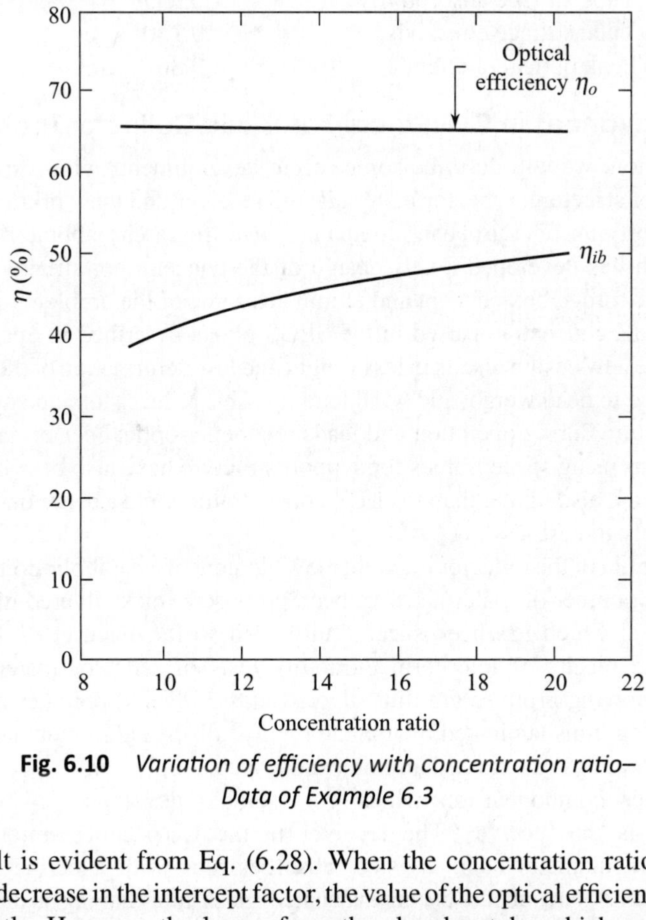

Fig. 6.10 *Variation of efficiency with concentration ratio–Data of Example 6.3*

This result is evident from Eq. (6.28). When the concentration ratio increases without a decrease in the intercept factor, the value of the optical efficiency changes very slightly. However, the losses from the absorber tube which are inversely proportional to C, decrease and hence the collection efficiency increases.

6.4.7 A Parabolic Trough Concentrating Collector

It will be of interest to the reader to note the dimensions and characteristics of the collector modules used in the 50 MW solar thermal power plant set up by Godawari Green Energy Ltd. [12] in Rajasthan. As mentioned in Section 2.3.4, this is the largest parabolic trough plant in India. The data is as follows:

Aperture of parabolic trough	: 5.76 m
Focal length	: 1.850 m
Length of one loop	: 600 m
Total reflecting area	: 392 400 m^2
Reflecting surface	: Curved mirror panels
Reflectivity	: 0.935
Glass cover transmissivity	: 0.955
Vacuum in annular space	: 0.1 Pa
Absorber tube O.D.	: 0.070 m
Tube surface absorptivity	: 0.955
Tube surface emissivity	: 0.120
Peak optical efficiency	: 0.86

6.4.8 Advances in Cylindrical Parabolic Collector Technology

In this section, we will describe some recent developments with respect to the concentrator structure, reflector materials, the receiver and the working fluid.

A consortium of European companies and research laboratories called EuroTrough has developed a new design of the concentrator structure. This is based on the torque box concept and eliminates some of the problems associated with the LUZ concentrator used in the SEGS plants described in Section 2.3.4 [13,14]. The new design results in less weight and less deformation of the collector structure due to dead weight and wind loading. This reduces torsion and bending of the structure during operation and leads to a better optical performance. Two other designs using space frames for support structure have also been developed [14]. These are also lighter than the LUZ concentrator, yield a better performance and are likely to cost less.

The material of the reflector is another key element of a parabolic concentrator. Although a number of materials have been proposed, thick silvered glass is the only material which has been successfully used so far. Fend *et al.* [15] have reported the results of long-term durability tests of reflector materials. The materials showing promise are thin silvered glass, silvered polymer films with UV-screening films laminated on top and highly polished aluminum sheets with protective coatings of oxidized alumina and a special polymer.

The other component on which considerable developmental work has been done is the receiver. The receiver in the LUZ concentrator consists of a 0.07 m diameter stainless steel absorber tube with a cermet selective surface coating surrounded by an evacuated glass tube having a diameter of 0.115 m. Glass-to-metal seals and metal bellows are provided at the two ends to take care of differential thermal expansion. Although this receiver was highly efficient, it experienced high failure rates for a number of reasons. These were identified, corrective measures were adopted and the failure rates were reduced significantly. However, one reason, the failure of the glass-to-metal seal remained as an issue. This failure was believed to be caused by the exposure of the seal to concentrated solar flux. Improved designs have therefore been developed by making modifications which protect the glass-to-metal seal from concentrated flux and reduce the stress levels at the seal. The Universal Vacuum (UVAC) receiver manufactured by the Solel Solar Systems, Israel is one of the improved designs developed. It has an internal reflective shield to protect the glass-to-metal seal from concentrated flux during low-sun-angle condition. It also has a better selective surface coating (α = 0.95-0.96, ε_p = 0.15) [13]. The molybdenum-alumina coating is stable and durable in an evacuated environment at high temperatures up to 350°C [16]. However, it has limited stability in air. Recently, the National Aerospace Laboratories (Bengaluru) has developed a selective carbide/nitride/oxide-based coating having α and ε_p values of 0.936 and 0.15 respectively. Prolonged testing shows that it is stable in air even at a temperature of 600°C [17].

Recently, Archimedes Solar Energy in Italy have begun manufacturing receiver tubes. These tubes are being used in an application in which heat is being transferred to a molten salt entering the solar field at 290°C and leaving at 550°C. This requires that the receiver tube and its selective coating be stable upto 580°C. The selective coatings are molybdenum – silicon oxide ($Mo\text{-}SiO_2$) or tungsten – alumina ($W\text{-}A1_2O_3$). For ($Mo\text{-}SiO_2$), values reported for α and ε_p are 0.94 and 0.13 at 580°C, while for ($W\text{-}A1_2O_3$), the values are 0.93 and 0.14 at 550°C. The coatings have also shown good stability when exposed to air for long periods at temperatures up to 450°C [18,19].

From 2007 onwards, a group of research institutes, laboratories and companies in Switzerland have collaborated to develop the design of a parabolic trough concentrating solar power plant with many unique features [20, 21]. The overall objective has been to make the solar power generated in such plants more cost effective. This has been achieved by introducing the following innovative features:

1. Air is used as the heat transfer fluid instead of thermic fluids. Thus the maximum operating temperature is not restricted to 400°C and can be raised to about 600°C. The efficiency of the thermodynamic cycle is thereby increased.
2. The traditional metal frame used for the parabolic concentrator is replaced by a concrete frame on which an inflated metallized polymer film is used as the reflector material.
3. Since air is the working fluid, the absorber tube of the receiver is a large horizontal pipe having a diameter of 30 or 40 cm and acting like a cavity-type receiver. It is well insulated and has window slits along the length through which the concentrated solar radiation enters, gets absorbed and heats the flowing air.

4. The use of air as the working fluid enables the use of a packed bed of rocks as the thermal storage.

The first pilot plant with this technology has been constructed in Ait Baha, Morocco in 2014. It has three collectors, 211m long and having an aperture of 9 m. The thermal output is 3 MW.

In recent years, parabolic trough systems in which steam is generated directly in the solar field have also been tested and developed. A good review of the subject has been made by Hirsch *et al.* [22]. The main motivation for having *Direct Steam Generation* (DSG) plants is to overcome the temperature limit of 400°C imposed by the use of thermic fluids. As stated earlier, with higher temperatures, the efficiency of the thermodynamic cycle increases.

The main technical questions which have to be tackled are the issues which arise due to the two-phase flow of water and steam in the long horizontal absorber tubes of the receiver. The two-phase flow results in non-uniform circumferential and axial cooling along the inner surface of the absorber tube. This coupled with non-uniform circumferential heating due to concentrated solar radiation falling on the outer surface leads to temperature gradients which are difficult to handle.

A number of test demonstration facilities have been set up starting with the DISS facility in Almeria, Spain [23]. These have led to a proof of concept, an analysis of various modes of operation and also an evaluation of the pressure drops occurring in such situations. The first commercial plant has also been built and went into operation in 2012 [24]. It has 14 evaporator loops and 7 superheated loops and delivers steam at 30 bar, 300°C to a 5 MW steam turbine. It is expected that the next generation of DSG parabolic trough solar power plants will be able to generate steam at higher temperatures and pressures. Progress has also been made in developing DSG linear Fresnel reflector power plants.

LINEAR FRESNEL REFLECTOR CONCENTRATING COLLECTOR 6.5

6.5.1 Description

The linear Fresnel reflector (LFR) concentrating collector has been described briefly in Section 2.2. It is made up of a number of reflector-rows placed parallel to each other at the ground level and a long horizontal receiver mounted at a height. The receiver remains stationary and the reflector-rows track the movement of the sun so that reflected rays from the reflectors are focused on the receiver. The reflectors are usually slightly curved to reduce the mouth size of the receiver so that the heat losses from the receiver are reduced. Each reflector is suitably mounted on the ground and connected to an appropriate tracking mechanism. The reflector-rows are spaced apart for reducing shading and blocking. Shading occurs when one reflector blocks the incident sun rays falling on an adjacent reflector. Blocking occurs when one reflector blocks the reflected rays from another reflector directed towards the receiver. These effects lead to the non-utilisation of a part of the aperture of a reflector resulting in energy loss. Higher the spacing, lower is the effect of shading and blocking. This results in a higher land usage and hence cost. Hence, a compromise is to be made on the land cost against shading and blocking losses.

The receiver may consist of a single absorber tube with a black coating placed in a long cavity (Fig. 6.11). Sometimes a row of tubes is used. Usually receivers have a transparent glass opening facing the reflector-rows. The side and back walls of the cavity act as secondary reflectors and are insulated to reduce thermal loss. In Fig. 6.11, the side and back walls have the shape of a compound parabolic concentrator (CPC) (discussed in Section 6.6). Thus, the radiation absorbed in the absorber tube is the sum of radiation coming directly from the reflector-rows and the radiation coming after secondary reflections in the CPC. The working fluid flowing through the absorber tube is usually water. Depending upon the application, the water may or may not be converted into steam. The use of water eliminates the need for a heat exchanger and reduces the cost of the system. Synthetic oils such as Dowtherm A, Therminol VP1, etc, are also used as working fluids. Efforts are being made to use a molten salt, which can also serve as a storage medium, as the working fluid.

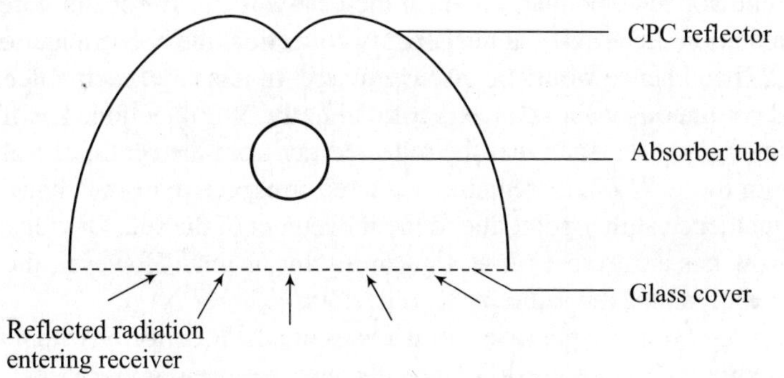

Fig. 6.11 *Cross-section of a typical receiver of LFR collector*

Mills and Morrison [25] have proposed an interesting modification to a LFR collector. They have suggested that if more than one receiver is used for a given field of reflector-rows in a LFR collector, the reflector-rows would have the option of directing reflected radiation to any close-by receiver. This would allow a reduction in the distance between reflector-rows. Yet it would be possible to avoid shading and blocking. The closely packed reflector-rows would result in lower land use. Further, the variable reflector orientation would allow the use of a lower receiver height. Such a collector has been called a Compact Linear Fresnel Reflector (CLFR) collector. Figure 6.12 shows the schematic cross-section of a CLFR collector with two receivers.

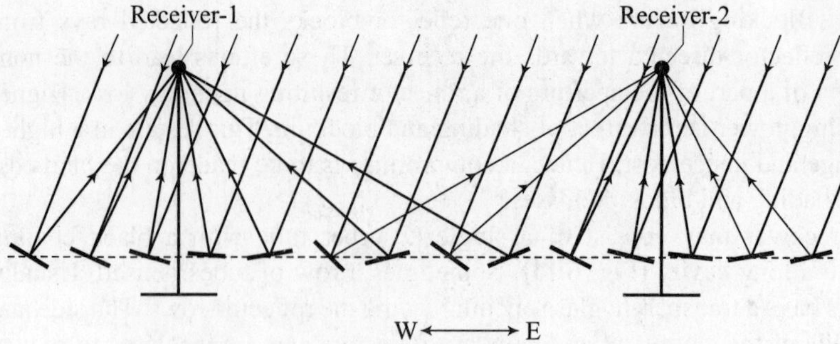

Fig. 6.12 *Schematic cross-section of a CLFR collector*

It may be mentioned that in a LFR collector, the reflector-rows are close to the ground level. Consequently, they experience less wind load compared to reflectors at a height. As a result, lighter structures can be adopted for their support. Besides, the cleaning of ground-based reflectors becomes easy. Thus, the LFR has some in-built economic advantages.

6.5.2 Tracking and Orientation

Like a cylindrical parabolic collector, a LFR concentrating collector is oriented with its reflector-rows pointed either in the east-west (E-W) or the north-south (N-S) orientation. In terms of annual energy collection, the N-S orientation is the best [26, 27] and hence would be discussed here. In this case, each reflector-row is tracked continuously about an axis parallel to the N-S direction. The tilt angle of a reflector is chosen such that the reflected ray from the centre of a reflector, projected in the E-W plane, remains unaltered irrespective of the change in the angle of incidence at that point due to the movement of the sun. Obviously, each reflector-row has a different tilt at a given instant of time. However, the rate of change of tilt remains the same for all reflectors.

Figure 6.13 shows the plan view of the axes of a LFR collector with n number of reflector-rows. The centre to centre distance between any adjacent pair of reflector-rows is the pitch (p). It may be constant or may vary for a collector. We will consider a constant pitch arrangement. When n is even, $[n/2]$ reflector-rows will be deployed on either side of the receiver axis. In case n is odd, $[(n-1)/2]$ reflector-rows will be deployed on either side of the receiver and the middle reflector-row will be placed just beneath the receiver. Let the extreme reflector-row lying on the negative side of the Y-axis be numbered as the 1st reflector-row. Then the extreme reflector-row lying on the positive side of the Y-axis would be numbered as the nth reflector-row.

In order to focus solar beam radiation on the receiver, an appropriate tilt angle of a reflector-row needs to be found out at any instant. We will derive an expression for the tilt of a reflector-row. Let us consider the ith row, the tilt angle being denoted as β_i. To find an expression for β_i, let us consider a ray incident on the ith reflector-row at any instant (Fig. 6.14). Let BO and OD be the projections of the incident and reflected rays on the E-W plane (plane perpendicular to the LFR axis) respectively. *OC* is the normal to the reflector-aperture and *OA* is the vertical. It is obvious that the angle *AOC* is equal to β_i. Let ψ denote the angle, in the E-W plane, between the vertical and the projection of the incident sun ray on the E-W plane. Let Ω_i be the angle in the E-W plane between the vertical and the projection of the reflected ray of ith reflector-row on the E-W plane. In Fig. 6.14, *AOB* is ψ and *AOD* is Ω_i.

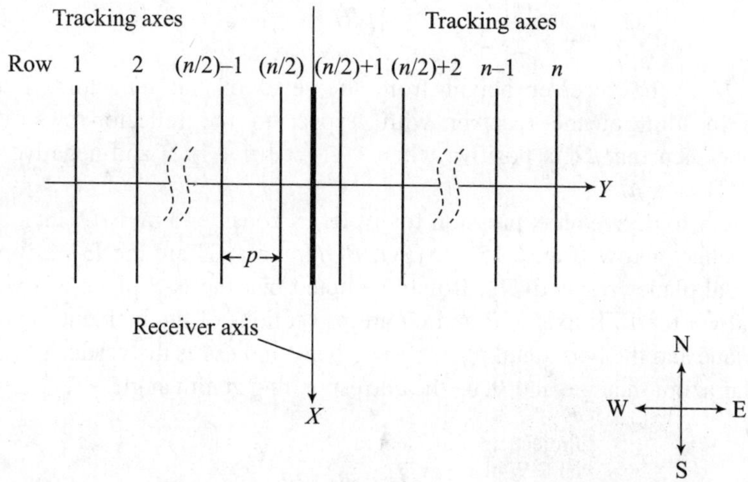

Fig. 6.13 *Plan view of LFR collector with an even number of reflector-rows*

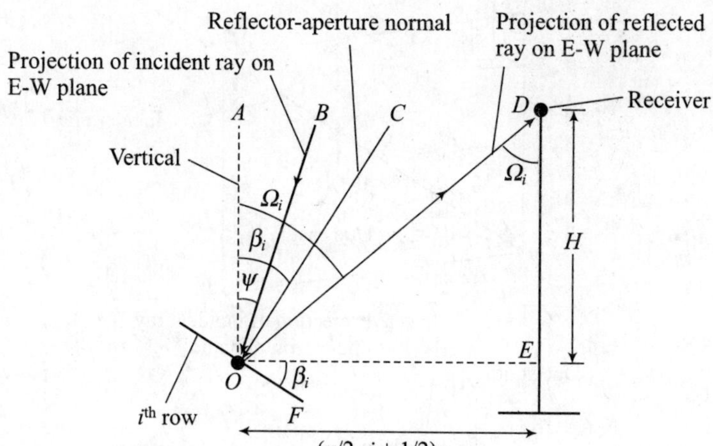

Fig. 6.14 *Calculation of tilt angle β_i of ith row*

Using the laws of reflection, it is seen that

$$\beta_i - \psi = \Omega_i - \beta_i$$

or

$$\beta_i = (\Omega_i + \psi)/2 \tag{6.51}$$

It is possible that the projections of the incident and reflected rays on the E-W plane may lie on the other side of the vertical depending on the location of the sun and the position as well as orientation of the reflector-row. It is therefore desirable to write the tilt angle β_i in the following form:

$$\beta_i = |(\Omega_i + \psi)/2| \tag{6.52}$$

Considering the right-angled triangle *DEO*, one can write

$$\Omega_i = \tan^{-1}\left[\left(\frac{n}{2} - i + \frac{1}{2}\right)p/H\right] \tag{6.53}$$

where H is the receiver height from the level of the reflectors. Because of the location of the receiver with respect to the reflector-rows, it can easily be seen that Ω_i is positive when $1 \leq i < [(n + 1)/2]$ and negative when $[(n + 1)/2] < i \leq n$.

In order to derive an expression for ψ, let us consider a ray, *JO*, incident on the ith reflector-row (Fig. 6.15). *OABG*, *BGIJ* and *OGI* are the E-W, N-S and horizontal planes respectively. It may be noted that the N-S plane is a vertical plane along the LFR axis. *OB* and *OI* are projections of the incident ray on the E-W plane and the horizontal plane respectively and *OA* is the vertical. Let γ_s be the solar azimuth angle and θ_z be the corresponding zenith angle.

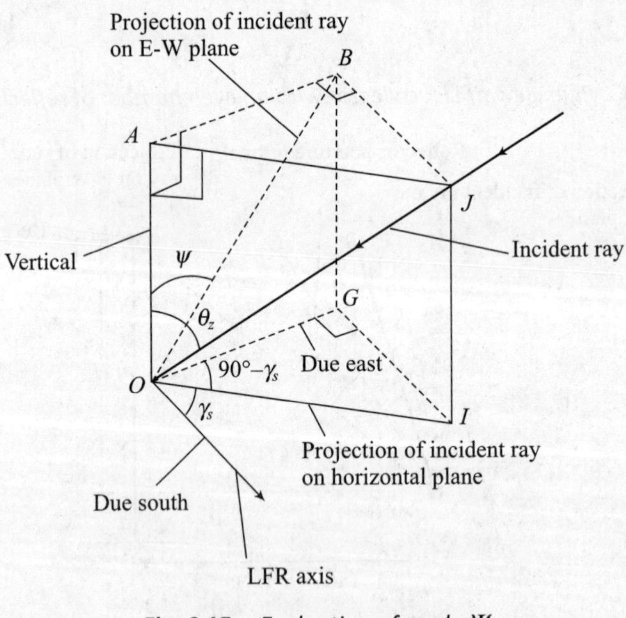

Fig. 6.15 *Evaluation of angle Ψ*

From triangle *OAB*, $\tan \psi = AB/AO$ (6.54)

From triangle *OAJ*, $\tan \theta_z = AJ/AO = OI/AO$ (6.55)

From triangle *OGI*, $\sin \gamma_s = OG/OI = AB/OI$ (6.56)

Combining Eqs (6.54) – (6.56)

$$\tan \psi = \tan \theta_z \sin \gamma_s$$

$$\psi = \tan^{-1} [\tan \theta_z \sin \gamma_s] \tag{6.57}$$

It may be noted that ψ is defined in such a way that it is positive when $0° < \gamma_s < 180°$, negative when $0° > \gamma_s > -180°$ and zero when $\gamma_s = 0°$ and 180°.

The values of Ω_i and ψ obtained from Eqs (6.53) and (6.57) respectively are substituted in Eq. (6.52) in order to obtain the tilt angle β_i.

Apart from the tilt, the orientation of a reflector-row, i.e. the surface azimuth angle γ_i, is needed for estimating the solar radiation incident on it. In the present case, because of the N-S orientation of the LFR axis, the surface azimuth angle of any row is simply given by

$$\begin{aligned}\gamma_i &= +90°, \text{ if } (\Omega_i + \psi) > 0, \text{ i.e. the reflector is facing due east}\\ &= 0°, \text{ if } (\Omega_i + \psi) = 0, \text{ i.e. the reflector is horizontal}\\ &= -90°, \text{ if } (\Omega_i + \psi) < 0, \text{ i.e. the reflector is facing due west}\end{aligned} \tag{6.58}$$

With the tilt and orientation of a reflector-row known, the amount of solar radiation incident on its aperture at any instant can be calculated using Eq. (3.5).

6.5.3 Performance Analysis

Following Sharma *et al.* [27], we now consider the performance analysis of a linear Fresnel reflector concentrating collector. The collector consists of n parallel rows of reflectors and a fixed receiver. The width of each reflector-row is w and its length is L. The pitch of the reflector field is p and is constant for the collector. The receiver consists of a compound parabolic concentrator (CPC) as a secondary reflector, a flat glass cover at the mouth and an absorber tube. The absorber tube has an outer diameter D_o and is of the same length L as the reflector-rows. We wish to derive an expression for the heat gain rate of the collector at a specified location on a given day at a given time.

The collector is operated as per the tracking mode described in Section 6.5.2. The following assumptions are made in order to simplify the analysis.

(a) The surface of each reflector is flat. This is a reasonable assumption because the small curvature of the mirror has no effect on shading calculations and has a very small effect on blocking.

(b) Since the sun's rays would be incident at an angle, some of the reflected rays reach the receiver a little beyond the length of the reflector. We assume that the receiver is long enough to intercept such reflected rays. For the purpose of analysis, this extra length is not considered and it is assumed that the radiation flux is the same all along the length of the receiver.

(c) The temperature drop across the absorber tube is negligible.

As mentioned, some portion of the aperture area of a reflector-row cannot receive solar radiation due to shading caused by a neighbouring row and hence would not contribute to focussing of radiation on the receiver. This area will keep on varying as the sun's position changes during the day. The calculation of this area is needed to estimate the solar radiation incident on the reflector-row. Similarly, due to blocking of reflected radiation by a neighbouring row, a part of the aperture area may not contribute to focussing of radiation on the receiver. Expressions will be derived for the shaded and blocked areas in order to calculate the effective aperture area.

❑ *Shaded Area* It may be noted first that no reflector row can shade the nth reflector-row when the solar azimuth angle lies between 0° and 180° (Fig. 6.13). Similarly, there will be no shading on the 1st reflector-row when the solar azimuth angle lies between 0° and –180°. If $L_{s,i}$ and $w_{s,i}$ are the length and width of the shaded area ($A_{s,i}$) of the ith reflector-row in the remaining cases, then $A_{s,i}$ can be written as

$$A_{s,i} = \begin{cases} 0 & \text{if } 0° \le \gamma_s \le 180° \text{ and } i = n \\ 0 & \text{if } 0° > \gamma_s > -180° \text{ and } i = 1 \\ L_{s,i}\, w_{s,i} & \text{for all other cases} \end{cases} \tag{6.59}$$

$L_{s,i}$ can be calculated using Fig. 6.16, which shows the plan view of two successive reflector-rows represented by two parallel lines; each line corresponds to the axis of rotation of the respective row. $L_{s,i}$ of the ith reflector-row is given by

$$L_{s,i} = L - p\,|\tan(\gamma_s - 90°)| \tag{6.60}$$

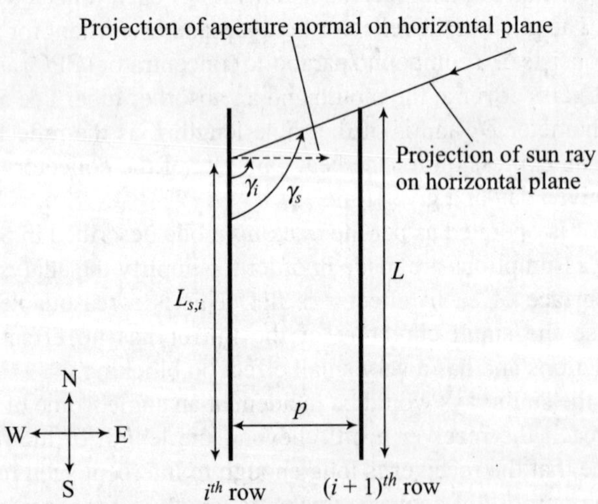

Fig. 6.16 *Plan view of successive rows*

It is observed that for typical values of p and L used in practice, the value of $p|\tan(\gamma_s - 90°)|$ is very small compared to L [27,28]. Thus, we can write $L_{s,i} = L$ without loss of accuracy.

Figure 6.17 is a sketch showing the loss of reflector-aperture area in the ith row due to shading by the $(i+1)^{th}$ row. This happens when $0° \le \gamma_s \le 180°$. When $0° > \gamma_s > -180°$, shading on the ith reflector-row occurs due to the $(i-1)^{th}$ reflector-row. Thus, the analysis for finding the shaded width of the ith reflector-row can be divided into the following eight cases on the basis of the values of the solar azimuth angle and surface azimuth angles of successive reflector-rows.

$0° \le \gamma_s \le 180°$

Case (i) $\quad \gamma_i = 90°, \quad \gamma_{i+1} = 90°$

Case (ii) $\quad \gamma_i = 90°, \quad \gamma_{i+1} = -90°$

Case (iii) $\quad \gamma_i = -90°, \quad \gamma_{i+1} = 90°$

Case(iv) $\quad \gamma_i = -90°, \quad \gamma_{i+1} = -90°$

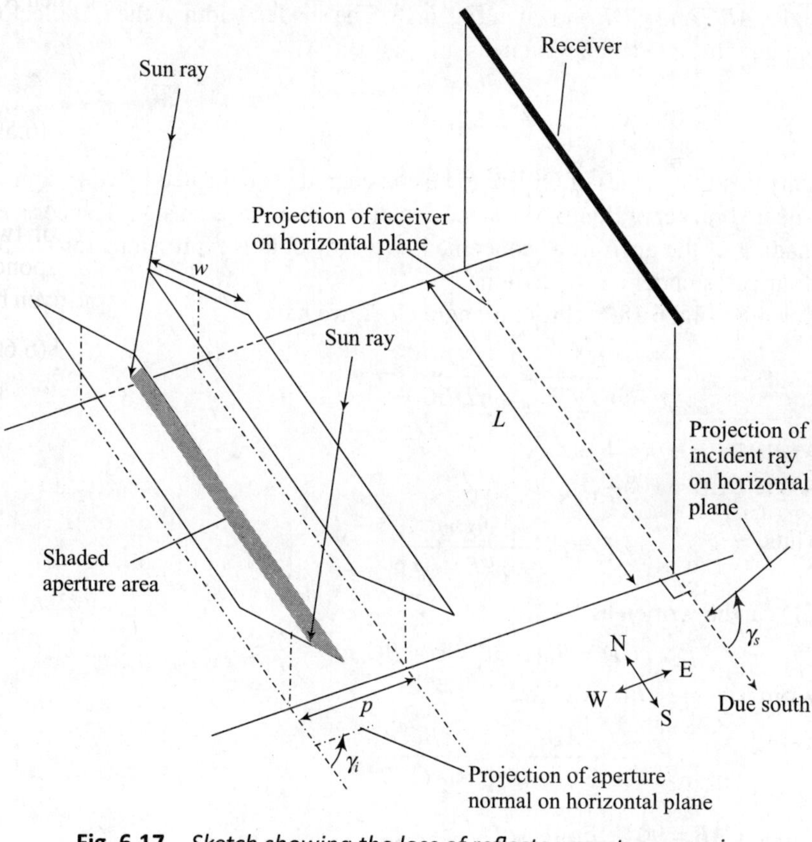

Fig. 6.17 *Sketch showing the loss of reflector-aperture area in i^{th} row due to shading caused by $(i+1)^{th}$ row*

$0° > \gamma_s > -180°$

Case (v) $\quad \gamma_i = 90°, \quad \gamma_{i-1} = 90°$

Case (vi) $\quad \gamma_i = 90°, \quad \gamma_{i-1} = -90°$

Case (vii) $\quad \gamma_i = -90°, \quad \gamma_{i-1} = 90°$

Case (viii) $\quad \gamma_i = -90°, \quad \gamma_{i-1} = -90°$

It may be noted that Cases (iii) and (vi) are not feasible and need not be considered further. In Case (iii), the ith row faces west, while the $(i + 1)^{th}$ row faces east and $0° \le \gamma_s \le 180°$. In Case (vi), the ith row faces east, while the $(i - 1)^{th}$ row faces west and $0° > \gamma_s > -180°$. One can see from Eq. (6.52) that such combinations of tilt angles (and hence surface azimuth angles) of adjacent rows cannot focus reflected radiation at any time on the receiver.

The remaining six cases are depicted in Fig. 6.18, which shows the aperture planes of two successive reflector-rows when seen along the tracking axis. Figures 6.18(a) - (f) respectively refer to Cases (i), (ii), (iv), (v), (vii) and (viii). It will be evident from Figs 6.18(c) and (d), Cases (iv) and (v), that because of the orientations of two neighbouring reflector-rows, there would be no shading at all. For the remaining four cases, the shaded widths can be calculated using the triangles *ABC* and *CDE* shown in Fig. 6.18. The shaded width of the ith reflector-row at any instant can be written as

$$w_{s,i} = \left[\frac{w}{2} - DE\right]^{+} \tag{6.61}$$

Only positive values of $w_{s,i}$ are to be considered, as indicated by a '+' sign on the square bracket of Eq. (6.61), since negative values have no significance as far as shading of the aperture is concerned. We now obtain expressions for *DE* for the four cases under consideration.

Consider Fig. 6.18(a). From triangle *CDE*, we have

$$\frac{DE}{\sin DCE} = \frac{CD}{\sin DEC}$$

Angle $DCE = 90° - \psi$
Angle $DEC = 90° - \beta_i + \psi$

Thus, $$DE = CD\frac{\cos\psi}{\cos(\beta_i - \psi)}$$

CD can be written as

$$CD = BD - BC = p - BC$$

From triangle *ABC*, we have

$$\frac{AB}{\sin(90° - \psi)} = \frac{BC}{\sin CAB}$$

Angle $CAB = 90° - \beta_{i+1} + \psi$

Thus, $$BC = \frac{w\cos(\beta_{i+1} - \psi)}{2\cos\psi}$$

and $$CD = \frac{2p\cos\psi - w\cos(\beta_{i+1} - \psi)}{2\cos\psi}$$

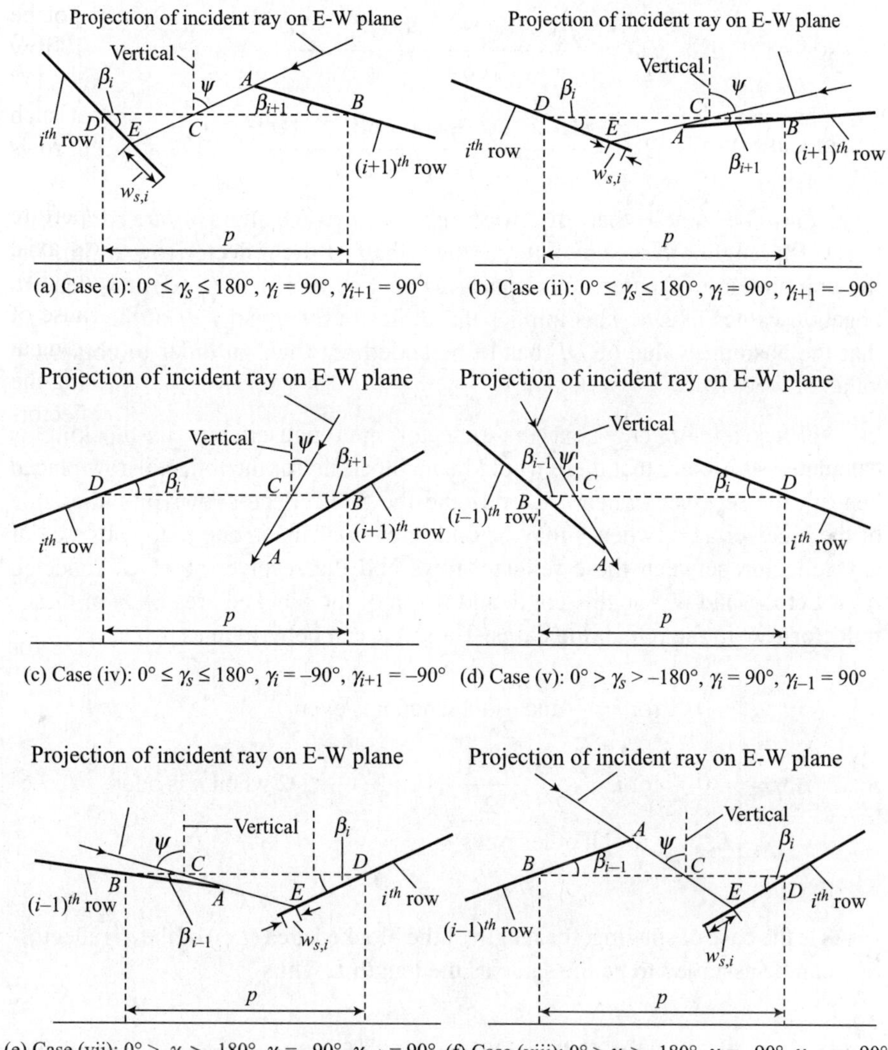

Fig. 6.18 *View of successive rows as seen along the tracking axis showing shaded widths*

For Case (i),
$$DE = \frac{2p\cos\psi - w\cos(\beta_{i+1} - \psi)}{2\cos(\beta_i - \psi)} \tag{6.62}$$

Similarly, using Fig. 6.18(b), (e) and (f), the following expressions are obtained for DE for the remaining cases:

For Case (ii),
$$DE = \frac{2p\cos\psi - w\cos(\beta_{i+1} + \psi)}{2\cos(\beta_i - \psi)} \tag{6.63}$$

For Case (vii), $$DE = \frac{2p\cos|\psi| - w\cos(\beta_{i-1} + |\psi|)}{2\cos(\beta_i - |\psi|)} \tag{6.64}$$

For Case (viii), $$DE = \frac{2p\cos|\psi| - w\cos(\beta_{i-1} - |\psi|)}{2\cos(\beta_i - |\psi|)} \tag{6.65}$$

It may be noted that the way the geometrical figures are drawn in Fig. 6.18, length *DE* is shown in the lower half of the reflector-row width. But the combination of parameters appearing in Eqs (6.62) to (6.65) may lead to negative values of *DE*. This implies that *E* lies in the upper half of the row and that the absolute value of *DE* has to be added onto *w*/2 in order to obtain the value of $w_{s,i}$.

❑ *Blocked Area* Next the reflector area unused due to blocking is calculated. It is clear that there would be no blocking for the reflector-row placed beneath the receiver (when *n* is odd) and the two reflector-rows on either side of the receiver axis (when *n* may be odd or even). This is due to the absence of any reflector between these reflector-rows and the receiver to block reflected rays. Let $L_{b,i}$ and $w_{b,i}$ be the length and width of the blocked area ($A_{b,i}$) of the ith reflector-row in the remaining cases. Thus, $A_{b,i}$ can be written as

$$A_{b,i} = \begin{cases} 0 & \text{for } i = \frac{n}{2} \text{ and } \frac{n}{2} + 1 \text{ when } n \text{ is even} \\ 0 & \text{for } i = \frac{n-1}{2}, \frac{n-1}{2} + 1 \text{ and } \frac{n-1}{2} + 2 \text{ when } n \text{ is odd} \\ L_{b,i} w_{b,i} & \text{for all other rows} \end{cases} \tag{6.66}$$

As is the case of shading, the length of the blocked area ($L_{b,i}$) of the ith reflector-row can be assumed to be the same as the length *L*. Thus

$$L_{b,i} = L \tag{6.67}$$

Figure 6.19 is a sketch showing the loss of reflector-aperture area in the ith row due to blocking by the $(i + 1)^{th}$ row. Blocking of the ith row occurs due to the $(i + 1)^{th}$ row for $1 \le i < n/2$ when *n* is even and for $1 \le i < [(n-1)/2]$ when *n* is odd. Blocking of the ith row occurs due to the $(i - 1)^{th}$ row for $[(n/2) + 1] < i \le n$ when *n* is even and for $[(n-1)/2 + 2] < i \le n$ when *n* is odd. Thus, the analysis for finding the blocked width of the ith reflector-row can be split into the following eight cases on the basis of the position of the row and the surface azimuth angle of successive rows.

Position of ith row: $1 \le i < n/2$ when *n* is even or $1 \le i < [(n-1)/2]$ when *n* is odd

Case (i) $\gamma_i = 90°$, $\gamma_{i+1} = 90°$

Case (ii) $\gamma_i = 90°$, $\gamma_{i+1} = -90°$

Case (iii) $\gamma_i = -90°, \quad \gamma_{i+1} = 90°$

Case (iv) $\gamma_i = -90°, \quad \gamma_{i+1} = -90°$

Position of ith row: $[(n/2) + 1] < i \le n$ when n is even or $[(n-1)/2 + 2] < i \le n$ when n is odd

Case (v) $\gamma_i = 90°, \quad \gamma_{i-1} = 90°$

Case (vi) $\gamma_i = 90°, \quad \gamma_{i-1} = -90°$

Case (vii) $\gamma_i = -90°, \quad \gamma_{i-1} = 90°$

Case (viii) $\gamma_i = -90°, \quad \gamma_{i-1} = -90°$

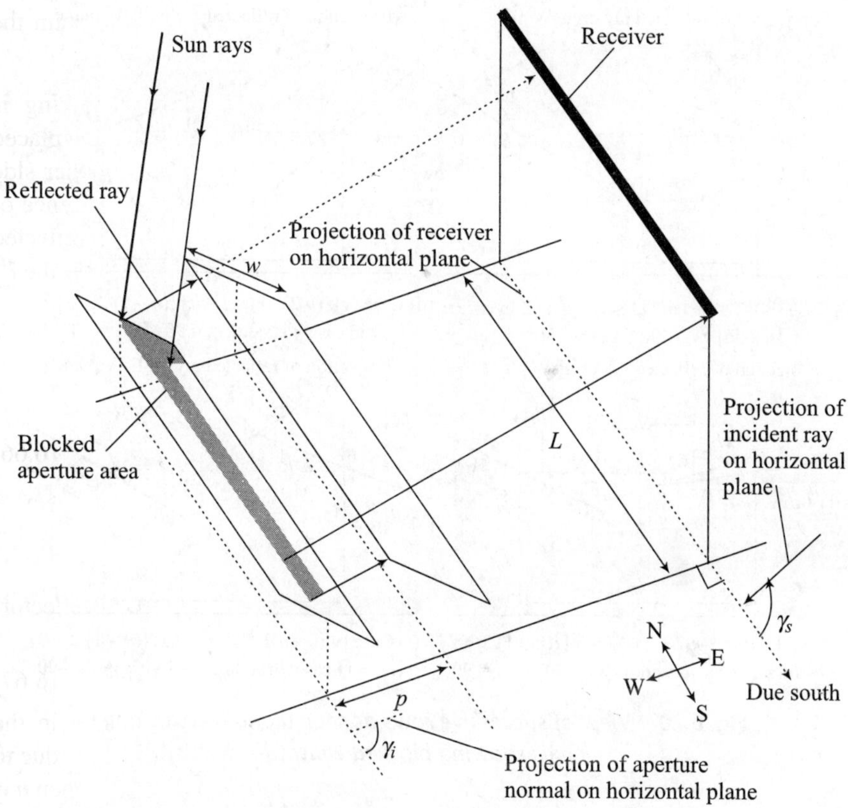

Fig. 6.19 *Sketch showing loss of reflector-aperture area in i^{th} row due to blocking by $(i + 1)^{th}$ row*

Obviously, Cases (iii) and (vi) are not feasible as these combinations of surface azimuth angles would not lead to focussing of radiation on the receiver. The remaining six cases are depicted in Fig. 6.20, which shows the aperture planes of two successive reflector-rows as seen along the tracking axis. Figure 6.20(a) – (f) respectively refer to Cases (i), (ii), (iv), (v), (vii) and (viii). It is seen from Figs 6.20(c) and (d) that there will be no blocking because of the orientations of the rows. For the remaining four cases, blocking can occur. The expression for the blocked width of the ith reflector-row can be written as follows (Fig. 6.21):

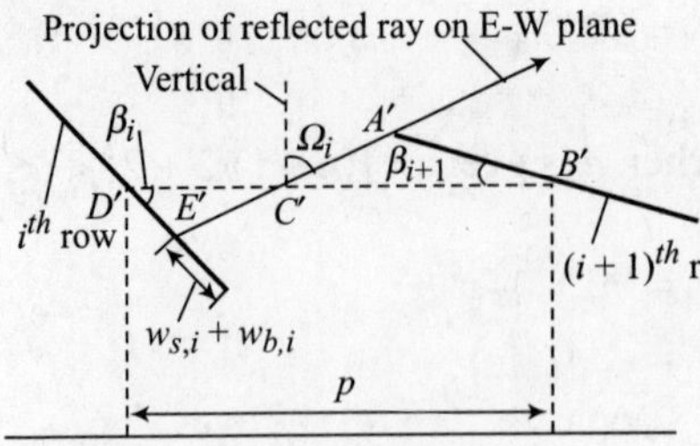

(a) Case (i): $1 \le i < n/2$ [$1 \le i < (n-1)/2$ in case n is odd], $\gamma_i = 90°$, $\gamma_{i+1} = 90°$

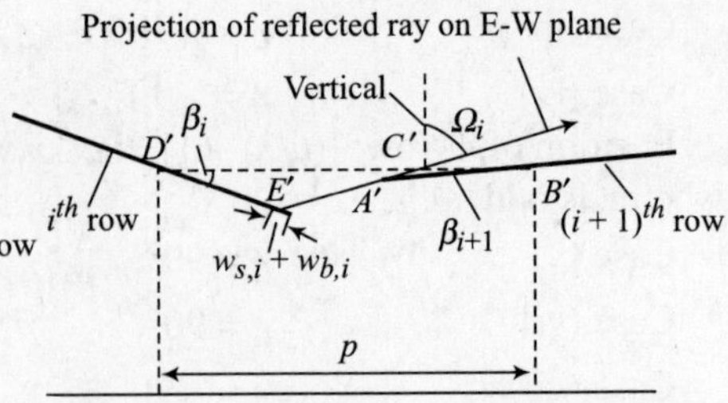

(b) Case (ii): $1 \le i < n/2$ [$1 \le i < (n-1)/2$ in case n is odd], $\gamma_i = 90°$, $\gamma_{i+1} = -90°$

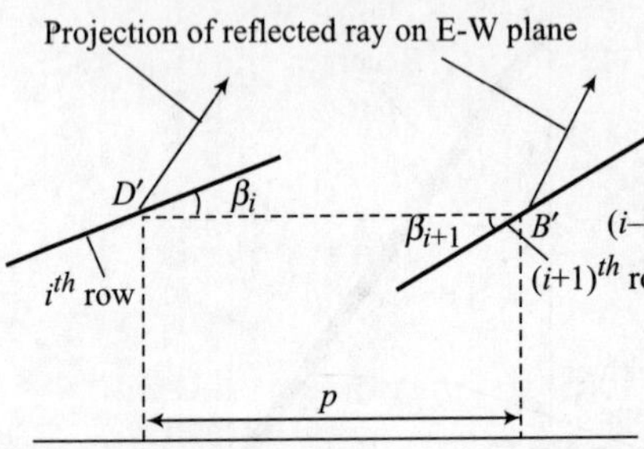

(c) Case (iv): $1 \le i < n/2$ [$1 \le i < (n-1)/2$ in case n is odd], $\gamma_i = -90°$, $\gamma_{i+1} = -90°$

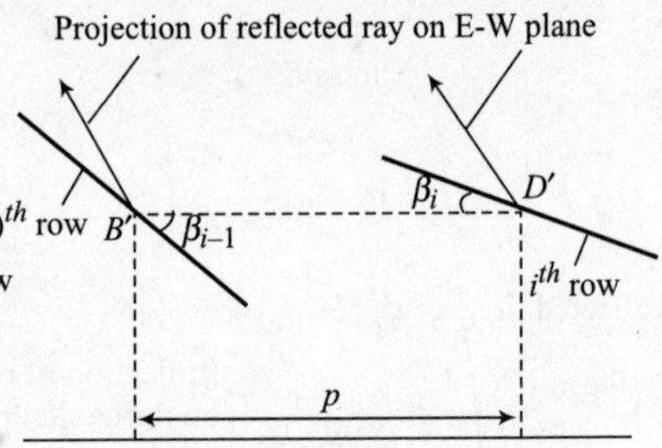

(d) Case (v): $(n/2 + 1) < i \le n$ [$\{(n-1)/2 + 2\} < i \le n$ in case n is odd], $\gamma_i = 90°$, $\gamma_{i-1} = 90°$

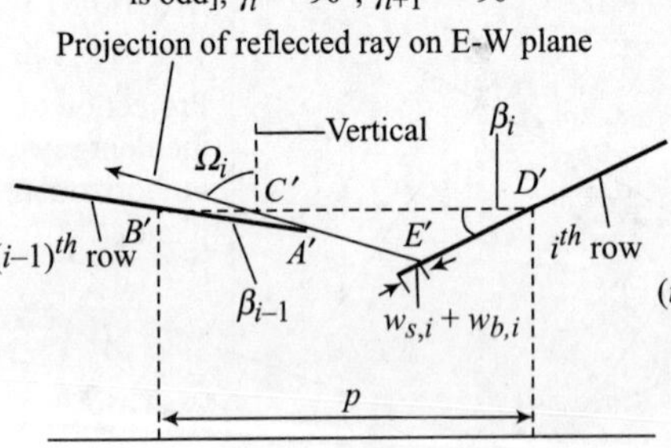

(e) Case (vii): $(n/2 + 1) < i \le n$ [$\{(n-1)/2 + 2\}$ $< i \le n$ in case n is odd], $\gamma_i = -90°$, $\gamma_{i-1} = 90°$

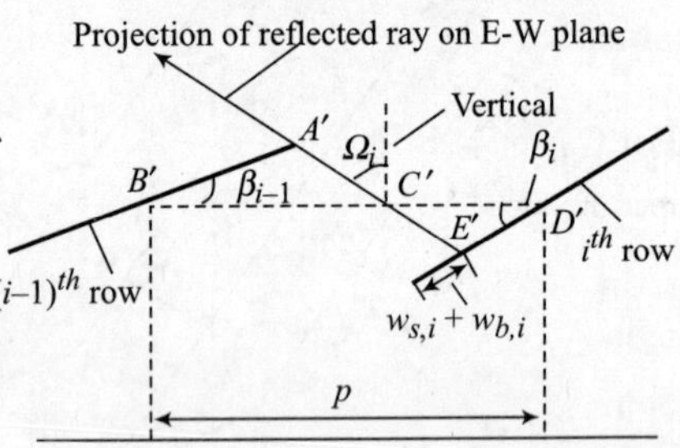

(f) Case (viii): $(n/2 + 1) < i \le n$ [$\{(n-1)/2 + 2\}$ $< i \le n$ in case n is odd], $\gamma_i = -90°$, $\gamma_{i-1} = -90°$

Fig. 6.20 *View of successive rows as seen along the tracking axis showing blocked widths*

$$w_{b,i} = \left[\frac{w}{2} - D'E' - w_{s,i}\right]^+ \tag{6.68}$$

Only positive values of $w_{b,i}$ are to be considered, as indicated by a '+' sign on the square bracket of Eq. (6.68), since negative values have no significance as far as blocking of aperture is concerned. Using the sine law of trigonometry for triangles $A'B'C'$ and $C'D'E'$ and following the same steps as outlined in the case of shading, we can show that for Case (i) in Fig. 6.20(a)

$$D'E' = \frac{2p\cos\Omega_i - w\cos(\beta_{i+1} - \Omega_i)}{2\cos(\beta_i - \Omega_i)} \tag{6.69}$$

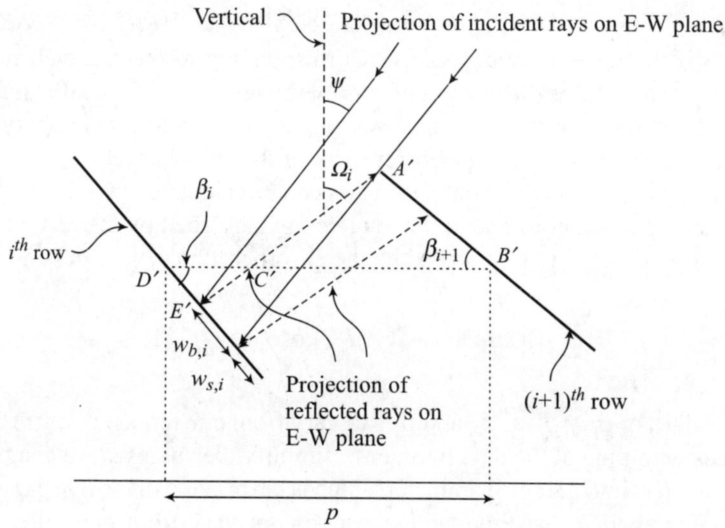

Fig. 6.21 *View of successive rows as seen along the tracking axis showing shaded and blocked widths*

Similarly, using Fig. 6.20(b), (e) and (f), the following expressions are obtained for $D'E'$:

For Case (ii), $$D'E' = \frac{2p\cos\Omega_i - w\cos(\beta_{i+1} + \Omega_i)}{2\cos(\beta_i - \Omega_i)} \tag{6.70}$$

For Case (vii), $$D'E' = \frac{2p\cos|\Omega_i| - w\cos(\beta_{i-1} + |\Omega_i|)}{2\cos(\beta_i - |\Omega_i|)} \tag{6.71}$$

For Case (viii), $$D'E' = \frac{2p\cos|\Omega_i| - w\cos(\beta_{i-1} - |\Omega_i|)}{2\cos(\beta_i - |\Omega_i|)} \tag{6.72}$$

It may be noted that the way the geometrical figures are drawn in Fig. 6.20, length $D'E'$ is shown in the lower half of the reflector-row width. But the combination of parameters appearing in Eqs (6.69) to (6.72) may lead to negative values of $D'E'$. This implies that E' lies in the upper half of the row and that the absolute value of $D'E'$ has to be added onto $w/2$ in order to obtain the value of $w_{b,i}$.

❑ *Useful Heat Gain Rate* The effective (useful) area of the ith row from which reflected radiation reaches the receiver

$$= wL - A_{s,i} - A_{b,i} \tag{6.73}$$

Thus, the effective incident radiation on the ith row $= (wL - A_{s,i} - A_{b,i})\, I_{bn}\cos\theta_i$ and the rate at which heat is absorbed by the absorber tube due to radiation reflected from the ith row

$$= [(wL - A_{s,i} - A_{b,i})\, I_{bn} \cos\theta_i]\, \rho_i\, \gamma_{int,i}\, \tau_i\, \rho_{ei}\, \alpha_i \qquad (6.74)$$

where ρ_i is the reflectivity of the row. Corresponding to the reflected radiation from the ith row, $\gamma_{int,i}$ is the intercept factor of the receiver, τ_i is the transmissivity of the transparent cover of the receiver, ρ_{ei} is the effective reflectivity of the secondary reflector and α_i is the absorptivity of the absorber tube.

The contribution of all the reflector-rows can be calculated in a similar fashion and combined together to obtain the total energy absorbed by the receiver. Thus the rate at which energy is absorbed by the absorber tube (q_{abs}) can be written as

$$q_{abs} = \sum_{i=1}^{n} [(wL - A_{s,i} - A_{b,i}) I_{bn} \cos\theta_i] \rho_i\, \gamma_{\text{int},i}\, \tau_i\, \rho_{ei}\, \alpha_i$$

The values of ρ_i, $\gamma_{int,i}$, τ_i, ρ_{ei} and α_i will vary from one reflector-row to another as well as with time of the day. In order to simplify the analysis, we combine the five parameters into a single parameter which is the product of the five parameters. Further, we assume that this parameter does not vary from one reflector-row to another and has a constant value. This single parameter is called the optical efficiency of the concentrator and will be denoted by the symbol η_o. Thus

$$q_{abs} = \eta_o \sum_{i=1}^{n} [(wL - A_{s,i} - A_{b,i}) I_{bn} \cos\theta_i] \qquad (6.75)$$

A part of the absorbed energy would be lost from the absorber tube. If q_l is the rate at which heat is lost from the absorber tube, then the useful heat gain rate q_u can be written, under steady-state conditions, as

$$q_u = q_{abs} - q_l = \eta_o \sum_{i=1}^{n} [(wL - A_{s,i} - A_{b,i}) I_{bn} \cos\theta_i] - q_l \qquad (6.76)$$

❑ *Expression for Heat Loss Rate* If U_l is the average overall loss coefficient of the entire absorber tube at an average temperature of T_{pm}, then the heat loss rate from the absorber tube can be written as

$$q_l = U_l (T_{pm} - T_a) \pi D_o L$$

$$= \frac{U_l}{C} (T_{pm} - T_a) nwL \qquad (6.77)$$

where C (= $nwL / \pi D_o L$) is the concentration ratio of the concentrator. T_a is the ambient temperature.

Based on experimental data, empirical correlations have been developed for calculating the heat loss from the absorber tube. For the receiver shown in Fig. 6.11, Morin *et al.* [29] have obtained the following expression for calculating q_l:

$$q_l = [a_1 (\overline{T}_f - T_a)^2 + a_2 (\overline{T}_f - T_a)]\, nwL \qquad (6.78)$$

In Eq. (6.78), q_l is expressed as a function of the temperature difference $(\bar{T}_f - T_a)$ where $\bar{T}_f$ is the arithmetic mean of the inlet and outlet fluid temperature and T_a is the ambient temperature. The constants a_1 and a_2 are given by

$$a_1 = 0.0007819 \text{ W/(m}^2\text{-K}^2) \text{ and } a_2 = -\,0.028164 \text{ W/(m}^2\text{-K)}$$

Equation (6.78) is based on data collected for an absorber tube diameter of 0.14 m, a secondary reflector having an opening of 0.5 m, and a mean fluid temperature lying between 25 and 500°C. It assumes that the absorber surface temperature is equal to the mean fluid temperature.

Two more receivers used in LFR collectors are shown in Fig. 6.22. Figure 6.22(a) shows a receiver with an evacuated glass tube around the absorber tube. A CPC secondary reflector is on the back side. Figure 6.22(b) shows a trapezoidal cavity receiver with four absorber tubes. Other types of trapezoidal receivers have also been investigated and heat loss correlations for them are available in the literature [30–32]. Moghimi *et al.* [33] have proposed an optimization technique for obtaining the geometrical parameters of the receiver shown in Fig. 6.22(b) when a set of operating conditions are specified.

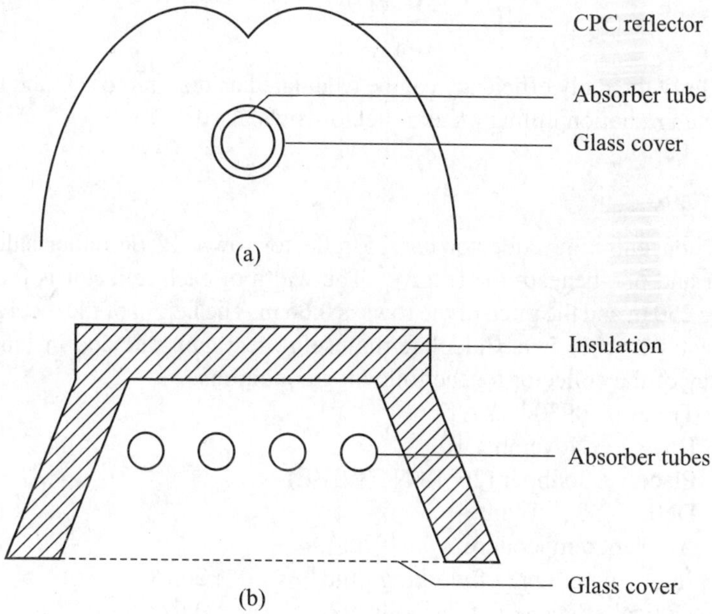

Fig. 6.22 *LFR receivers (a) Receiver with an evacuated annular space around the absorber tube and a CPC secondary reflector (b) Trapezoidal cavity receiver with four absorber tubes*

❑ *Useful Heat Gain* The useful heat gain of the collector over a time interval from t_1 to t_2 is given by

$$Q_u = \int_{t_1}^{t_2} q_u \, dt \tag{6.79}$$

For numerical calculations, it is useful to divide the time period t_1 to t_2 into N number of equal intervals. In each interval Δt, the values of all parameters are assumed to be constant. Then the integration is converted into a summation and we have

$$Q_u = \sum_{k=1}^{N} q_{u,k} \, \Delta t \tag{6.80}$$

where $q_{u,k}$ is the value of q_u at time $t = t_k$ and $t_k = t_1 + k\Delta t - \Delta t/2$

❑ *Efficiency* The instantaneous collection efficiency of the collector based on beam radiation is given by

$$\eta_{ib} = \frac{q_u}{wL\, I_{bn} \sum_{i=1}^{n} \cos\theta_i} \tag{6.81}$$

Similarly the daily efficiency can be calculated as the ratio of Q_u and the total daily beam radiation falling on all reflector-rows.

Example 6.4

A LFR concentrating collector has 25 reflector-rows, 12 on either side of the receiver and one beneath the receiver. The width of each reflector is 0.6 m, the length is 250 m and the pitch of the rows is 0.66 m. The height of the receiver from the reflector level is 5 m. Calculate the useful heat gain rate and instantaneous efficiency of the collector for the following conditions:

Time : 0930 h (LAT)
Day : November 10
Place : Jodhpur (26°18′N 73°01′E)
DNI : 818 W/m^2
Ambient temperature : 19°C
Inlet temperature of working fluid : 200°C
Optical efficiency of the collector : 0.62
Fluid flow rate through the absorber : 6.6 kg/s
Specific heat of working fluid : 2.154 kJ/kg-K

Assume that the receiver is of the type shown in Fig. 6.11 and that the correlation given in Eq. (6.78) for calculating the heat loss rate is valid.

Solution

On November 10, $n = 314$. From Eq. (3.4),

$$\delta = 23.45 \sin [(284 + 314) \times 360/365] = -17.913°$$

At 0930 h(LAT), $\omega = 37.5°$. Substituting in Eq. (3.7), we have

$$\cos \theta_z = \sin 26.3° \sin (-17.913°) + \cos 26.3° \cos (-17.913°) \cos 37.5°$$

$$= 0.5405$$

$$\theta_z = 57.28°$$

From Eq. (3.12),

$$\cos \gamma_s = [\cos 57.28° \sin 26.3° - \sin (-17.913°)] / (\sin 57.28° \cos 26.3°)$$

$$= 0.7253$$

$$\gamma_s = 43.51°$$

From Eq. (6.57),

$$\tan \psi = \tan 57.28° \sin 43.51° = 1.0716$$

$$\psi = 46.98°$$

We will show the detailed calculations only for the 1st reflector-row. From Eq. (6.53),

$$\tan \Omega_1 = [\{(25/2) - 1 + (1/2)\} \times 0.66] / 5 = 1.584$$

$$\Omega_1 = 57.74°$$

Therefore, from Eq. (6.52),

$$\beta_1 = (57.74° + 46.98°)/2 = 52.36°$$

Similarly, values of Ω and β can be calculated for all rows. The values are given in Table 6.4. It can be seen from the values of Ω and ψ that rows 1–21 face east and rows 22–25 face west.

For calculating the widths of the shaded area in rows 1–20, Fig. 6.18(a) is to be used, for the 21st row, Fig. 6.18(b) is to be used and for rows 22–24, Fig. 6.18(c) is to be used. It may be noted that the 25th row would not get shaded being the extreme row on the east and the sun being on the east.

Thus, for the 1st row, from Eq. (6.62),

$$DE = [2 \times 0.66 \times \cos 46.98° - 0.6 \times \cos (51.21° - 46.98°)] / [2 \times \cos (52.36° - 46.98°)]$$

$$= 0.152 \text{ m}$$

From Eq. (6.61),

$$w_{s,1} = (0.6/2) - 0.152 = 0.148 \text{ m}$$

Shaded area, $A_{s,1} = 0.148 \times 250 = 37.00 \text{ m}^2$

Table 6.4 Performance of a LFR concentrating collector - Example 6.4

Row	Ω (°)	β (°)	Incident radiation – no shading or blocking (kW)	w_s (m)	w_b (m)	Useful area (m^2)	Effective Incident radiation (kW)
1	57.74	52.36	96.786	0.148	0.098	88.50	57.104
2	55.44	51.21	96.943	0.149	0.075	94.00	60.744
3	52.85	49.92	97.080	0.149	0.051	100.00	64.650
4	49.91	48.45	97.176	0.150	0.025	106.25	68.917
5	46.56	46.77	97.207	0.150	0	112.50	72.981
6	42.74	44.86	97.141	0.149	0	112.75	73.051
7	38.38	42.68	96.934	0.147	0	113.25	73.151
8	33.42	40.20	96.528	0.144	0	114.00	73.285
9	27.83	37.41	95.854	0.140	0	115.00	73.458
10	21.60	34.29	94.833	0.134	0	116.50	73.666
11	14.79	30.89	93.397	0.125	0	118.75	73.896
12	7.52	27.25	91.500	0.114	0	121.50	74.122
13	0	23.49	89.151	0.100	0	125.00	74.314
14	–7.52	19.73	86.418	0.083	0	129.25	74.446
15	–14.79	16.10	83.423	0.064	0	134.00	74.505
16	–21.60	12.69	80.310	0.043	0	139.25	74.495
17	–27.83	9.57	77.214	0.022	0	144.50	74.434
18	–33.42	6.78	74.243	0	0	150.00	74.243
19	–38.38	4.30	71.461	0	0	150.00	71.461
20	–42.74	2.12	68.903	0	0	150.00	68.903
21	–46.56	0.21	66.578	0	0	150.00	66.578
22	–49.91	1.46	64.479	0	0	150.00	64.479
23	–52.85	2.93	62.590	0	0	150.00	62.590
24	–55.44	4.23	60.892	0	0	150.00	60.892
25	–57.74	5.38	59.366	0	0.015	146.25	57.857
			2096.407				1738.222

Note: Reflector-rows 1-21 are facing east. Reflector-rows 22-25 are facing west.

For calculating the widths of the blocked areas in rows 1–11, Fig. 6.20(a) is to be used; for rows 15–21, Fig. 6.20(d) is to be used; for the 22nd row, Fig. 6.20(e) is to be used and for rows 23–25, Fig. 6.20(f) is to be used. There would be no blocking for rows 12–14. As explained in Section 6.5.3, there would be no blocking for the situations shown in Fig. 6.20(d).

Thus, for the 1st reflector-row, from Eq. (6.69),

$$D'E' = [2 \times 0.66 \times \cos 57.74° - 0.6 \times \cos (51.21° - 57.74°)] / [2 \times \cos (52.36° - 57.74°)]$$

$$= 0.054$$

From Eq. (6.68),

$$w_{b,1} = (0.6/2) - 0.054 - 0.148 = 0.098 \text{ m}$$

Blocked area, $A_{b,1} = 0.098 \times 250 = 24.50 \text{ m}^2$

Useful area of the 1st row, Eq. (6.73),

$$= 0.6 \times 250 - 37.00 - 24.50 = 88.50 \text{ m}^2$$

The solar radiation incident on the row is $I_{bn} \cos \theta_1$

From Eq. (3.5), we have

$$\cos \theta_1 = \cos \theta_z \cos \beta_1 + \cos \delta \sin \omega \sin \beta_1$$

$$= \cos 57.28° \cos 52.36° + \cos (-17.913°) \sin 37.5° \sin 52.36°$$

$$= 0.7888$$

Solar radiation incident on the 1st row if there was no shading and blocking

$$= 0.818 \times 0.7888 \times 0.6 \times 250 = 96.786 \text{ kW}$$

Solar radiation incident on the useful area of the 1st row

$$= 0.818 \times 0.7888 \times 88.50 = 57.104 \text{ kW}$$

Similar calculations have been carried out for all reflector-rows and the values obtained are given in Table 6.4. It can be seen that the solar radiation received by all the reflector-rows if there was no shading and blocking = 2096.407 kW and that incident on the useful area = 1738.222 kW. Hence, the power absorbed by the absorber

$$= 1738.222 \times 0.62 = 1077.698 \text{ kW}$$

In order to calculate q_l, we need the value of $\overline{T}_f$. Since this is not known, an iterative calculation is necessary. We assume the value of $\overline{T}_f$ and proceed.

First iteration: Assume $\overline{T}_f = 215°\text{C}$

From Eq. (6.78), we get

$$q_l = [0.0007819(215 - 19)^2 - 0.028164(215 - 19)] \times 25 \times 0.6 \times 250 = 91.940 \text{ kW}$$

From Eq. (6.76),

$$q_u = 1077.698 - 91.940 = 985.758 \text{ kW}$$

$$= \dot{m}\, C_p\, \Delta T$$

$$\Delta T = 985.758/(6.6 \times 2.154) = 69.34 \text{ °C}$$

New value of $\overline{T}_f = 200 + (69.34/2) = 234.67 \text{ °C}$

Second iteration:

Repeating the calculations with $\overline{T}_f$ = 234.67 °C, we get

$$q_l = 113.605 \text{ kW}$$

$$q_u = 964.093 \text{ kW}$$

$$\Delta T = 67.82 \text{ °C}$$

$$\overline{T}_f = 233.91 \text{ °C}$$

Third iteration:

With this new value of $\overline{T}_f$, we get

$$q_l = 112.724 \text{ kW}$$

$$q_u = 964.974 \text{ kW}$$

$$\Delta T = 67.88 \text{ °C}$$

$$\overline{T}_f = 233.94 \text{ °C}$$

This value is close to the value of $\overline{T}_f$ obtained after the 2nd iteration. Repeating the calculations once more, we get the final values of q_l and q_u

$$q_l = 112.760 \text{ kW}$$

$$q_u = 964.938 \text{ kW}$$

Instantaneous efficiency, from Eq. (6.81),

$$\eta_{ib} = 964.938/2096.407 = 0.460$$

6.5.4 Parametric Study of Collector Performance

We now study the effect of two geometric parameters, the height of the receiver and the pitch of the reflector-rows, on the performance of a LFR concentrating collector.

❑ *Effect of Height* The effect of increasing the height of the receiver is analysed for the LFR of Example 6.4. The values of height are increased from 5 m to 7 m in steps of 1 m, while the pitch is kept constant at 0.66 m. The shaded area, the blocked area and the effective incident solar radiation of each reflector-row are presented in Table 6.5. The total incident radiation for no shading and blocking, total effective incident radiation, the corresponding useful heat gain rate and efficiency are also given in the table. It is seen that the blocked area as well as the number of rows affected by blocking decreases as H increases. This is to be expected. As far as shading is concerned, it is seen that the magnitude of the shaded area changes marginally with increase in receiver height and there is a small increase in the number of shaded rows. This is because shading is mainly dependent on the elevation of sun. It can be seen that the effective incident radiation on the reflector-rows increases due to less blocking resulting in a higher useful heat gain rate. The instantaneous efficiency also increases.

Table 6.5 LFR concentrating collector – Effect of height of receiver (Pitch = 0.66 m)

Row	*H* = 5 m			*H* = 6 m			*H* = 7 m		
	Shaded area (m²)	Blocked area (m²)	Effective incident radiation (kW)	Shaded area (m²)	Blocked area (m²)	Effective incident radiation (kW)	Shaded area (m²)	Blocked area (m²)	Effective incident radiation (kW)
1	37.00	24.50	57.104	37.35	12.78	64.633	37.43	3.25	70.839
2	37.25	18.75	60.744	37.42	7.35	68.166	37.40	0	72.971
3	37.25	12.75	64.650	37.42	1.55	71.953	37.29	0	73.005
4	37.50	6.25	68.917	37.34	0	72.997	37.09	0	73.049
5	37.50	0	72.981	37.14	0	73.052	36.77	0	73.104
6	37.25	0	73.051	36.78	0	73.125	36.29	0	73.173
7	36.75	0	73.151	36.22	0	73.218	35.63	0	73.255
8	36.00	0	73.285	35.40	0	73.334	34.74	0	73.352
9	35.00	0	73.458	34.26	0	73.471	33.59	0	73.461
10	33.50	0	73.666	32.73	0	73.625	32.12	0	73.578
11	31.25	0	73.896	30.75	0	73.787	30.32	0	73.696
12	28.50	0	74.122	28.30	0	73.942	28.17	0	73.809
13	25.00	0	74.314	25.37	0	74.076	25.65	0	73.907
14	20.75	0	74.446	21.96	0	74.175	22.78	0	73.985
15	16.00	0	74.505	18.13	0	74.233	19.59	0	74.036
16	10.75	0	74.495	13.97	0	74.249	16.13	0	74.061
17	5.50	0	74.434	9.54	0	74.229	12.44	0	74.060
18	0	0	74.243	4.94	0	74.181	8.58	0	74.039
19	0	0	71.461	0.26	0	74.114	4.61	0	74.001
20	0	0	68.903	0	0	71.910	0.59	0	74.952
21	0	0	66.578	0	0	69.730	0	0	73.234
22	0	0	64.479	0	0	67.712	0	0	72.336
23	0	0	62.590	0	0	65.854	0	0	68.557
24	0	0	60.892	0	0	64.150	0	0	66.896
25	0	3.75	57.857	0	0	62.590	0	0	65.352
Total incident radiation – no shading and blocking (kW)			2096.407			2122.997			2142.821
Total effective incident radiation (kW)			1738.222			1786.506			1812.708
Useful heat gain rate (kW)			964.938			993.703			1009.310
Instantaneous efficiency (%)			46.0			46.8			47.1

❑ *Effect of Pitch* The effect of increasing the pitch is also analysed for the LFR of Example 6.4. The values of pitch are increased from 0.60 m (the minimum possible) to 0.84 m in steps of 0.12 m, while the height is kept constant at 5 m. For the three values of pitch, viz. 0.6 m, 0.72 m and 0.84 m, values of the shaded area, the blocked area and the effective incident radiation for each reflector-row are presented in Table 6.6. The total incident radiation for no shading and blocking, total effective incident radiation, the corresponding

Table 6.6 LFR concentrating collector – Effect of pitch of reflector-row (Receiver height = 5 m)

Row	$p = 0.6$ m			$p = 0.72$ m			$p = 0.84$ m		
	Shaded area (m^2)	Blocked area (m^2)	Effective incident radiation (kW)	Shaded area (m^2)	Blocked area (m^2)	Effective incident radiation (kW)	Shaded area (m^2)	Blocked area (m^2)	Effective incident radiation (kW)
1	47.50	16.59	55.529	26.56	32.52	58.542	5.40	50.09	60.607
2	47.60	11.59	58.771	26.80	26.52	62.383	5.75	43.28	64.904
3	47.66	6.20	62.283	27.00	19.91	66.639	6.08	35.68	69.745
4	47.64	0.41	66.068	27.14	12.65	71.346	6.38	27.17	75.204
5	47.52	0	66.387	27.19	4.71	76.523	6.61	17.66	81.351
6	47.26	0	66.460	27.09	0	79.644	6.71	7.10	88.238
7	46.80	0	66.557	26.77	0	79.745	6.59	0	92.926
8	46.09	0	66.683	26.14	0	79.883	6.12	0	93.070
9	45.05	0	66.837	25.06	0	80.072	5.11	0	93.282
10	43.61	0	67.016	23.40	0	80.309	3.33	0	93.573
11	41.70	0	67.207	20.99	0	80.579	0.49	0	93.929
12	39.28	0	67.393	17.74	0	80.850	0	0	92.064
13	36.34	0	67.552	13.58	0	81.078	0	0	89.151
14	32.90	0	67.666	8.58	0	81.226	0	0	85.621
15	29.03	0	67.725	2.86	0	81.281	0	0	81.730
16	24.82	0	67.732	0	0	79.458	0	0	77.771
17	20.37	0	67.695	0	0	76.115	0	0	73.981
18	15.77	0	67.627	0	0	72.952	0	0	70.504
19	11.13	0	67.540	0	0	70.037	0	0	67.397
20	6.51	0	67.445	0	0	67.397	0	0	64.661
21	1.97	0	67.349	0	0	65.030	0	0	62.268
22	0	0	66.180	0	0	62.919	0	0	60.178
23	0	2.58	63.194	0	0	61.039	0	0	58.352
24	0	7.24	59.571	0	0	59.366	0	0	56.749
25	0	11.81	56.235	0	0	57.872	0	0	55.338
Total incident radiation – no shading and blocking (kW)			2110.706			2082.644			2056.868
Total effective incident radiation (kW)			1630.702			1812.283			1902.594
Useful heat gain rate (kW)			900.863			1009.056			1062.837
Instantaneous efficiency (%)			42.7			48.5			51.7

useful heat gain rate and efficiency are also given in the table. It is seen that as spacing increases, the shading decreases. This is to be expected. Consequently, the total effective incident radiation and the useful heat gain rate increase resulting in higher instantaneous efficiency.

On the other hand, blocking increases as the pitch increases. The farther the reflector-row from the receiver, the higher is the tilt required of the row so as to focus radiation on the receiver and hence the increase in blocking. It may be mentioned that the effect of blocking would reduce if the height of the receiver is increased.

With the increase in the value of the pitch, two opposing effects are observed from Table 6.6. While the shading decreases, the blocking increases. It is, therefore, expected that there would be an optimum value for the pitch for a given width of reflector-row and the height of the receiver for which the useful heat gain rate would be maximum. In order to examine this, calculations have been done for three more pitch values of 0.96 m, 1.08 m and 1.20 m. The corresponding values of total incident radiation for no shading and blocking, total effective incident radiation, useful heat gain rate and instantaneous efficiency are presented in Table 6.7. It is seen that the total effective incident radiation and the useful heat gain rate increase initially, pass through a maximum and then decrease with increase in the values of pitch.

Table 6.7 Variation of total effective incident radiation, useful heat gain rate and instantaneous efficiency with pitch

Pitch (m)	Total incident radiation – no shading and blocking (kW)	Total effective incident radiation (kW)	Useful heat gain rate (kW)	Instantaneous efficiency (%)
0.60	2110.706	1630.702	900.863	42.7
0.72	2082.644	1812.283	1009.056	48.5
0.84	2056.868	1902.594	1062.837	51.7
0.96	2033.394	1911.574	1068.184	52.5
1.08	2012.078	1904.793	1064.147	52.9
1.20	1992.711	1894.025	1057.736	53.1

It is also observed from Table 6.6 that shading and blocking are less for reflector-rows close to the receiver. On the other hand, rows located far from the receiver, experience higher shading and blocking. This has led to the variable pitch concept. It implies that the pitch of rows should be gradually increased from the central region to the extreme ends. As a consequence, the overall shading and blocking will decrease resulting in higher useful area and an increase in the heat gain rate of collector.

The preceding calculations are based on the assumption that the value of optical efficiency (η_o) does not change. This is not quite correct. With changes in the values of tilts of reflector-rows, η_o would change and hence the numerical values of various parameters presented in Table 6.7 would get affected. Also, the calculations have been made only for one instant of time. If the calculations are carried out on a daily or annual basis, the changes in blocking effect are not found to be significant with changes in spacing. On the other hand, the cosine effect dominates. This effect refers to the fact that the aperture of the reflector-rows is not normal to the direction of the sun's rays. However, the total effective incident radiation and useful heat gain rate would still pass through a maximum and there exists an optimum value of the pitch for a given width of reflector-rows and height of the receiver [27].

COMPOUND PARABOLIC COLLECTOR (CPC) 6.6

Like the flat-plate collector with plane reflectors, the compound parabolic concentrating collector is also a non-imaging device. It has a large acceptance angle and requires only intermittent tracking. The usefulness of the geometry of the CPC for solar energy collection was noted by Winston [34] and it has been the subject of considerable attention. However, no significant commercial development has taken place.

6.6.1 Geometry of CPC

The geometry of an ideal two-dimensional CPC is shown in Fig. 6.23. The concentrator consists of two segments *AB* and *DC* which are parts of two parabolas 1 and 2. *AD* is the aperture of width *W*, and *BC* is the absorber surface of width *b*. The axes of two parabolas are oriented to each other at an angle in such a manner that the point *C* is the focus of parabola 1 and point *B* is the focus of parabola 2. Tangents drawn to the parabolas at points *A* and *D* are parallel to the axis of the CPC.

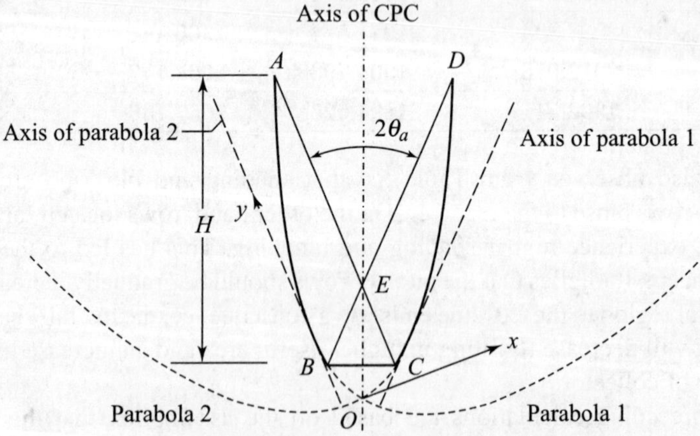

Fig. 6.23 *Geometry of a compound parabolic concentrating collector*

The acceptance angle of the CPC is the angle *AED* ($2\theta_a$), made by the lines obtained by joining each focus to the opposite aperture edge. The concentration ratio is given by $C = (W/b)$. It can be shown that $(W/b) = (1/\sin\theta_a)$ and that this concentration ratio is the maximum possible for the acceptance angle $2\theta_a$.

Using the *x-y* coordinate system shown in Fig. 6.23, with origin *O* at the vertex of parabola 2, it is easy to show that the equation for parabola 2 is

$$y = \frac{x^2}{2b(1+\sin\theta_a)} \tag{6.82}$$

where the focal length $OB = (1 + \sin\theta_a)\, b/2$. The coordinates of the end points of the segment *CD* are as follows,

Point *C*: $x = b\cos\theta_a$ $\quad y = \frac{b}{2}(1-\sin\theta_a)$

Point *D*: $x = (b + W)\cos\theta_a$ $\quad y = \frac{b}{2}(1-\sin\theta_a)\left(1+\frac{1}{\sin\theta_a}\right)^2$

The height-to-aperture ratio of the concentrator is given by

$$\frac{H}{W} = \frac{1}{2}\left(1+\frac{1}{\sin\theta_a}\right)\cos\theta_a = \frac{1}{2}(1+C)\left(1-\frac{1}{C^2}\right)^{1/2} \tag{6.83}$$

The surface area of the concentrator is obtained by integrating along the parabolic arc. Rabl [35] has shown that the ratio of the surface area of the concentrator to the area of the aperture is given by the expression

$$\frac{A_{con}}{WL} = \sin\theta_a(1+\sin\theta_a)\left[\frac{\cos\theta_a}{\sin^2\theta_a} + \ln\left\{\frac{(1+\sin\theta_a)(1+\cos\theta_a)}{\sin\theta_a[\cos\theta_a + (2+2\sin\theta_a)^{1/2}]}\right\} - \frac{\sqrt{2}\cos\theta_a}{(1+\sin\theta_a)^{3/2}}\right] \tag{6.84}$$

For values of concentration ratio greater than 3, it can be shown that the following simple expression (which predicts values to an accuracy better than 5 per cent) may be used in place of Eq. (6.84).

$$\frac{A_{con}}{WL} = 1 + C \tag{6.85}$$

Rabl has also shown that the average number of reflections *m* undergone by all radiation falling within the acceptance angle, before reaching the absorber surface, is given by the expression

$$m = \frac{1}{2\sin\theta_a}\left(\frac{A_{con}}{WL}\right) - \frac{(1-\sin\theta_a)(1+2\sin\theta_a)}{2\sin^2\theta_a} \tag{6.86}$$

where the value of (A_{con}/WL) is to be calculated from Eq. (6.84). Thus, the effective reflectivity of the concentrator surface is given by

$$\rho_e = \rho^m \tag{6.87}$$

where ρ_e = effective reflectivity, and

ρ = reflectivity value for a single reflection.

Example 6.5

A compound parabolic collector, 1 m long, has an acceptance angle of 20°. The absorber surface of the collector is flat and has a width of 10 cm. Calculate the concentration ratio, the aperture, the height and the surface area of the concentrator.

Solution

Concentration ratio $C = 1/\sin 10° = 5.76$

Therefore, aperture $W = 5.76 \times 10 = 57.6$ cm

From Eqs (6.83) and (6.85),

$$\frac{H}{W} = \frac{1}{2}\left(1 + \frac{1}{\sin 10°}\right)\cos 10° = 3.328 \quad \text{and} \quad \frac{A_{con}}{WL} = 1 + 5.76 = 6.76$$

Therefore, height $H = 3.328 \times 57.6 = 191.7$ cm

Surface area of concentrator,

$$A_{con} = 6.76 \times 0.576 \times 1 = 3.90 \text{ m}^2$$

The results of Example 6.5 show that compared to a cylindrical parabolic collector, a CPC is very deep and requires a large concentrator area for a given aperture. Fortunately, however, it has been shown that a large portion of the top of a CPC can be removed with negligible loss in performance. Thus, in practice, a CPC is generally truncated (reduced in height) by about 50 per cent in order to reduce its cost. A detailed study on the effects of truncation has also been carried out by Rabl [35].

6.6.2 Tracking Requiremesnts

A two-dimensional CPC is usually oriented with its length parallel to the horizontal E-W direction and aperture plane sloping towards the south. As stated earlier, only intermittent tracking is required, the frequency of adjustment depending upon the concentration ratio. Thus, for example, for a concentration ratio of 10, the acceptance angle is 11.5° and a tracking adjustment may be needed once every few days in order to ensure collection for 7 or 8 hours every day. On the other hand, for a lower concentration ratio of 5, the acceptance angle is 23.1° and a tracking adjustment may be needed only once in a month or two months.

In order to calculate the tracking requirement, we will use the solar geometry equations of Chapter 3. The apparent plane of motion of the sun does not coincide with the E-W plane passing through the axis of the CPC. Hence the sun 'rises' and 'falls' with respect to the CPC, an extreme position being attained at solar noon. It is thus necessary to calculate the solar 'swing' for the time period of the day for which collection is to be done. Referring to Fig. 6.24, assume that *OG* is a vertical stick whose shadow in the horizontal plane is *GE*. *EFGH* is a rectangle in the horizontal plane, with *EF* and *GH* being E-W lines. Thus angle *OEG* is the solar altitude angle α_a. The projection of this angle in a vertical north-south plane, i.e. angle *OFG,* may be called the *solar elevation angle.* We will denote it by the symbol α_v. The change in the angle α_v over the specified time period is the *solar swing.*

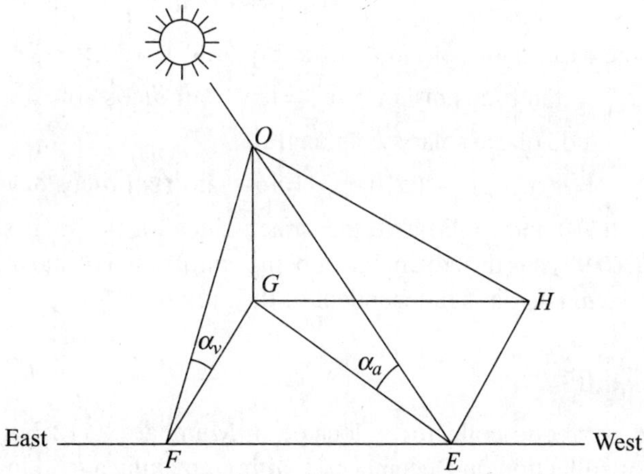

Fig. 6.24 *Calculation of solar swing*

From Fig. 6.24,

$$\tan \alpha_v = \frac{OG}{FG} = \left(\frac{OG}{OE}\right) \Big/ \left(\frac{EH}{OE}\right) = \frac{\sin \alpha_a}{\cos OEH}$$

Substituting expressions for sin α_a and cos *OEH* from Eqs (3.7) and (3.9) respectively, we have

$$\tan \alpha_v = \frac{\sin \phi \sin \delta + \cos \phi \cos \delta \cos \omega}{\sin \phi \cos \delta \cos \omega - \cos \phi \sin \delta} \tag{6.88}$$

At noon, when $\omega = 0$, we have,

$$\tan \alpha_v = \frac{\cos (\phi - \delta)}{\sin (\phi - \delta)} = \cot (\phi - \delta)$$

$$\text{Therefore, } (\alpha_v)_{\omega = 0} = \frac{\pi}{2} - (\phi - \delta) \tag{6.89}$$

The solar swing angle over a time period corresponding to an hour angle $+\omega_t$ to $-\omega_t$ is the change in α_v from the time corresponding to the angle $\pm\, \omega_t$ to solar noon. Thus the magnitude of the solar swing angle

$$= \left| \left[\frac{\pi}{2} - (\phi - \delta) \right] - \tan^{-1} \left[\frac{\sin \phi \sin \delta + \cos \phi \cos \delta \cos \omega_t}{\sin \phi \cos \delta \cos \omega_t - \cos \phi \sin \delta} \right] \right| \tag{6.90}$$

Equation (6.90) has been expressed in a neater form by Rabl [1]. He defines a solar elevation angle α'_v measured with reference to the equatorial plane rather than the horizontal plane. α'_v is related to α_v by the equation

$$\alpha'_v = \alpha_v - \left(\frac{\pi}{2} - \phi\right) \tag{6.91}$$

$$\text{Thus, } \tan \alpha'_v = -\cot(\alpha_v + \phi) = -\left[\frac{\cos\phi - \tan\alpha_v \sin\phi}{\tan\alpha_v \cos\phi + \sin\phi}\right]$$

Substituting the expression for $\tan \alpha_v$ from Eq. (6.88), we get

$$\tan \alpha'_v = \tan\delta/\cos\omega \quad \text{or} \quad \alpha'_v = \tan^{-1}(\tan\delta/\cos\omega) \tag{6.92}$$

Thus, the magnitude of the solar swing angle

$$= |(\alpha'_v)_{\omega=0} - (\alpha'_v)_{\omega=\omega_t}| = |\delta - \tan^{-1}(\tan\delta/\cos\omega_t)| \tag{6.93}$$

Equations (6.90) and (6.93) give the same values for the solar swing angle. However, Eq. (6.93) has the advantage of being simpler and of showing explicitly that the solar swing angle is independent of the latitude.

Example 6.6

A compound parabolic collector is located in Mumbai (19.12°N) and is to be used for 8 h of collection on December 21 without making a tracking adjustment during the day. Calculate the minimum acceptance angle required for the collector, its concentration ratio and its orientation.

Solution

First we calculate the solar swing angle on December 21 from 0800 h to 1200 h (LAT). Substituting $\delta = -23.45°$, $\omega_t = 60°$ in Eq. (6.93),

$$\text{Solar swing angle} = |-23.45° - \tan^{-1}[\tan(-23.45°)/\cos 60°]|$$

$$= |-23.45° - (-40.94°)| = 17.5°$$

The minimum acceptance angle required for the collector is obviously equal to the solar swing angle. Thus,

$2\theta_a = 17.5°$ and concentration ratio $C = (1/\sin 8.75°) = 6.57$.

The slope of the collector aperture plane would have to be adjusted such that the sun's rays enter parallel to the axis of the upper parabolic segment at 0800 h. This will ensure that the sun's rays at noon enter parallel to the axis of the lower parabolic segment. Now from Eq. (6.89),

$$(\alpha_v)_{\omega=0} = \frac{\pi}{2} - (19.12° + 23.45°) = 47.43°$$

Therefore, angle made by the axis of the lower parabolic segment with the horizontal = 47.43°.

Angle made by axis of CPC with the horizontal

$$= 47.43° - \theta_a = 38.68°$$

It should be noted that the solar swing angle is maximum on the solstice days, viz. June 21 and December 21, and equal to zero on the equinox days, viz. March 21 and September 21. Thus a collector having an acceptance angle of 17.5° would give 8 or more hours of collection on all the days of the year without requiring a diurnal tracking adjustment.

Equation (6.93) can also be used for calculating the collection time per day for a CPC with a given acceptance angle or for calculating the number of adjustments required in a year for a CPC with a given acceptance angle and a required minimum collection time per day. This calculation has been done by Rabl [1]. The procedure adopted was as follows: On June 21, the collector is assumed to be oriented with its axis pointing at an angle θ_a above the solar noon altitude angle. It is left in this position until the day when the collection time falls below the minimum specified value. On this day, it is adjusted with its axis again pointing at an angle θ_a above the solar noon altitude angle. The procedure is repeated over a time span of one year in order to obtain the number of adjustments per year.

6.6.3 Performance Analysis

Consider a compound parabolic collector having an aperture W, length L and acceptance angle $2\theta_a$ (Fig. 6.25). The absorber surface has a width b. The heat collected at the absorber surface is transferred to a fluid flowing through N tubes, having an outer diameter D_o and an inner diameter D_i, attached to the bottom side. The fluid enters at a temperature T_{fi}, leaves at a temperature T_{fo}, and has a mass flow rate $\dot{m}$. The aperture, which is covered with a transparent sheet, is assumed to be sloping south at such an angle that the beam radiation incident on it is within the acceptance angle of the collector. It is also assumed that the length of the absorber surface is a little more than the length L so that it intercepts all the reflected radiation.

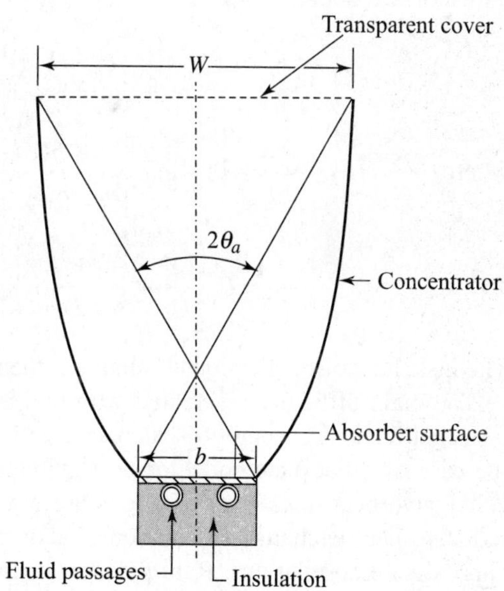

Fig. 6.25 *Compound parabolic collector with flat absorber surface*

We first derive an expression for the flux S absorbed at the absorber surface. Because of its large acceptance angle, a CPC accepts both beam and diffuse radiation. The beam radiation flux falling on the aperture plane is $I_b r_b$, while the diffuse radiation flux within the acceptance angle[¶] is given by (I_d/C). Thus, the total effective flux entering the aperture plane is $[I_b r_b + (I_d/C)]$ and

$$S = \left[I_b r_b + \frac{I_d}{C}\right] \tau \rho_e \alpha \qquad (6.94)$$

where τ = transmissivity of the cover,
ρ_e = effective reflectivity of the concentrator surface for all radiation, and
α = absorptivity of the absorber surface.

The values of τ, ρ_e and α are assumed to be the same for beam and diffuse radiation. It is to be noted that the flux S is based on the area of the aperture.

In order to obtain an expression for the useful heat gain rate, we take an energy balance on an elementary slice dx of the absorber surface at a distance x from the inlet. This yields the equation

$$dq_u = \left[S - \frac{U_l}{C}(T_p - T_a)\right] W\,dx \qquad (6.95)$$

Proceeding along lines similar to those adopted for the cylindrical parabolic collector, we obtain

$$q_u = F_R W L \left[S - \frac{U_l}{C}(T_{fi} - T_a)\right] \qquad (6.96)$$

where
$$F_R = \frac{\dot{m} C_p}{b U_l L}\left[1 - \exp\left\{-\frac{F' b U_l L}{\dot{m} C_p}\right\}\right] \qquad (6.97)$$

and
$$\frac{1}{F'} = U_l \left[\frac{1}{U_l} + \frac{b}{N \pi D_i h_f}\right] \qquad (6.98)$$

The instantaneous collection efficiency is then given by Eq. (6.30).

The main difficulty associated with the use of Eq. (6.96) is that the overall loss coefficient U_l cannot be calculated with the same degree of accuracy as in the case of a flat-plate collector or a cylindrical parabolic collector. Adequate convective heat transfer correlations are not available for the purpose and the radiative heat exchange calculation is also more complicated. Based on some approximate correlations, Rabl [35] has estimated the value of U_l for different values of the absorber plate temperature, plate emissivity and the concentration ratio. These values are given in Table 6.8. It is seen that they vary from 4 to 19.4 W/m^2-K. Values of U_l for situations differing from those given in Table 6.8 may be obtained by interpolation.

[¶] The derivation of this expression is not given here. It is based on the use of the reciprocal relation between shape factors for radiative heat exchange.

Table 6.8 Overall loss coefficient in a compound parabolic concentrator (Units: W/m^2-K)

T_{pm} (°C)	$C = 1.6$		4.0		8.0	
	$\varepsilon_p = 0.1$	0.9	0.1	0.9	0.1	0.9
110	4.0	8.2	5.6	10.5	6.9	11.8
210	5.0	11.8	7.8	16.0	10.4	19.4

Apart from the flat one-sided absorber surface shown in Fig. 6.25, CPCs can also be designed for other absorber surfaces. Two shapes, (a) tubular, and (b) tubular with longitudinal fins, are shown in Fig. 6.26. Unlike the flat absorber surface, these shapes have the advantage of not having a back side through which heat can be lost.

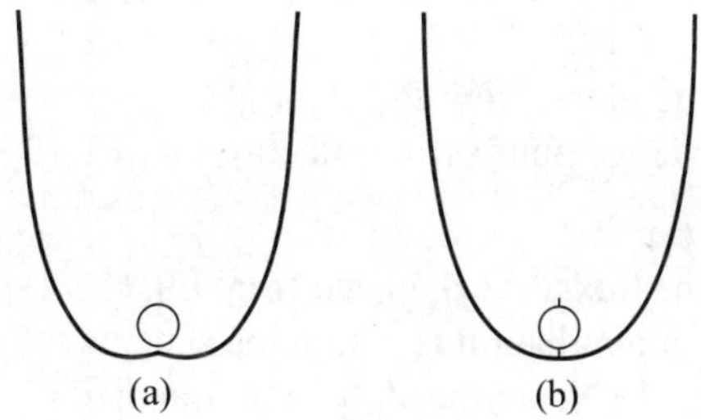

Fig. 6.26 *Other absorber shapes used in CPCs: (a) Tubular, (b) Tubular with longitudinal fins*

Example 6.7

A CPC is mounted on a horizontal E-W axis and oriented with its aperture plane sloping at an angle of 40°. The concentration ratio of the collector is 6.5, the width of its absorber plate is 6 cm and its length is 2 m. The collector is used for heating a fluid (C_p = 2.35 kJ/kg-K) which enters at a temperature of 130°C. Calculate the exit temperature of the fluid and the instantaneous collection efficiency for the following situation:

Location of collector	:	New Delhi (28.58°N)
Date	:	November 5
Time	:	1100 h (LAT)
I_g	:	0.735 kW/m^2
I_d	:	0.162 kW/m^2
Number of tubes	:	2
Tube outer diameter	:	18 mm
Tube inner diameter	:	14 mm
Transmissivity of glass cover	:	0.89
Reflectivity of concentrator	:	0.87
Absorptivity of absorber surface	:	0.94
Overall loss coefficient	:	10.5 W/m^2-K
Heat transfer coefficient on inside of absorber tube	:	230 W/m^2-K
Mass flow rate of fluid	:	1.25 kg/minute
Ambient temperature	:	21°C

Solution

Geometry of Collector:

Half acceptance angle $\theta_a = \sin^{-1}(1/6.5) = 8.85°$

Therefore acceptance angle = 17.70°

Aperture width $W = 0.06 \times 6.5 = 0.39$ m

Substituting the value of θ_a in Eq. (6.84), we get

$$\frac{A_{con}}{WL} = 0.1538 \times 1.1538(41.7459 + 1.7832 - 1.1275) = 7.524$$

Therefore, from Eq. (6.86),
average number of reflections $m = 1.12$

Flux S:

The flux S is calculated from Eq. (6.94). Before doing so, we must check that the beam radiation is within the acceptance angle of the collector.

On November 5, $n = 309$ and $\delta = -16.55°$. Therefore, from Eq. (6.88), the solar elevation angle is given by,

$$\tan \alpha_v = \frac{\sin 28.58° \sin(-16.55°) + \cos 28.58° \cos(-16.55°) \cos 15°}{\sin 28.58° \cos(-16.55°) \cos 15° - \cos 28.58° \sin(-16.55°)}$$

$$= 0.9766$$

$$\alpha_v = 44.32°$$

Since the slope of the aperture plane is 40°, the axis of the CPC makes an angle of 50° with the horizontal. Thus, beam radiation having a solar elevation angle between the limits of (50° ± 8.85°), i.e. between 58.85° and 41.15°, would be accepted by the collector. In the present problem, the beam radiation is within these limits.

Now, from Eq. (3.41), tilt factor r_b

$$= \frac{\sin(-16.55°) \sin(28.58° - 40°) + \cos(-16.55°) \cos 15° \cos(28.58° - 40°)}{\sin 28.58° \sin(-16.55°) + \cos 28.58° \cos(-16.55°) \cos 15°}$$

$$= 1.4243$$

Therefore, $$S = \left(573 \times 1.4243 + \frac{162}{6.5}\right) 0.89 \times 0.87^{1.12} \times 0.94$$

$$= 602.3 \text{ W/m}^2$$

Useful Heat Gain Rate:

From Eqs (6.96) – (6.98),

$$\frac{1}{F'} = 10.5\left[\frac{1}{10.5} + \frac{0.06}{2 \times \pi \times 0.014 \times 230}\right] = 1.0311$$

$$F' = 0.9698$$

$$\frac{\dot{m}C_p}{bU_lL} = \frac{1.25}{60} \times \frac{2350}{0.06 \times 10.5 \times 2} = 38.856$$

$$F_R = 38.856[1 - \exp(-0.9698/38.856) = 0.9578$$

Useful heat gain rate $q_u = 0.9578 \times 0.39 \times 2 \times \left[602.3 - \frac{10.5}{6.5}(130 - 21)\right]$

$$= 318.4 \text{ W}$$

Exit Fluid Temperature and Collection Efficiency:

The exit fluid temperature is calculated from the equation

$$(1.25/60) \times 2350 \times (T_{fo} - 130) = 318.43$$

$$T_{fo} = 136.50°\text{C}$$

From Eq. (6.30)

$$\eta_i = \frac{318.4}{\left[573 \times 1.4243 + \frac{162(1 + \cos 40°)}{2}\right] 0.39 \times 2} = 0.4256$$

PARABOLOID DISH COLLECTOR 6.7

The principle of working of a paraboloid dish collector system has been described in Sections 2.2 and 2.3 along with a few details of some commercial systems. Mancini *et al.* [36] have presented a review of these systems. Here we present descriptions of the concentrator and receiver in the collector assemblies of three systems.

Science Applications International Corp. and STM Power, Inc., USA have developed the SunDish system with a peak output of 22.9 kW. The paraboloid concentrator of the system is made up of 16 round, stretched membrane mirror facets, each 3.2 m diameter, which are mounted on a truss structure. The total reflecting area is 117.2 m^2 with a reflectivity of 0.95. The truss is attached to an azimuth/elevation drive on top of a pedestal. A distinguishing feature of the concentrator is that the facets are attached to the truss in a staggered manner so that there are gaps between the facets. This helps to reduce the wind load. The focal length of the concentrator is 12 m. The receiver is a cavity with an aperture having a diameter of 38 cm. The cavity contains the heater head of a Stirling engine in the shape of a truncated cone on which the insolation falls directly. The intercept factor for the radiation reflected through the receiver aperture is 0.90 and the concentration ratio is 2500. A four-cylinder Stirling engine is used. Burners are provided in the receiver to use other fuel for running the engine when solar energy is not available.

The 'EuroDish' system was developed under a joint-venture project between the European Community and German/Spanish industries and research institutions from 1998 to 2001. The concentrator consists of a glass-fibre composite shell onto which glass mirrors (reflectivity = 0.94) are bonded with an adhesive. It is 8.5 m in diameter and the total glass area is 60 m^2. The concentrator dish is supported by a space frame ring truss and the whole concentrator is suspended on a space frame turntable rolling on six wheels. A cavity-type receiver with an aperture diameter of 15 cm is kept at the focus (focal length = 4.5 m) and is directly attached to the cylinder heads of a Stirling engine. The engine is coupled to an induction generator which has produced a peak output of 8.5 kW. Subsequently an improved system called 'Envirodish' was developed in 2004. This yielded a peak output of 11 kW and had an overall efficiency of 22.5 per cent.

The third system described here is the SES system, which has a peak output of 25.3 kW. The concentrator is approximately paraboloid and is made from 82 reflecting glass facets with a total area of 91.0 m^2. The reflectivity of the glass is 0.91. The diameter of the aperture of the receiver is 20 cm and the collector has an intercept factor of 0.97. Like other systems, the temperature in the receiver is around 700°C. The prime mover is a 4-cylinder Stirling engine operating at 1800 rpm.

In spite of the disappointing growth in production of parabolidal dish systems for power generation, new ideas continue to be presented in the literature. Li and Dubowsky [37] have suggested that the concentrator be formed from several thin flat, flexible metal strips shaped like petals. One side of the strips would have a highly reflective surface. Thin layers are attached to the non-reflecting back side of each strip. These layers have optimized shapes so that each petal forms into a parabola of the desired curvature when the ends are pulled towards each other by thin tension cables. The authors claim that the concept would permit flat mirror strips to be efficiently packed and assembled at site leading to a substantial reduction in the initial cost.

The receiver is the other important component of a paraboloid dish system. It is difficult to design because it has to receive a high heat flux (of the order of 10^6 W/m^2 and at about 700°C), absorb it and transfer the energy to the working fluid of the engine. A number of new ideas have therefore been tried out. Three suggestions are as follows:

1. Heat pipes using liquid metals like sodium have been used to transfer the heat from the receiver to the engine head.
2. Hybrid receivers which can absorb solar energy as well as energy from some other sources like a fossil fuel or bio-gas have been developed. The development of such receivers enables power to be developed in the system at all hours.
3. The use of volumetric receivers has also been experimented with. These receivers are similar to those described in Section 6.8 for use in central receiver power systems.

Finally, it should be mentioned that the paraboloid dish collector has been effectively used for providing process heat in industries which require heat at temperatures up to 250°C. For demonstrating such a use, a paraboloid dish collector (Plate 6.2) has been installed at the Mahanand Dairy in Latur, Maharashtra for providing heat for pasteurization of 20000 – 25000 litres of

Plate 6.2 *Paraboloid dish collector of 160 m^2 aperture area installed at Mahanand Dairy, Latur, Maharashtra. (Courtesy: Clique Developments Ltd., Mumbai)*

milk per day. The dish is octagonal in shape and uses glass mirror facets to concentrate the solar radiation. The aperture area is 160 m^2 and the focal length is 8 m. The facets are fixed in such a manner that no facet falls in the shadow cast by the receiver and its supporting structure on the dish. The total reflector area is 123 m^2. The receiver is a cylindrical cavity with helical M S tubes fixed on the inner surface. The diameter of the cavity is 0.8 m and it has a skirt of 1.26 m diameter at the open end. The concentration ratio is 128. Pressurised water at 20 bar is used as the working fluid. The dish provides process heat at 160 to 180°C. About 1.6×10^6 kJ of heat energy are delivered on a typical day [38]. A few such systems, having a total aperture area of about 5000 m^2, have been installed in India. The aperture areas of the dishes installed vary from 34 m^2 to 169 m^2.

CENTRAL RECEIVER COLLECTOR 6.8

The principle of working of a central receiver power plant was described in Section 2.3.4. Some details of a number of operational plants were also given. The building of such a plant was first suggested by Baum *et al.* [39]. Based on their calculations, they indicated the possibility of erecting an installation in sunny regions of the USSR, to produce 11 to 13 t of steam per hour at 30 atm and 400°C. The optical system was calculated to consist of 1293 mirrors of dimensions 3 m × 5 m. These heliostats were proposed to be mounted on carriages which moved on rails in arcs around the tower.

A number of small pilot plants were built by Francia [40] in Italy in the period 1965 to 1967. In one of these, he collected 50 kW of energy. After a break of a few years, the design of central receiver collector systems again attracted attention and a number of demonstration plants ranging in capacity from 0.5 to 10 MWe were built. Central receiver power plants have been operated commercially since 2007.

The two major components making up a central receiver collector are the heliostats and the receiver. These will now be described.

6.8.1 Heliostats

The heliostats form an array of circular arcs around the central tower. They intercept, reflect and concentrate the solar radiation on to the receiver. The array is served by a tracking control system which ensures that each heliostat focuses beam radiation towards the receiver during collection. A two-axis tracking mode involving adjustments to the surface azimuth angle and the slope of the heliostat is usually adopted. In addition, when solar radiation is not being collected, the control system orients the heliostats in a safe direction so that the receiver is not damaged.

As stated earlier, the 10 MWe plant at Barstow was the largest of the pilot plants built. It was operated for six years, from 1982 to 1988 as Solar One and again after modifications from 1996 to 1999 as Solar Two. Solar One had a field of 1818 heliostats positioned all around a central tower of height 80 m. Each heliostat was an assembly of 12 slightly concave glass mirrors mounted on a support structure and geared drive that could be controlled for azimuth as well as elevation. The total reflective area of each heliostat was 39.3 m^2. A rear view sketch is shown in Fig. 6.27. The 12-mirror panels in each heliostat were 1 m × 3 m in size and were made from 3 mm low iron float glass. When clean, the heliostats had an average reflectivity of 0.903. However, dirt accumulated due to exposure to the environment, reduced the average reflectivity to 0.82. Solar Two had an additional 108 heliostats each having an area of 95 m^2.

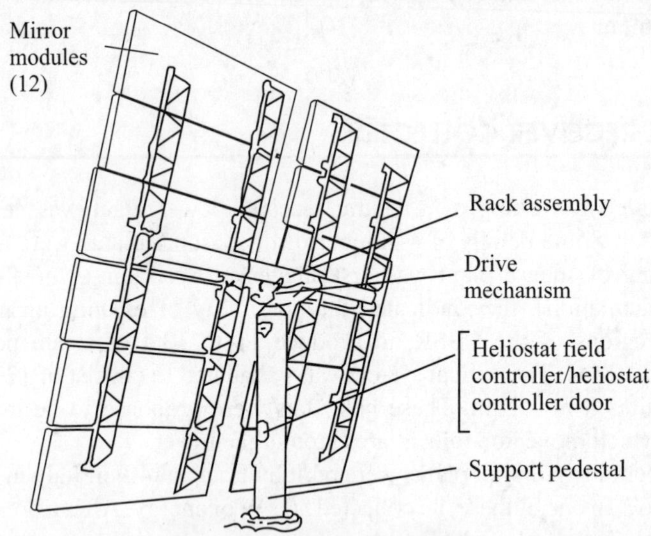

Fig. 6.27 *A heliostat [41]*

Degradation of the mirrored surfaces of the heliostats and low availability of the heliostats are major issues in the development of central receiver systems. Thus an increase in the reflectivity and in the reliability of the heliostat system is important. In order to achieve these goals, many new concepts have been tried. For example, larger size glass-mirror heliostats having areas of 150 m^2 and reflectivity values up to 0.94 have been built. Also a new type of heliostat using a stretched membrane has been developed. In this heliostat, the reflector is a silvered polymer film laminated to a thin metal foil which is stretched over a large-diameter metal ring. The reflectivity of this surface has been measured to be 0.92. Because of its simplicity and reduced weight, a stretched-membrane heliostat would be about 30 per cent less costly than a glass-mirror design. Various improved versions of reflective polymeric films have been developed. However, the ability to reflect solar flux onto the receiver from the furthest receiver (the distance can be as large as 1600 to 1700 m) is an issue. Recently, Ho *et al.* [42] have tested a commercially available polymer film, Solar Mirror

Film 1100, made by the 3M company. Based on one year of testing, the authors have reported that the decrease in the solar reflectivity of the film due to soiling and outdoor exposure was not significantly different from that of silvered glass. The peak flux resulting from reflections from the film and from silvered-glass was also measured at distances of 1700 m. The results showed that the mean of the measured peak fluxes generated by the film and the silvered glass were statistically similar.

Deposition of dust on the reflecting surface of heliostats is a major concern and more so in view of the large number of reflectors. Mazumdar *et al.* [43] have proposed an innovative way of automatic removal of the dust by using a thin transparent electrodynamic screen. It consists of thin rectangular, transparent conducting electrodes (50 – 100 μm in width). The gap between electrodes are typically 700 μm. When voltage is applied across the electrodes, an electrodynamic field is created. This lifts the dust particles from the surface, thus cleaning it. Prototype testing has been carried out by the authors. Exhaustive investigations including economic analysis will need to be carried out before large scale application is considered.

Heliostats contribute 40 per cent to the total cost of central receiver plants and it is clear that significant heliostat cost reduction is needed in order to ensure that central receiver plants become cost competitive [44]. A numbers of heliostat designs that could potentially lower the cost are under development. Pfahl [45] has surveyed these designs and identified the merits and demerits of the proposed features. The cost of a heliostat can be divided into six parts. These are the reflector (25 per cent), the support (15 per cent), the foundation (15 per cent), controls (5 per cent), drives (30 per cent) and installation (10 per cent). The on-going investigations on cost reduction reported in the literature have been categorized by Pfahl into the above-mentioned groups and he has made recommendations on the promising concepts. Blackmon [46] has developed a model to calculate heliostat cost taking into account most of the above-mentioned groups. Coventry and Pye [47] have reviewed the cost-reduction efforts by considering three angles. These are manufacturing and assembly of the heliostat, heliostat size and wind load analysis. They have also identified a series of promising design concepts.

The heliostat field layout has also been a topic of intense investigation. A number of concepts have been proposed. The two most common layouts have been the surround field layout and the polar field layout. The first one refers to the case where heliostats are arranged all around the tower. Examples of power plants using this arrangement are Solar Two, Gemasolar and Ivanpah. In the second case, heliostats are arranged only on one side of the tower. Examples of such plants are PS10 and PS20. The actual layout suitable at a place has to be identified based on a number of factors such as blocking, shading, cosine effect, atmospheric attenuation, tower height, spread of image at the receiver, receiver type, latitude, plant size and land cost. A number of computer codes have been developed for designing heliostat field layout. Mutuberria *et al.* [48]

have recently compared the codes for the 'surround field'. They concluded that the Campo code [49] and the model proposed by Noone *et al.* [50] yield heliostat field layouts having high field efficiency and high annual energy collection.

6.8.2 Receiver

The receiver is the most complex part of the collection system. The main factor influencing its design is its ability to accept the large and variable heat flux which results from the concentration of the solar radiation by the heliostats. This flux has to be transferred to the receiver fluid. The value of the heat flux can range from 100 to 1000 kW/m^2 and this results in high temperatures, high thermal gradients and high stresses in the receiver. The value depends on the concentration ratio and varies with the season and the day. It also varies over the surface of the receiver. For these reasons, attention has to be given to the absorber shape, the heat transfer fluid and the materials used for construction. Receivers can be classified into three broad categories: tubular liquid receivers, volumetric air receivers and falling solid particle receivers. Out of these, the development of tubular liquid receivers has matured and been adopted commercially. Volumetric air receivers have been in the development stage for many years. However, the technology has not yet been commercialized. Only a demonstration power tower plant with a volumetric air receiver has been in operation in Germany since 2009[51]. Falling solid particle receivers are still in the development stage. It may be mentioned that falling film receivers and tubular gas receivers are also being studied [52].

Tubular Liquid Receivers

There are two types of tubular receiver designs: the external type and the cavity type. The external receiver (Fig. 6.28) is usually cylindrical in shape. The solar flux is directed onto the outer surface of the cylinder consisting of a number of panels and is absorbed by the receiver fluid flowing through closely spaced tubes fixed on the inner side. On the other hand, in a cavity receiver, the solar flux enters through one or more small apertures in an insulated enclosure. The cavity contains a suitable tube configuration through which the receiver fluid flows. The geometry of the cavity is such that it maximizes the absorption of the entering

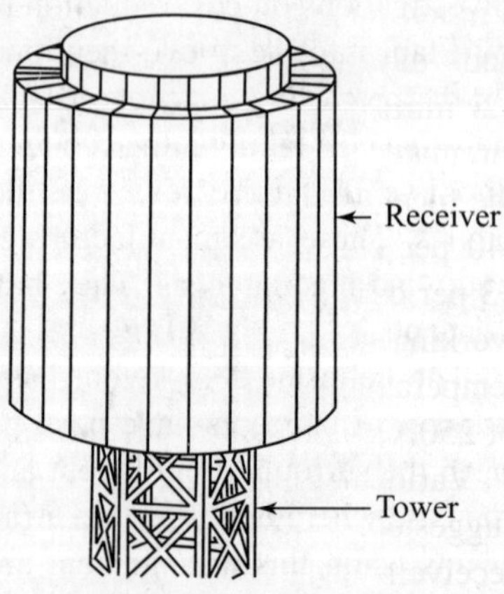

Fig. 6.28 *External receiver*

radiation, minimizes heat losses by convection and radiation to the ambient and at the same time accommodates the heat exchanger that transfers the radiant energy to the receiver fluid. Both types of receivers have their advantages and disadvantages. The external type has a very wide acceptance angle, while the cavity type has a small acceptance angle. On the other hand, the cavity type traps the solar flux more effectively and consequently has a higher efficiency than the external type.

As described earlier in Chapter 2, Solar One had an external type, tubular receiver in which water was heated directly and converted to superheated steam. The receiver was a cylinder, 7 m in diameter and 13.5 m in height, made up of 24 vertical panels painted black. Tubes made from Incoloy 800 (0.6 cm I.D., 1.25 cm O.D.) were fixed on the inside. The receiver was located on a tower 80 m high and produced steam at 102 bar and 510°C. The receiver had an annual efficiency of 0.69, which was rather low. In order to achieve a higher efficiency and reduce the failure rate, the receiver for Solar Two was re-designed to use molten nitrate salt as the heat transfer fluid instead of water/steam. With a molten salt, the receiver could operate with higher incident solar fluxes. Consequently the size of the receiver was reduced resulting in smaller thermal losses. In addition, the molten salt was essentially at atmospheric pressure, therby permitting the use of thinner walled tubes in the receiver. An annual efficiency of 0.88 was attained.

Although the use of molten nitrate salts has yielded a higher efficiency, some issues have also arisen. Nitrate salts starts decomposing at temperature above 600°C. They become chemically unstable and produce oxide ions, which are highly corrosive. Alternate liquids such as chlorides, fluorides and carbonates have been considered. Each one has certain merits and some demerits. Another point worth mentioning is that, at high temperatures, the tube materials carrying hot liquid experience thermal fatigue. Thus, attention is focused on the choice of containment materials.

A cavity type receiver was used in the 2 MWe Themis demonstration power plant in France [53]. The receiver was almost cubic having dimensions of 4 m × 4 m × 3.5 m (Fig. 6.29). The inner walls had tubes (diameter: 18 mm, thickness: 1.5 mm) in which the heat transfer fluid was circulated. The aperture of the receiver was in the focal plane of the heliostat field. Hitec salt (40 per cent $NaNO_2$, 7 per cent $NaNO_3$ and 53 per cent KNO_3 by weight) was used as the working fluid. The typical value of the outlet temperature of the fluid was 430°C for an inlet of 250°C.

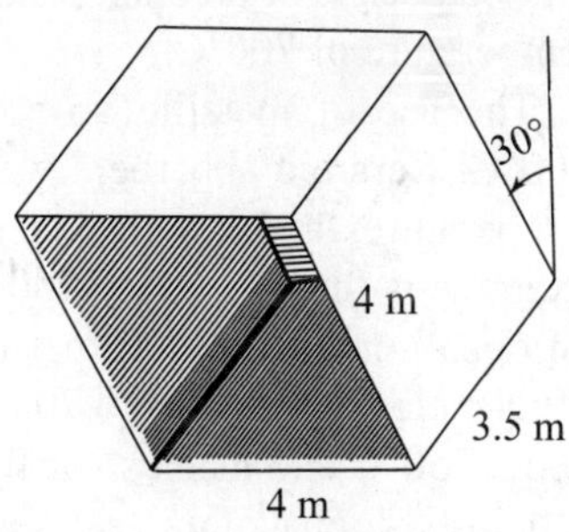

Fig. 6.29 *Cavity receiver [53]*

Various types of cavity receivers have been suggested in the literature and investigations on the heat loss behavior of such receivers have been undertaken [54, 55]. These investigations help in designing more efficient receivers.

An interesting suggestion has been made recently to use a dual-receiver for a central receiver power plant, one on top of the other [56]. The one on top would be an external type of receiver and would serve to generate saturated steam, while the one below would be a cavity type of receiver and would serve as a superheater. A case study of a 11 MW solar power plant has been conducted to demonstrate that the dual-receiver concept could produce superheated steam at 107 bar and 515°C.

Volumetric Air Receivers

In the nineties, suggestions were made that air could be used as the heat transfer fluid in the receiver. As a result, cavity type, volumetric receivers in which tubes are not used for circulating the fluid were developed. Highly porous materials like metallic wire meshes, ceramics or honeycomb structures are used to absorb the concentrated solar radiation in these receivers. Because of the porous nature, the radiation penetrates and is absorbed deep within its interior. A fan sucks the air through the absorbent pores and hot air is obtained. Since the absorption of radiation takes place over a certain depth and not at the surface, thermal losses are reduced. In addition, the heat transfer area is more as compared to a tubular surface. As a result, a higher quantity of heat can be transferred, resulting in a reduction of receiver size. Further, the absorption of solar radiation as well as convective heat transfer to the working fluid occur at the same surface; this reduces the thermal stress on the material. Another advantage of using volumetric receivers is that they have provided opportunities to integrate the solar power plant with gas turbines for higher plant efficiency. This results in reduced heliostat area and a reduction in the investment cost [57].

Volumetric receivers can be of the open-loop or closed-loop type. An excellent review has been provided by Avila-Marin [51] of the designs developed and tested till 2011. A few will be described in the following paragraphs.

An open-loop volumetric air receiver using a metallic absorber was developed under the TSA project [58]. The absorber was made up of "coiled knit-wire knitted packs" of inconel. It had an efficiency of 0.85 and delivered an air outlet temperature of 700°C.

The use of metallic absorbers restricted the maximum air temperature to 700°C. Ceramic absorbers were therefore developed in order to obtain higher temperatures and a series of open-loop receivers using them were tested. An example is the SOLAIR 3000 receiver [59] which used recrystallized SiC with an open porosity of 49.5 per cent as the absorber material. With this receiver, efficiencies in the range 0.70 to 0.75 were obtained under nominal conditions of about 750°C and mean solar fluxes in the range of 370 – 520 kW/m^2.

Closed-loop volumetric absorbers are also being developed with a view to integrating the use of solar energy into a combined cycle with gas turbines. In this case, the high pressure operation makes it necessary to build a transparent window through which the concentrated solar radiation enters the receiver cavity. An example of this application is the REFOS project [60]. In this case, the absorber consisted of several layers of the heat-resistant wire screens (Inconel 600). It was placed in a pressure vessel having a domed quartz window which could withstand a pressure upto 19.5 bar. A secondary concentrator fixed outside the quartz window helped to improve the optical efficiency of the system. The project goals of delivering air at temperature of 800°C and a receiver efficiency of 0.80 were achieved.

The SOLGATE project [61] was initiated as a follow-up with the main goal of providing pressurized air at 1000°C. A solar receiver cluster consisting of three modules connected in the series was built. The first module was a low temperature receiver, a tubular coil being attached to the back side of the secondary concentrator on which the solar radiation was being focused. The second module in the cluster was similar to the one used in the REFOS project, while the third module was similar to the second module excepting for the fact that the metal-wire mesh absorber was replaced by a ceramic absorber made of SiC. The tests showed that the receiver cluster was able to produce 1000°C air to drive a gas turbine.

Falling Solid Particle Receivers

Falling solid particle receivers are being developed to increase the operating temperature of central receiver collectors and the collection efficiency. Solid particles falling under gravity within a cavity receiver serve as the heat collecting medium. The particles are stored in an insulated tank and heat is transferred to the fluid (air or steam) to be used for the power cycle. The particles are typically ceramic based and are stable at high temperatures. Thus the receivers are capable of yielding outlet temperatures upto 1000°C.

A number of analytical and experimental laboratory studies have been conducted on solid particle receivers. However, only one preliminary investigation has been conducted with concentrated solar radiation [62]. Since the receiver design was not optimized, a thermal efficiency of only 50 per cent was achieved and the particles were heated by only about 250°C. However, the principle of working was established. Recently, Christian and Ho [63] have presented the system design of a prototype 1 MW solid particle receiver. Field tests will be performed on the receiver and the solid particles will be maintained at a temperature of 700°C while the system is running.

6.8.3 Analysis

Consider a central receiver collector consisting of N mirrors** each of area A_m and span w covering a ground area A_g around the tower. The mirrors have to be laid out in such a manner that incident or reflected radiation associated with one heliostat is not blocked by a neighbouring heliostat. As a result, they have to be spaced apart and cover only a fraction of the ground area. If the fraction of the ground area covered is ψ, then

$$NA_m = \psi A_g \tag{6.99}$$

In most central receiver collector designs, the value of ψ is around 0.4.

Let the receiver's absorber surface (on which the radiation is focussed) have an area A_p and be located on top of a tower of height H. Taking an energy balance on the absorber, we obtain the following expression for the useful heat gain rate,

** The mirrors could be circular in which case $A_m = \pi w^2/4$, or square in which case $A_m = w^2$.

$$q_u = I_b\left[\sum_{j=1}^{N} r_{bj}\right]\rho\gamma_{int}\,\alpha A_m - U_l A_p (T_{pm} - T_a) \tag{6.100}$$

We assume ρ, γ_{int} and α to be constants.

Defining an average tilt factor $(r_b)_{av} = \dfrac{1}{N}\displaystyle\sum_{j=1}^{N} r_{bj}$ (6.101)

and using Eq. (6.99), we have

$$\begin{aligned} q_u &= I_b (r_b)_{av}\psi A_g\,\rho\gamma_{int}\alpha - U_l A_p (T_{pm} - T_a) \\ &= \psi A_g\left[I_b (r_b)_{av}\,\rho\gamma_{int}\,\alpha - \frac{U_l}{C}(T_{pm} - T_a)\right] \end{aligned} \tag{6.102}$$

where $C = (NA_m / A_p) = (\psi A_g / A_p)$

In order to obtain expressions for the size of the external absorber and the concentration ratio, we assume that the mirror field is circular with the tower at the centre. The rim angle, i.e. the angle made by the line joining the absorber and the outermost mirror with the vertical is taken to be ϕ_r (Fig. 6.30). Thus the distance between the outermost mirror and the absorber is $(H/\cos\phi_r)$. If the mirrors are flat, it follows that the size of the image at the absorber is given by

$$L_i = \frac{H}{\cos\phi_r}(\theta_s + \theta_e) + w \tag{6.103}$$

where θ_s is the angle subtended by the sun at the earth and θ_e is the total angular error associated with the reflection due to factors like mirror surface imperfections and mirror orientation. If the mirrors are suitably dished, then the spread of the image due to the mirror span could be eliminated, and the size L_i is obtained by putting $w = 0$ in Eq. (6.103).

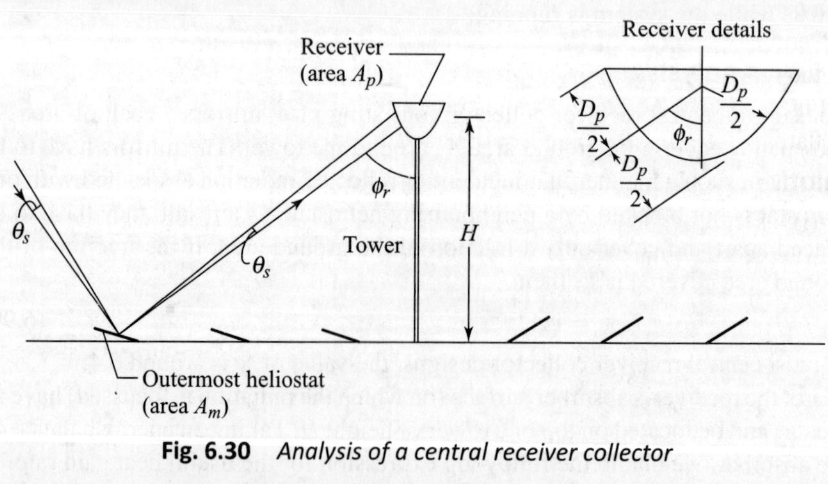

Fig. 6.30 *Analysis of a central receiver collector*

Vant-Hull and Hildebrandt [64] have suggested that an appropriate shape for the absorber could be a spherical segment with a conical section as shown in Fig. 6.30. The area of such a shape is given by

$$A_p = \frac{\pi}{2} D_p^2 \left(1 + \sin\phi_r - \frac{\cos\phi_r}{2}\right) \tag{6.104}$$

where D_p is the diameter of the sphere as well as the apparent height of the absorber when viewed from the outermost mirror. Equating D_p with the image size given in Eq. (6.103), we have

$$A_p = \frac{\pi}{2}\left\{\frac{H}{\cos\phi_r}(\theta_s + \theta_e) + w\right\}^2 \left(1 + \sin\phi_r - \frac{\cos\phi_r}{2}\right) \tag{6.105}$$

Thus, the concentration ratio for this shape is given by

$$C = \frac{N A_m}{\frac{\pi}{2}\left\{\frac{H}{\cos\phi_r}(\theta_s + \theta_e) + w\right\}^2 \left(1 + \sin\phi_r - \frac{\cos\phi_r}{2}\right)}$$

$$= \frac{\psi \pi H^2 \tan^2\phi_r}{\frac{\pi}{2}\left\{\frac{H}{\cos\phi_r}(\theta_s + \theta_e) + w\right\}^2 \left(1 + \sin\phi_r - \frac{\cos\phi_r}{2}\right)} \tag{6.106}$$

For suitably dished mirrors, in which the spread of the image due to mirror span is eliminated, the concentration ratio is obtained by putting $w = 0$ in Eq. (6.106). Expressions similar to Eq. (6.106) can be obtained for other absorber shapes.

Example 6.8

In a central receiver collector, the height of the tower is 150 m, the rim angle is 50° and the diameter of the mirrors is 4.5 m. Find the size of the image formed by the outermost mirror at the receiver, the area of the absorber (if it is as shown in Fig. 6.30) and the corresponding concentration ratio. Assume that mirrors are (i) flat, and, (ii) dished. Take $\psi = 0.38$ and $\theta_e = 0.002$ radians.

Solution

(i) From Eqs (6.103) and (6.104),

$$L_i = \frac{150}{\cos 50°}\left(\frac{32\pi}{60 \times 180} + 0.002\right) + 4.5 = 7.14 \text{ m}$$

$$A_p = \frac{\pi}{2} \times 7.14^2 \left(1 + \sin 50° - \frac{\cos 50°}{2}\right) = 115.67 \text{ m}^2$$

Thus, $$C = \frac{0.38 \times \pi \times 150^2 \times \tan^2 50°}{115.67} = 330$$

(ii) If the mirrors are dished,

$$L_i = \frac{150}{\cos 50°}\left(\frac{32\pi}{60 \times 180} + 0.002\right) = 2.64 \text{ m}$$

$$A_p = \frac{\pi}{2} \times 2.64^2 \left(1 + \sin 50° - \frac{\cos 50°}{2}\right) = 15.82 \text{ m}^2$$

Thus, $$C = \frac{0.38 \times \pi \times 150^2 \times \tan^2 50°}{15.82} = 2412$$

6.8.4 Alternative Concepts

We now discuss two alternative concepts. One is the multi-tower concept and the other, the beam-down concept.

Danielli *et al.* [65] have proposed a modification to the central receiver power plant in order to improve the optical efficiency. They suggest that the central tower be replaced by a number of towers of smaller height distributed across the heliostat field. The heliostats would still have two-axis tracking and each heliostat would be directed suitably to an appropriate tower depending upon the time and the day of the year. By adopting this modification, the effect of factors like the distance to be covered by a reflected ray, the cosine effect, the intercept factor and the blocking of reflected rays by an adjacent heliostat is reduced. The receivers on the towers are linked to each other by connecting pipes so that the fluid to be heated moves from receiver to receiver. In order to minimize the heat losses, the connecting pipes arc well insulated. The modification suggested by Danielli *et al.* is in many ways similar to the concept of CLFR, mentioned in Section 2.2.2.

Another modification suggested for the central receiver power station is to use the beam-down concept. In such a station, a hyperbolic shaped reflector would be located at the top of the tower instead of the receiver. The reflector would re-direct the concentrated radiation coming from the heliostats downwards where it is captured by the receiver placed at the ground level. In this way, the cost of the power tower and the long heat transport system is reduced. Heat losses from the piping carrying the heat transfer fluid up and down are also reduced. A disadvantage associated with the beam-down concept is that the reflected radiation received at the ground level tends to spread out. There are other issues also. These have been discussed by Vant-IIull [66].

PROBLEMS

6.1 Obtain the expression

$$\psi = (\pi + \beta + 2\phi - 2\delta)/3$$

for the correct inclination of a south-facing specular reflector fixed on the top edge of a flat-plate collector (as shown in Fig. 6.2) and having the same dimensions as the flat-plate collector.

6.2 Compare the values of the tilt factor for beam radiation incident on a flat-plate collector with the tilt factors obtained for north and south-facing reflectors fixed to it. Given the following data:

Location of collector-reflector array	:	Chennai (13.00°N)
Slope of collector	:	13° facing south
Date	:	Jun. 21 and Dec. 21
Time	:	1200 h (LAT)

Assume that the reflectors are inclined at angles given by Eq. (6.6) and by Problem 6.1.

6.3 A cylindrical parabolic collector is operated in Kolkata (22.65°N, 88.45°E) in tracking mode II. Calculate the variation in the slope of the aperture plane from 0800 to 1600 h (LAT) on June 21.

6.4 A cylindrical parabolic collector is located in Pune (18.53°N) and operates in tracking mode II on May 1. Calculate the values of the slope of the aperture plane from 0600 to 1200 h (LAT) at hourly intervals and the corresponding angle of incidence. Calculate also the time at sunrise.

6.5 A cylindrical parabolic concentrating collector is operating in tracking mode III. Calculate the slope of the aperture plane and the tilt factor for beam radiation for the following conditions:

Location	:	21° 06′ N, 79° 03′ E
Date	:	November 13
Time	:	1500 h (LAT)

6.6 Calculate the overall loss coefficient for an evacuated glass tube cylindrical parabolic focussing collector with the following data:

Absorber tube:	Outer diameter	:	6.5 cm
	Inner diameter	:	6.0 cm
Glass cover:	Outer diameter	:	15.8 cm
	Inner diameter	:	15.0 cm
Aperture		:	1.90 m
Length of concentrator		:	3.50 m
Emissivity of absorber tube surface		:	0.22
Emissivity/absorptivity of glass		:	0.88
Average temperature of absorber tube		:	200°C
Ambient temperature		:	20°C
Wind velocity		:	1.5 m/s

6.7 Use Eq. (6.38) to calculate h_w in Example 6.2. Hence re-calculate the overall loss coefficient. Compare the answers with the values obtained in the text.

6.8 Use Eqs (6.45–6.48) to calculate the glass cover temperature and the overall loss coefficient for the data of Example 6.3. Take h_w = 34.119 W/m^2-K. Compare the answers with the values obtained by the iterative procedure.

6.9 In order to check the sensitivity of the overall loss coefficient to the value of h_w, repeat the calculation of Problem 6.8 with a value of h_w which is 10 per cent higher. What is the per cent change in the value of U_l?

6.10 A cylindrical parabolic focussing collector is used for heating a thermic fluid (C_p = 2.2 kJ/kg-K) which enters with a temperature of 160°C. The concentrator has an aperture of 1.8 m and a length of 3 m. The absorber tube has an inner diameter of 2.8 cm and outer diameter of 3.2 cm and has a concentric glass cover around it. Given that:

Specular reflectivity of concentrator surface	:	0.82
Intercept factor	:	0.91
$(\tau\alpha)_b$	:	0.8
Beam radiation incident normally on aperture plane	:	556 W/m^2
Diffuse radiation incident on aperture plane	:	152 W/m^2
Overall loss coefficient	:	9.5 W/m^2-K
Convective heat transfer coefficient on inside of absorber tube	:	325 W/m^2-K
Ambient temperature	:	27°C
Mass flow rate of fluid	:	360 kg/h

Calculate the useful heat gain rate, the exit temperature of the fluid and the instantaneous efficiency.

6.11 A cylindrical parabolic concentrating collector is operating under tracking mode II. It is used for heating a thermic fluid (C_p = 2.3 kJ/kg-K), which enters the receiver at a temperature of 125°C with a flow rate of 330 kg/h. The following data are available:

Location	:	21° 06′ N, 79° 03′ E
Date	:	October 21
Time	:	1400 h (LAT)
Concentrator width	:	2 m
Concentration length	:	6 m
Absorber outer diameter	:	8.1 cm
Absorber inner diameter	:	7.5 cm
Collector overall loss coefficient	:	6.2 W/m^2-K
Convective heat transfer coefficient on the inside surface of the absorber tube	:	205 W/m^2-K
Intercept factor	:	0.94
Reflectivity of the concentrating surface	:	0.86
$(\tau\alpha)_b$ of the absorber and cover assembly	:	0.81
Ambient temperature	:	23°C
Global solar radiation on horizontal surface	:	681 W/m^2
Diffuse solar radiation on horizontal surface	:	136 W/m^2

Calculate the exit temperature of the fluid, the useful heat gain rate and the instantaneous efficiency of the concentrating collector.

6.12 Calculate the performance of the cylindrical parabolic collector of Example 6.3 for the situation when (i) the absorber tube surface is selective (ε_p = 0.15), (ii) the annulus is evacuated, and (iii) the fluid inlet temperature is 180°C. All other data remain the same.

6.13 Study the effect of having two glass covers instead of one on the performance of the cylindrical parabolic collector of Example 6.3. Assume that the outer glass cover has an outer diameter of 8.0 cm and an inner diameter of 7.2 cm, while the dimensions of the inner glass cover are unchanged. All other data remain the same.

Calculate the value of η_{ib} for inlet temperatures of 120, 150 and 180°C and plot the variation. It should be noted that the curve obtained with two glass covers will intersect the curve with one glass cover (Fig. 6.8, curve I).

6.14 Study the effect of varying the annular gap between the absorber tube and the glass cover on the performance of the cylindrical parabolic collector of Example 6.3. Keep all data the same and calculate the efficiency for two situations: (i) D_{co} = 5.5 cm, D_{ci} = 4.8 cm, and (ii) D_{co} = 7.1 cm, D_{ci} = 6.4 cm. Compare the results with those obtained in Example 6.3. Is there an optimum annular gap?

6.15 Derive Eq. (6.69).

6.16 A LFR concentrating collector has 13 reflector-rows, 6 on either side of the receiver and one beneath the receiver. The width of each reflector is 1 m, the length is 300 m and the pitch of the rows is 1.1 m. The height of the receiver from the reflector plane is 5 m. Calculate the useful heat gain rate and instantaneous efficiency of the collector for the following conditions:

Time : 0930 h (LAT)
Day : December 1
Place : Nagpur (21°06′N, 79°03′E)
DNI : 678 W/m^2
Ambient temperature : 20°C
Inlet temperature of working fluid : 220°C
Optical efficiency of the collector : 0.60
Fluid flow rate through the absorber : 6.7 kg/s
Specific heat of working fluid : 2.181 kJ/kg-K

Assume that the receiver is of the type shown in Fig. 6.11 and that the correlation given in Eq. (6.78) for calculating the heat loss rate is valid.

6.17 Calculate the daily performance of the LFR concentrating collector of Example 6.4. The hourly DNI and ambient temperatures are as follows:

Time (LAT)	7 – 8	8 – 9	9 – 10	10 – 11	11 – 12
DNI (W/m^2)	534	728	818	853	865
T_a(°C)	16.7	19.1	21.9	25.4	27.9
	12 – 13	13 – 14	14 – 15	15 – 16	16 – 17
	858	828	787	696	510
	29.3	29.8	29.8	29.5	28.8

Given: The specific heat of the working fluid changes with temperature as follows:

Temperature (°C)	210	220	230	240	250
C_p (kJ/kg-K)	2.075	2.110	2.128	2.154	2.181

6.18 Derive an expression for the useful heat gain rate of a parabolic dish collector having a cylindrical absorber at its focus. The diameter and length of the absorber on the outside are D_o and L_o, and on the inside, D_i and L_i. Assume that the heat transfer fluid changes its phase as it passes through the absorber and consequently remains at a constant temperature T_{fi}.

6.19 An untruncated CPC has a half-acceptance angle of 36° and an absorber width of 10 cm. Calculate the concentration ratio, the aperture, the height, and the surface area of the concentrator per metre length.

6.20 A CPC has an acceptance angle of 7.5°. Find the maximum collection period possible on April 15 without making a tracking adjustment.

6.21 Solve Example 6.8 by assuming that the absorber has the shape of a hemispherical bowl.

6.22 A central receiver collector system consists of 1800 heliostats, each 6.5 m × 6 m in size. The height of the central tower is 80 m and the rim angle is 81°. Find the size of the image formed by the outermost arc of heliostats at the receiver. The receiver is a vertical cylinder whose height is two times the diameter. Calculate the size of the receiver and the concentration ratio. Assume that the angular error associated with the heliostats does not exceed 0.0055 radians and that the heliostats are suitably dished.

REFERENCES

1. Rabl, A. 1976. Comparison of solar concentrators. *Solar Energy*, 18: 93.
2. McDaniels, D.K., Lowndes, D.H., Mathew, H., Reynolds, J. and Gray, R. 1975. Enhanced solar energy collection using reflector-solar thermal collector combinations. *Solar Energy,* 17: 277.
3. Grassie, S.L. and Sheridan, N.L. 1977. The use of planar reflectors for increasing the energy yield of flat-plate collectors. *Solar Energy,* 19: 663.
4. Raithby, G.D. and Hollands, K.G.T. 1975. A general method of obtaining approximate solutions to laminar and turbulent free convective problems. *Advances in Heat Transfer,* 11: 265.
5. Hilpert, R. 1933. Warmeabgabe von geheizen Drahten und Rohren. *Forsch Gebiete Ingenieurw,* 4: 220.
6. Churchill, S.W. and Bernstein, M. 1977. A correlating equation for forced convection from gases and liquids to a circular cylinder in cross flow. *Journal of Heat Transfer, Trans. ASME,* 99: 300.
7. Hong, S.W. and Bergles, A.E. 1976. Augmentation of laminar flow heat transfer in tubes by means of twisted tape inserts. *Journal of Heat Transfer, Trans. ASME,* 98: 251.
8. Date, A.W. and Singham, J.R. 1972. Numerical prediction of friction and heat transfer characteristics of fully developed laminar flow in tubes containing twisted tapes. *ASME Paper* No. 72-HT-17.
9. Mullick, S.C. and Nanda, S.K. 1989. An improved technique for computing the heat loss factor of a tubular absorber. *Solar Energy,* 42: 1.
10. Burkholder, F. and Kutscher, C. 2009. Heat loss testing of Schott's 2008 PTR 70 parabolic trough receiver. *Technical Report NREL/TP-550-*45633, National Renewable Energy Laboratory, Denver, USA.
11. Kelkar, K.M. 1982. Performance analysis of a cylindrical parabolic collector. *B. Tech. Thesis,* Department of Mechanical Engineering, IIT Bombay.
12. Dubey, A. 2016. Personal communication, Godawari Green Energy Ltd., Raipur, India.
13. Price, H, Luepfert, E., Kearney, D., Zarza, E., Cohen, G., Gee, R. and Mahoney, R. 2002. Advances in parabolic trough solar power technology. *Journal of Solar Energy Engineering, Trans. ASME,* 124: 109.

14. Schiel, W., Schweitzer, A., Kracht, O. and Hunt, B. 2006. Collector development for parabolic trough power plants at Schlaich Bergermann und partner. *Proc. of Solar PACES,* June 20 – 23, Seville, Spain.
15. Fend, T., Hoffschmidt, B., Jorgensen, G., Kuester, H., Krueger, D., Pitz-Pall, R., Rietbrock, P. and Riffelmann, K. 2003. Comparative assessment of solar concentrator materials. *Solar Energy,* 74: 149.
16. Lanxner, M. and Elgat, Z. 1990. Solar selective absorber coating for high service temperatures produced by plasma sputtering. *Proc. Photo-Opt. Instrum. Engr.* 1272: 240.
17. Barshilia, H.C. 2016. Personal communication, CSIR-National Aerospace Laboratories, Bengaluru, India.
18. Selvakumar, N. and Barshilia, H.C. 2012. Review of physical vapor deposited (PVD) spectrally selective coatings for mid- and high-temperature solar thermal applications. *Solar Energy Materials and Solar Cells*, 98: 1.
19. Raccurt, O., Disdier, A., Bourdon, D., Donnola, S., Stollo, A. and Gioconia, A. , 2015, Study of the stability of a selective solar absorber coating under air and high temperature conditions. *Energy Procedia*, 69: 1551.
20. Good, P., Zanganeh, G., Ambrosetti, G., Barbato, M.C., Pedretti, A. and Steinfeld, A. 2014. Towards a commercial parabolic trough CSP system using air as heat transfer fluid. *Energy Procedia*, 49: 381.
21. Zanganeh, G, Ambrosetti, G., Pedretti, A., Zavattoni, S., Barbato, M., Good, P., Haselbacher, A. and Steinfeld, A. 2014. A 3 MWth parabolic trough CSP plant operating with air at up to 650°C. *Proc. IEEE*, 978-1-4799-7336-1/14.
22. Hirsch, T., Feldhoff, J.F., Hennecke, K. and Pitz-Paal, R. 2014. Advancements in the field of direct steam generation in linear solar concentrators – a review. *Heat Transfer Engineering*, 35: 258.
23. Zarza, E., Valenzuela, L., Leon, J., Weyers, H.D., Eickhoff, M., Eck, M. and Hennecke, K. 2002. The DISS project: direct steam generation in parabolic trough systems. Operation and maintenance experience and update on project status. *Journal of Solar Energy Engg., Trans. ASME*, 124: 126.
24. Kruger, D., Kruger, J., Pandian, Y., O'Connell, B., Feldhoff, J.F., Karthikeyan, R., Hempel, S., Muniasamy, K., Hirsch, T., Eickhoff, M. and Hennecke, K. 2012. Experiences with direct steam generation at the Kanchanaburi solar thermal power plant. *Proc. 18th SolarPACES Conference*, Marrakech, Morocco.
25. Mills, D.R. and Morrison, G.L. 2000. Compact linear Fresnel reflector solar thermal power plants. *Solar Energy*, 68: 263
26. Sharma, V.M., Nayak, J.K. and Kedare, S.B. 2013. Shading and available energy in a parabolic trough concentrator field. *Solar Energy*, 90: 144.
27. Sharma, V.M. 2014. Effects of shading and blocking in line focusing solar concentrator fields. *Ph.D. Thesis*, Department of Energy Science and Engineering, IIT Bombay.
28. Sharma, V.M., Nayak, J.K. and Kedare, S.B. 2015. Effects of shading and blocking in linear Fresnel reflector field. *Solar Energy*, 113: 114.
29. Morin, G., Dersch, J., Platzer, W., Eck, M. and Haberle, A. 2012. Comparison of linear Fresnel and parabolic trough collector power plants. *Solar Energy*, 86: 1.
30. Sahoo, S.S., Singh, S. and Banerjee, R. 2012. Analysis of heat losses from a trapezoidal cavity used for linear Fresnel reflector system. *Solar Energy*, 86: 1313

31. Larsen, S.F., Altamirano, M. and Hernandez, A. 2012. Heat loss of a trapezoidal cavity absorber for linear Fresnel reflector solar concentrator. *Renewable Energy*, 39: 198

32. Manikumar, R. and Arasu, A.V. 2014. Heat loss characteristics study of a trapezoidal cavity absorber with and without plate for a linear Fresnel reflector solar concentrator system. *Renewable Energy*, 63: 98

33. Moghimi, M.A., Craig, K.G and Meyer, J.P. 2015. Optimization of a trapezoidal cavity absorber for the linear Fresnel reflector. *Solar Energy*, 119: 343.

34. Winston, R. 1974. Principles of solar concentrators of a novel design. *Solar Energy*, 16: 89.

35. Rabl, A. 1976. Optical and thermal properties of compound parabolic concentrators. *Solar Energy*, 18: 497.

36. Mancini,T, Heller, P., Butler, B., Osborn, B., Schiel, W., Goldberg, V., Buck, R., Diver, R., Andraka, C. and Moreno, J. 2003. Dish-Stirling systems: an overview of development and status. *Journal of Solar Energy Engineering, Trans. ASME*, 125: 135.

37. Li, L. and Dubowsky, S. 2011. A new design approach for solar concentrating parabolic dish based on optimized flexible petals. *Mechanism and Machine Theory*, 46: 1536.

38. Kedare, S.B., Nayak, J.K. and Paranjape, A.D. 2006. Development, installation and evaluation of large scale concentrating solar collector for medium temperature industrial thermal applications, *Final Report of Project No. 15/ST/2002*, Ministry of New and Renewable Energy, Government of India, New Delhi.

39. Baum, V.A., Aparasi, R.R. and Garf, B.A. 1957. High power solar installations. *Solar Energy*, 1:6.

40. Francia, G. 1968. Pilot plants of solar steam generating stations. *Solar Energy*, 12: 51.

41. Bartel, J.J. and Skvarna, P.E. 1984. 10 MWe solar thermal central receiver pilot plant. *Journal of Solar Energy Engineering, Trans. ASME*, 106: 50.

42. Ho, C.K., Sment, J., Yuan, J. and Sims, C.A. 2013. Evaluation of a reflective polymer film for heliostats. *Solar Energy*, 95: 229.

43. Mazumdar, M.K., Horenstein, M.N., Stark, J.W., Hudelson, J.N., Sayyah, A., Joglekar, N., Yellowhair, J. and Bolts, A. 2014. Self-cleaning solar mirrors using electrodynamic dust shield: prospects and progress. *Proc. ASME 2014, 8th International Conf. on Energy Sustainability*, June 30 – July 2, Boston, USA.

44. Kolb, G.J., Ho, C.H., Mancini, T.R. and Gary, J.A. 2011. Power tower technology roadmap and cost reduction plan. *SANDIA report SAND2011-2419*, Albuquerque, USA.

45. Pfahl, A. 2014. Survey of heliostat concepts for cost reduction. *Journal of Solar Energy Engg, Trans. ASME*, 136: 014501-1.

46. Blackmon, J.B. 2013. Parametric determination of heliostat minimum cost per unit area. *Solar Energy*, 97: 342.

47. Coventry, J. and Pye, J. 2014. Heliostat cost reduction - where to now? *Energy Procedia*, 49: 60.

48. Mutuberria, A., Pascual, J., Guisado, M.V. and Mallor, F. 2015. Comparison of heliostat field layout design methodologies and impact on power plant efficiency. *Energy Procedia*, 69: 1360.

49. Collado, F.J. and Guallar, J. 2013. A review of optimized design layouts for solar power tower plants with *Campo* code. *Renewable and Sustainable Energy Reviews*, 20: 142.

50. Noone, C.J., Torrilhon, M. and Mitsos, A. 2012. Heliostat field optimization; a new computationally efficient model and biomimetic layout. *Solar Energy*, 86: 792.

51. Avila-Marin, A. 2011. Volumetric receivers in solar thermal power plants with central receiver system technology: a review. *Solar Energy*, 85: 891.

52. Ho, C.K. and Iverson, B.D. 2014. Review of high-temperature central receiver designs for concentrating solar power. *Renewable and Sustainable Energy Reviews*, 29: 835.

53. Bonduelle, B. and Cazin-Bourguignon, A.M. 1986. in *Proc. of the 3rd International Workshop on Solar Thermal Central Receiver Systems Vol I: Design, construction and operation of CRS-plants (M. Becker, ed.) June 23-27, 1986*, Konstanz, Germany. Themis receiver: thermal losses and performance, pp. 273-282

54. Prakash, M. 2009. Analysis of convective loss from a solar cavity receiver. *Ph.D. Thesis*, Department of Energy Science and Engineering, IIT Bombay.

55. Jilte, R.D. 2014. Investigations on heat losses of solar cavity receivers of different shapes. *Ph. D. Thesis*, Department of Energy Science and Engineering, IIT Bombay.

56. Luo, Y., Du, X. and Wen, D. 2015. Novel design of central dual-receiver for solar power tower. *Applied Thermal Engineering*, 91: 1071.

57. Romero, M., Buck, R. and Pacheco, J.E. 2002. An update on solar central receiver systems, projects and technologies. *Journal of Solar Energy Engineering, Trans. ASME*, 124: 98.

58. Meinecke, W. Cordes, S. and Merten, I. 1994. PHOEBUS technology program solar air receiver (TSA) - final report on test evaluation. *DLR(MD-ET)*, Koln, Germany.

59. Tellez, F., Romero, M., Heller, P., Valverde, A., Reche, J.F., Ulmer, S. and Dibowski, G. 2002. Thermal performance of "solair 3000 kWh" ceramic volumetric solar receiver. *Proc.* 11th *SolarPACES, International Symposium on Concentrated Solar Power and Chemical Energy, Technologies*, Zurich, Switzerland.

60. Buck, R., Braeuning, T., Denk, T., Pfaender, M., Schwarzboezl, P. and Tellez, F. 2002. Solar-hybrid gas turbine-based power tower systems (REFOS). *Journal of Solar Energy Engineering, Trans. ASME*, 124: 2.

61. Heller, P., Pfaender, M., Denk, T., Tellez, F., Valverde, A., Fernandez, J. and Ring, A. 2004. Test and evaluation of a solar powered gas turbine system. *Proc. 12th SolarPACES, International Symposium on Concentrated Solar Power and Chemical Energy, Technologies*, Oaxaca, Mexico.

62. Siegel, N.P., Ho, C.K., Khalsa, S.S. and Kolb, G.J. 2010. Development and evaluation of a prototype solid particle receiver: on-sun testing and model validation. *Journal of Solar Energy Engg., Trans. ASME*, 132: 2.

63. Christian, J. and Ho, C. 2014. Alternative designs of a high efficiency, north-facing, solid particle receiver. *Energy Procedia*, 49: 314.

64. Vant-Hull, L. and Hildebrandt, A.F. 1976. Solar thermal power system based on optical transmission. *Solar Energy*, 18:31.

65. Danielli, A., Yatir, Y. and Mor, O. 2011. Improving the optical efficiency of a concentrated solar power field using a concatenated micro-tower configuration. *Solar Energy*, 85: 931.

66. Vant-Hull, L. 2014. Issues with beam-down concepts. *Energy Procedia*, 49; 257.

Thermal Energy Storage

CHAPTER 7

INTRODUCTION 7.1

This chapter deals with the description and analysis of various thermal energy storage systems. The chapter commences with an introductory section in which various situations requiring the use of a storage system are described and the basic methods of storing thermal energy are discussed. This is followed by a detailed description of sensible heat storage systems. Various materials, their properties and the range of temperatures over which they can be used for storage are mentioned. Analyses of a few situations are given and testing procedures described. Finally, latent heat storage systems and thermochemical storage systems are briefly described.

GENERAL FEATURES 7.2

The intermittent, variable and unpredictable nature of solar radiation generally leads to a mismatch between the rate and time of collection of solar energy and the load needs of a thermal application. As a result it is often necessary to use a storage system in between. The storage system stores energy when the collected amount is in excess of the requirement of the application and discharges energy when the collected amount is inadequate. The size of a storage system is largely determined by the specific purpose for which it is used. Three situations are shown in Fig. 7.1.

In Fig. 7.1 (a), the time interval during the day over which the energy is required is essentially the same as the time of collection. However, a storage system is needed because there is some mismatch between the amount of energy required and the amount collected at any instant. The storage system in such a situation has to store energy only for short intervals of time and is relatively small in size. It is called a 'buffer storage'. In Fig. 7.1 (b), the load demand shown extends over all 24 hours, whereas the collection takes place only during the sunshine hours. As a result, a system larger than a buffer storage having the capacity to store energy for a day or two is required. Such a system is called a 'diurnal storage'. Figures 7.1(a) and (b) are both 'short-term' storage systems. In contrast, Fig. 7.1 (c) illustrates a situation in which the storage system stores energy during the summer when the collection is in excess of the demand, and delivers the excess energy in winter when the collection is less than the demand. A large 'long-term' storage system is required for such a situation, since excess energy has to be stored over a period of months. Such a system is also called an 'annual storage'.

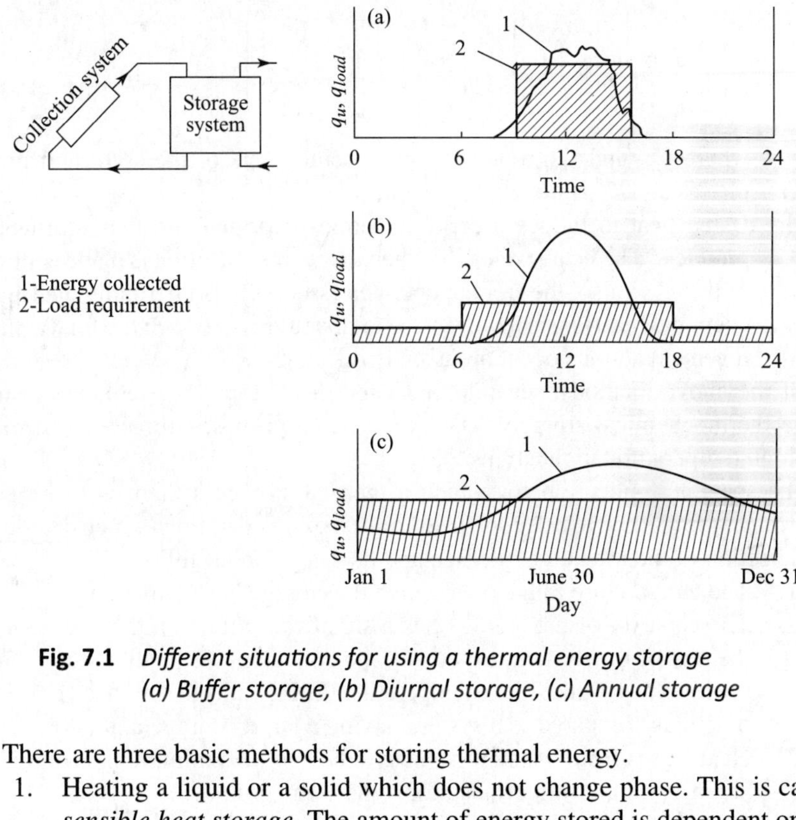

Fig. 7.1 *Different situations for using a thermal energy storage (a) Buffer storage, (b) Diurnal storage, (c) Annual storage*

There are three basic methods for storing thermal energy.

1. Heating a liquid or a solid which does not change phase. This is called *sensible heat storage*. The amount of energy stored is dependent on the temperature change of the material and can be expressed in the form

$$E = m\int_{T_1}^{T_2} C_p \, dT \tag{7.1}$$

where m is the mass and C_p, the specific heat. T_1 and T_2 represent the lower and upper temperature levels between which the storage operates. The difference $(T_2 - T_1)$ is referred to as the temperature swing.

2. Heating a material which undergoes a phase change (usually melting). This is called *latent heat storage.* The amount of energy stored in this case depends upon the mass and the latent heat of fusion of the material. Thus,

$$E = m\lambda \tag{7.2}$$

where λ is the latent heat of fusion. In this case, the storage operates isothermally at the melting point of the material. If isothermal operation at the phase change temperature is difficult, the system operates over a range of temperatures T_1 to T_2 which includes the melting point. Then sensible heat contributions have to be considered and the amount of energy stored is given by

$$E = m\left[\left\{\int_{T_1}^{T^+} C_{ps}\, dT\right\} + \lambda + \left\{\int_{T^+}^{T_2} C_{pl}\, dT\right\}\right] \tag{7.3}$$

where C_{ps} and C_{pl} represent the specific heats of the solid and liquid phases and T^+ is the melting point.

3. Using heat to induce a certain chemical reaction and then storing the products. The heat is released when the reverse reaction is made to occur. In this case also, the storage operates essentially isothermally during the chemical reactions. However, the temperatures at which the forward and reverse reactions occur are usually different.

Of the above methods, sensible and latent heat storage systems are in use, while thermochemical storage systems are being proposed for use in medium and high temperature applications.

The specific application for which a thermal storage system is to be used determines the method to be adopted. Some of the considerations which determine the selection of the method of storage and its design are as follows.

1. The temperature range over which the storage has to operate.
2. The capacity of the storage has a significant effect on the operation of the rest of the system, especially the collectors. A smaller storage unit operates at a higher mean temperature. This results in a reduced collector output as compared to a system having a larger storage unit.
3. Heat losses from the storage have to be kept to a minimum. Heat losses are of particular importance for 'long-term' storage.
4. Cost of the storage unit. This includes the initial cost of the storage medium, the containers and insulation, and the operating cost.

Other considerations include the suitability of materials used for the container, the means adopted for transferring the heat to and from the storage, and the power requirements for these purposes.

SENSIBLE HEAT STORAGE 7.3

In the case of sensible heat storage systems, energy is stored or extracted by heating or cooling a liquid or a solid which does not change its phase during the process. A variety of substances have been used in such systems. These include liquids like water, heat transfer oils and certain inorganic molten salts, and solids like rocks, pebbles and refractories. In the case of solids, the material is invariably in the porous form and heat is stored or extracted by the flow of a gas or a liquid through the pores or voids.

The choice of the substance used largely depends on the temperature level of the application, water being used for temperatures below 100°C and refractory bricks being used for temperatures around 1000°C. Sensible heat storage systems are simpler in design than latent heat or thermochemical storage systems. However, they suffer from the disadvantage of being bigger in size. For this reason, an important criterion in selecting a material for sensible heat storage is its (ρC_p) value. A second disadvantage associated with sensible heat systems is that they cannot store or deliver energy at a constant temperature. We shall first take up for consideration the various materials used.

7.3.1 Materials for Low Temperature Storage

❑ *Liquids* Water is the most commonly used medium in a sensible heat storage system at temperature below 100°C. Most solar water-heating and space-heating systems use hot water storage tanks located either inside or outside the buildings, or underground. The sizes of the tanks used vary from a few hundred litres to a few thousand cubic metres. An approximate thumb rule followed for fixing the size is to use about 75 to 100 litres of storage per square metre of collector area.

Water storage tanks are made from a variety of materials like steel, concrete and fibreglass. The tanks are suitably insulated with glass wool, mineral wool or polyurethane. The thickness of insulation used is large and ranges from 10 to 20 cm. Because of this, the cost of the insulation represents a significant part of the total cost and means to reduce this cost have to be explored. Shelton [1] has shown that in an underground tank, the insulating value of the earth surrounding the tank may be adequate and this could provide the bulk of the insulation thickness required. However, it may take as much as one year for the earth around a large storage tank to reach a steady state by heating and drying, and a considerable amount of energy may be required for this purpose.

If the water is at atmospheric pressure, the temperature is limited to 100°C. It is possible to store water at temperatures a little above 100°C by using pressurized tanks. This has been done in a few instances.

In order to reduce the cost of water storage systems, an alternative way, which has been examined for very large systems, is to make use of naturally occurring underground aquifers which already contain water. In such systems, the need for building a storage tank is eliminated. For storing energy, hot water is pumped into the aquifer through an injection well. At the same time, cold ground water is displaced through another well. For withdrawing energy, the reverse procedure is followed. These systems are used for long-term storage purposes. Xu *et al.* [2] have described a few such installations. Since the investment required is a series of openings for injecting and withdrawing water, the storage costs for such systems are low (Fig. 7.2). However, the geological conditions at the site are the deciding factors for using aquifer-based storage systems. There should be no or low natural ground flow. The quality of water available in the aquifer should be fit for direct use. Since heat insulations cannot be used, the problem of heat loss is also an issue in such systems.

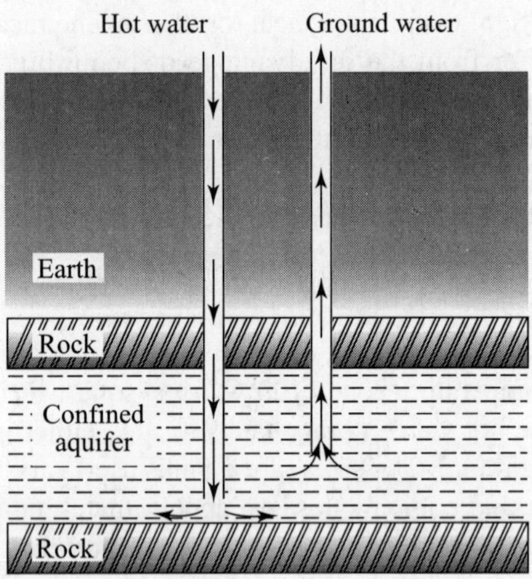

Fig. 7.2 *Schematic diagram of the underground aquifer storage concept*

❑ *Solids* For temperatures upto 100°C, energy can be stored in rocks or pebbles packed in insulated vessels. This type of storage is used very often in conjunction with solar air heaters. It is simple in design and relatively inexpensive. Typically, the characteristic size of the pieces of rock used varies from 1 to 5 cm. An approximate thumb rule followed for sizing is to use 300 to 500 kg of rock per square metre of collector area for space-heating applications.

Undisturbed earth or processed earth has also been used in space-heating applications, but for long-term storage. In such storage systems, the heat transfer with the air takes place through a network of plastic heat exchange tubes buried in the earth and through which the air flows.

7.3.2 Materials for Intermediate and High Temperature Storage

❑ *Liquids* Synthetic oils are used in sensible heat storage systems for intermediate temperatures ranging from 100 to 300°C. Some of the US brands which are used for this purpose are Dowtherm A and Therminol VP1. One of the Indian brands which would be suitable is Servotherm. The main problem associated with the use of synthetic oils is that they tend to degrade with time. The degradation is particularly serious if they are used above their recommended temperature limit. The use of oils presents safety problems since there is a possibility of ignition above their flash point. For this reason, it is recommended that they be used in systems with an inert gas cover. A further limitation to the use of these oils is their cost. For this reason, they can be considered for use only in small storage systems.

A few molten inorganic salts have been considered for high temperatures (300°C and above). One is an eutectic mixture of 40 per cent $NaNO_2$, 7 per cent $NaNO_3$ and 53 per cent KNO_3 (by weight) and is available under the trade name of Hitec. Hitec has a low melting point of 145°C and can be used up to a temperature of 425°C. Above this temperature, decomposition and oxidation begin to take place.

A promising molten salt which has been used in solar thermal power plants is 'solar salt', a binary mixture containing 60 per cent $NaNO_3$ and 40 per cent KNO_3 (by weight). Solar salt has a melting point of 220°C and can be used upto a temperature of 585°C. Based on experimental data, Bauer *et al.* [3] have developed the following empirical correlations for calculating the values of ρ, C_p, and k for solar salt as a function of temperature:

$$\rho\,(\text{kg/m}^3) = 2106.0 - 0.66795\,T \tag{7.4}$$

$$C_p\,(\text{kJ/kg-K}) = 1.5404 + 3.0924 \times 10^{-5}\,T \tag{7.5}$$

$$k\,(\text{W/m-K}) = 0.3804 + 3.452 \times 10^{-4}\,T \tag{7.6}$$

In Eqs (7.4)–(7.6), temperature T is to be substituted in degrees Centigrade.

Fernandes *et al.* [4] have developed a quaternary molten nitrate salt with the following composition by weight: 20 per cent $NaNO_3$, 60 per cent KNO_3, 10 per cent $LiNO_3$ and 10 per cent $Ca(NO_3)_2$. It is claimed that this salt exhibits better physical and chemical properties than 'solar salt'. In particular its melting point is 132°C which is significantly lower than the melting point of 220°C of 'solar salt'. The reduction is attributed to the presence of lithium nitrate ($LiNO_3$). Other inorganic mixtures with lithium compounds have also been studied for sensible heat storage [5].

Apart from molten salts, liquid metals have been considered for use at high temperatures. Liquid sodium was used as the storage fluid of a demonstration 0.5 MW solar thermal electric power station working on the central receiver concept in Spain in 1981.

Properties of some of the liquids used for sensible heat storage are given in Table 7.1.

Table 7.1 Properties of some sensible heat storage liquids

	Water[#]	Servotherm[#] Light	Servotherm[#] Medium	Hitec[*]	Sodium[+]
Highest temperature for use (°C)	100	~ 150	~ 250	425	760
Density ρ (kg/m^3)	958.4	867	880	1760	847
Specific heat C_p (kJ/kg-K)	4.22	2.24	2.22	1.51	1.27
ρC_p (kJ/m^3-K)	4044	1942	1954	2653	1079
Thermal conductivity (W/m-K)	0.683	0.126	0.123	0.54	68.3

[#]Properties at 100°C; [*]Properties at 400°C; [+]Properties at 450°C

❏ *Solids* Refractory materials like magnesium oxide (magnesia), aluminium oxide (alumina) and silicon oxide are suitable for high-temperature sensible heat storage. Bricks made of magnesia have been used in many countries for many years for storing heat. They are available in the form of devices with electric heater elements embedded in the bricks. The heat is stored at night (when the electricity rates are low) by switching on the electric heaters and is supplied during the day for space-heating purposes by allowing air to pass through the devices.

Properties of some solids used for sensible heat storage are given in Table 7.2. It will be seen that the (ρC_p) and k values for the refractory materials are higher than those for rocks or pebbles, making them more suitable from the point of view of compactness of the storage vessel. However, they are also many times more costly.

Table 7.2 Typical properties of some sensible heat storage solids

	Rocks	Pebbles	Magnesium oxide	Aluminium oxide	Silicon oxide
Density ρ (kg/m^3)	2245	1350	3575	4000	2600
Specific heat C_p (kJ/kg-K)	0.81 (0.71–0.92)	0.90	1.06	1.02	1.26
ρC_p (kJ/m^3-K)	1818	1215	3790	4080	3276
Thermal conductivity (W/m-K)	0.13	0.85	10.5	6.3	2.3

Finally it should be noted that combinations of liquid and solid sensible heat storage media are also used. In such cases, the solid is porous or granular with the liquid filling the hollow spaces. Generally the liquid is also the heat transfer fluid.

7.3.3 Analysis of a Liquid Storage Tank

❑ *Well-mixed Situation* Consider the insulated liquid-storage tank shown in Fig. 7.3 receiving energy (when available) from an array of collectors and discharging energy (when needed) to a load for use in an application. We assume that the liquid in the tank is always well-mixed and consequently is at a uniform temperature T_l, which varies only with time. An energy balance on the tank yields the following equation.

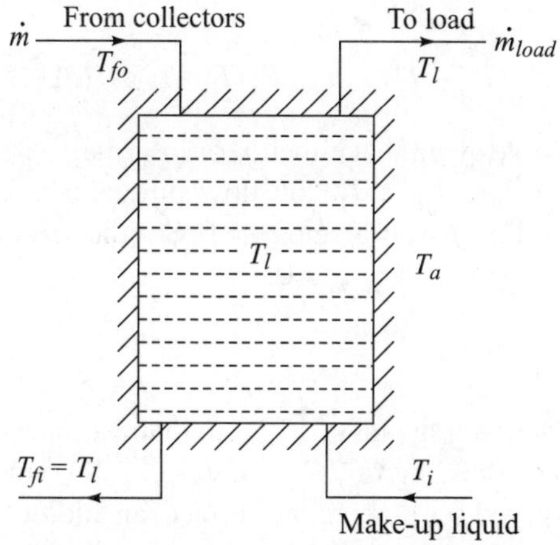

Fig. 7.3 *Analysis of a well-mixed sensible heat liquid storage tank*

$$\left[(\rho VC_p)_l + (\rho VC_p)_t\right]\frac{dT_l}{dt} = q_u - q_{load} - (UA)_t(T_l - T_a) \tag{7.7}$$

where $(\rho VC_p)_l$ represents the heat capacity of the liquid in the tank, $(\rho VC_p)_t$ the heat capacity of the tank material, q_u the rate of useful heat gain received from the collectors, q_{load} the rate at which energy is being discharged to the load, $(UA)_t$ the product of the overall heat transfer coefficient and surface area of the tank, and T_a the ambient temperature around the tank. The heat capacity term $(\rho VC_p)_t$ is likely to be of importance for small-sized tanks only. For large tanks, its value may be negligible in comparison to $(\rho VC_p)_l$. Denoting the sum of the two heat capacities by the symbol $(\rho VC_p)_e$ and integrating differential equation (7.7)* under the assumption that q_u, q_{load} and T_a are constants, we get, subject to the initial condition $t = 0$, $T_l = T_{li}$,

$$\frac{q_u - q_{load} - (UA)_t(T_l - T_a)}{q_u - q_{load} - (UA)_t(T_{li} - T_a)} = \exp\left[-\frac{(UA)_t t}{(\rho VC_p)_e}\right] \tag{7.8}$$

The assumption that q_u, q_{load} and T_a are constants is valid if the time interval of the integration is kept reasonably small (say 1 hour or less). Equation (7.8) can

* Equation (7.7) could also be solved by a finite-difference technique similar to the one used in Example 7.3.

be used to determine the variation of the temperature T_l with time if the variation of q_u and q_{load} is given.

Alternatively, the values of q_u and q_{load} may be calculated from the inlet and outlet temperatures and flow rates as

$$q_u = \dot{m}C_p(T_{fo} - T_{fi}) = \dot{m}C_p(T_{fo} - T_l) \tag{7.9}$$

and

$$q_{load} = \dot{m}_{load}\, C_p\,(T_l - T_i) \tag{7.10}$$

Equation (7.7) then becomes

$$(\rho VC_p)_e \frac{dT_l}{dt} = \dot{m}C_p(T_{fo} - T_l) - \dot{m}_{load}\, C_p\,(T_l - T_i) - (UA)_t(T_l - T_a) \tag{7.11}$$

It is important to note again that either q_u or q_{load} may be zero at any particular time. For example, when $T_{fo} < T_l$, the flow through the collectors would be stopped and we would have $\dot{m} = 0$ and $q_u = 0$. Similarly if no energy is required on the load side, $\dot{m}_{load} = 0$ and $q_{load} = 0$.

Example 7.1

A cylindrical hot water storage tank, 1.7 m in diameter and 2.1 m high, is made from a steel plate ($\rho = 7800$ kg/m^3, $C_p = 0.46$ kJ/kg-K) 6 mm thick. Apart from the mass of steel required for making the surface, an additional 200 kg of steel is required in the form of angles, etc. for strengthening the tank, which is insulated all round with glass wool insulation 20 cm thick ($k = 0.040$ W/m-K). The initial temperature of the water in the tank is 50.0°C at 0700 h in the morning on a particular day and the variation of q_u and T_a up to 2000 h is as shown.

Time	7–8	8–9	9–10	10–11	11–12	12–13	13–14
q_u (kJ/h)	18660	37496	54890	60070	69890	70120	62205
T_a (°C)	17.8	21.9	25.1	27.4	29.1	30.2	30.9
		14–15	15–16	16–17	17–18	18–19	19–20
		55490	41070	19195	0	0	0
		31.2	31.3	30.8	28.9	26.0	24.0

The load requirement is such that energy is continuously withdrawn from the tank at a constant rate of 27 000 kJ/h for 18 hours a day starting at 0500 h. Assuming that the water in the tank is always well-mixed, calculate the variation of its temperature.

Solution

We shall first determine the values of $(\rho VC_p)_e$ and $(UA)_t$.

$$\text{Internal volume of the tank} = \frac{\pi}{4} \times 1.7^2 \times 2.1 = 4.767 \text{ m}^3$$

$$(\rho VC_p)_l = 971.8 \times 4.767 \times 4.195 = 19432 \text{ kJ/K}$$

$$\text{Mass of steel used} = \left\{(\pi \times 1.7 \times 2.1 + 2 \times \frac{\pi}{4} \times 1.7^2) \times 0.006 \times 7800\right\} + 200$$

$$= 937.0 \text{ kg}$$

Therefore, $(\rho V C_p)_t = 937.0 \times 0.46 = 431$ kJ/K

Hence, $(\rho V C_p)_e = 19432 + 431 = 19\,863$ kJ/K

In calculating the value of $(UA)_t$, we shall assume that the flow of heat through the cylindrical surface and through the flat top and bottom is one-dimensional. We shall also assume that the thermal resistance offered by the insulation dominates and that the other thermal resistances to the flow of heat at the inner and outer surfaces can be neglected by comparison. The overall heat transfer coefficient U_1 (based on the inner diameter of the tank) for the flow of heat through the cylindrical surface is given by

$$\frac{1}{U_1} = \frac{r_1}{k_i} \ln \frac{r_2}{r_1}$$

where r_1 is the radius of the tank, r_2 is the radius with the insulation added on, and k_i is the thermal conductivity of the insulation. Substituting, we have,

$$\frac{1}{U_1} = \frac{0.85}{0.04} \ln \frac{1.05}{0.85} = 4.49$$

Therefore, $U_1 = 0.223$ W/m^2-K

Similarly, the overall heat transfer coefficient U_2 for the flow of heat through the flat surfaces is given by

$$U_2 = \frac{0.04}{0.2} = 0.2 \text{ W/m}^2\text{-K}$$

Hence, $(UA)_t = (0.223 \times \pi \times 1.7 \times 2.1) + (0.2 \times 2 \times \frac{\pi}{4} \times 1.7^2) = 3.409$ W/K

We now apply Eq. (7.8) over a one hour interval of time. Substituting the given values, we have from 0700 to 0800 h,

$$\frac{18\,660 - 27\,000 - 3.409 \times 3.6(T_l - 17.8)}{18\,660 - 27\,000 - 3.409 \times 3.6(50.00 - 17.8)} = \exp\left[-\frac{3.409 \times 1 \times 3600}{19\,863 \times 1000}\right]$$

Hence $(T_l)_{0800\,h} = 49.56$°C

Continuing in the same way, we obtain the following variation for T_l up to 2000 h.

Time (h)	8	9	10	11	12	13	14
T_l (°C)	49.56	50.07	51.46	53.11	55.25	57.41	59.16
		15	16	17	18	19	20
		60.58	61.27	60.86	59.48	58.10	56.72

The results are plotted in Fig. 7.4. It is seen that from 0700 to 0800 h, the energy withdrawn exceeds the useful heat gain and that the water temperature decreases. Thereafter, the useful heat gain (whose variation is typical of a clear sunny day) exceeds the energy withdrawn up to 1600 h. As a result, the tank water temperature starts increasing steadily and touches a peak around 1600 h. After this the useful heat gain decreases to zero and since energy is still being withdrawn, the tank water temperature decreases. From the load demand, it is clear that if the calculations were continued over a 24 hour period, the tank water temperature would be around 50°C at 0700 h on the next day. This indicates that there is an approximate matching between the useful heat gain and the energy withdrawn over a day and that the purpose of the storage tank in the present case is only to take care of the short term mismatch between supply and demand of energy over a day.

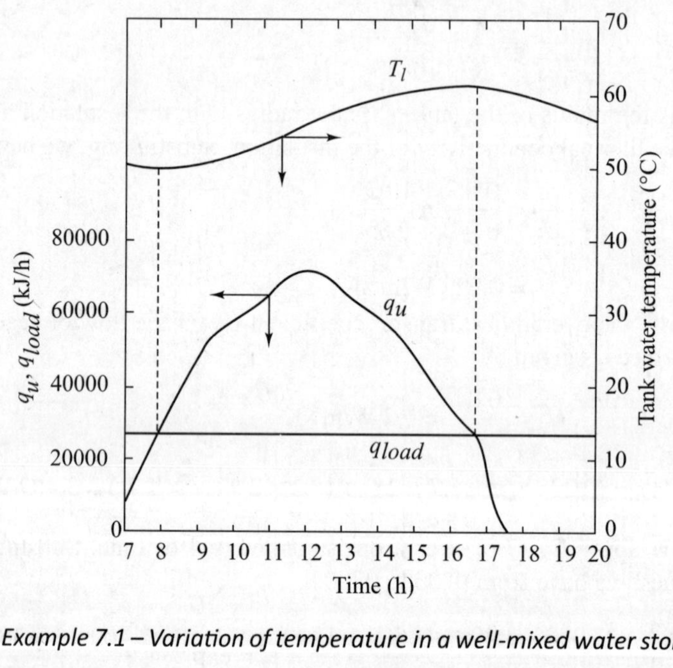

Fig. 7.4 *Example 7.1 – Variation of temperature in a well-mixed water storage tank*

It should also be noted from Fig. 7.4 that the tank temperature fluctuates by about 12°C. The magnitude of this fluctuation essentially depends upon the volume of water in the tank. If the volume of the water were to be doubled, the fluctuation would be approximately halved.

Kulkarni *et al.* [6] have recently proposed a new methodology for designing a solar water heating system with a flat-plate collector array and a well-mixed storage tank. They have introduced the concept of a design space to identify all possible designs based on the specifications of a given situation. The design space is shown on a collector area vs. storage volume diagram and can be obtained for different values of the solar fraction (fraction of annual energy supplied by the solar system). A particular design within the design space can be chosen based on further considerations.

The preceding analysis has been performed under the assumption that the liquid flowing through the collectors and to the load and that stored in the tank is the same, and that one stream mixes with the other. This may not always be the case. For example, in certain cases, it may be necessary from the point of view of corrosion to pass specially treated water mixed with chemical inhibitors through the collector circuit. It may also be necessary to add antifreeze compounds to the water passing through the collectors if temperatures below 0°C are likely to be encountered. In such cases, the fluid streams have to be separated and heat has to be transferred from the collector circuit liquid to the storage tank liquid through a heat exchanger. The heat exchanger may be a separate unit external to the storage tank or may be in the form of a coil immersed in the storage tank. We will now analyse the second situation shown schematically in Fig. 7.5.

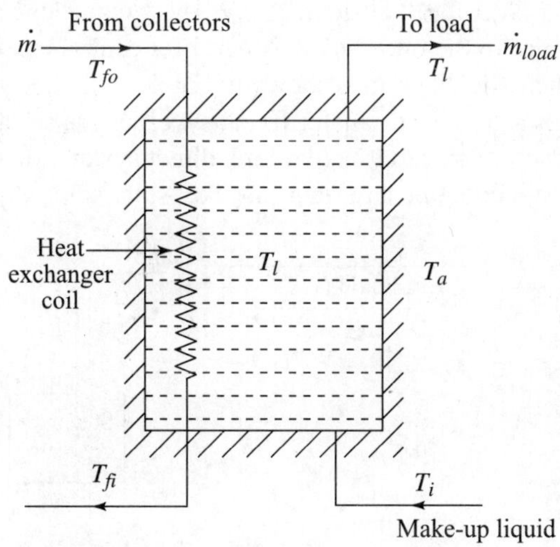

Fig. 7.5 *Analysis of a well-mixed sensible heat liquid storage tank with an immersed heat exchanger coil*

For an immersed coil, it is easy to show from heat exchange theory that

$$\frac{T_{fo} - T_{fi}}{T_{fo} - T_l} = 1 - \exp\left[-(UA)_e / \dot{m}\, C_p\right] \tag{7.12}$$

where $(UA)_e$ is the product of the overall heat transfer coefficient and the area for the heat exchanger coil. Thus, the expression for the rate of useful heat gain becomes

$$q_u = \dot{m}\, C_p (T_{fo} - T_{fi}) = \dot{m}\, C_p (T_{fo} - T_l)\,[1 - \exp\{-(UA)_e / \dot{m}\, C_p\}] \tag{7.13}$$

Substituting Eq. (7.13) for q_u and Eq. (7.10) for q_{load} into the energy balance Eq. (7.7) we get

$$(\rho VC_p)_e \frac{dT_l}{dt} = \dot{m}C_p(T_{fo} - T_l)[1 - \exp\{-(UA)_e / \dot{m}C_p\}]$$

$$- \dot{m}_{load} C_p (T_l - T_i) - (UA)_t (T_l - T_a) \qquad (7.14)$$

In the above equation, the specific heats of the two liquid streams have been assumed to be the same. In case they are different, they would need to be distinguished in an appropriate manner.

It is important to realize that the use of a heat exchanger for a given application not only increases the complexity of the set-up but also raises the temperature level of operation of the collector array by a few degrees. Since collector efficiency decreases with increase in temperature, more collector area is required for meeting the given energy demand.

❑ *Well-mixed Two-tank Situation* We now consider a well-mixed two-tank situation similar to the one shown as part of a central receiver power plant in Fig. 2.24. A schematic diagram is shown in Fig. 7. 6. In the system shown, the same fluid operates as the heat transfer fluid as well as the storage medium. For this reason, it is also referred to as a two-tank direct system. In a central receiver power plant, the fluid could be a molten salt like 'solar salt'.

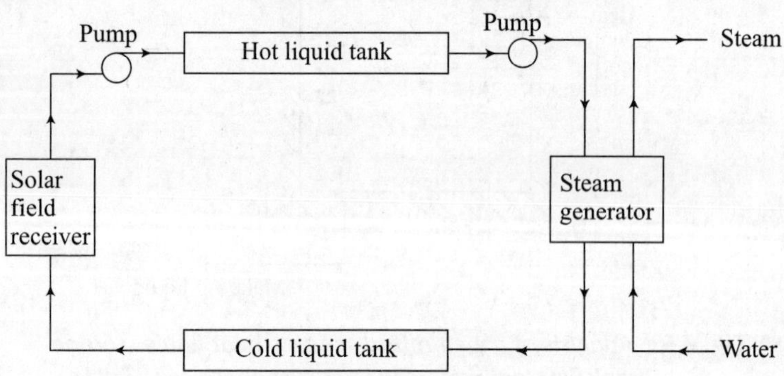

Fig. 7.6 *Two-tank direct sensible heat storage system*

Example 7.2

A two-tank direct heat storage system (Fig. 7.6) is used in a 1 MWe central receiver power plant. Solar salt is the heat transfer fluid and the heat storage medium, and the temperatures of the hot and cold molten salt in the two tanks, are 565 and 290°C respectively. The energy conversion efficiency of the electric power generation cycle is 40 per cent. Heat energy is collected only during the sunshine hours at the following rate:

Time (h)	0800–0900	0900–1000	1000–1100	1100–1200	1200–1300
Heat energy collected (kWh)	2000	5700	7900	9400	10 500
Time (h)	1300–1400	1400–1500	1500–1600	1600–1700	
Heat energy collected (kWh)	11 000	9000	6000	3000	

However, it is required that the electricity be generated at a constant rate of 1 MW continuously for 24 hours a day. Thus the thermal energy storage has to meet the mismatch between the heat energy collected and the heat energy required for energy conversion to electricity. Estimate the size of the hot and cold molten salt tanks.

Solution

Let $\dot{m}_1$ (in kg/s) be the rate at which the cold molten salt is pumped from the cold tank to the hot tank via the solar receiver in the central tower. $\dot{m}_1$ will have a positive value which changes during the sunshine hours from 8 am to 5 pm when heat energy is collected. It will be zero at all other times.

From 8 to 9 am, we have

$$\dot{m}_1 \times C_p \times (565 - 290) = 2000$$

From Eq. (7.5), $C_p = 1.55$ kJ/kg-K

Thus $\dot{m}_1 = 2000/(1.55 \times 275) = 4.692$ kg/s

Let $\dot{m}_2$ (in kg/s) be the rate at which hot molten salt is pumped from the hot tank to the cold tank via the steam generator of the energy conversion cycle.

Since the energy conversion efficiency is 40 per cent, the heat energy required per hour on a continuous basis = 1000/0.4 = 2500 kWh

Thus at all times of the day

$$\dot{m}_2 \times 1.55 \times (565 - 290) = 2500$$

$$\dot{m}_2 = 5.865 \text{ kg/s}$$

The rate at which the mass of molten salt increases in the hot tank from 8 to 9 am $= \dot{m}_1 - \dot{m}_2$

$= (4.692 - 5.865) = -1.173$ kg/s

The negative sign indicates that the mass of liquid in the hot tank has decreased from 8 to 9 am by 1.173 × 3600 = 4222.8 kg

Continuing in the same way from hour to hour, we obtain the following results from 8 am to 5 pm.

Time (h)	8–9	9–10	10–11	11–12	12–13	13–14	14–15	15–16	16–17
Energy collected (kWh)	2000	5700	7900	9400	10 500	11 000	9000	6000	3000
$\dot{m}_1$ (kg/s)	4.692	13.372	18.533	22.052	24.633	25.806	21.114	14.076	7.038
$\dot{m}_2$ (kg/s)	5.865	5.865	5.865	5.865	5.865	5.865	5.865	5.865	5.865
$\dot{m}_1 - \dot{m}_2$ (kg/s)	–1.173	7.507	12.668	16.187	18.768	19.941	15.249	8.211	1.173

For the remaining period from 5 pm in the evening to 8 am in the morning, we have

$$\dot{m}_1 = 0,\ \dot{m}_2 = 5.865 \text{ kg/s and } \dot{m}_1 - \dot{m}_2 = -5.865 \text{ kg/ s}$$

From the results it is clear that the level of liquid in the hot tank will be at its lowest at 9 am. It will keep on rising till 5 pm. Thereafter through the evening and night, the level will keep on falling till 9 am as energy stored in the tank is used for electricity generation.

The mass of liquid added to the hot tank from 9 am to 5 pm

$$= (7.507 + 12.668 + 16.187 + 18.768 + 19.941 + 15.249 + 8.211 + 1.173) \times 3600$$

$$= 99.704 \times 3600 = 358\,934.4 \text{ kg}$$

From Eq. (7.4), density of solar salt at 565°C is given by

$$\rho = 2106.0 - (0.66795 \times 565) = 1728.6 \text{ kg/m}^3$$

Volume of liquid corresponding to the mass added from 9 am to 5 pm

$$= 358\,934.4/1728.6$$

$$= 207.6 \text{ m}^3$$

Assuming that the volume of the hot tank is made about 20 per cent more as a margin of safety, volume of tank = 1.2 × 207.6 = 249.1 m^3

This corresponds to a cylindrical tank 6 m in diameter and 8.8 m high.

As far as the cold tank is concerned, the mass of liquid taken away from 9 am to 5 pm = Amount added to the hot tank

$$= 358\,934.4 \text{ kg}$$

However the volume of this liquid is less because the density of solar salt at 290°C is more than that at 565°C

Density of solar salt at 290°C = 2106.0 – (0.66795 × 290) = 1912.3 kg/m^3

Corresponding volume = 187.7 m^3

With a margin of safety of 20 per cent, this corresponds to a cylindrical tank 6 m in diameter and 8 m high.

It is worth comparing the above results for tank size with data available for the Gemasolar central receiver plant (Table 2.3). The plant generates 19.9 MWe and has a two-tank thermal storage system using solar salt. The system stores enough excess energy during the day to generate the rated electrical capacity for 15 hours without any solar input. Thus the characteristics of the storage at Gemasolar are similar to the storage analyzed in this example. The hot and cold storage tanks at Gemasolar are 23 m in diameter and 14 m in height. This is equivalent to 292 m^3 per MWe, a number close to the value of 250 m^3 per MWe obtained in this example.

❑ *Thermocline Systems* Thermocline storage systems have been proposed as an alternative to the two-tank storage to reduce cost. The system consists of a single storage tank where hot fluid from the solar field receiver is pumped into the top of the tank and colder fluid from the bottom part of the tank is taken to the solar field receiver. Thermal stratification is created within the tank and is preserved by buoyancy effects. Such a system is called as a *thermocline* storage.

Thermal stratification is obviously desirable if the temperature difference across the solar receiver is significant. If the line taking hot liquid to the load (as in Fig. 7.3) is at the required high temperature and is located appropriately near the top of the storage tank, then only the liquid near this exit port has to be at that temperature. The rest of the liquid in the storage tank can be at a lower temperature at all times. As a result, heat losses from the tank are reduced. A second advantage is that the collector operates at a lower average temperature level and consequently delivers a higher collection efficiency.

An insulation baffle can also be used to keep the hot fluid and the cold fluid separated from each other. The baffle can move freely up and down inside the tank. In place of an insulation baffle, filler material can also be inserted in the tank. The filler material acts as the primary heat storage material and reduces the overall need for a relatively higher cost working liquid. It also helps in maintaining the thermal gradient by reducing the natural convection within the liquid. Thermoclines of this type are called as dual-media storage systems. Various types of filler materials such as silica sand, quartzite rock, taconite, basalt, granite and marble have been investigated and their merits and demerits have been reported [7 – 10]. Galione *et al.* [11] have proposed a new concept of using a packed bed of different layers of solid and PCM filler materials. However, testing has not yet been done.

Some of the considerations which need to be kept in mind for the selection of the filler material are as follows.

(i) It should be inexpensive, non-hazardous and readily available.
(ii) It should have a high heat capacity.
(iii) The material needs to be compatible with heat transfer fluid being used.

The analysis of a thermocline storage tank is very complex due to the presence of filler material. We will discuss here the analysis of a thermally stratified liquid storage tank without filler material. Even that analysis is complicated by the fact that the temperature profile has to be solved for as a function of time. In addition, in most practical situations, the problem is not one-dimensional in nature and involves consideration of the heat conduction through the walls of the storage tank. A simplified analysis is possible by assuming that the storage tank consists of a number of well-mixed sections at different temperatures and writing down energy balance equations for each section. Consider the simplified situation in which the storage tank is assumed to consist of only two well-mixed sections (Fig. 7.7) at temperatures T_{l1} and T_{l2}. Writing energy balances on each of the sections, we obtain

$$(\rho V C_p)_{e1} \frac{dT_{l1}}{dt} = \dot{m}\, C_p(T_{fo} - T_{l1}) - \dot{m}_{load}\, C_p(T_{l1} - T_{l2}) - (UA)_{t1}(T_{l1} - T_a) \quad (7.15)$$

$$(\rho V C_p)_{e2} \frac{dT_{l2}}{dt} = \dot{m}\, C_p(T_{l1} - T_{l2}) - \dot{m}_{load}\, C_p(T_{l2} - T_i) - (UA)_{t2}(T_{l2} - T_a) \quad (7.16)$$

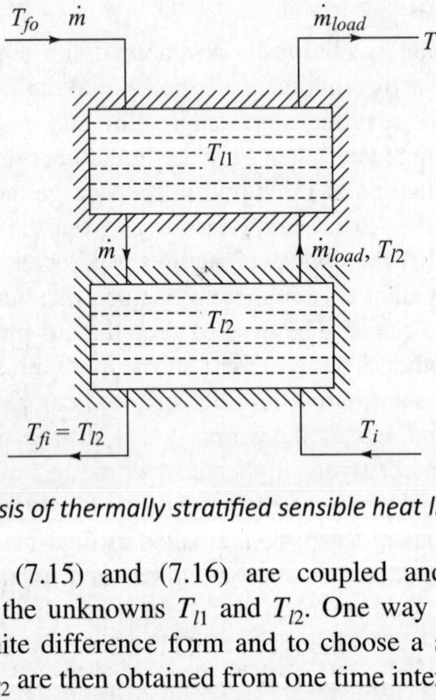

Fig. 7.7 *Analysis of thermally stratified sensible heat liquid storage tank*

Differential Eqs (7.15) and (7.16) are coupled and have to be solved simultaneously for the unknowns T_{l1} and T_{l2}. One way of solving them is to express them in finite difference form and to choose a suitable time interval. Values of T_{l1} and T_{l2} are then obtained from one time interval to the next.

Equations (7.15) and (7.16) have been written under the assumption that the flow from the collectors can enter the storage tank only at the top of section 1. This occurs when $T_{fo} > T_{l1}$. When $T_{fo} < T_{l1}$ no flow occurs.

In case the tank is also provided with an inlet between the two sections, three possibilities exist.

1. Flow $\dot{m}$ enters at the top of section 1 when $T_{fo} > T_{l1}$.
2. Flow $\dot{m}$ enters between sections 1 and 2 when $T_{l1} > T_{fo} > T_{l2}$.
3. There is no flow, and consequently no energy to be stored, when $T_{fo} < T_{l2}$.

Equations (7.15) and (7.16) can be easily modified to account for this situation.

Example 7.3

The temperatures in a hot water stratified tank are to be determined by assuming that the tank consists of two equal well-mixed sections with inlets at the top and in-between the two sections. The following data is given.

Hour	$\dot{m}$(kg/h)	T_{fo} (°C)	$\dot{m}_{load}$ (kg/h)	T_a (°C)
1200–1300	2000	80	250	22
1300–1400	2000	80	240	23
1400–1500	2000	76	220	22

It is also given that,

(1) Mass of water in the tank = 5000 kg

(2) $(UA)_{t1} = (UA)_{t2} = 30$ kJ/h-°C

(3) Make up water enters at 20°C at the same rate as the rate of withdrawal to the load,

(4) At 1200 h, $T_{l1} = 70$°C, $T_{l2} = 65$°C

Calculate the values of T_{l1} and T_{l2} at 1500 h.

Solution

Neglecting the value of the heat capacity term $(\rho VC_p)_t$, Eqs (7.15) and (7.16) can be written in a finite difference form as follows:

$$(\rho VC_p)_{l1}\left(\frac{T_{l1,f} - T_{l1,i}}{\Delta t}\right) = \dot{m}C_p\left(T_{fo} - \frac{T_{l1,f} + T_{l1,i}}{2}\right)$$

$$- \dot{m}_{load}C_p\left(\frac{T_{l1,f} + T_{l1,i}}{2} - \frac{T_{l2,f} + T_{l2,i}}{2}\right) - (UA)_{t1}\left(\frac{T_{l1,f} + T_{l1,i}}{2} - T_a\right) \quad (7.17)$$

$$(\rho VC_p)_{l2}\left(\frac{T_{l2,f} - T_{l2,i}}{\Delta t}\right) = \dot{m}C_p\left(\frac{T_{l1,f} + T_{l1,i}}{2} - \frac{T_{l2,f} + T_{l2,i}}{2}\right)$$

$$- \dot{m}_{load}C_p\left(\frac{T_{l2,f} + T_{l2,i}}{2} - T_i\right) - (UA)_{t2}\left(\frac{T_{l2,f} + T_{l2,i}}{2} - T_a\right) \quad (7.18)$$

In these equations, $T_{l1,f}$ and $T_{l1,i}$ represent the final and initial values of T_{l1} over the time interval Δt. A similar meaning is attached to the symbols $T_{l2,f}$ and $T_{l2,i}$. The symbols T_{fo}, T_a and T_i now represent average values over the time interval Δt.

We take $\Delta t = 1$ h and note the presence of the inlet in between the two sections while substituting the given data from one hour to the next.

Time interval 1200–1300 h: $T_{l1,i} = 70$°C; $T_{l2,i} = 65$°C.

Substituting into Eq. (7.17) we get

$$2500 \times 4.19 \times (T_{l1,f} - 70) = 2000 \times 4.19\left(80 - \frac{T_{l1,f} + 70}{2}\right)$$

$$- 250 \times 4.19\left(\frac{T_{l1,f} + 70}{2} - \frac{T_{l2,f} + 65}{2}\right) - 30\left(\frac{T_{l1,f} + 70}{2} - 22\right)$$

or $$3628.6\, T_{l1,f} - 125\, T_{l2,f} = 264\,282 \tag{7.19}$$

Similarly, substituting into Eq. (7.18), we get

$$-1000\, T_{l1,f} + 3628.6\, T_{l2,f} = 164\,300 \tag{7.20}$$

Solving Eqs (7.19) and (7.20),

$$T_{l1,f} = 75.10°\text{C};\ T_{l2,f} = 65.98°\text{C}$$

Time interval 1300–1400 h: Now $T_{l1,i} = 75.10°\text{C}$; $T_{l2,i} = 65.98°\text{C}$.

Substituting into Eqs (7.17) and (7.18),we get

$$3623.6\, T_{l1,f} - 120\, T_{l2,f} = 271\,457$$

and $$-1000\, T_{l1,f} + 3623.6\, T_{l2,f} = 170\,882$$

Solving $$T_{l1,f} = 77.18°\text{C};\ T_{l2,f} = 68.46°\text{C}$$

Time interval 1400–1500 h: Now $T_{l1,i} = 77.18°\text{C}$; $T_{l2,i} = 68.46°\text{C}$.

Since $T_{l1,i} > T_{fo} > T_{l2,i}$, the flow from the collectors will bypass section 1 and enter in between sections 1 and 2. Thus, the first term of the right hand side of Eq. (7.17) will be zero and in the first term of the right hand side of Eq. (7.18), we would put T_{fo} instead of $(T_{l1,f} + T_{l1,i})/2)$. Substituting, we get

$$2613.6\, T_{l1,f} - 110\, T_{l2,f} = 191\,875$$

and $$3613.6\, T_{l2,f} = 251\,469$$

Solving, $$T_{l1,f} = 76.34°\text{C};\ T_{l2,f} = 69.59°\text{C}$$

7.3.4 Analysis of a Packed-bed Storage

We now take up for analysis a packed-bed storage unit as shown in Fig. 7.8. The unit is packed with rocks, pebbles or bricks through which air is circulated. Hot air from solar air heaters is usually passed down through the bed when sensible heat is to be stored in the particulate solid, while cold air from the load is circulated upwards when heat is to be extracted from the solid. Unlike a liquid storage tank, the two processes cannot be executed simultaneously.

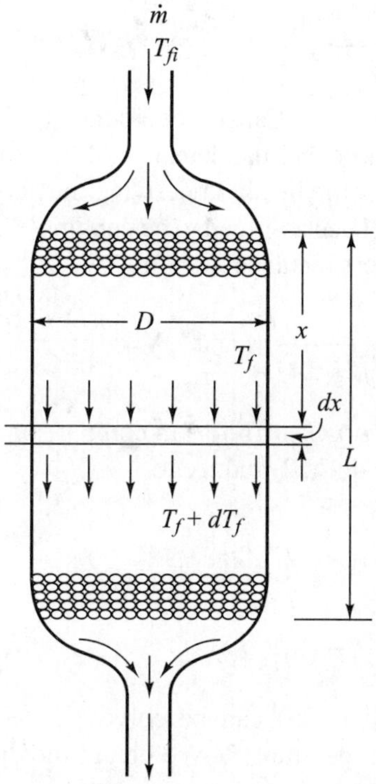

Fig. 7.8 *Analysis of a packed-bed storage unit*

The transient heat transfer analysis which follows is due to Schumann [12]. Consider a packed-bed unit of length L and diameter D packed with solid having an equivalent spherical diameter d and a void fraction ε. The mass flow rate of the air is $\dot{m}$ and it enters with a constant temperature T_{fi}. For the purposes of analysis, it is assumed that,

(1) The bed material has infinite thermal conductivity in the radial direction and zero conductivity in the axial flow direction,

(2) The heat transfer coefficient does not vary with time and place inside the bed, and

(3) The bed is semi-infinite in the direction of the flow.

Considering separate energy balances on the bed material and air in a slice dx of the bed across which the temperature of the solid changes from T_s to $(T_s + dT_s)$ and the temperature of the air changes from T_f to $(T_f + dT_f)$, we have

$$(1-\varepsilon)\rho_s C_{ps}\frac{\partial T_s}{\partial t} = h_v(T_f - T_s) \tag{7.21}$$

Hence, $$\varepsilon \rho_f C_{pf} \frac{\partial T_f}{\partial t} + \frac{4\dot{m}C_{pf}}{\pi D^2} \frac{\partial T_f}{\partial x} = h_v(T_s - T_f) \tag{7.22}$$

where h_v is the volumetric heat transfer coefficient (in W/m^3-K) defined per unit volume of the bed, ρ_s and ρ_f are the densities of the solid and fluid, and C_{ps} and C_{pf} are the respective specific heats. In deriving the above equations, heat losses to the surroundings have been assumed to be negligible. Defining a dimensionless time τ and a dimensionless distance X as follows,

$$\tau = \frac{h_v t}{\rho_s C_{ps}(1-\varepsilon)} \quad \text{and} \quad X = \frac{\pi D^2 h_v x}{4\dot{m}C_{pf}} \tag{7.23}$$

and neglecting the term $\varepsilon \rho_f C_{pf} (\partial T_f/\partial t)$ in comparison to the other two terms in Eq. (7.22), Eqs (7.21) and (7.22) reduce to

$$\frac{\partial T_s}{\partial \tau} = (T_f - T_s) \tag{7.24}$$

$$\frac{\partial T_f}{\partial X} = (T_s - T_f) \tag{7.25}$$

Equations (7.24) and (7.25) can be solved if the solid is assumed to be initially at a uniform temperature T_i. We obtain the dimensionless temperature distributions

$$\frac{T_s - T_i}{T_{fi} - T_i} = 1 - e^{-(X+\tau)} \sum_{n=0}^{\infty} X^n M_n(X\tau) \tag{7.26}$$

and $$\frac{T_f - T_i}{T_{fi} - T_i} = 1 - e^{-(X+\tau)} \sum_{n=1}^{\infty} X^n M_n(X\tau) \tag{7.27}$$

where $$M_n(X\tau) = \sum_{k=0}^{\infty} \frac{(X\tau)^k}{k!(k+n)!}$$

Values of $(T_s - T_i)/(T_{fi} - T_i)$ and $(T_f - T_i)/(T_{fi} - T_i)$ have been computed from Eqs (7.26) and (7.27) for $0 \le X \le 20$ and $0 \le \tau \le 30$ and are given in Tables 7.3 and 7.4 so that they can be used easily.

Table 7.3 Values of $(T_s - T_i)/(T_{fi} - T_i)$ in a packed-bed storage

τ	$X \rightarrow 0$	1	2	3	4	5	6	7	8	9	10	12	14	16	18	20
0	0.000	0.000	0.000	0.000	0.000	0.000	0.000	0.000	0.000	0.000	0.000					
1	0.632	0.346	0.183	0.094	0.047	0.023	0.011	0.006	0.003	0.001	0.001	0.000				
2	0.865	0.606	0.397	0.247	0.148	0.086	0.049	0.027	0.015	0.008	0.004	0.001	0.000			
3	0.950	0.775	0.585	0.417	0.283	0.185	0.117	0.072	0.044	0.026	0.015	0.005	0.001	0.000	0.000	
4	0.982	0.877	0.730	0.573	0.428	0.307	0.212	0.143	0.093	0.059	0.037	0.014	0.005	0.002	0.001	0.000
5	0.993	0.934	0.831	0.702	0.565	0.436	0.325	0.234	0.164	0.112	0.074	0.031	0.012	0.005	0.002	0.001
6	0.998	0.966	0.898	0.800	0.682	0.559	0.442	0.338	0.251	0.181	0.128	0.060	0.026	0.011	0.004	0.002
7	0.999	0.983	0.940	0.870	0.776	0.667	0.554	0.446	0.349	0.265	0.197	0.101	0.048	0.022	0.009	0.004
8		0.991	0.966	0.918	0.847	0.757	0.655	0.551	0.450	0.358	0.277	0.156	0.081	0.039	0.018	0.008
9			0.981	0.949	0.898	0.827	0.741	0.646	0.548	0.453	0.365	0.222	0.124	0.065	0.032	0.015
10			0.990	0.969	0.934	0.880	0.811	0.728	0.638	0.545	0.455	0.297	0.178	0.100	0.052	0.026
12				0.989	0.974	0.946	0.905	0.851	0.784	0.707	0.625	0.459	0.312	0.197	0.116	0.065
14					0.990	0.978	0.956	0.925	0.882	0.827	0.763	0.616	0.462	0.324	0.213	0.131
16						0.991	0.981	0.965	0.940	0.905	0.861	0.746	0.608	0.465	0.334	0.226
18							0.992	0.984	0.971	0.951	0.924	0.843	0.732	0.601	0.467	0.342
20								0.993	0.987	0.977	0.961	0.909	0.828	0.720	0.596	0.468
22									0.994	0.989	0.981	0.950	0.896	0.815	0.710	0.591
24											0.991	0.974	0.940	0.883	0.803	0.701
26												0.987	0.967	0.930	0.872	0.792
28													0.983	0.960	0.921	0.861
30													0.991	0.978	0.953	0.912

Table 7.4 Values of $(T_f - T_i)/(T_{fi} - T_i)$ in a packed-bed storage

τ	$X \to 0$	1	2	3	4	5	6	7	8	9	10	12	14	16	18	20
0	1.000	0.368	0.135	0.050	0.018	0.007	0.003	0.001								
1	1.000	0.654	0.394	0.225	0.123	0.066	0.034	0.017	0.009	0.004	0.002					
2	1.000	0.817	0.604	0.415	0.270	0.169	0.102	0.060	0.034	0.019	0.011	0.003	0.001			
3	1.000	0.906	0.753	0.583	0.427	0.298	0.200	0.130	0.082	0.051	0.031	0.011	0.003	0.001		
4	1.000	0.953	0.852	0.717	0.572	0.435	0.318	0.224	0.154	0.102	0.067	0.026	0.010	0.003	0.001	
5	1.000	0.977	0.914	0.815	0.693	0.564	0.441	0.333	0.243	0.173	0.120	0.054	0.022	0.009	0.003	0.001
6	1.000	0.989	0.951	0.883	0.788	0.676	0.558	0.446	0.345	0.259	0.189	0.095	0.044	0.019	0.008	0.003
7	1.000	0.995	0.973	0.928	0.858	0.766	0.662	0.554	0.449	0.354	0.272	0.149	0.075	0.035	0.016	0.007
8	1.000	0.997	0.985	0.957	0.907	0.837	0.749	0.651	0.550	0.452	0.362	0.216	0.119	0.060	0.029	0.013
9			0.992	0.974	0.941	0.888	0.819	0.735	0.643	0.547	0.455	0.293	0.173	0.095	0.049	0.024
10			0.996	0.985	0.963	0.926	0.872	0.803	0.723	0.635	0.545	0.375	0.237	0.139	0.076	0.039
12				0.995	0.986	0.969	0.940	0.899	0.844	0.778	0.704	0.541	0.385	0.254	0.157	0.091
14					0.995	0.988	0.974	0.952	0.919	0.876	0.822	0.688	0.538	0.392	0.268	0.172
16						0.996	0.989	0.979	0.961	0.935	0.901	0.803	0.676	0.535	0.399	0.280
18							0.996	0.991	0.982	0.968	0.948	0.884	0.788	0.666	0.533	0.404
20								0.996	0.992	0.985	0.974	0.935	0.869	0.774	0.658	0.532
22									0.997	0.994	0.988	0.966	0.923	0.855	0.762	0.650
24											0.995	0.983	0.957	0.912	0.843	0.752
26												0.992	0.977	0.949	0.901	0.830
28													0.988	0.972	0.940	0.891
30													0.994	0.985	0.966	0.932

Lof and Hawley [13] have suggested the following correlation for calculating the value of the volumetric heat transfer coefficient h_v, which is required for evaluating the parameters X and τ.

$$h_v = 650\,(G/d)^{0.7}\ \text{W/m}^3\text{-K} \tag{7.28}$$

where G is the superficial mass velocity based on the cross-sectional area of the bed ($= 4\dot{m}/\pi D^2$) in kg/s-m^2 and d is the average diameter of the bed material in metres. Subsequently, based on extensive experimental data, Chandra and Willits [14] have obtained the dimensionless correlation

$$(h_v\, d^2/k_f) = 1.45\,(\text{Re}_d)^{0.7} \tag{7.29}$$

where $\text{Re}_d (= G\,d/\mu_f)$ is the Reynolds number of the flow based on the characteristic dimension d. Equation (7.29) is valid for the range $100 < \text{Re}_d < 1000$ and is recommended for use.

It should be noted that Eqs (7.26) and (7.27) are valid regardless of whether $T_i < T_{fi}$ or $T_i > T_{fi}$. In the first case, the bed heats up and energy is stored, while in the second case, the reverse occurs.

The pressure drop across a packed-bed storage unit is also of importance since large volumes of fluid are being handled. Dunkle and Ellul [15] have suggested the correlation

$$\Delta p = \frac{LG^2}{\rho_f\, d}[21 + 1750/\text{Re}_d] \tag{7.30}$$

Later, Chandra and Willits [14] have suggested the correlation

$$\Delta p = \frac{LG^2}{\rho_f d}\varepsilon^{-2.6}[1.7 + 185/\text{Re}_d] \tag{7.31}$$

Equation (7.31) is valid for the range $0.33 < \varepsilon < 0.46$ and $1 < \text{Re}_d < 1000$. It is recommended for use in view of the fact that it includes the effect of void fraction on the pressure drop.

Example 7.4

A packed-bed storage unit, 1 m in height and 0.70 m in diameter, is filled with rock pieces ($\rho_s = 2800$ kg/m^3, $C_{ps} = 0.90$ kJ/kg-K) having an average diameter of 2 cm. The void fraction is 0.35. Initially the bed is at a uniform temperature of 25°C everywhere. Air heated to a temperature of 70°C in solar air heaters starts flowing in with a flow rate of 0.4 kg/s. Find the temperature distributions in the bed after (i) 5 minutes, and (ii) 10 minutes. Calculate also the energy stored in the bed material as a fraction of the maximum amount which can be stored and the pressure drop across the bed.

Solution

We first calculate the value of the volumetric heat transfer coefficient.

$$\text{Mass velocity } G = \frac{0.4}{(\pi \times 0.7^2/4)} = 1.039\ \text{kg/s-m}^2$$

The properties of air at 70°C are

$\rho_f = 1.029$ kg/m^3 $\qquad k_f = 0.0297$ W/m-K

$\mu_f = 20.6 \times 10^{-6}$ N-s/m^2 $\qquad C_{pf} = 1.009$ kJ/kg-K

$$\text{Re}_d = Gd/\mu_f = (1.039 \times 0.02)/20.6 \times 10^{-6} = 1009$$

Therefore, from Eq. (7.29)

$$h_v = 0.0297 \times 1.45 \times (1009)^{0.7}/(0.02)^2 = 13\,639 \text{ W/m}^3\text{-K}$$

(i) At $t = 5$ minutes,

$$\tau = \frac{13\,639 \times 5 \times 60}{2800 \times 0.9 \times (1 - 0.35)} \times \frac{1}{1000} = 2.50$$

We will determine the values of T_s and T_f at $x = 0, 0.25, 0.5, 0.75$ and 1 m.

At $x = 0.25$ m,

$$X = \frac{13\,639 \times \pi \times 0.7^2 \times 0.25}{4 \times 0.4 \times 1.009 \times 1000} = 3.25$$

Thus, for the values of x chosen, we have

$$X = 0, 3.25, 6.50, 9.75 \text{ and } 13.00$$

Corresponding to these values of X and for $\tau = 2.50$, the values of $(T_s - T_i)/(T_{fi} - T_i)$ and $(T_f - T_i)/(T_{fi} - T_i)$ can be read by interpolation from Tables 7.3 and 7.4 and the values of T_s and T_f can be calculated. The results are plotted in Fig. 7.9.

(ii) At $t = 10$ minutes, $\tau = 5.00$. The calculation proceeds in the same manner as in part (i). The results are again plotted in Fig. 7.9. The trends exhibited by the temperature distributions are as expected. It is seen that temperature decreases in the direction of the flow and that there is a point of inflection in each temperature distribution.

The energy stored in the bed material at any instant of time

$$= \int_0^L (1 - \varepsilon)\frac{\pi}{4} D^2 \rho_s C_{ps} (T_s - T_i)\, dx$$

The maximum amount of energy which can be stored

$$= (1 - \varepsilon)\frac{\pi}{4} D^2 L \rho_s C_{ps} (T_{fi} - T_i)$$

Thus, the fraction of the maximum stored

$$F = \left\{\int_0^L (T_s - T_i)\, dx\right\} \Big/ L(T_{fi} - T_i) \qquad (7.32)$$

From Fig. 7.9, we obtain by numerical integration

$$F = 0.20 \text{ for } t = 5 \text{ minutes} \quad \text{and} \quad F = 0.38 \text{ for } t = 10 \text{ minutes}$$

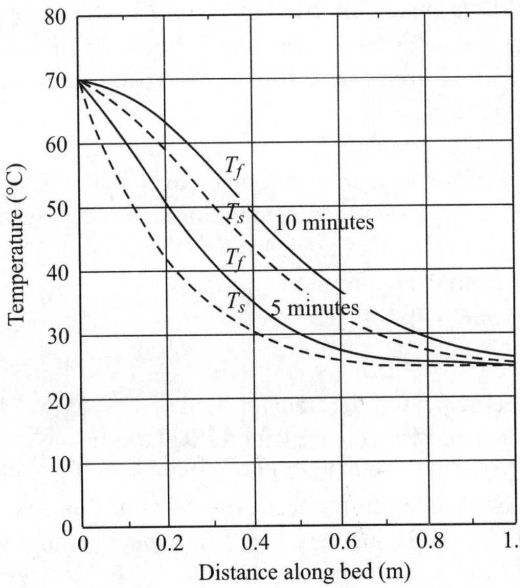

Fig. 7.9 *Example 7.4 – Temperature distributions in a packed-bed storage*

Using Eq. (7.31) and the properties of air at 70°C, we have

$$\Delta p = \frac{1 \times 1.039^2}{1.029 \times 0.02} \times 0.35^{-2.6}[1.7 + 185/1009] \text{ N/m}^2$$

$$= 1514 \text{ N/m}^2 = 15.4 \text{ cm of water}$$

This pressure drop is rather high and may not be acceptable in an actual application. In order to reduce it to an acceptable value, the air flow rate would have to be reduced by a factor of 4 or 5. This would also result in a reduction in the value of h_v. Consequently the rate at which energy is stored in the solid would also reduce (see Problem 7.6).

The solutions just given are valid for the situation in which the solid is initially at a uniform temperature T_i and the fluid entering the bed is also at a constant temperature T_{fi}. If these conditions are not satisfied, it becomes difficult to obtain an analytical solution and recourse has to be taken to a numerical method. The governing differential Eqs (7.24) and (7.25) are expressed in a suitable finite-difference form by considering the bed to consist of a finite number of slices of width Δx and considering finite time intervals of Δt. Sets of simultaneous linear equations at each instant of time are obtained and these are solved in order to obtain the temperature distributions.

7.3.5 Testing Procedures

Certain standard procedures suggested for the testing of thermal storage devices [16,17] are now considered. The set-ups used are similar to those described earlier for liquid flat-plate collectors (Section 4.13) and solar air heaters (Section

5.5), with the liquid flat-plate collector or solar air heater being replaced by the thermal storage device under test using either a liquid or air as the transfer fluid. The tests to be performed are the heat loss test and two charge-discharge tests for determining the storage capacity.

In the heat loss test, the transfer fluid is passed through the storage device at a fixed mass flow rate ($\dot{m}$) and at an inlet temperature (T_{fi}) 25°C above the ambient air temperature (T_a). After steady state conditions are achieved, the difference between the inlet and outlet fluid temperature is measured. The heat loss factor $(UA)_t$ *is* calculated from the expression

$$(UA)_t = \frac{\dot{m}C_p(T_{fi} - T_{fo})}{(T_{fi} - T_a)} \tag{7.33}$$

where C_p is the specific heat of the transfer fluid.

In the charge-discharge test, the transfer fluid at a constant temperature is first passed through the device and the device brought to a uniform initial temperature T_i. Then the flow is adjusted to the test flow rate ($\dot{m}$) and the temperature of the transfer fluid is suddenly increased in a step-wise manner to $T_i + \Delta T_i$. The difference between the temperature of the transfer fluid entering and leaving the device is continuously recorded over a specified time period t_p as charging takes place. The temperature of the transfer fluid is maintained at $(T_i + \Delta T_i)$ until the exit temperature of the transfer fluid reaches a steady value. Thereafter with the same mass flow rate, the temperature of the transfer fluid is suddenly decreased in a step-wise manner back to the initial value T_i. The difference between the temperature of the transfer fluid entering and leaving the device is again continuously recorded over the same time period t_p as discharging takes place. The charge-discharge cycle is performed for two test conditions. If the transfer fluid is a liquid, the value of ΔT_i is taken as 15°C for both test conditions, while the values of t_p are taken to be 2 and 4 hours. If the transfer fluid is air, the corresponding values are $\Delta T_i = 35$°C, and $t_p = 2$ and 4 hours. The variation of the inlet and exit temperature of the transfer fluid during a typical charge-discharge cycle is shown in Fig. 7.10.

The value of T_i, the initial temperature of the storage device, is chosen by considering the intended temperature range of the device during use, while the value of the mass flow rate maintained during charging and discharging is calculated from the expression

$$\dot{m} = \frac{\text{TSC}}{t_p C_p \Delta T_i} \tag{7.34}$$

where, TSC = theoretical storage capacity of the device for a temperature change from T_i to $(T_i + \Delta T_i)$.

From the data recorded, the charge and discharge capacities of the device are obtained by integration as follows,

$$C_c = \dot{m}C_p \int_0^{t_p} (T_{fi} - T_{fo})dt - (UA)_t\, t_p \left[\frac{T_{fi} + \overline{T}_{fo}}{2} - T_a\right] \tag{7.35}$$

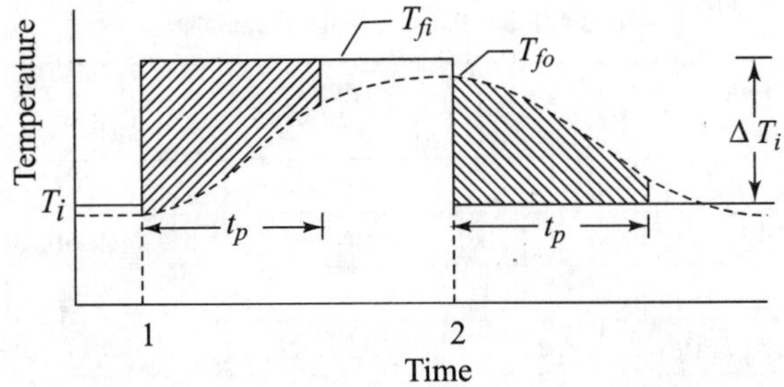

Fig. 7.10 *Temperature variations during a charge-discharge test cycle. Time instants 1 and 2 correspond to steady state situations. The shaded areas are measures of C_c and C_d*

$$C_d = \dot{m} C_p \int_0^{t_p} (T_{fo} - T_{fi})\, dt \tag{7.36}$$

The standard recommends that the performance of the thermal storage device should be judged by the values of C_c and C_d calculated from Eqs (7.35) and (7.36) along with plots of dimensionless temperature $(T_{fo} - T_{fi})/\Delta T_i$ vs. time obtained for the two conditions specified.

LATENT HEAT STORAGE 7.4

In a latent heat storage system, heat is stored in a material when it melts and extracted from the material when it freezes. Compared to sensible heat storage, latent heat storage systems exhibit the characteristic of having a high energy density leading to more compact systems. A number of materials called phase change materials (PCMs) have been investigated from the point of view of their suitability for solar energy applications.

The properties required for a *phase change material* are as follows,

1. A melting point in the temperature range of the application for which it is being considered,
2. A high value of the latent heat of fusion,
3. A small volume change during the phase change, and
4. A negligible amount of supercooling or superheating for the phase change to occur.

These properties should be stable and should not degrade after repeated cycling involving melting and solidification. In addition, it is desirable that the thermal conductivity of the material should be high in both phases, that it should have a low vapour pressure at the temperature of use, and that it should be non-corrosive.

Table 7.5 lists a few PCMs along with the values of their melting points and latent heats of fusion. An exhaustive listing has been compiled by Zalba *et al.* [18] and Xu *et al.* [19]. PCMs may be considered under the following groupings: (1) Organic materials, (2) Inorganic materials.

Table 7.5 Phase change materials

Substance	Formula	Melting point (°C)	Heat of fusion (kJ/kg)
1. Organic materials:			
Paraffin wax Pll6E	–	42–50	209
Capric acid	–	32	153
2. Inorganic materials:			
(i) Hydrated salts			
Calcium chloride hexahydrate	$CaC1_2.6H_2O$	30	168
Sodium sulphate decahydrate	$Na_2SO_4.10H_2O$	32	241
Magnesium nitrate hexahydrate	$Mg(N0_3)_2.6H_2O$	89	167
(ii) Other compounds/eutectics			
Ice	H_2O	0	335
Sodium nitrate	$NaNO_3$	310	173
Sodium hydroxide	NaOH	320	159
Lithium carbonate/sodium carbonate	Li_2CO_3/Na_2CO_3 (44%: 56% by weight)	496	370
Lithium carbonate/potassium carbonate	Li_2CO_3/K_2CO_3 (35%:65% by weight)	505	345

The most suitable PCMs under the category of 'organic materials' are paraffin waxes. They are readily available with melting points in the range of 40° to 60°C. This makes them suitable for space heating and cooling applications. The properties of one typical paraffin wax are given in Table 7.5. It will be noted that it does not have a well-defined melting point. In addition to paraffin waxes, a wide variety of other organic materials such as fatty acids, esters, alcohols, eutectics, etc. have been tried as PCMs. Their melting points are in the range of 7 to 187°C and heats of fusion in the range of 42 to 250 kJ/kg. Fatty acids have melting points suitable for heating applications and have heat of fusion values comparable to those of paraffins. They exhibit excellent melting-freezing characteristics without any supercooling. However, they are costlier than paraffins. The properties of one typical fatty acid (capric acid) are given in Table 7.5.

Among inorganic materials, a number of hydrated salts have been developed as PCMs. Most of them have transition temperatures in the range of 10° to 100°C. The properties of three of them, viz. calcium chloride hexahydrate, sodium sulphate decahydrate and magnesium nitrate hexahydrate are listed in Table 7.5. It is of interest to note that sodium sulphate decahydrate (Glauber's salt) was the first PCM to be studied for solar applications. This salt undergoes the hydration-dehydration reaction at 32°C. It was studied by Telkes [20] for a space-heating application.

$$\mathrm{Na_2SO_4 \cdot 10H_2O} \underset{\text{Energy extraction}}{\overset{\text{Energy storage}}{\rightleftharpoons}} \mathrm{Na_2SO_4 + 10H_2O} \tag{7.37}$$

Hydrated salts are attractive materials due to their high volumetric storage density (~350 MJ/m^3), relatively high thermal conductivity (~ 0.5 W/m-K) and moderate cost compared to paraffin waxes. The main difficulty encountered with them is that their thermal performance degrades with repeated cycling due to phase segregation. This difficulty is generally overcome by mixing certain additives.

Some inorganic compounds and eutectics are also used as PCMs. Ice is a good PCM if the energy is to be stored/extracted at 0°C. For high temperature applications, PCMs developed include sodium nitrate, sodium hydroxide and a eutectic of lithium carbonate and potassium carbonate. Their properties are also given in Table 7.5.

An advantage associated with a latent heat storage system is that it is more compact than a sensible heat system. However, the heat exchange system for transferring energy between the working fluid (which is usually a gas) and the storage material is always more complex. This is due to the fact that during the heat extraction process, the storage material first solidifies at the heat transfer surface. Hence, the thermal resistance to the flow of heat keeps on increasing as heat is extracted from the storage. In order to prevent the thermal resistance from becoming too large, the storage material is placed in long thin containers and the gas is passed through narrow spaces between the tubes. Alternatively the storage material is in the spaces between the tubes and the gas is passed through the tubes. Both these arrangements are shown in Fig. 7.11.

Some novel arrangements have also been proposed or tried. In one system, the phase change material is contained in a bundle of long cylindrical tubes which is rotated at a low speed. This helps to keep the storage material mixed and it is claimed that solidification begins at a number of nuclei inside the liquid and not on the tube walls alone.

Various techniques have been tested or proposed in order to increase the heat transfer area, to eliminate chemical reactions of PCMs with outside materials and to control the changes in the storage material volume as phase change occurs [18, 21]. A large improvement in the heat transfer rate is obtained by encapsulating the PCM in small spherical capsules to form a packed bed storage unit. High pressure drop of the heat transfer fluid passing through the bed and high initial cost of encapsulation are the major drawbacks. If the size of the PCM along with its encapsulation is in the range of 1 to 2 cm, then the encapsulation is called macro-encapsulation. On the other hand, micro-encapsulation refers to a process in which tiny particles or droplets of PCM are used. They are surrounded by a coating or embedded in a matrix to form a capsule. Micro-encapsulation has the advantage of providing higher surface area to volume ratio. While significant development has taken place on encapsulation technology of low-temperature PCMs, there are technical challenges associated with high temperature PCMs because very often they are inorganic salts, which are corrosive to most encapsulating materials [19]. It is worth noting that no commercial high temperature storage system using a PCM has been built so far.

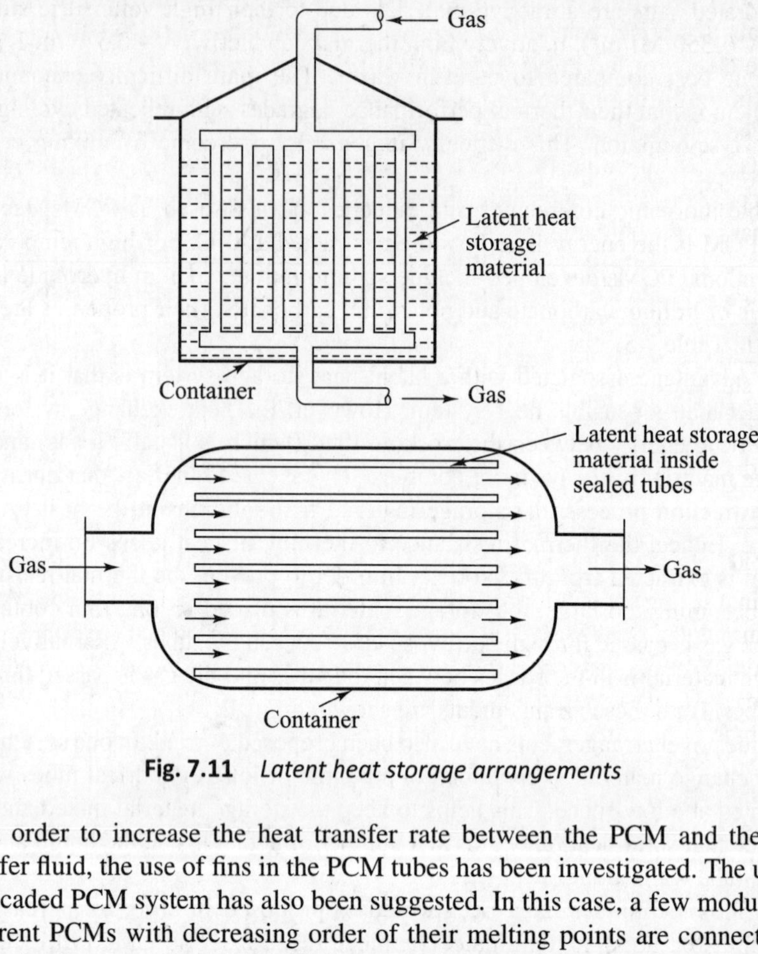

Fig. 7.11 *Latent heat storage arrangements*

In order to increase the heat transfer rate between the PCM and the heat transfer fluid, the use of fins in the PCM tubes has been investigated. The use of a cascaded PCM system has also been suggested. In this case, a few modules of different PCMs with decreasing order of their melting points are connected in series. In the case of a single PCM system, the temperature of the heat transfer fluid decreases in the direction of flow as it transfers heat to the PCM. Thus, the heat flux transferred to the PCM gradually decreases. However, in the case of the cascaded configuration, even if the temperature of the heat transfer fluid decreases, the difference of temperature between the PCM and the fluid can be maintained more or less constant by proper selection of PCM melting temperature. This results in maintaining a constant heat flux between the fluid and the PCMs. Liu *et al.* [9] and Zhang *et al.* [22] have presented excellent reviews on latent heat storage investigations.

Efforts have also been made to increase the thermal conductivity of the PCM. Addition of highly thermal conductive particles, use of carbon fibre brushes, insertion of a metal matrix, etc. have been tried [19].

Extensive work has been carried out on the use of PCMs in the outer walls, roofs and floors of buildings for passive space-heating applications [23, 24]. The PCMs have been impregnated in wall boards or absorbed in the pores of building materials like brick, concrete or gypsum plaster. In order to avoid any possible adverse effects of the PCM on the properties of the construction materials, the PCM (in the form of particles) is sometimes encapsulated in a thin polymeric film. The micro-capsules are then mixed with the building material. Table 7.6 presents a list of some PCMs investigated for space-heating applications for human comfort. It is seen that their melting points lie in the range of 18 to 22°C. This makes them suitable for such applications.

Table 7.6 PCMs for space-heating applications in buildings [23]

Type	Melting point (°C)	Heat of fusion (kJ/kg)
Butyl stearate	19	140
Capric-lauric acid (45-55%)	21	143
Hexadecane	18	236
Heptadecane	22	214
Propyl palmitate	19	186

An interesting application utilizing PCMs in India is the temperature control of telecommunication sheds [25]. Typically these sheds are thermally well-insulated and have dimensions of about 3 m × 3 m × 2 m. The sheds contain electronic equipment used for cell phone transmission purposes and are located on the roofs of tall buildings next to transmitting towers. The electronic equipment generates about 2 to 2.5 kW of heat and the temperature in the shed is generally controlled to a value around 30°C by an air-conditioning unit. In the event of non-availability of electricity (either from the mains or a diesel-generator stand-by set), power is available from charged batteries to run the electronic equipment but not the air-conditioning unit. During this period, temperature control is achieved with the help of a PCM. About 300 kg of the PCM kept in sealed heavy density polyethylene panels is stacked in suitable racks in the shed. The PCM is usually a hydrated inorganic material. The heat rejected by the electronic equipment is absorbed by the PCM in solid form and causes it to melt slowly. Consequently the temperature in the shed is maintained at a value about 2°C higher than the melting point of the PCM. Usually the temperature can be controlled in this manner for about 5 or 6 hours.

THERMOCHEMICAL STORAGE 7.5

In a thermochemical storage system, the solar energy to be stored is used to produce a certain endothermic chemical reaction and the products of the reaction are stored. When the energy is required to be released, the reverse exothermic reaction is made to take place. Both reactions take place at different temperatures, the forward reaction occurring at a higher temperature than the reverse reaction. Thermochemical storage systems are suitable for medium or high temperature applications only. They have some inherent advantages. Typically, their volumetric energy densities are about 5 times the values obtained in latent heat storage systems and about 10 times the values in sensible heat storage systems. Also, the products of the chemical reactions can be stored for any length of time and can be transported to any distance without thermal loss.

The reactions are shown schematically in Fig. 7.12. Chemicals A and B are converted through the forward reaction into products X and Y.† This reaction occurs because of the solar heat input and may or may not need a catalyst. The products X and Y are generally stored at room temperature. This is an advantage since the storage tanks do not have to be insulated. However, in case a catalyst is not needed, the products of the forward reaction must be separated at the temperature at which the reaction occurs in order to prevent the reverse reaction. The products must then be stored separately. In case a catalyst is required, separation may not be required.

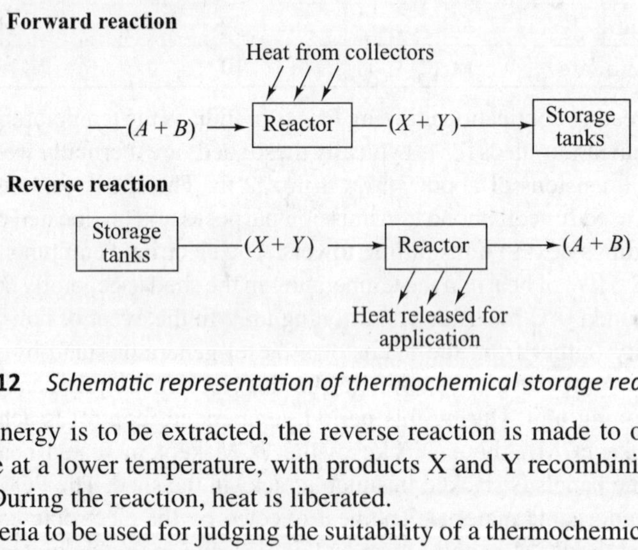

Fig. 7.12 *Schematic representation of thermochemical storage reactions*

When energy is to be extracted, the reverse reaction is made to occur. This takes place at a lower temperature, with products X and Y recombining to form A and B. During the reaction, heat is liberated.

The criteria to be used for judging the suitability of a thermochemical reaction for a solar application are as follows:

1. The forward reaction should occur in the temperature range of the solar collectors used.
2. The reverse reaction should occur in the temperature range in which heat is to be extracted.
3. The energy absorbed per unit volume of the products stored should be as large as possible and the products should be in the liquid form. These factors help in reducing the size of the storage tanks.
4. The reactions in both directions should be fast and completely reversible with no side reactions which may produce contaminants.

† The number of chemicals involved may be more or less than two.

In addition, it is preferable that the chemicals required should be readily available at low cost and should be easy to handle.

A number of chemical reactions have been examined from the standpoint of their suitability based on the above criteria. Pardo *et al.* [26] have reviewed the literature on thermochemical energy storage systems. They have compiled data on the volumetric energy densities of a number of possible reactions and the temperature at which the forward reaction takes place. A few of these have been found to be suitable for CSP application because of the work done so far, the cost and the temperature range [26]. The details of four such reactions are given in Table 7.7.

Table 7.7 Thermochemical storage reactions

Reaction	Phase	Temperature of forward reaction (°C)	Temperature of reverse reaction (°C)	Volumetric energy density (MJ/m^3)
$MgH_2 \rightleftharpoons Mg + H_2$	Solid → Solid + Gas	380	230	2088
$PbCO_3 \rightleftharpoons PbO + CO_2$	Solid → Solid + Gas	450	300	1091
$Ca(OH)_2 \rightleftharpoons CaO + H_2O$	Solid → Solid + Gas	450	25–400	1573
$2NH_3 \rightleftharpoons N_2 + 3H_2$	Liquid → Gas	450 (Pressure 9–150 bar)	450 (Pressure 10–300 bar)	2682

The first reaction in Table 7.7 involves the endothermic decomposition of magnesium hydride into magnesium and hydrogen in the forward step and the exothermic recombination of magnesium and hydrogen to form magnesium hydride in the reverse reaction. This material has attracted attention due to the fact that it has a very high capacity of hydrogen storage. The Max-Planck Institute has tested its performance in a thermochemical solar plant, after carrying out extensive investigations by making various prototypes [27]. A number of disadvantages exist with this system. Although magnesium can be stored as a solid, hydrogen has to be stored as a gas under pressure (50 – 100 bar). Besides, the chemical reaction is slow. In order to improve the rate of the chemical reaction, it becomes necessary to use nickel powder or iron powder as doping agents.

The second reaction also involves the endothermic decomposition of a solid material (lead carbonate) into a solid (lead oxide) and a gas (carbon dioxide). This reaction does not require any catalyst. It also has a number of drawbacks. The reversibility of the reaction is poor and the products are toxic. In addition, there is a requirement to store carbon dioxide gas. So far, enough experimental work has not been done on this reaction.

The third reaction involves the hydration/dehydration of calcium oxide. A number of investigators have worked on the feasibility of using this chemical in a thermo-chemical storage system for a concentrated solar power plant. The advantages associated with this reaction are that no catalyst is required, the reaction is reversible and there is no by-product. Also, the products are readily separated and are non-toxic. However, there is considerable agglomeration and sintering and large change in volume. Also, the material has a low thermal conductivity.

The last reaction listed in Table 7.7 is the classical Haber-Bosch process involving the decomposition and synthesis of ammonia. This process has to be adapted for thermochemical storage. A considerable amount of work has been done in this direction over many years at the Australian National University [28]. The main advantage of the process is that there is no side reaction. The disadvantages are that the operating pressures are high, catalysts are needed for both the forward and the reverse reaction, and hydrogen and nitrogen have to be stored as gases.

Recently Prieto *et al.* [29] have reviewed three types of redox reactions, viz. sulphur-based cycles, metal oxide cycles and Perovskite-form structures. They conclude that though the cycles show promise, yet none seems to be adequate in all aspects. The above review shows that despite the inherent advantages associated with thermochemical storage, the development of engineering systems has proceeded slowly. It has been essentially limited so far to laboratory and pilot experiments. There are still many technical issues to be sorted out. Further, the economic viability of thermochemical storage systems needs to be established.

PROBLEMS

7.1 A well-mixed water storage unit contains 3000 kg of water and is provided with auxiliary heating of 3 kW as shown in Fig. 7.13. Data for the useful heat gain from the collectors, ambient temperature and rate of withdrawal to the load on a particular day from 0300 to 1200 h are as follows:

Hour	q_u (kJ/h)	T_a (°C)	$\dot{m}_{load}$ (kg/h)
0300 – 0400	0	16	200
0400 – 0500	0	16	200
0500 – 0600	0	18	220
0600 – 0700	0	20	250
0700 – 0800	4 000	22	260
0800 – 0900	15 000	24	260
0900 – 1000	30 000	26	230
1000 – 1100	50 000	28	220
1100 – 1200	70 000	30	220

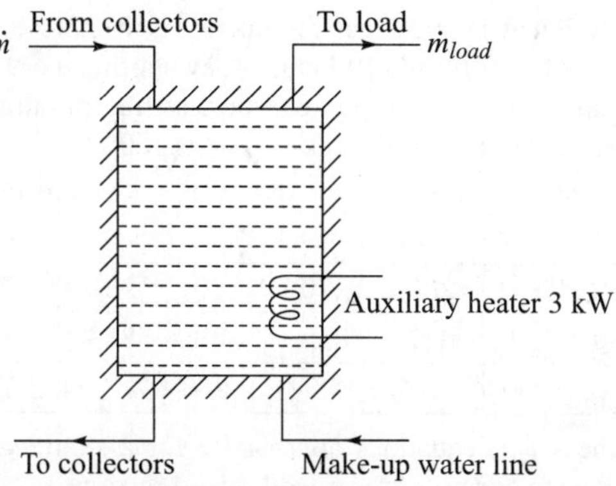

Fig. 7.13 *Problem 7.1*

Assume: (i) The auxiliary heater switches on when the temperature in the tank falls below 45°C. (ii) Make-up water at 24°C enters at the same rate as the rate of withdrawal to the load. (iii) $(UA)_t = 60$ kJ/h-°C.

Calculate the hourly variation of T_l from 0300 to 0800 h starting with the value of 49.5°C at 0300 h. At what time does the auxiliary heater switch on?

7.2 For a well-mixed liquid storage tank, it is given that (i) the useful heat gain rate can be approximated by a half-sine wave $q_u = A_1 \sin(\pi t/t_1)$ where A_1 and t_1 are constants, (ii) q_{load} = a constant, (iii) the ambient temperature T_a is constant, and (iv) at $t = 0$, the stored liquid is at a temperature T_{l1}. Show that the temperature variation in the tank is given by

$$(T_t - T_a) = \frac{\pi c_2 t_1}{(\pi^2 + c_1^2 t_1^2)}\left[\frac{c_1 t_1}{\pi}\sin\frac{\pi t}{t_1} - \cos\frac{\pi t}{t_1} + e^{-c_1 t}\right] + \left[T_{li} - T_a + \frac{c_3}{c_1}\right]e^{-c_1 t} - \frac{c_3}{c_1}$$

where $$c_1 = \frac{(UA)_t}{(\rho V C_p)_e};\ c_2 = \frac{A_1}{(\rho V C_p)_e};\ c_3 = \frac{q_{load}}{(\rho V C_p)_e}$$

7.3 The volume of a storage tank used for storing hot oil is 18 m^3. The tank is well insulated and its (UA) value is 3.85 W/K. The initial temperature of the oil in the tank is 280°C at 0900 h in the morning on a particular day. The variation of heat input to the tank and ambient temperature over a 24 h period is as follows:

Time (h)	0-1	1-2	2-3	3-4	4-5	5-6	6-7	7-8	8-9	9-10
q_u(MJ/h)	0	0	0	0	0	0	0	0	0	1550
T_a (°C)	29.1	28.4	27.8	27.2	26.7	26.3	27.1	29.3	31.8	34.1
	10-11	11-12	12-13	13-14	14-15	15-16	16-17	17-18	18-19	19-20
	1900	2050	2230	2245	1950	1520	1400	0	0	0
	35.7	36.9	37.7	38.3	38.5	38.7	38.0	37.5	35.9	34.1
	20-21	21-22	22-23	23-24						
	0	0	0	0						
	32.6	31.5	30.7	29.9						

The load requirement is such that the energy is withdrawn from the tank at a constant rate of 1440 MJ/h for 10 hours a day starting at 0900 h. Assuming that the oil in the tank is always well-mixed, calculate the variation of its temperature. Given that the temperature of the make-up oil is 260°C.

The density and specific heat of the oil used vary with temperature in the following manner:

Temperature (°C)	260	270	280	290	300	310	320
Density (kg/m^3)	857	848	838	828	817	806	796
Specific heat (kJ/kg-K)	2.207	2.234	2.260	2.287	2.314	2.341	2.369

7.4 Write down the set of equations appropriate for a stratified water storage tank which is assumed to consist of three well-mixed sections.

7.5 Use the data of Example 7.2 with the assumption that the hot water tank consists of three equal well-mixed sections. Take T_{l1} = 70°C, T_{l2} = 67.5°C and T_{l3} = 65°C at 1200 h and find their values at 1300 h.

7.6 Solve Example 7.4 by taking the air flow rate to be 0.10 kg/s and time t = 45 minutes.

7.7 (a) A cylindrical hot water storage tank has a diameter of 1.1 m and a height of 2.16 m. It is insulated on all faces with glass wool insulation 10 cm thick, having a thermal conductivity of 0.042 W/m-K. Calculate the heat loss factor $(UA)_t$ for the tank.

(b) A heat loss test is conducted on the tank in which the flow rate of water is 0.11 kg/s, the inlet water temperature is 54.0°C and the ambient air temperature is 29.0°C. Under steady-state conditions, the average difference between the inlet and outlet water temperature is measured to be 0.32°C. Calculate the heat loss factor $(UA)_t$ from the experimental data. Compare the value obtained with the value calculated in part (a). Comment on possible reasons for the difference between the two values.

7.8 A hot water storage tank has a volume of 2.0 m^3. For conducting a charge-discharge test according to the standard code, it is first brought to a uniform initial temperature of 53.0°C. Then the flow is adjusted to 0.273 kg/s and the temperature of the inlet water is increased stepwise by 15.3°C to 68.3°C. The temperature of the leaving water is measured at regular time intervals over a period of 2 hours and the following readings are recorded.

t (minute)	0 to 78	84	90	96	102	108	114	120
T_{fo} (°C)	53.0	53.3	54.9	57.5	60.0	62.5	64.9	67.0

Plot the dimensionless temperature $(T_{fo} - T_{fi})/\Delta T_i$ against time and calculate the charge capacity of the device. Given that the heat loss factor (as obtained from a heat loss test) is 6.22 W/K and that the ambient air temperature is 24.5°C. Calculate also the ratio of the charge capacity to the theoretical storage capacity for the step change of 15.3°C.

REFERENCES

1. Shelton, J. 1975. Underground storage of heat in solar heating systems. *Solar Energy*, 17: 138.
2. Xu, J., Wang, R.Z. and Li, Y. 2014. A review of available technologies for seasonal thermal energy storage. *Solar Energy*, 103: 610.
3. Bauer, T., Pfleger, N., Breidenbach, N., Eck, M., Laing, D. and Kaesche, S. 2013. Material aspects of solar salt for sensible heat storage. *Applied Energy*, 111: 1114.
4. Fernandez, A.G., Ushak, S., Galeguillos, H. and Perez, F.J. 2015. Thermal characterization of an innovative quaternary molten nitrate mixture for energy storage in CSP plants. *Solar Energy Materials and Solar Cells*, 132: 172.
5. Cabeza, L.F., Gutierrez, A., Barreneche, C, Ushak, S., Fernandez, A.G., Ines Fernandez, A. and Grageda, M. 2015. Lithium in thermal energy storage: a state-of-the-art review. *Renewable and Sustainable Energy Reviews*, 42: 1106.
6. Kulkarni, G.N., Kedare, S.B. and Bandyopadhyay, S. 2007. Determination of design space and optimization of solar water heating systems. *Solar Energy*, 81: 958.
7. Pacheco, J.E., Showalter, S.K. and Kolb, W.J. 2002. Development of a molten-salt thermocline thermal storage system for parabolic trough plants. *Journal of Solar Energy Engg., Trans ASME*, 124: 153.
8. Kuravi, S., Trahan, J., Goswami, D.Y., Rahman, M.M. and Stefanakos, E.K. 2013. Thermal energy storage technologies and systems for concentrating solar power plants. *Progress in Energy and Combustion Science*, 39: 285.
9. Liu, M., Steven Tay, N.H., Bell, S., Belusko, M., Jacob, R., Will, G., Saman, W. and Bruno, F. 2016. Review on concentrating solar power plants and new developments in high temperature thermal energy storage technologies. *Renewable and Sustainable Energy Reviews*, 53: 1411.
10. Grirate, H., Agalit, H., Zari, N., Elmachaouri, A., Molina, S. and Couturier, R. 2016. Review: experimental and numerical investigation of potential filler materials for thermal oil thermocline storage. *Solar Energy*, 131: 260.
11. Galione, P.A., Perez-Segarra, C.D., Rodriguez, I., Torras, S. and Rigola, J. 2015. Multi-layered solid-PCM thermocline thermal storage for CSP. Numerical evaluation of its application in a 50 MWe plant. *Solar Energy*, 119: 134.
12. Schumann, T.E.W. 1929. Heat transfer: A liquid flowing through a porous prism. *Journal Franklin Institute*, 208: 405.
13. Lof, G.O.G. and Hawley, R.W. 1948. Unsteady-state heat transfer between air and loose solids. *Industrial and Engineering Chemistry*, 40: 1061.
14. Chandra, P. and Willits, D.H. 1981. Pressure drop and heat transfer characteristics of air-rock bed thermal storage systems. *Solar Energy*, 27:547.
15. Dunkle, R.V. and Ellul, W.H.J. 1972. Randomly-packed particulate bed regenerators and evaporative coolers. *Trans. Institution of Engineers (Australia)*, MC 8, 117.

16. ASHRAE Standard 94-77. 1977. Methods of testing thermal storage devices based on thermal performance. *American Society of Heating, Refrigerating and Air-conditioning Engineers*, USA.
17. Hill, J.E., Kelly, G.E. and Peavy, B.A. 1977. A method of testing for rating thermal storage devices based on thermal performance. *Solar Energy*, 19: 721.
18. Zalba, B., Marin, J.M., Cabeza, L.F. and Mehling, H. 2003. Review on thermal energy storage with phase change: materials, heat transfer analysis and applications. *Applied Thermal Engineering*, 23: 251.
19. Xu, B., Li, P. and Chan, C. 2015. Application of phase change materials for thermal energy storage in concentrated solar thermal power plants: a review to recent developments. *Applied Energy*, 160: 286.
20. Telkes, M. 1955. *Solar Heat Storage*, Solar Energy Research, University of Wisconsin Press, USA.
21. Farid, M.M., Khudhair, A.M., Razack, S.A.K. and Al-Hallaj, S. 2004. A review on phase change energy storage: materials and applications. *Energy Conversion and Management*, 45: 1597.
22. Zhang, H., Baeyens, J., Caceres, G., Degreve, J. and Lv, Y. 2016. Thermal energy storage: recent developments and practical aspects. *Progress in Energy and Combustion Science*, 53:1.
23. Zhang, Y., Zhou, G., Lin, K., Zhang, Q. and Di, H. 2007. Application of latent heat thermal energy storage in buildings: state-of-the-art and outlook. *Building and Environment*, 42: 2197.
24. Khudhair, A.M. and Farid, M.M. 2004. A review on energy conservation in building applications with thermal storage by latent heat using phase change materials. *Energy Conversion and Management*, 45: 263.
25. Budhi, D. 2007. Personal communication, Green Hi-Tech Energy (P) Ltd. (http://www.ghel.in)
26. Pardo, P., Deydier, A., Anxionnaz-Minvielle, Z., Rouge, S., Cabassud, M. and Cognet, P. 2014. A review on high temperature thermochemical heat energy storage. *Renewable and Sustainable Energy Reviews*, 32: 591.
27. Felderhoff, M. and Bogdanovic, B. 2009. High temperature metal hydrides as heat storage materials for solar and related applications. *International Journal of Molecular Sciences*, 10: 325.
28. Dunn, R., Lovegrove, K. and Burgess, G. 2012. A review of ammonia-based thermochemical energy storage for concentrating solar power. *Proc. IEEE*, 100(2): 391.
29. Prieto, C, Cooper, P., Ines Fernandez, A. and Cabeza, L.F. 2016. Review of technology: thermochemical energy storage for concentrating solar power plants. *Renewable and Sustainable Energy Reviews*, 60: 909.

Photovoltaic Conversion

CHAPTER 8

INTRODUCTION 8.1

The focus thus far has been essentially on the direct thermal route for utilizing solar energy. We will now take up for consideration photovoltaic conversion, which is the other direct method for utilization.

The devices used in photovoltaic conversion are called solar cells. When solar radiation falls on these devices, it is converted directly into DC electricity. The principal advantages associated with solar cells are that they have no moving parts, require little maintenance, and work quite satisfactorily with beam or diffuse radiation. Also, they are readily adapted for varying power requirements because a cell is like a 'building block'. The main factor limiting their use is their high cost. Clearly, increasing the efficiency of solar cells and containing production costs are crucial factors in the future.

Significant developments have taken place in the past few years. New types of cells have been developed, innovative manufacturing processes introduced, conversion efficiencies of existing types increased, costs reduced and the volume of production steadily increased. The annual world production of photovoltaic devices in 2015 was about 57 GWp* and the cumulative installation at the end of 2015 was 242 GWp [1]. The corresponding production in India in 2014 was about 1.6 GWp [2] and the cumulative installation at the end of 2015 was 4943 MWp [3]. In the near future solar cells may become important sources of power for providing electrical energy through the grid as well as for localized use through stand-alone systems. Various materials have been used and many types of solar cells have been developed. However, about 93 per cent of the current commercial production of solar cells is from single crystal and multicrystalline silicon only [1].

---

*The term 'peak watt' or 'watt-peak' (Wp) is used for the power produced by a photovoltaic device at a radiation level of 1000 W/m^2, a standardised spectral distribution of AM1.5 and a cell temperature of 25°C.

DESCRIPTION AND PRINCIPLE OF WORKING 8.2

8.2.1 The Single Crystal Silicon Cell

The first solar cells were made in the 1950s from single crystal silicon. Even today silicon is the material generally used for making most cells. Single crystal silicon cells are thin wafers about 180 – 220 μm in thickness, sliced from a single crystal of p-type doped silicon (Fig. 8.1). A shallow junction (about 0.3 μm) is formed at one end by diffusion of the n-type impurity. An anti-reflection coating of silicon nitride, having a thickness of about 0.075 μm is applied on the top surface. On the front side, metallic contacts having a finger and busbar structure are placed to permit sunlight to go through. Generally, these contact grids are made by a screen printing technique using aluminium-silver paste at a high temperature. The finger grids are about 70 to 100 μm wide with a pitch of about 2 mm. Typically, two to three busbars are used, each one is approximately 1.5 mm wide. On the back side, the contact is of aluminium and completely covers the surface. Aluminium, not only provides the rear side contact, but also forms a layer called as the back surface field (BSF). A typical cell develops a voltage of 0.5 - 0.7 V and a current density of 30 - 40 mA/cm^2. In order to obtain higher voltages and currents, individual cells are fixed side by side on a suitable backup board and connected in series and parallel to form a module. The cells are encapsulated in a thin transparent material in order to protect them from the environment and support them in the module. A number of modules are interconnected to form an array. Earlier the cells used to be circular in shape with diameters ranging from 6 to 15 cm. Now, they are rectangular in shape, resulting in more compact modules.

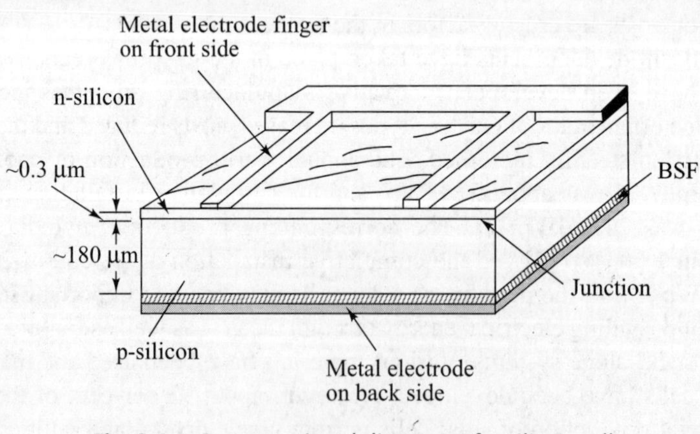

Fig. 8.1 *Cross-sectional diagram of a silicon cell*

Apart from single crystal silicon, silicon solar cells are now also made in large numbers from multicrystalline silicon and amorphous silicon.

8.2.2 Principle of Working of a Solar Cell

Two important steps are involved in the principle of working of a solar cell. These are:

1. Creation of pairs of positive and negative charges (called electron-hole pairs) in the solar cell by absorbed solar radiation.
2. Separation of the positive and negative charges by a potential gradient within the cell.

For the first step to occur, the cell must be made of a material which can absorb the energy associated with the photons of sunlight. The energy *(E)* of a photon is related to the wavelength (λ) by the equation

$$E = hc/\lambda \tag{8.1}$$

where h = Planck's constant = 6.62×10^{-27} erg-s
and c = velocity of light = 3×10^8 m/s.
Substituting these values, we get

$$E = 1.24/\lambda \tag{8.2}$$

where E is in electron-volts (eV) and λ is in μm.

Materials suitable for absorbing the energy of the photons of sunlight are semiconductors like silicon (Si), cadmium telluride (CdTe), gallium arsenide (GaAs), etc. In a semiconductor, the electrons occupy one of two energy bands, the valence band and the conduction band. The valence band has electrons at a lower energy level and is fully occupied, while the conduction band has electrons at a higher energy level and is not fully occupied. The difference between the minimum energy of electrons in the conduction band and the maximum energy of the electrons in the valence band is called the *band gap energy* E_g. Photons of sunlight having energy E greater than the band gap energy E_g are absorbed in the cell material and excite some of the electrons.* These electrons jump across the band gap from the valence band to the conduction band leaving behind holes in the valence band. Thus electron-hole pairs are created.

The electrons in the conduction band and the holes in the valence band are mobile. They can be made to flow through an external circuit (thereby executing the second step of the photovoltaic effect) if a potential gradient exists within the cell. In the case of silicon, the potential gradient is obtained by making the cell as a sandwich of two types of silicon, p-type and n-type. Silicon of p-type is 'doped' with some trivalent atoms like those of boron, while silicon of n-type is 'doped' with some pentavalent atoms like those of phosphorus. The n-type silicon has excess electrons, while the p-type has excess holes. When these materials are joined together, excess electrons from the n-type diffuse to recombine with the holes in the p-type. Similarly, excess holes from the p-type diffuse to the n-type. As a result, close to the junction, the n-type material becomes positively charged, while the p-type is negatively charged. This creates a 'built-in' potential (Fig. 8.2) at the junction. The consequent electric field is adequate to separate

*The excess energy $(E - E_g)$ is lost as heat.

the electrons and holes, and cause a direct electric current to flow in the external load, under sunlight.

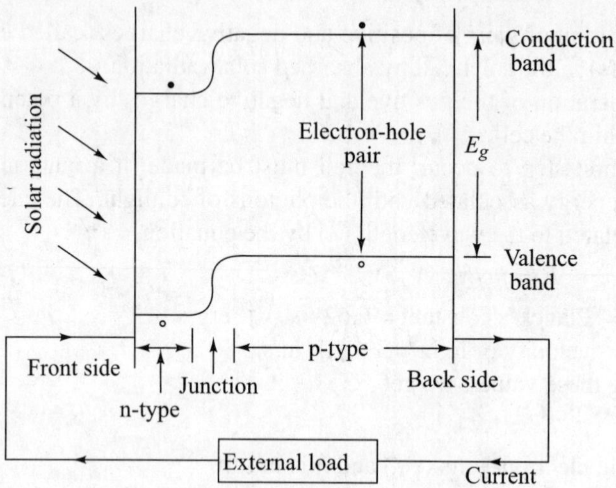

Fig. 8.2 *Principle of working of a solar cell*

In a silicon cell, the junction is a thin region separating the n-type and p-type portions. Since the basic material is all silicon, such a junction is more specifically called a *homojunction.* Solar cells can also be made from dissimilar materials. For example, in one type of solar cell, a layer of cadmium sulphide is deposited on a layer of cadmium telluride. The junction formed along the contact between the two materials is then called a *heterojunction.*

PERFORMANCE CHARACTERISTICS OF A SOLAR CELL 8.3

8.3.1 Current-voltage Characteristic

The performance of a solar cell is displayed by plotting its current-voltage *(I-V)* characteristic. A typical characteristic curve is shown in Fig. 8.3. It is seen that the current first stays more or less constant as the voltage increases. It then drops rapidly to zero. The intercepts of the curve on the *x*-axis and *y*-axis are called the open circuit voltage V_{oc} and the short circuit current I_{sc} respectively. The power delivered by the cell is also shown in Fig. 8.3 plotted as a function of the voltage. It is seen that the power increases more or less linearly from zero, goes through a maximum value and then decreases again to zero when the voltage corresponds to the open-circuit voltage. The values of current and voltage at which the maximum power occurs are denoted by the symbols I_m and V_m. A power electronic converter with an appropriate controller is used to track this maximum power point. This device is called a *maximum power point tracker*

(MPPT). The ratio $(I_m V_m / I_{sc} V_{oc})$ is called the *fill factor (FF)* of the cell. It is a measure of how closely the *I-V* characteristic of the cell approaches being a rectangle. Its value obviously lies between 0 and 1. Good silicon solar cells have *FF* values greater than 0.75.

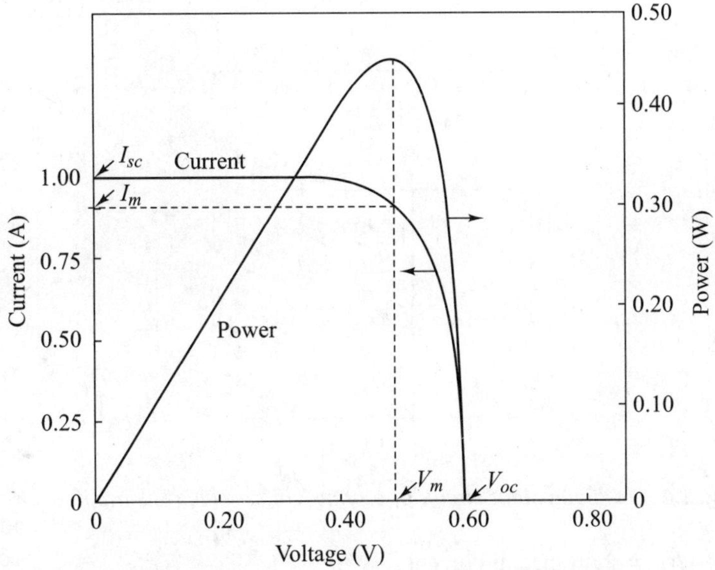

Fig. 8.3 *Typical current-voltage and power-voltage characteristics of a solar cell*

In order to analyze the performance of a cell, it is convenient to draw its equivalent circuit, shown in Fig. 8.4(a). The circuit consists of a current-source supplying current I_{sc}. R_{sh} and R_s are the internal shunt and series resistances respectively. I_j is the current at the p-n junction and is known as the junction current. I_{sh} is the current through the shunt resistance. I is the current passing through a load of resistance R_L and having a voltage drop of V across it. Usually the shunt resistance is much greater than the external load resistance, so that most of the current passes through the load. Again, the internal series resistance is much less than that of the load, so that a negligible amount of power is dissipated internally. Figure 8.4(b) shows a simplified equivalent circuit of a solar cell taking account of the above two factors. Based on this circuit, the load current (I), at fixed values of the temperature and solar radiation, can be shown to be related to the voltage (V) by the expression

$$I = I_{sc} - I_o\,[\exp(eV/mkT) - 1] \tag{8.3}$$

where I_o = reverse saturation or dark current

e = charge of an electron = 1.609×10^{-19} J/V

m = ideality factor

k = Boltzmann constant = 1.381×10^{-23} J/K

T = absolute temperature (in K) and

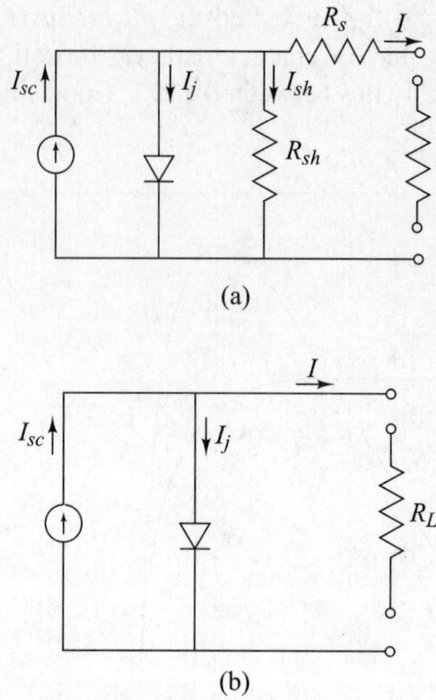

Fig. 8.4 *(a) Equivalent circuit of a photovoltaic cell, (b) Simplified circuit*

I_{sc} = short circuit current

I_o represents the current that flows from the p side to the n side in the dark. The ideality factor m characterizes the recombination caused due to defects in cells. It is equal to 1 for an ideal cell and usually lies between 1 and 2 for actual cells.

For an open circuit, $I = 0$ and $V = V_{oc}$. We have from Eq. (8.3)

$$0 = I_{sc} - I_o\,[\exp\,(eV_{oc}/kT) - 1]$$

or

$$V_{oc} = \frac{kT}{e}\ln(I_{sc}/I_o + 1) \tag{8.4}$$

The power P from the photovoltaic cell is given by

$$P = IV$$

$$= [I_{sc} - I_o\,\{\exp\,(eV/kT) - 1\}]V \tag{8.5}$$

Equation (8.5) exhibits the feature of the power passing through a maximum as described earlier. In order to find the location of the maximum, we differentiate Eq. (8.5) with respect to V and equate it to zero. We get

$$(1 + eV_m/kT)\exp\,(eV_m/kT) = 1 + \frac{I_{sc}}{I_o} \tag{8.6}$$

where V_m = voltage at which the power is maximum. Equation (8.6) is an implicit equation for V_m and the value of V_m has to be found out by trial-and-error.

The load current I_m corresponding to maximum power can be determined by substituting Eq. (8.6) in Eq. (8.3). We get

$$I_m = \frac{(eV_m/kT)}{(1 + eV_m/kT)}(I_{sc} + I_o) \tag{8.7}$$

The maximum power output P_m of the cell is

$$P_m = \frac{(eV_m^2/kT)}{(1 + eV_m/kT)}(I_{sc} + I_o) \tag{8.8}$$

Example 8.1

A photovoltaic cell has an open circuit voltage of 0.6 V and a short circuit current of 350 A/m^2 at a cell temperature of 40°C. Calculate the voltage and current density that maximizes the power of the cell. What would be the corresponding maximum power output per unit cell area? What is the fill factor of the cell?

Solution

From Eq. (8.4), we have

$$\frac{I_{sc}}{I_o} = \exp(eV_{oc}/kT) - 1$$

$$= \exp\left(\frac{1.602 \times 10^{-19} \times 0.6}{1.381 \times 10^{-23} \times 313}\right) - 1 = 4.544 \times 10^9$$

The reverse saturation current density

$$\frac{I_o}{A_c} = \frac{I_{sc}/A_c}{I_{sc}/I_o} = \frac{350}{4.544 \times 10^9} = 7.702 \times 10^{-8} \text{ A/m}^2$$

From Eq. (8.6),

$$\left(1 + \frac{1.602 \times 10^{-19}\, V_m}{1.381 \times 10^{-23} \times 313}\right)\exp\left(\frac{1.602 \times 10^{-19}\, V_m}{1.381 \times 10^{-23} \times 313}\right) = 1 + 4.544 \times 10^9$$

or, $(1 + 37.062\, V_m) \exp(37.062\, V_m) = 1 + 4.544 \times 10^9$

Solving by trial-and-error, we obtain

$$V_m = 0.519 \text{ V}$$

From Eq. (8.7), current density at maximum power

$$\frac{I_m}{A_c} = \left[\frac{(1.602 \times 10^{-19} \times 0.519)/(1.381 \times 10^{-23} \times 313)}{1 + \left(\dfrac{1.602 \times 10^{-19} \times 0.519}{1.381 \times 10^{-23} \times 313}\right)}\right](350 + 7.702 \times 10^{-8})$$

$$= 332.7 \text{A/m}^2$$

Maximum power output per unit cell area

$$= 332.7 \times 0.519 = 172.7 \text{ W/m}^2$$

$$FF = \frac{P_m \text{(per unit cell area)}}{V_{oc} I_{sc} \text{ (per unit cell area)}} = \frac{172.7}{0.6 \times 350} = 0.822$$

Figure 8.5 shows a set of typical *I-V* curves obtained for a solar cell maintained at a constant temperature and subjected to two different solar radiation levels. It is seen that the cell power output increases with the solar radiation. For a given voltage, the changes in the short circuit current are essentially proportional to the changes in solar radiation. Thus, the conversion efficiency does not change much with the radiation level.

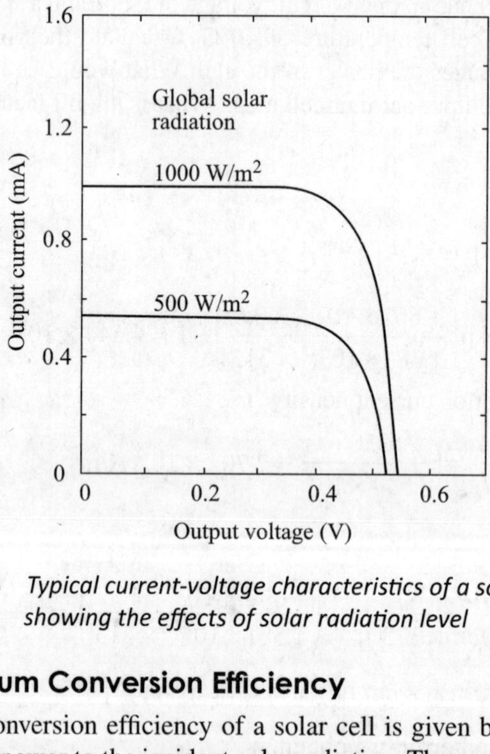

Fig. 8.5 *Typical current-voltage characteristics of a solar cell showing the effects of solar radiation level*

8.3.2 Maximum Conversion Efficiency

The maximum conversion efficiency of a solar cell is given by the ratio of the maximum useful power to the incident solar radiation. Thus

$$\eta_{max} = \frac{I_m V_m}{I_T A_c} = \frac{(FF) I_{sc} V_{oc}}{I_T A_c} \tag{8.9}$$

where I_T = incident solar flux and A_c = area of the cell.

For an efficient cell, it is desirable to have high values of fill factor, short circuit current and open circuit voltage. From solid state physics theory, expressions can be derived for each of these quantities. The expressions show that high values of I_{sc} are obtained with low band gap materials, while high values of V_{oc} and *FF* are possible with high band gap materials. Thus if theoretical values of η_{max}

are calculated for different values of E_g, it is obvious that a maximum value would be obtained at some value of E_g. Figure 8.6 shows the results of such calculations for homojunction cells*. These calculations are done on the basis of a few idealisations. Curve I is obtained by assuming that extra-terrestrial radiation (AM0) is incident on the cell. Essentially, similar curves differing only in detail would be obtained if the idealisations were changed or if spectral values corresponding to terrestrial conditions were assumed. Thus, for example, curve II is obtained by assuming the incident global radiation to be AM1.5 under a clear sky (1000 W/m^2).

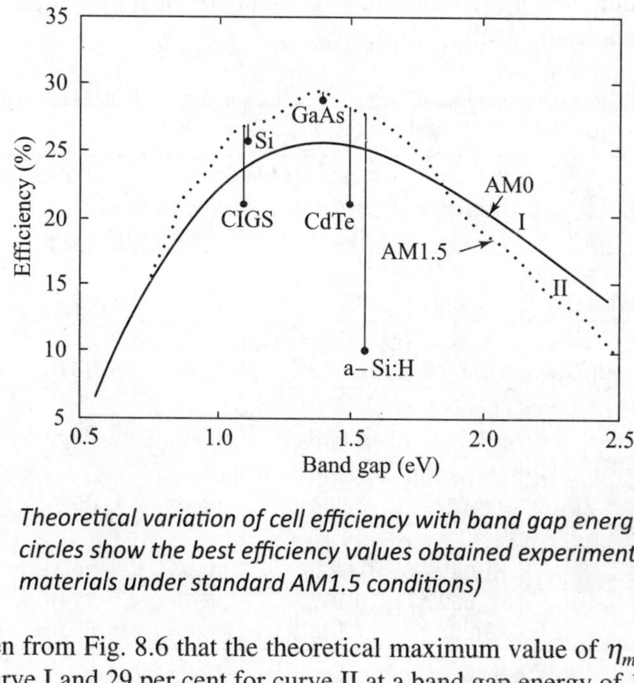

Fig. 8.6 *Theoretical variation of cell efficiency with band gap energy (The solid circles show the best efficiency values obtained experimentally for five materials under standard AM1.5 conditions)*

It is seen from Fig. 8.6 that the theoretical maximum value of η_{max} is 26 per cent for curve I and 29 per cent for curve II at a band gap energy of 1.4 eV. It is also seen that ideal solar cell materials should have band gaps in the range of 1.1 to 1.7 eV. Many semiconductors like Si, GaAS and CdTe have band gap energies in this range.

8.3.3 Actual Efficiency Values

For a variety of reasons, actual values of efficiency attained in practice are lower than the theoretical maximum efficiency values. However, sustained research and development (R&D) efforts have slowly yielded higher values and, in some cases, efficiencies close to the theoretical values have been obtained in laboratory scale experiments. Tables 8.1 and 8.2 summarise the values of efficiencies reported in

*The efficiency of a heterogeneous cell can also be approximately obtained from Fig. 8.6 by reading the value of η_{max} corresponding to the smaller of the band gap energies of the two materials making up the cell.

the literature for a few important types of cells and modules respectively [4].

The values given in Tables 8.1 and 8.2 are the highest reported up to 2016 for the type mentioned and for the areas specified. The following observations can be made:

(i) Solar cells and modules are being essentially made from silicon (single crystal, multicrystalline and amorphous), gallium arsenide (GaAs), copper indium gallium diselenide (CIGS) and cadmium telluride (CdTe).

(ii) Efficiency values obtained for cells are higher than those obtained for corresponding laboratory modules. As will be seen later from Section. 8.4, these values in turn are higher than values obtained with commercially available modules.

Table 8.1 Efficiency values reported for some solar cells. Measurements made under global AM 1.5 spectrum (1000 W/m^2) and at a cell temperature of 25°C [4].

Type	Area (cm^2)	Open circuit voltage (V)	Short-circuit current (mA/cm^2)	Fill factor (%)	Efficiency (%)
Silicon (Single crystal)	143.7	0.740	41.8	82.7	25.6
Silicon (Multicrystalline)	242.74	0.6678	39.8	80.0	21.3
Silicon (Amorphous)	1.001	0.896	16.36	69.8	10.2
GaAs (Multicrystalline)	4.011	0.994	23.2	79.7	18.4
Thin film:					
CdTe	1.0623	0.8759	30.25	79.4	21.0
CIGS	0.9927	0.757	35.7	77.6	21.0
GaAS	0.9927	1.122	29.68	86.5	28.8

Table 8.2 Efficiency values reported for some PV modules. Measurements made under global AM1.5 spectrum (1000 W/m^2) and at a cell temperature of 25°C [4].

Type	Area (cm^2)	Open circuit voltage (V)	Short circuit current (mA/cm^2)	Fill factor (%)	Efficiency (%)
Silicon (Single crystal)	11 562	53.4	6.32	81.6	23.8
Silicon (Multicrystalline)	15 349	41.53	9.299	77.4	19.5
Thin film:					
CdTe	7038.8	110.6	4.533	74.2	18.6
CIGS	808	47.6	0.408	72.8	17.5
GaAs	858.5	10.89	2.255	84.2	24.1

(iii) The highest efficiencies obtained for single crystal Si cells and for amorphous Si, CdTe, CIGS and GaAs thin film cells (reported in the last column of Table 8.1) are also plotted in Fig. 8.6 as solid circles. Vertical lines are drawn upwards from each solid circle to intersect the AM1.5 theoretical curve. The points of intersection represent the maximum efficiency attainable for the material. It can be seen that the highest values of 25.6 and 28.8 per cent obtained for single crystal Si and thin film GaAs cells are the best for all materials and are within 1 or 2 per cent of the maximum value predicted theoretically. On the other hand, the highest values of 10.2, 21.0 and 21.0 per cent obtained for amorphous Si, CdTe and CIGS cells are much lower than the theoretically predicted values.

It may be mentioned that, like solar thermal collectors, the performance of solar cells under outdoor conditions also decreases due to contamination of their surfaces. This aspect in PV systems is usually referred to as a *soiling effect.* It includes the effect of dust and other surface contaminations like soot, sand, salt, etc. [5-7]. The cleaning and mitigation approaches for Indian climatic conditions discussed earlier for flat-plate collectors in Section 4.10.7 are applicable to PV modules as well.

TYPES OF SOLAR CELLS 8.4

In this section, we describe different types of solar cells developed. Some of them have reached the commercial stage and are used extensively for the generation of electricity. As opposed to them, other solar cells are being developed and are still in the laboratory or R&D stage. We will first discuss solar cells which are commercially available. The efficiency values will be given and some of the production processes described. We will then briefly discuss other types of solar cells, which are promising.

Commercially available solar cells can be divided into two groups (i) cells based on crystalline silicon (Si) technology (or Si wafers), and (ii) cells based on thin film technology. The two groups are often categorized as first-generation and second-generation photovoltaic technology.

8.4.1 Crystalline Silicon Solar Cells

The crystalline silicon solar cell is the most developed and widely used cell. This is because extensive work had been done on the processing of silicon due to its prior use in the microelectronics industry. In addition, the material was available in abundance and was non-toxic. Silicon was originally developed as a single crystal cell. However, the multicrystalline form has gradually taken over due to its cost effectiveness. We will discuss both types of cells and also indicate the current advancements.

❑ *Singe Crystal Silicon* For single crystal silicon, an efficiency value as high as 25.6 per cent has been obtained in a cell under laboratory conditions. The highest value reported for a laboratory module is 23.8 per cent. However, for commercially available modules, the efficiency ranges between 15 and 18 per cent. The following are the specifications and performance details of a typical single crystal silicon commercial module tested under standard conditions (global solar radiation 1000 W/m^2, AM 1.5 spectrum and cell temperature 25°C) [8]:

Module size	: 163.30 cm × 98.22 cm
Module weight	: 18.2 kg
Cell size	: 15.6 cm × 15.6 cm
Number of cells	: 60
Peak power	: 285 Wp
Maximum voltage	: 31.5 V
Open circuit voltage	: 39.8 V
Short circuit current	: 9.98 A
Module efficiency	: 17.76 per cent

In order to understand some of the technical developments that have taken place for manufacturing solar cells, it will be useful to describe the conventional process of making single crystal silicon cells. A block diagram showing the main steps in the manufacturing process of a single crystal silicon solar cell is given in Fig. 8.7. The usual source of silicon is silica (silicon dioxide). In order to produce silicon, silica is melted and a measured amount of carbon is added. As a result, carbon dioxide is formed and silica is reduced to relatively pure silicon called metallurgical grade (MG) silicon. This grade contains about 1 per cent impurities and is fairly inexpensive. However, this grade of silicon is not suitable for making solar cells. A much purer grade is needed. In the past, till about the year 2000, the production of solar cells was relatively small. The silicon needed for making them was obtained from the pure silicon left-over or rejected by the electronics industry which uses semi-conductor grade (SeG) silicon. SeG-silicon is one of the purest forms of silicon with impurities being less than 1 part per billion. The process for producing it is to treat MG-Si with hydrochloric acid

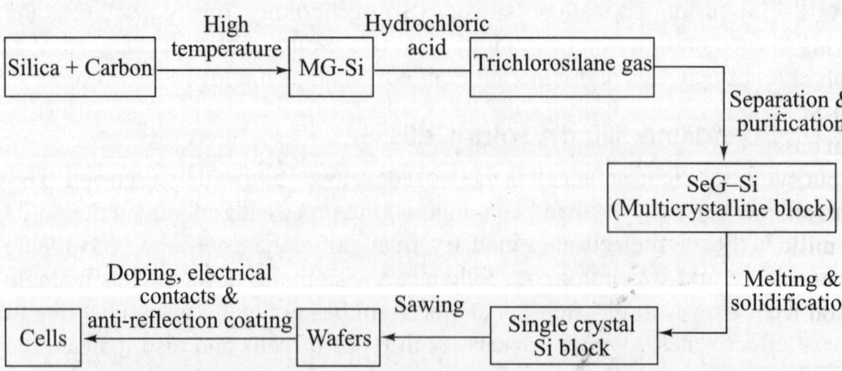

Fig. 8.7 *Block diagram showing manufacturing process for a single crystal silicon solar cell from SeG-silicon*

to get trichlorosilane ($SiHCl_3$), dichlorosilane ($SiHCl_2$) or silane (SiH_4) gas. The gas is condensed and fractionally distilled to yield solid polycrystalline SeG-silicon having a small grain size (less than 1 mm). This silicon is melted and cooled in a controlled manner to yield ingots of single crystal SeG-silicon. The ingots obtained are in the form of long cylindrical blocks, 6 to 15 cm in diameter. The above methods for converting MG-Si to SeG-Si and then obtaining single crystal ingots are energy consuming and are also very expensive. Typically the cost of the silicon increases by a factor of about one hundred.

Next, the single crystal silicon* block is sliced into a number of wafers (about 200-μm thick) by sawing. The standard process has been to use very thin saws so as to reduce the loss of material associated with the sawing process. Recently, a new type of saw, called the multi-wire saw has been developed to reduce the sawing losses further. Before junction formation, wafers undergo a texturing process in order to reduce reflection from the surface. The p-type wafers are then passed through a furnace containing phosphorus vapour so that phosphorus atoms diffuse a short distance through one face of the wafers, thereby forming n-type silicon up to a depth of about 0.3 microns. After this, an anti-reflection coating is laid on the front side followed by fixing of metallic contacts on both the front and rear sides. The rear side contact also forms a layer of the back surface field. Finally, the wafer is encapsulated in a weather-resistant transparent coating.

❑ *Multicrystalline Silicon* With the increasing production of solar PV cells and the need for reducing their cost, it became necessary to develop separate processes for manufacturing the silicon. These processes were based on the following observations.

1. Researchers realized that the efficiency of solar PV cells made from multicrystalline Si was not much lower than those made from single crystal Si if the grain size was large (around 1 cm).
2. It was noted that circular silicon PV cells were at a disadvantage compared to square cells because they did not use the area of the modules on which they were attached efficiently.

As a consequence, different processes of crystallization were developed for making SeG-Si. These yielded rectangular ingots with large grain sizes and were much cheaper than the traditional process of forming long cylindrical blocks of SeG-Si.

In parallel, it was also noted that efficient solar cells can be fabricated even if the purity level is not as high as in SeG-Si. This led to the development of solar grade silicon (SoG-Si), which is less costly. SoG-Si has impurity levels of parts per million (ppm), rather than parts per billion as in SeG-Si. Several refining processes have been developed in the last few years for the production of SoG-silicon from MG-silicon. These include acid leaching, directional solidification

---

*The silicon is usually of p-type because it is doped with measured amounts of boron during the purification process.

and vacuum refining. For more details on the processes used for producing different grades of silicon as well as processes for manufacturing solar cells, the reader is referred to Solanki [9].

Today, most of the current commercial modules are made from multicrystalline silicon. Multicrystalline wafers are prepared from large-grained multicrystalline ingots. The grains are generally much larger than the wafer thickness. Consequently, the grain boundaries do not interfere much with the flow of electrons. Thus, the cells yield efficiencies not very inferior to those obtained with single crystal cells and have the advantage of lower cost. Cell efficiencies up to 21.3 per cent have been obtained (Table 8.1).

Multicrystalline silicon solar cells have a great potential for further development in terms of increased cell efficiency and reduced cost. As noted earlier, they can be manufactured in square shapes and hence, when used in modules, a more complete utilization of the module area is realized. Consequently, the lower efficiency of the cell is compensated at the module level. The highest value reported for a laboratory module is 19.5 per cent (Table 8.2). For commercially available modules, the efficiency ranges between 14.1 and 16.7 per cent in India.

❑ *Advances in Crystalline Silicon Cell Technology* Several innovations in manufacturing processes are being developed and implemented by industry to increase the efficiency of crystalline silicon solar cells. At the same time, extensive R&D work is going on to improve the performance of different components of the conventional cell. These include advances in doping concentration at the surfaces of the cell, alternative top metal contacts, better anti-reflection coatings, alternative bulk material (n-type silicon instead of p-type silicon), alternative surface passivation techniques, etc. Some of the research initiatives have also led to the development of advanced silicon solar cells. These include passivated emitter solar cells (PESC), buried contact solar cells, rear point contact solar cells, passivated emitter and rear cells (PERC), passivated emitter and rear locally diffused (PERL) cells, heterojunction solar cells with amorphous silicon, metal wrap-through (MWT) cells, emitter wrap-through (EWT) cells, etc. These developments are yet to reach the commercial stage. Detailed descriptions of these cells are beyond the scope of this book. For more details, the reader is referred to Ramanujam *et al.* [10], Solanki [9] and Glunz *et al.* [11].

8.4.2 Thin Film Solar Cells

Thin film solar cell technology was developed with the prime objective of reducing the cost of wafer-based photovoltaic modules. Apart from requiring less quantity of materials, thin film modules can be made in large sizes and some of them can be laid on thin flexible substrates allowing their mounting on curved surfaces. However, the cell conversion efficiencies are lower than those of wafer-based cells. We will discuss four types of thin film solar cells, viz. cadmium telluride, copper indium gallium diselenide, amorphous silicon and gallium arsenide.

❑ *Cadmium Telluride* Cadmium telluride (CdTe) is a compound semiconductor. Its band gap of 1.45 eV is close to the optimum value which can yield the maximum theoretical efficiency (Fig. 8.6). Besides, it has a high absorption coefficient for photons having energy greater than the band gap [9]. Consequently, material of small thickness is required. Many efforts have been made towards the large-scale production of modules. A laboratory efficiency as high as 21.0 per cent has been achieved for a CdTe cell (Table 8.1). A schematic diagram of a typical cell is shown in Fig. 8.8. A heterojunction is formed between a layer of p-type CdTe, 1.5 to 8 μm thick, and n-type cadmium sulphide (CdS), 0.05 to 0.3 μm thick. A 0.1 to 0.3 μm transparent layer of tin oxide known as transparent conducting oxide (TCO), acts as the front contact while an extremely thin layer of copper and gold forms the back contact. The typical efficiency of a commercial module is 16.7 per cent (120 Wp, 7200 cm^2 area) [12]. It may be mentioned that research on nano-structured CdTe cells is being carried out in order to increase the efficiency [10]. There are some concerns about the toxic effects of cadmium. Investigations, however, have shown that these issues can be handled effectively [13].

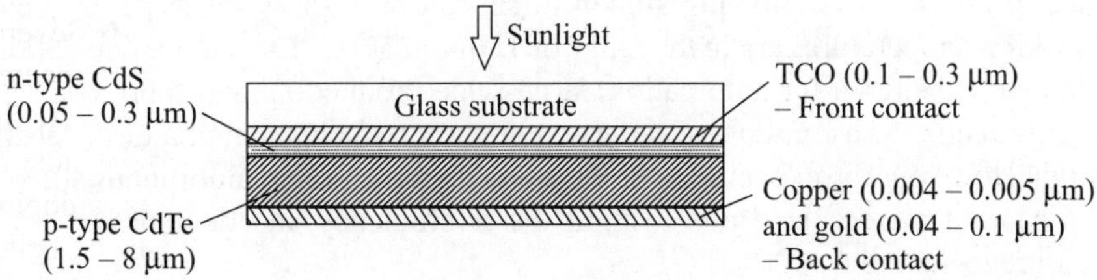

Fig. 8.8 *Schematic diagram of a CdTe cell*

❑ *Copper Indium Gallium Diselenide* Among the chalcopyrite compounds, the copper indium gallium diselenide (CIGS) solar cell has the highest efficiency. Thus rapid progress has been made in the last few years with thin film multicrystalline cells made from CIGS. A laboratory efficiency of 21.0 per cent has been reported (Table 8.1). A schematic diagram of a typical CIGS cell is shown in Fig. 8.9. The substrate on which the cell is based is sodalime glass. The back contact is a thin 1 μm film of molybdenum. The junction is formed by chemical deposition of a thin 0.05 μm cadmium sulphide (CdS) layer on a CIGS film (2 to 4 μm thick). Like CdTe, a heterojunction is formed. A thin layer of zinc oxide, which acts as a

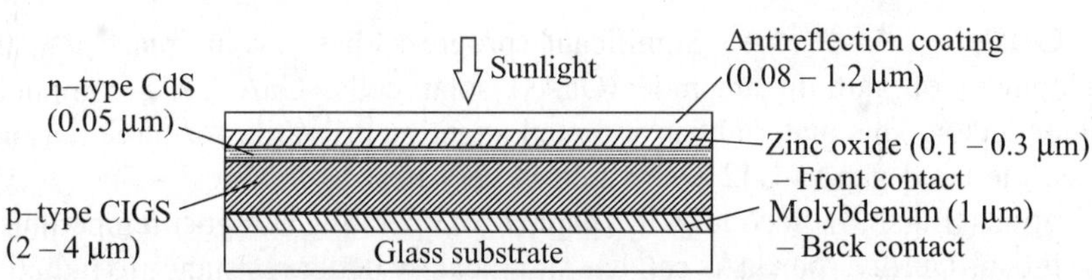

Fig. 8.9 *Schematic diagram of a CIGS cell*

TCO, is deposited on the top in addition to an antireflection coating. Unlike amorphous silicon, the properties of CIGS are stable. Several manufacturing processes have been developed for increasing the cell efficiency.

Commercial modules are available in the range of 150 to 170 Wp with a module area of 125.7 cm × 97.7 cm, the module efficiencies ranging from 12.2 to 13.8 per cent [14]. It may be noted that CIGS solar cells have been used for building integrated photovoltaic (BIPV) applications and the possibility of making semi-transparent CIGS solar cells is being explored [15].

❑ *Amorphous Silicon* The amorphous form of silicon was considered for making thin film solar cells because of the high cost of making crystalline silicon. However, its electrical properties were not suitable. On the other hand, thin film cells based on a hydrogenated alloy of amorphous silicon have been successfully commercialized. This alloy is represented by the symbol a-Si:H. Typically, a cell consist of three layers, viz. p-type, undoped and n-type amorphous silicon. It is manufactured by a chemical vapour deposition technique. In this technique, large vacuum chambers are filled with mixtures of silane gas, hydrogen and small quantities of doping agents (diborane for p-type, phosphine for n-type). The silane gas dissociates allowing silicon to deposit on a glass substrate. Typical deposition temperatures are in the range of 150° to 250°C. The vapour deposition technique is suitable for automation, large-scale production, and fabrication of large size cells. As a consequence, the potential for reducing manufacturing costs is high. However, the expectation of large-scale production of amorphous silicon cells has not materialized due to their lower efficiency and degradation with exposure to sunlight.

The conversion efficiency of amorphous silicon cells is lower than single crystal and multicrystalline silicon cells. Efficiencies around 10.2 per cent have been obtained in the laboratory, while the highest efficiencies reported with commercially available modules are about 8 per cent [10].

The other problem associated with amorphous silicon is that the performance of the cell is found to degrade by 10 to 20 per cent with exposure to sunlight. Thus an initial efficiency of 8 per cent is likely to stabilize at a value between 6.5 and 7 per cent after about one year of operation. Therefore, these cells are not much used for photovoltaic power production. On the other hand, they are commonly used in devices that require much less power e.g., calculators, watches, etc. It may be mentioned that semi-transparent a-Si:H solar cells with tunable colours are being developed for BIPV applications [15].

❑ *Gallium Arsenide* Significant progress has taken place in the development of gallium arsenide (GaAs) solar cells. GaAs is a compound semiconductor. This material has attracted attention primarily for three reasons. Firstly, the band gap is 1.42 eV, which is very near to the ideal value for PV applications (Fig. 8.6). Secondly, it is capable of operating at higher temperatures than silicon. Thirdly, the GaAs cell has significantly better resistance to radiation than silicon cells. In view of these merits, GaAs solar cells have high efficiencies and have been used extensively for space applications whenever solar power is needed.

Apart from being made in the thin film form, GaAs cells have been made in the single crystal, multicrystalline, and multijunction forms. The highest laboratory efficiency of thin film GaAs single junction solar cells has been reported to be 28.8 per cent, where as for multicrystalline GaAs solar cells, the highest efficiency is 18.4 per cent (Table 8.1). The highest module efficiency of a thin film GaAs solar cell made in a laboratory is reported to be 24.1 per cent (Table 8.2). Because of these high values, the GaAs solar cell is preferred in concentrated PV applications (Section 8.6.8), where it is generally used in multijunction solar cells.

8.4.3 Multijunction Solar Cells

In order to improve the efficiency of solar cells, a strategy that has been adopted is to stack cells one above the other. These type of cells are called multijunction solar cells. The basic idea behind such a concept is that two or more materials with different band gaps can absorb the solar spectrum more efficiently than a single material. Thus the cells could be fabricated as double junction cells (also called tandem cells) or multijunction cells. There are many types of multijunction cells reported in the literature, but we will describe only a few.

Because of the low efficiency of the amorphous silicon solar cell, double as well as triple junction cells have been made. Commercially, the most successful ones are stacked cells using amorphous silicon with micro- or nano-crystalline silicon. Such two junction cells are called tandem micro morph cells and are used for commercial photovoltaic power production. Table 8.3 presents efficiencies of a few dual-junction and triple-junction amorphous silicon cells. Some cells are of the laboratory scale and have small areas. Commercial production of large-area modules has also begun.

Table 8.3 Efficiency values of multijunction amorphous silicon cells. Measurements made under global AM1.5 spectrum (1000 W/m^2) and at a cell temperature of 25°C [16–18].

Type of cell	Area (cm^2)	Efficiency (%)
Dual junction (a-Si:H/nc-Si:H)	14 300	10.35
Triple junction (a-Si:H/a-SiGe:H/nc-Si:H)	399.8	11.8
Triple junction (a-Si:H/μc-Si:H/μc-Si:H)	1.043	13.6

N.B. nc = nano-crystalline; μc = micro-crystalline

Another successful development of multijunction solar cells has been through the use of compound semiconductors from group III and V elements of the periodic table. We have already discussed the gallium arsenide (GaAs) cell earlier as a single junction cell. A number of two-, three- and four-junction cells have been developed along with GaAs. Apart from the advantages associated with the GaAs cell which were mentioned earlier in Section 8.4.2, it is found that the crystal composition of compounds of III–V group elements can be altered to suit the required band gap in multijunction cells. Table 8.4 presents efficiencies of a few multijunction cells and modules based on III–V compounds.

Table 8.4 Efficiency values of multijunction III-V cells and modules. Measurements made under global AM1.5 spectrum (1000 W/m^2) and at a cell temperature of 25°C [4].

Type	Area (cm^2)	Open circuit voltage (V)	Short-circuit current (mA/cm^2)	Fill factor (%)	Efficiency (%)
Cell					
GaInP/GaAs	0.999	2.538	14.18	87.7	31.6
InGaP/GaAs/InGaAs	1.047	3.065	14.27	86.7	37.9
Module					
InGaP/GaAs/InGaAs	968	23.95	1.556	93.6	31.2

It may be mentioned that the multijunction cells of III–V compounds are primarily designed to be used with solar concentrators. While they are routinely used for providing power in space applications, some terrestrial power generation has also been attempted. We discuss these multijunction cells under concentrated photovoltaic (CPV) systems in Section 8.6.8.

8.4.4 Other Solar Cells

We will now describe four types of solar cells which are still in the R&D stage. They are the perovskite solar cell, the organic solar cell, the dye sensitized solar cell and the kesterite solar cell.

Two other types are also worth mentioning. They are the quantum dot solar cell and the bifacial solar cell. Quantum dot solar cells are made up of semiconductors of nanocrystal size having dimensions smaller than the Bohr exciton radius for that material. Substantial progress has been made in the development of this type of cell. Zheng *et al.* [19] provide a review of the current progress. However, significant research and developmental investigations are needed before these cells can be commercialized. Another direction where R&D work is in progress is to develop bifacial solar cells, which absorb both incident and reflected radiation. Guerreo-Lemus *et al.* [20] provide a technological review of such cells.

❑ *Perovskite Solar Cells* Perovskite solar cells have attracted attention as emerging thin film solar cells. The term 'perovskite' refers to a large group of compounds having a crystal structure similar to a mineral called perovskite named in honour of the Russian mineralogist L.A. Perovski. The mineral is composed of calcium titanate ($CaTiO_3$). Compounds having a crystal structure similar to $CaTiO_3$ are called perovskites and represented as ABX_3, where A and B represent cations, and X is an anion. Out of a large number of structural variations of perovskites which are possible, it is found that organometallic halides are the most suitable for solar cell applications. They have very high light absorbing properties and have satisfactory carrier transportation properties. Besides, these compounds can be processed from solutions at low temperatures and their production processes can be readily scaled up for mass-scale production. Methyl ammonium lead trihalides ($CH_3NH_3PbX_3$; X = Cl, Br or I) are the most commonly studied perovskite materials. A variety of other materials have also investigated.

Perovskite thin film solar cells have undergone very rapid development over the last few years with the efficiency increasing from 3.8 per cent to more than 20 per cent. They have emerged as the potential new generation solar cells. The typical structure of a planar perovskite solar cell is shown in Fig. 8.10. It consists of a layer of fluorine-doped tin oxide (FTO) glass substrate. This acts as a conducting layer. Over it, a layer of titanium oxide (TiO_2) is deposited as a hole-blocking layer. A layer of perovskite is deposited over TiO_2 followed by a layer of hole transporting material (HTM). The last layer is the cathode, usually made from gold.

The architecture of the cells, materials of different components and device developments are experiencing constant variations and improvements. The maximum efficiency obtained for a perovskite solar cell thus far has been reported to be 22.1 per cent. This is for a cell of area 0.046 cm^2 and with V_{oc} = 1.105 V, I_{sc} = 24.97 mA/cm^2 and FF = 80.3 per cent [4]. For the current status and issues, the reader is referred to Qin *et al.* [21], Park [22], Petrovic *et al.* [23], Liu *et al.* [24] and Habibi *et al.* [25].

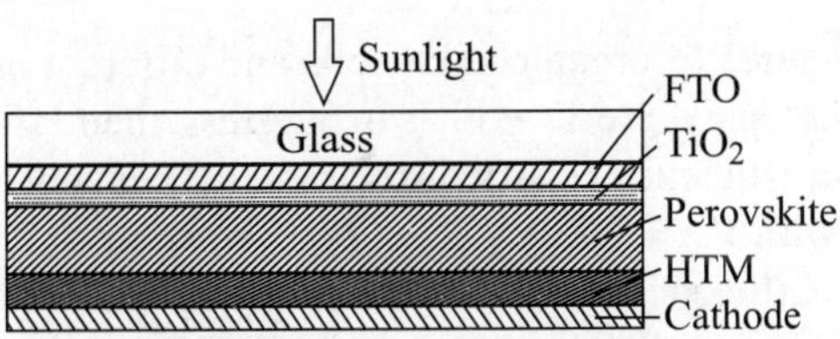

Fig. 8.10 *Schematic architecture of a planar perovskite solar cell*

While the developments on perovskite solar cells are quite encouraging, there are a number of issues which need to be addressed before commercialization takes place. These include stability and reproducibility. Another factor is the use of lead halides in perovskite solar cells. Since lead is toxic, its effect on the environment is an issue. Efforts are being undertaken to address this concern by replacing lead with tin. Finally, it may be mentioned that semi-transparent perovskite solar cells are being developed for building integrated PV applications. The possibility of making tandem solar cells having a perovskite solar cell as the bottom layer and a thin film crystalline silicon solar cell as the top layer, is also being explored [24].

❑ *Organic Solar Cells* Organic solar cells (OSC) or organic photovoltaic cells (OPV) are thin film cells that use organic semiconductors. Various types of organic materials in the form of monomers, oligomers as well as polymers have been tried. The attraction of these cells lies in the fact that they could be scalable, flexible and of light weight. In addition, they can be made from materials which can be processed in solution and have the potential to be of low cost.

Figure 8.11 shows a schematic diagram of a typical organic solar cell having conventional architecture. It consists of a layer of indium tin oxide (ITO) coated on a glass substrate. This acts as the anode of the cell. A buffer layer usually of poly(ethylenedioxythiophene) blended with poly(styrenesulphonate) (PEDOT:PSS) is deposited on the surface of the ITO film. This enhances the hole transfer to the ITO electrode. The next component of the cell is the layer of an organic semiconductor. It is usually a mixture of conjugated polymers and fullerene derivatives e.g., poly(3-hexylthiophene) (P3HT) and [6,6]-phenyl C61 butyric acid methyl ester (PCBM). There is a second buffer layer of lithium fluoride (LiF) to enhance electron transport. This is followed by a metal cathode of aluminium.

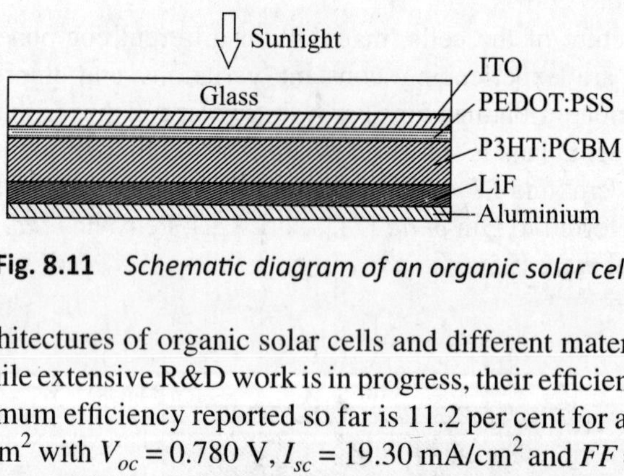

Fig. 8.11 *Schematic diagram of an organic solar cell*

Various architectures of organic solar cells and different materials have been developed. While extensive R&D work is in progress, their efficiencies have been low. The maximum efficiency reported so far is 11.2 per cent for a cell having an area of 0.992 cm^2 with V_{oc} = 0.780 V, I_{sc} = 19.30 mA/cm^2 and FF = 74.2 per cent [4]. The maximum efficiency obtained with a module having an area of 802 cm^2 is 8.7 per cent [4]. Small OPV modules are available commercially and are used for applications such as building integrated photovoltaics. However, research efforts are necessary for scaling up for large scale deployment [26].

In addition to having low efficiencies, organic solar cells also have stability issues. For the current status, challenges and innovations needed, the reader is referred to Kaur *et al.* [27], Liu *et al.* [28] and Wang *et al.* [29].

❑ *Dye Sensitized Solar Cells* Dye sensitized solar cells (DSSC) are photo-electrochemical cells. The cell consists of a photoanode, sensitizer, electrolyte and a counter electrode. The sensitizer is embedded in the photoanode and can be comprised of inorganic dye molecules, organic dye molecules or metal-organic dye molecules. Incident light is absorbed by the dye molecules creating electron-hole pair. While electrons are injected into the photoanode, the holes migrate via the electrolyte to the counter electrode. The electrons travel through the photoanode to the external load to reach the counter electrode, thus generating electricity.

Figure 8.12 shows a schematic diagram of a typical DSSC. It consists of a transparent conducting oxide (TCO) layer of fluorine-doped tin oxide deposited on a glass or plastic substrate. The TCO layer is used to collect electrons. It is followed by a nanocrystalline oxide semiconductor layer of titanium dioxide (TiO_2) deposited on the TCO layer. Next, a layer of dye (sensitizer) is deposited on the surface of the TiO_2 film. The dye molecules penetrate the pores of TiO_2 resulting in effective absorption of photons. The next component of the cell is an electrolyte solution containing iodide/tri-iodide redox ions. Finally, a layer of platinum coated on TCO substrate is used as an electrode. So far a large number of materials have been developed for various components of DSSCs and widely varying performances have been obtained.

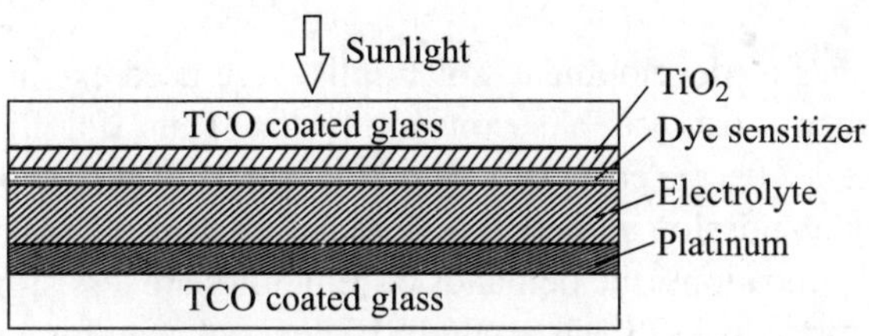

Fig. 8.12 *Schematic diagram of a DSSC*

Dye sensitized solar cells have attracted attention because of their low cost potential and high photon to electricity conversion efficiency. The maximum efficiency reported so far is 11.9 per cent for a cell of area 1.005 cm^2 with $V_{oc} = 0.774$ V, $I_{sc} = 22.47$ mA/cm^2 and $FF = 71.2$ per cent [4].

In order to make an impact, the efficiency of DSSCs needs to be increased. Electrolyte leakage and degradation of sealants, electrolytes and dyes are other issues that need to be sorted out so that the reliability and durability of these cells are improved before commercialization. The status of the R&D activities on DSSCs and improvements needed are presented by Sugathan *et al.* [30]. Shalini *et al.* [31] have provided a review of natural dye sensitized solar cells.

❑ *Kesterite Solar Cells* Recently rapid progress has been made with thin film solar cells made out of two synthetic compounds, copper zinc tin sulphide (Cu_2ZnSnS_4) and copper zinc tin selenide ($Cu_2ZnSnSe_4$). They are denoted by the acronyms CZTS and CZTSe respectively. These compounds are referred to as 'kesterite' because of their structural similarity to a naturally occurring mineral with the formula $Cu_2(Zn,Fe)SnS_4$ having the same name. In the naturally occurring mineral, zinc and iron atoms share the same lattice sites with zinc dominating.

The optical and electronic properties of CZTS and CZTSe are similar to CdTe and CIGS. However, the advantage with them is that they do not contain toxic elements like cadmium or rare-earth elements like indium. Rapid progress has taken place and the efficiency obtained for a CZTS cell having an area of 1.067 cm^2 is reported to be 7.6 per cent with $V_{oc} = 0.6585$ V, $I_{sc} = 20.43$ mA/cm^2 and $FF = 56.7$ per cent. Similarly, a maximum efficiency of 9.8 per cent has been obtained with a CZTSe cell having an area of 1.115 cm^2 with $V_{oc} = 0.5073$ V, $I_{sc} = 31.95$ mA/cm^2 and $FF = 60.2$ per cent [4]. There are, however, device defects which need to be sorted out before large scale fabrication and commercialization can take place. The reader is referred to Ramanujam *et al.* [10] for the current status and a technological review of these cells.

COSTS 8.5

In the early stages of development, solar cells were used exclusively for space applications where cost was a secondary factor. Thus, for the first Mariner spacecraft in the 1960s, the cost was several thousand dollars per peak watt, while for the later Skylab mission in the 1970s, the cost was around $100 per peak watt. For terrestrial applications, the demands of reliability are less stringent. Because of this fact, the price in 1973 was around $10 per peak watt for a module. Figure 8.13 shows the variation in module cost from 1994 to 2015 [32, 33]. It can be seen that the average module cost has decreased from about $5.20 per Wp in 1994 to $0.65 per Wp in 2015, a drop of 87.5 per cent.

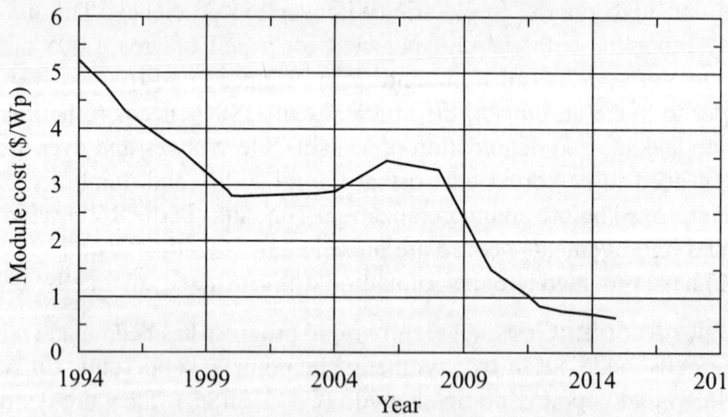

Fig. 8.13 *Yearwise change in the average cost of PV modules (Adapted from Solar Cell Central [32])*

There are a number of reasons that have contributed to this rapid reduction in cost. The most important reason is the decrease in the cost of making multicrystalline silicon. This has taken place due to the large expansion in the manufacturing capacity of this material globally. Secondly, the efficiencies of solar cells and hence modules, have been increasing year after year due to sustained R&D efforts as well as the introduction of innovative manufacturing techniques. Because of these combined initiatives, the usage of various materials in a cell has significantly decreased causing a reduction in cost. Apart from these facts, the volume of production has increased due to high global demand. Thus economics of scale as well as stiff competition among the suppliers has contributed to the reduction in module cost.

It is instructive to discuss the cost of a complete solar PV system. As is mentioned later in Section 8.6.2, the PV system essentially consists of arrays of PV modules and other hardware equipment called 'balance of system' (BOS). Thus the cost of a complete PV system can be split into three groups, viz. hardware cost, soft costs and other costs. The hardware cost includes modules, BOS, mounting structures, etc. The soft cost refers to the installation cost, land cost, etc., while the other cost refers to taxes. From this break-up, it can be seen that the cost of a PV system could also depend on the location of the installation. The module cost, however, contributes the largest share. According to the norms laid down by the Central Electric Regulatory Commission (CERC) of India, the module price in the financial year 2016-17 has been fixed to be $ 0.48 per peak watt, while the total installation cost of PV power is Rs 530.02 lakh per MW [34]. It may be mentioned that there are various schemes and types of subsidies available in India through the Ministry of New and Renewable Energy (MNRE) for PV system installation [3, 35].

PHOTOVOLTAIC SYSTEMS AND APPLICATIONS 8.6

8.6.1 Introduction

Photovoltaic systems can be broadly classified into two categories:

1. Stand-alone PV systems
2. Grid-connected PV systems.

As the name implies, a stand-alone system is one which is not connected to a power grid. It feeds a load which is often in watts or kilowatts directly. As opposed to this, the power generated in a grid-connected system is fed to the power grid and becomes part of the power generated by other energy sources. In 2006, the grid-connected and stand-alone PV power generation capacity in India was only 2.74 and 1.86 MWp, respectively. The impressive growth in recent years can be seen from the fact that upto June 2016, the total PV power generation capacity was 7585 MWp. Out of this, the stand-alone capacity was 325 MWp and grid-connected capacity was 7260 MWp.

A stand-alone system generally has an energy storage unit in the form of a battery as a component, while a grid-connected system usually does not need one. Apart from being stand-alone or connected to the grid, PV systems are also described by other characteristics. They can be (i) regulated or unregulated,

(ii) supply a DC load, an AC load or both types of loads, (iii) have a battery storage or no storage. We will describe a few configurations in detail from Section 8.6.3 onwards.

8.6.2 Balance of System Components

A PV system usually has some other components which come between the PV modules and the load or between the PV modules and the power grid. These are referred to as 'balance of system' components. The following are the commonly used components:

1. Electric batteries
2. DC to DC converters
3. Inverters (DC to AC converters).

Electric batteries are used for storing electrical energy generated by PV modules and supplying the energy for later use against a load requirement. Batteries are generally needed in stand-alone systems. The batteries used are storage batteries which can be charged and recharged. The common batteries are lead-acid batteries, lithium-ion batteries and nickel-cadmium (Ni-Cd) batteries. A battery is made up of a number of electrochemical cells, each of which generates a voltage associated with the electrochemistry of the chemicals in it. For example, the Ni-Cd cell generates a voltage of 1.2 V DC. Thus if the voltage required is 24 V DC, 20 cells would need to be connected in series. Apart from the voltage which it supplies, a battery is characterized by its capacity and its efficiency. The *capacity* is expressed in ampere-hours and is obtained by multiplying the current (in amperes) supplied by the battery to feed a load and the number of hours for which it can supply this current till it is fully discharged. Rechargeable batteries are available in capacities ranging from a few A-h to a few thousand A-h. The *efficiency* of a battery is the ratio of the energy discharged by a fully charged battery to the energy supplied to it during the complete charging process. It is usually expressed as a percentage. We can consider the efficiency of a battery to be the product of two efficiencies, a charging efficiency and a discharging efficiency. Typically, a rechargeable battery might have a charging efficiency around 95 per cent and a discharging efficiency which is a few per cent lower. It should be mentioned that the capacity of a battery multiplied by its voltage gives the energy discharged by a fully charged battery.

In practice, the charge stored in a battery is not used up fully during discharging. This practice helps to protect the battery and extends its life. The extent to which a battery is allowed to discharge is called the Depth of Discharge (DoD). Tolerable values of DoD range from 60 to 90 per cent. This is implemented with the help of charge controllers.

DC to DC converters are used to convert DC voltage from one level to another. They serve as an interface between the PV modules and the load. The converter is referred to as 'buck' type if it steps down the voltage, 'boost' type if it steps up the voltage, and 'buck boost' type if it can perform either function. With suitable modifications, DC to DC converters can also be used as *battery charge controllers* and as maximum power point trackers (MPPT) in solar PV systems. As charge controllers, they regulate the charging and discharging process of a battery. The controller senses the voltage of the battery and disconnects the supply from the battery to the load during the discharging process if the battery has reached the allowable DoD. Similarly, it disconnects the battery from the PV modules during charging when it senses that overcharging is occurring. The other function of serving as a MPPT can be visualized if one refers to Fig. 8.5 which shows the *I-V* curves obtained as the incident solar radiation varies. With suitable control action of a DC to DC converter, a MPPT can be devised which adjusts to the maximum power point on the *I-V* characteristic through the day as electrical energy is charged into the battery. Now-a-days, a single unit having all the three facilities, viz. DC to DC conversion, battery charge control and MPPT, is available. Typically, the efficiency of well-designed DC to DC converters would be greater than 95 per cent, while the efficiencies of MPPTs and charge controllers would be more than 90 per cent.

Inverters are DC to AC converters. Since PV modules generate direct current, it often becomes necessary to use inverters in stand-alone systems. These inverters not only convert to AC but adjust to the voltage requirement of the load. In grid-connected PV systems, inverters are a must since power grids are invariably AC. Efficiencies of inverters are as high as 94 to 96 per cent.

8.6.3 Stand-alone System with Battery Storage and a DC Load

Figure 8.14 shows the configuration of a stand-alone system with battery storage and a DC load which generates power during the day. The electricity passes through a charge controller and is stored in a battery. The energy stored may be used during the day or night to operate a DC load. A simple example of an application of this system is a small capacity LED lamp connected to a PV panel which provides light at night to rurai huts in villages where there is no electric supply. The system looks like a small table lamp. Already a million such lamps have been provided at a subsidized price through a government-sponsored scheme in many states in India. The aim is to reach out to rural school-going children and provide some light at night. A description of the scheme and of the design of the solar lamp is given by Arora *et al.* [36]. The success of the scheme so far has provided the encouragement needed to expand it at a rapid rate. The children benefitted would otherwise not be able to study after sunset or would have to use kerosene lamps. The solar lamps are provided at a subsidized rate.

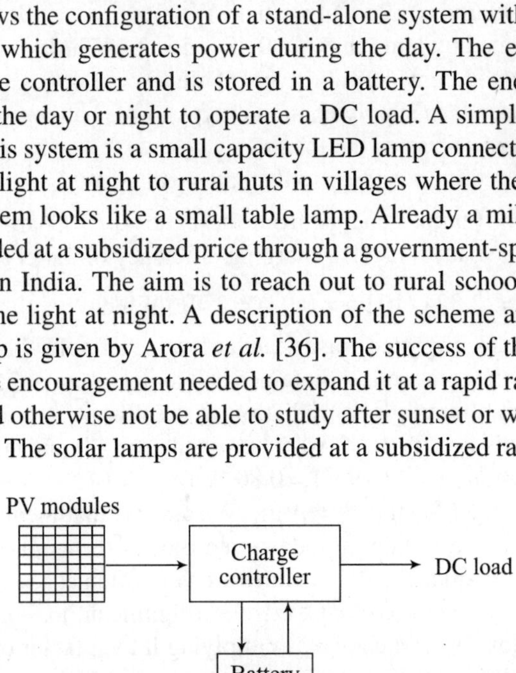

Fig. 8.14 *Schematic diagram of a stand-alone PV system with battery storage and a DC load*

It is worth noting that India consumed about 9 billion litres of kerosene in 2015 [37]. This was sold at a subsidized price of Rs 32 per litre. A very large number of rural households consume kerosene for lighting purposes [38]. It is certainly more desirable and cost effective to subsidize solar PV-LED lamps which provide better illumination and reduce indoor pollution.

Example 8.2

A solar PV-LED lamp operates with a voltage of 3.2 V DC and draws a current of 0.15 A. The lamp is required to operate for 5 hours every night. Assuming that the configuration of Fig. 8.14 is used, calculate the wattage and area of the PV module required which would collect enough energy through the day and operate satisfactorily at night through the year. Also calculate the capacity of the battery if the battery voltage is 2.4 V and the depth of discharge is 80 per cent.

Given:

(i) The daily global radiation varies from a minimum of 4 kWh/m^2 in the winter to a maximum of 7 kWh/m^2 in summer.

(ii) The PV module efficiency is 15 per cent, while the battery charging efficiency is 95 per cent, and the battery discharging efficiency is 93 per cent.

(iii) The LED driver efficiency is 85 per cent.

Solution

Power of LED lamp = 3.2 × 0.15 = 0.48 W

Energy consumed by lamp in 5 h = 0.48 × 5 = 2.4 Wh/day

Energy consumed considering driver efficiency = 2.4 /0.85 = 2.824 Wh/day

The PV module needs to supply enough energy to the battery, which in turn can supply the required energy to the load.

Thus the required input energy to the battery in a day

= (Output energy from the battery) / [(Charging efficiency of battery) × (Discharging efficiency of battery)]

= 2.824 / (0.95 × 0.93) = 3.196 Wh/day

The PV modules are rated for a standard value of solar global radiation of 1000 W/m^2 with the solar cells maintained at a temperature of 25°C. We will use the given daily minimum global radiation of 4 kWh/m^2 in our calculation and interpret it as equivalent to 1000 W/m^2 available for 4 hours per day. Thus, PV module wattage required = 3.196/4 = 0.80 Wp.

Let the area of the PV module = A m^2. We assume that the module efficiency of 15 per cent is valid through the day even though the global radiation varies (Fig. 8.5). However since the PV modules are very often at a temperature greater than the standard test value of 25°C, there is a significant loss of power. We will account for this loss in efficiency by multiplying it by a factor of 0.90. Thus

$$A \times 4 \times 1000 \times 0.15 \times 0.90 = 3.196$$

$$A = 5.92 \times 10^{-3}\ \text{m}^2 = 59.2\ \text{cm}^2$$

Capacity of battery = (Energy consumed)/(Battery voltage × Battery discharging efficiency × Depth of discharge)

= 2.824 /(2.4 × 0.93 × 0.8) A-h

= 1.58 A-h

It may be mentioned that in order to charge the battery safely, the difference in voltage between the PV module and the battery typically should not be greater than 1.4 V. This value is specified by the manufacturer of the battery.

8.6.4 Stand-alone System without Battery Storage and with an AC Load

A block diagram of a stand-alone system without battery storage and with an AC load is shown in Fig. 8.15. The configuration shown is a simple one. The PV panel generates power only during the day when adequate radiation is available. The DC power is converted to AC in an inverter and is used straightaway to drive an AC load. The power fed to the load is unregulated and it varies as the radiation level changes. Such systems are used widely in rural areas. They are simple, rugged, less costly and easy to maintain. No battery back-up is provided and as a result, maximum power point tracking cannot be implemented. Because of this fact and also because of the higher temperature attained by the PV modules, there is a loss of at least 20 per cent in the standard test conversion efficiency reported by the manufacturers.

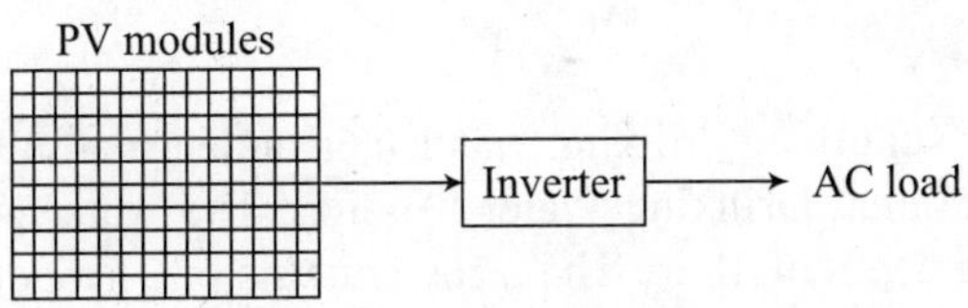

Fig. 8.15 *Schematic diagram of a stand-alone PV system without battery storage and connected to an AC load*

A diagram of a typical system for pumping water from a bore well is shown in Fig. 8.16. Such systems have been designed to pump water from depths of 20 to 100 metres and to supply 5000 to 50 000 litres per day. As seen in Fig. 8.16, a PV array supplies power through a DC-AC inverter to an electric motor coupled to a submersible pump. The pump is installed below the water level of the bore well. Its discharge is connected through a delivery pipe to a storage tank at ground level. Water can be withdrawn from the storage tank for use when required. Usually systems for pumping water do not have storage batteries and work only during the day when adequate solar radiation is available.

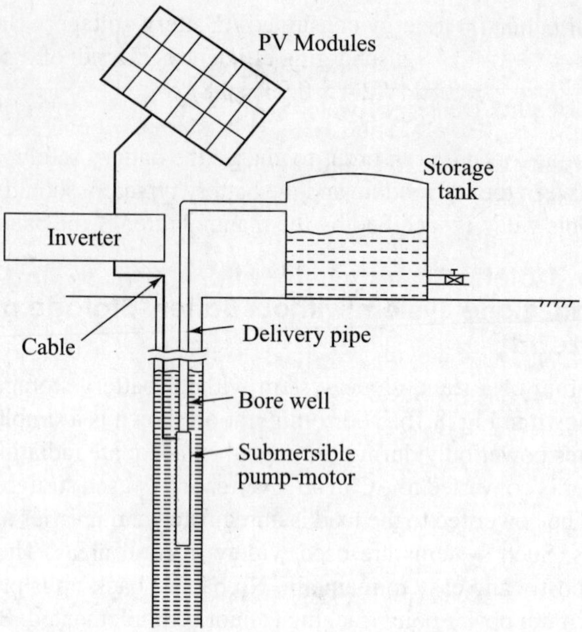

Fig. 8.16 *A photovoltaic water-pumping system*

Example 8.3

A PV system for supplying drinking water is installed in a village in Rajasthan as part of the National Drinking Water Mission. The water is pumped from a bore well, from a depth of 48 m. The array consists of 24 modules, each having an area of 0.5625 m^2 and an overall 'standard test' conversion efficiency of 14.1 per cent for global radiation. The conversion efficiency is based on the module area. It is given that the inverter efficiency is 94 per cent and the pump - motor set efficiency is 45 per cent. Calculate the water discharge rate at noon when the total radiation incident normally on the modules is 945 W/m^2.

Solution

We assume that the actual conversion efficiency of the PV modules is 80 per cent of the 'standard test' conversion efficiency.

Power output from array = Incident flux × Module area × Actual conversion efficiency

$$= 945 \times (24 \times 0.5625) \times (0.141 \times 0.80)$$

$$= 1439.0 \text{ W}$$

Power available for lifting water

$$= 1439.0 \times 0.94 \times 0.45 = 608.7 \text{ W}$$

Water discharge rate (Q)

$$= \frac{\text{Power available}}{\text{Depth} \times \text{Density of water} \times \text{Acceleration due to gravity}}$$

Taking the density of water to be 996 kg/m^3,

$$Q = 608.7/(48 \times 996 \times 9.81) \text{ m}^3\text{/s} = 0.001298 \text{ m}^3\text{/s} = 4672 \text{ litres/h}$$

8.6.5 Stand-alone System with Battery Storage and Supplying an AC Load

A schematic diagram of a regulated stand-alone system having a battery storage and supplying an AC load is shown in Fig. 8.17. The configuration shown is more sophisticated than the ones discussed in Section 8.6.3 and Section 8.6.4. A MPPT unit operates the PV array at an optimum power level. The power is then stored in a battery bank through a controller which regulates both the charging and discharging process from the batteries. The DC output of the battery bank is passed through an inverter and converted to the AC voltage required by the load. In case there is a DC load in addition to the AC load, some of the output from the batteries would bypass the inverter. We will analyze a 24 hour cycle for such a configuration through an example.

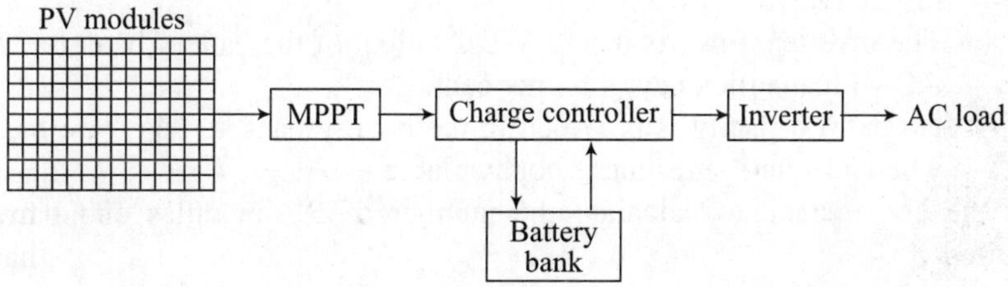

Fig. 8.17 *Schematic diagram of a stand-alone system with battery storage and connected to an AC load*

Example 8.4

A residential house has the following 220 V AC load requirement over a 24 hour day-night cycle:

Time (h)	0-1	1-2	2-3	3-4	4-5	5-6	6-7	7-8
Load (kWh)	0.05	0.05	0.05	0.05	0.05	0.05	0.15	0.35
	8-9	9-10	10-11	11-12	12-13	13-14	14-15	15-16
	0.60	0.50	0.40	0.35	0.35	0.35	0.35	0.45
	16-17	17-18	18-19	19-20	20-21	21-22	22-23	23-24
	0.60	0.80	0.90	1.00	0.80	0.50	0.30	0.10

It is proposed to supply the load with a roof-top PV array system connected to a battery storage bank and an inverter. The following solar radiation data is available for the location (latitude = 27°N):

Date: May 1

Time-LAT (h)	7	8	9	10	11	12	13	14	15	16	17	18
I_b (kWh/m²-h)	0.07	0.19	0.34	0.47	0.57	0.61	0.60	0.53	0.43	0.30	0.16	0.05
I_d (kWh/m²-h)	0.11	0.19	0.24	0.28	0.30	0.32	0.32	0.31	0.29	0.25	0.19	0.11

Given:

(i) The PV modules used are made from multicrystalline silicon and have a 'standard test' conversion efficiency of 14.4 per cent which does not change with the incident irradiance through the day, but is reduced by 15 per cent because of higher temperatures encountered by the modules.

(ii) The modules are tilted at an angle of 27° and face due south.

(iii) The size of each PV module is 1.667 m × 1.0 m.

(iv) The efficiency of the charge controller and MPPT together = 97 per cent.

(v) The battery bank is made up of 12 V DC batteries having a capacity of 150 A-h each. The batteries have a depth of discharge = 72 per cent and the efficiencies for charging and discharging are 94 and 92 per cent respectively.

(vi) The inverter converts the 12 V DC output of the battery bank to 220 V AC with an efficiency of 94 per cent.

(vii) No extra capacity is provided in the battery bank to take care of days when adequate sunshine is not available.

Use the above data, to calculate the number of PV modules and batteries required.

Solution

We assume that the values of I_b and I_d at a given time are valid for one hour preceding that time. Thus I_T, the solar radiation incident on one module in a specific hour in a day is $(I_b r_b + I_d r_d)$ and the radiation incident over a day is obtained by summing over all hours when it is available.

Sample calculation for the hour preceding 1100 h (LAT):

Since tilt is equal to the latitude in this case, Eq. (3.8) reduces to

$$\cos\theta = \cos\delta\cos\omega$$

From Eq. (3.4), $\delta = 14.9°$ on May 1

Thus $\cos\theta = \cos 14.9° \cos\omega = 0.966 \cos\omega$

From Eq. (3.7),

$$\cos\theta_z = \sin 27° \sin 14.9° + \cos 27° \cos 14.9° \cos\omega = 0.117 + 0.861 \cos\omega$$

From Eq. (3.41), tilt factor $r_b = \dfrac{0.966\cos\omega}{0.117 + 0.861\cos\omega}$

At 1100 h, $r_b = \dfrac{0.966 \cos 15°}{0.117 + 0.861 \cos 15°} = 0.983$

From Eq. (3.42),

tilt factor $r_d = (1 + \cos 27°) / 2 = 0.946$

Assuming that the reflected radiation is negligible, we have from Eq. (3.44)

$$I_T = 0.57 \times 0.983 + 0.30 \times 0.946 = 0.844 \text{ kW/m}^2$$

Repeating this calculation for all hours and summing up, we have radiation flux incident over the day from 0700 to 1800 h *(H_T)*

$= 0.155 + 0.347 + 0.547 + 0.721 + 0.844 + 0.906 + 0.893 + 0.807 + 0.679 + 0.501 + 0.298 + 0.104$

$= 6.802 \text{ kW/m}^2$

Output from one module over a day

$= 6.802 \times (1.667 \times 1.000) \times 0.144 \times 0.85$

$= 1.388$ kWh

Energy required by the AC load in the house during 24 hours

$= (0.05 \times 6 + 0.15 + 0.35 + 0.60 + 0.50 + 0.40 + 0.35 \times 4 + 0.45 + 0.60 + 0.80 + 0.90 + 1.00 + 0.80 + 0.50 + 0.30 + 0.10)$

$= 9.15$ kWh

Let the number of modules be *N*. Multiplying the output of the PV array by the efficiencies of the controller, the battery bank (for charging and discharging) and the inverter, and equating to the energy requirement of the house, we get

$$1.3881 \times N \times 0.97 \times 0.94 \times 0.92 \times 0.94 = 9.15$$

$$N = 8.36$$

We will recommend 9 modules.

Let the number of batteries required = *M*

Then

$$12 \times 150 \times 0.72 \times M \times 0.94 = 9.15 \times 10^3$$

$$M = 7.5$$

Rounding off, we require 8 batteries for storage.

Note: The system in Fig. 8.17 may be re-designed in the following situation: It is possible that adequate radiation is not available on a particular day and the PV system is required to meet the need for electricity for an additional 24 hours. In this case, the system would need to be provided with double the battery storage capacity. However, the capacity of the PV array would need to be increased by only 10 or 20 percent depending upon how often a day with poor availability of radiation is likely to be encountered.

8.6.6 Hybrid PV-Concentrated Solar Thermal Electric Power System

Hybrid electric power systems are systems in which one combines the output of two or more energy sources producing electricity separately. Because of the combination, it is possible to obtain the desired characteristics in the output in a more economical manner. In the context of this chapter, a hybrid power system implies that one of the energy sources is a PV system. A block diagram of a hybrid stand-alone regulated system in which electricity is generated by a PV array as well as a concentrated solar thermal power (CSP) plant is shown in Fig. 8.18.

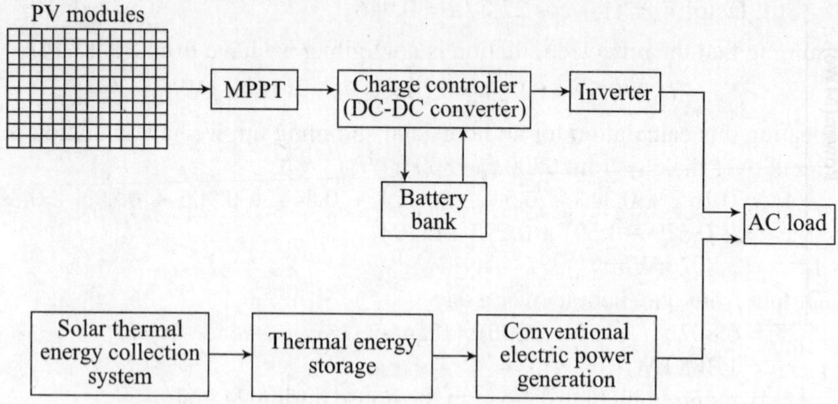

Fig. 8.18 *Block diagram of a PV-concentrated solar thermal hybrid stand-alone system*

The PV panels generate DC power during the day whenever adequate solar radiation is available. This DC power is maximized by a MPPT unit. Part of it is directly fed to an AC load through a charge controller (DC-DC converter) and an inverter, while the remaining power is fed through the charge controller to the battery bank and stored. It is withdrawn from the battery bank when required (usually when adequate solar radiation is not available in the day light hours) and fed to the DC–DC converter. The DC–DC converter is in turn connected to an inverter and the resulting AC power is connected to the load. Storing large amounts of electrical energy in battery banks is not economical. For this reason, in parallel there is a CSP plant with a thermal energy storage system which generates AC power through a conventional electricity generation system. This power is also fed to the load mostly at night. Because of the presence of storage systems, it is possible to control the AC output of the PV arrays and the CSP plant and obtain a total output from the hybrid system which matches the AC load demand on a continuous basis in the manner desired, day and night.

In order to understand the working of such a hybrid system better, let us assume that we have to design a system to meet a constant load requirement of 1 MWe of electric power on a continuous basis round the clock. Solar radiation is available for about 10 hours and the sky is clear. Because of this situation, the PV array will generate electricity for about 8 or 9 hours as shown by graph 1 in Fig. 8.19(a). It is sized so as to give an output of a little more than 1 MWe at its peak. Thus the energy output of the PV array will be about 8 to 9 MWh(e) and this output will be essentially used during the day light hours by the load of 1 MWe, the shaded rectangle in Fig. 8.19(a). The small mismatch between the energy generated and the energy requirement is adjusted by the charging and discharging of a battery storage unit.

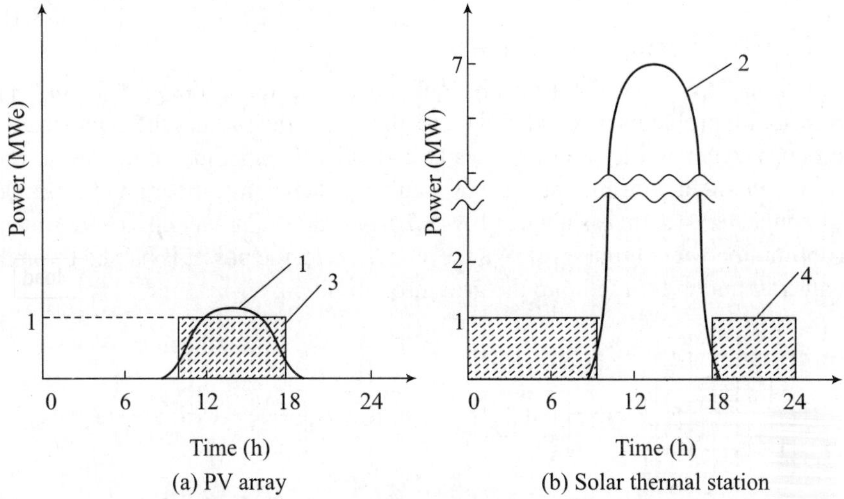

Fig. 8.19 *1 MWe hybrid PV-concentrated solar thermal power plant*
1. Electrical output of PV array during the day (MWe)
2. Thermal output of solar thermal plant collected during the day (MWth)
3. Load requirement of 1 MWe met for 8 hours by PV array
4. Load requirement of 1 MWe met for 16 hours by solar thermal plant

In contrast to the PV array, the solar thermal system is designed to collect energy at the rate of about 7 MW(th) during the day at its peak as shown in Fig. 8.19(b). In 8 or 9 hours this will amount to about 55 MWh(th) of heat. This heat is stored in the thermal energy storage unit. The electric power generation unit of the CSP plant will operate for about 16 hours in the evening, night and morning to generate the required electricity using the stored heat energy. Assuming that it has a power cycle efficiency of 30 per cent, it will generate 1 MWe and produce a total of approximately 16 MWh(e). This is shown by the shaded rectangles in Fig. 8.19(b). It is to be noted that there are many possible variations in the performance characteristics specified for such a plant. A detailed design would require consideration of these variations as well as the overall cost in order to arrive at an optimal solution.

An example of a solar hybrid power plant under construction is the Copiapo Project in the Atacama desert in Chile. The power station will comprise two 130 MW central receiver CSP plants with 14 hours of thermal storage capacity and a 150 MW PV plant. It is expected to be operational in 2019 and generate 1700 GWh annually on a continuous basis.

Finally, it would be useful to note that the hybrid power system described could be a stand-alone system or be connected to a power grid. If it is a stand-alone system, the AC output voltage should match the requirement of the load, where it is installed. On the other hand, if it is to be connected to the power grid, it would probably be necessary to pass the power through a transformer and to step up the voltage to the grid's requirement.

8.6.7 Grid-connected System

Most of the MW-sized PV arrays in India are connected to the grid (Plate 8.1). There are no problems associated with feeding their output into the grid because the power generated by PV arrays is still small in magnitude compared to the power from other conventional energy sources. A schematic diagram of a simple grid-connected system is shown in Fig. 8.20. A MPPT unit operates a PV array at its optimum power output level. Whenever available, the power is passed through an inverter and is then fed into the grid through filters.

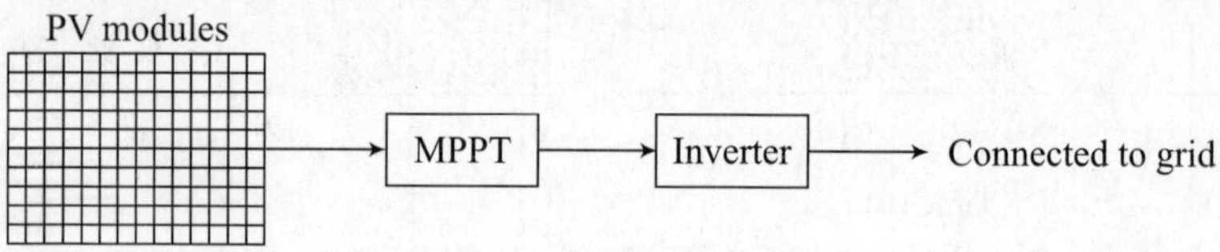

Fig. 8.20 *Schematic representation of a grid-connected PV system*

There can be many ways of specifying the requirements of a PV power plant from the design point of view. One way is to specify the capacity of the plant and the annual plant load factor (APLF). This defines the annual generation of a fixed quantity of energy by the plant and can be calculated by multiplying the capacity by the number of hours in a year and the APLF. For PV plants, the APLF generally ranges between 0.17 to 0.19.

Example 8.5

Consider a grid-connected PV power plant with a capacity 10 MW and an APLF of 0.17 to be installed at a place (latitude = 20° N) near Nagpur. Estimate the module area and land area required for the plant. The modules are tilted at an angle equal to the latitude of the location and face due south. The efficiencies of the PV module, MPPT and inverter are 14.5, 95 and 96 per cent respectively. The monthly average hourly global and diffuse solar radiation data for the location are given in Table 8.5.

Solution

Following the same procedure, as in Example 8.4, we calculate $\bar{H}_T$, the monthly average of daily radiation incident on the PV modules on the average day of each month (Section 3.6.1). The values obtained are as follows:

Month	J	F	M	A	M	J
$\bar{H}_T$ (kWh/m^2-day)	6.059	6.460	6.621	6.598	6.385	5.219
Month	J	A	S	O	N	D
$\bar{H}_T$ (kWh/m^2-day)	4.158	3.998	5.134	6.174	9.306	8.877

Plate 8.1 *Grid-connected PV plant in IIT Bombay.*

Table 8.5 Solar radiation data for Example 8.5

	Time-LAT (h)	6	7	8	9	10	11	12	13	14	15	16	17	18	19
J	$\bar{I}_g$	0	.015	.159	.360	.538	.667	.733	.732	.658	.527	.350	.148	.012	0
	$\bar{I}_d$	0	.011	.066	.110	.138	.148	.152	.149	.140	.129	.103	.060	.009	0
F	$\bar{I}_g$	0	.035	.216	.434	.618	.754	.821	.817	.739	.598	.413	.198	.029	0
	$\bar{I}_d$	0	.022	.085	.129	.153	.165	.173	.175	.166	.152	.125	.076	.018	0
M	$\bar{I}_g$	.001	.069	.270	.490	.680	.816	.884	.878	.800	.651	.462	.245	.057	0
	$\bar{I}_d$	.001	.044	.111	.153	.179	.189	.200	.205	.199	.182	.151	.100	.035	0
A	$\bar{I}_g$	.006	.117	.332	.546	.727	.860	.924	.917	.819	.660	.480	.283	.093	.003
	$\bar{I}_d$	.005	.075	.141	.183	.206	.224	.236	.239	.241	.223	.192	.134	.057	.002
M	$\bar{I}_g$	.016	.146	.355	.565	.738	.857	.921	.900	.805	.655	.485	.298	.118	.011
	$\bar{I}_d$	.014	.095	.165	.211	.240	.253	.264	.272	.271	.251	.209	.154	.079	.009
J	$\bar{I}_g$	.018	.126	.292	.465	.610	.714	.752	.728	.657	.530	.391	.237	.101	.013
	$\bar{I}_d$	.018	.096	.182	.248	.302	.338	.353	.352	.333	.281	.227	.153	.077	.011
J	$\bar{I}_g$	.012	.093	.223	.363	.470	.532	.560	.556	.528	.438	.330	.195	.079	.009
	$\bar{I}_d$	.011	.082	.176	.268	.343	.386	.406	.396	.363	.311	.241	.154	.067	.008
A	$\bar{I}_g$	.006	.079	.207	.344	.461	.515	.534	.532	.497	.415	.308	.185	.068	.004
	$\bar{I}_d$	.004	.068	.158	.254	.334	.375	.390	.374	.344	.287	.224	.139	.058	.003
S	$\bar{I}_g$	.002	.070	.238	.421	.571	.667	.712	.696	.622	.500	.359	.197	.056	.001
	$\bar{I}_d$	.002	.056	.147	.219	.272	.306	.329	.321	.286	.248	.195	.120	.040	.001
O	$\bar{I}_g$	0	.046	.231	.442	.614	.740	.796	.787	.712	.581	.410	.208	.038	0
	$\bar{I}_d$	0	.029	.097	.141	.168	.188	.204	.211	.198	.176	.135	.085	.024	0
N	$\bar{I}_g$	0	.021	.181	.388	.563	.691	.757	.751	.675	.542	.370	.166	.018	0
	$\bar{I}_d$	0	.014	.070	.108	.134	.146	.152	.156	.152	.134	.106	.066	.011	0
D	$\bar{I}_g$	0	.011	.144	.336	.510	.640	.705	.702	.634	.503	.333	.136	.010	0
	$\bar{I}_d$	0	.009	.056	.097	.120	.126	.128	.133	.130	.116	.092	.054	.007	0

N.B. Units of $\bar{I}_g$ and $\bar{I}_d$: kWh/m^2-h

Let the area of the PV modules = A m^2. Then using the module efficiency of 14.5 per cent, the MPPT efficiency of 95 per cent and the inverter efficiency of 96 per cent, the energy output of the PV modules in January

$$= A \times 6.059 \times 31 \times 0.145 \times 0.95 \times 0.96$$

Repeating this calculation for all months and adding, we have the annual energy output

$$= A \times 0.145 \times 0.95 \times 0.96 \times [6.059 \times 31 + 6.460 \times 28 + 6.621 \times 31 + 6.598 \times 30 + 6.385 \times 31 + 5.219 \times 30 + 4.158 \times 31 + 3.998 \times 31 + 5.134 \times 30 + 6.174 \times 31 + 9.306 \times 30 + 8.877 \times 31]$$

$$= 301.378\, A \text{ kWh}$$

Equating this output with the required output, we have

$$301.378\, A = 10 \times 10^3 \times 8760 \times 0.17$$

$$A = 49\,413 \text{ m}^2 = 4.94 \text{ hectares (ha)}$$

Assuming that the slope of the modules is equal to the latitude, the module area projected on a horizontal plane

$$= A \cos 20° = 4.94 \times 0.94 = 4.64 \text{ ha}$$

The above calculations have been done under the assumption that the module efficiency, which is measured at a temperature of 25°C, has a constant value of 14.5 per cent. In reality, the temperature of the modules in India is some times much above 25°C and consequently the efficiency decreases. This fact may need to be taken into account by using experimental data showing the decrease of efficiency with temperature or simply by using a factor as is done in Example 8.2. It could lead to an increase of 10 to 20 per cent in the value of the module area required.

In order to calculate the land area required for the plant, one has to take account of the fact that open land has to be provided between the rows of modules of a PV plant and also around them. This has to be done to (1) avoid the shade of one module falling on another module in an adjoining row, (2) provide easy vehicular access to the modules for cleaning and maintenance and (3) provide space for locating balance of system components. Typically for MW-sized plants, about 1 hectare per MW is provided for these purposes. Thus for the present 10 MW plant, the total land area required = 4.64 + 10 = 14.64 ha. Land area required per MW capacity = 1.46 ha.

8.6.8 Concentrated Photovoltaic Systems

A novel approach in PV conversion systems is making use of concentrators to focus the sunlight on high efficiency solar cells. The concentration is achieved by using either linear or circular Fresnel lenses or parabolic or paraboloid concentrators, which focus along a line or at a point. Concentration ratios ranging from 2 to 1000 are being used. For a given conversion efficiency and output power, the collection area required is the same whether the sunlight is concentrated or not concentrated. However with concentrators, most of the area is made up of reflecting surfaces or lenses, which are likely to be more amenable to low cost and high volume production. Thus, the cells which are expensive do not account for a major fraction of the total system cost. Because of the concentration, the cells become hot and the cell conversion efficiency decreases with increasing temperature. For this reason, some form of cell cooling mechanism is employed.

Figure 8.21 shows a schematic of a concentrated photovoltaic (CPV) system. It has four basic components, viz. concentrator, cell, heat sink and supporting structure with tracking facility. Various types of concentrators based on Fresnel lenses or parabolic/paraboloid reflectors have been developed specially for CPV applications. The reader is referred to Shanks *et al.* [39] for more information on the development of concentrators for CPV applications.

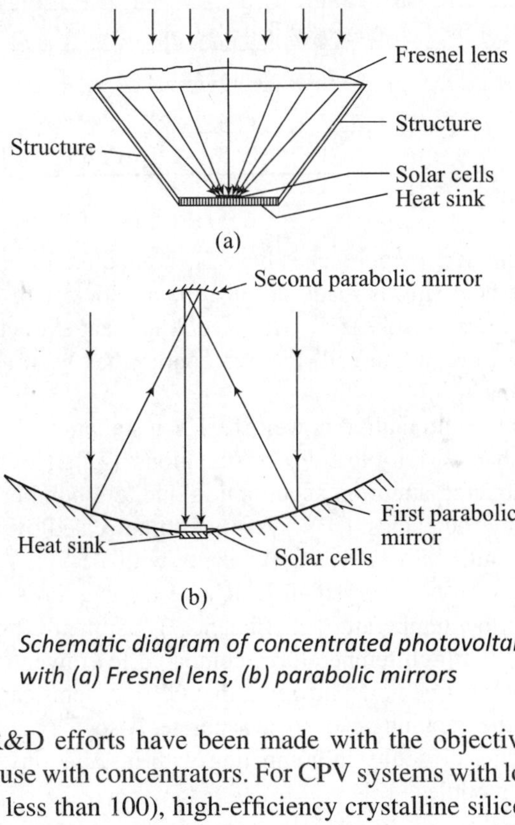

Fig. 8.21 *Schematic diagram of concentrated photovoltaic system with (a) Fresnel lens, (b) parabolic mirrors*

Significant R&D efforts have been made with the objective of developing special cells for use with concentrators. For CPV systems with low concentration ratios (typically less than 100), high-efficiency crystalline silicon solar cells are used. On the other hand, systems with high concentration ratios (between 300–1000) use III-V multijunction solar cells. Table 8.6 presents values of efficiencies of a few cells used for CPV applications. It is seen that the highest laboratory multijunction cell efficiency obtained is 46.0 per cent with an intensity of 508 suns, while the highest CPV module efficiency attained is 38.9 per cent with an intensity of 333 suns. More information on CPV cells is given by Philipps *et al.* [40] and Xing *et al.* [41].

Table 8.6 Efficiency values of concentrator PV cells and modules. Measurements made under global AM1.5 spectrum (1000 W/m^2) and at a cell temperature of 25°C [4].

Type	Area (cm^2)	Intensity* (suns)	Efficiency (%)
Single cell			
GaAs	0.0505	117	29.1
Si	1.0	92	27.6
Multijunction cell			
GaInP/GaAs	0.0520	508	46.0
GaInAsP/GaInAs	0.0520	508	46.0
GaInP/GaAs/ GaInAs/GaInAs	0.09709	234	45.7
InGaP/GaAs/InGaAs	0.1652	302	44.4
Modules			
Si	1875	79	20.5
Four junction	812.3	333	38.9

*1 sun = 1000 W/m^2

Both passive and active approaches have been adopted for cooling the cells. In the former, the heat sink is made up of a metal block with fins underneath for losing heat to the atmosphere. In active cooling, heat extraction is achieved by pumping a fluid below the cell. Passive cooling has been adopted in most commercial systems.

The current global cumulative power of CPV installations is only 360 MW, much less than that of flat-plate PV installations [42]. However, the CPV technology has attracted attention as an option due to a number of advantages. In addition to high efficiencies, CPV systems can be scaled over a wide range, i.e., from kW to multi-MW. Also, CPV plants with III-V multijunction cells are ideal for locations with very high DNI (greater than 2000 kWh/m^2-year), because they have low temperature coefficients. Thus, their efficiency does not reduce significantly at high temperatures compared to conventional crystalline silicon modules. It must be noted however that CPV systems cannot use diffuse radiation and that the tracking needs to be accurate. Also, they are expensive and the industry is facing difficulty in competing with installations using flat-plate crystalline silicon modules [42].

8.6.9 Satellite Solar Power Station

In 1968, Glaser [43, 44] made a novel suggestion for generating large amounts of electrical power using PV systems in space. He proposed the concept of a space solar power (SSP) station in which a satellite would be placed in a geosynchronous orbit far above the earth (Fig. 8.22). The satellite would consist of a very large array of photovoltaic modules. The DC power generated by these modules, when solar radiation falls on them, would be fed to microwave oscillators and large antennas would beam the output of the oscillators towards the earth. Antennas on the earth would receive the microwave energy, which would then be converted back to DC power. The principal advantage of this concept is that the array in space would not be subject to any day-night cycles if located properly, and that it would receive a much higher intensity of solar radiation than on earth.

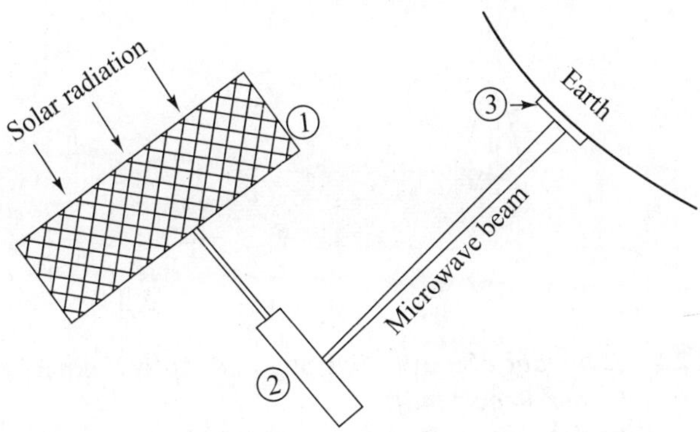

Fig. 8.22 *Satellite solar power station concept*
1. PV array
2. Converter and transmitting antenna
3. Receiving antenna

Various studies of the Glaser concept were conducted, a major one being a study conducted by the US Department of Energy and NASA in 1978 [45]. The study identified a reference 5 GW SSP station sketched in Fig. 8.23. The station consisted of an enormous 5 km × 10 km single platform structure to be assembled in space. On one side, the platform would be packed with PV module arrays and would be oriented by control systems to face the sun normally at all times. Single crystal silicon modules and gallium-aluminium-arsenide modules were considered as the promising options then. The DC power generated by the PV arrays would be converted by DC-RF amplifiers to a high power density microwave beam (1 km diameter) of suitable frequency and amplitude for transmission. In turn, the beam would be directed by a transmitting antenna towards a fixed location on the earth. The three systems, the PV array, DC-RF amplifiers and the transmitting antenna would be fixed to the platform structure and would in effect constitute a three-layered sandwich. In the process of transmission, the microwave beam would be spread out so that it would reach the earth as a low power density beam about 13 km in diameter. It would be intercepted by a receiving antenna, reconverted to DC and then AC power, and finally connected to the electricity grid. The report calculated that the complete power station would have an overall efficiency of about 7 per cent. The study showed that although the project was technically feasible, the estimated cost would be exorbitant. The scheme was, therefore, shelved for the time being.

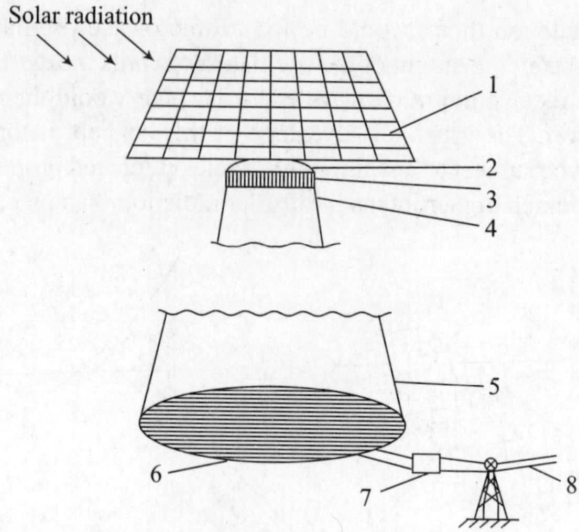

Fig. 8.23 *Schematic of 5 GW SSP station. (Adapted from Reference System Report [46])*
1. Platform for PV array (5 km x 10 km)
2. DC–RF amplifier
3. Transmitting antenna
4. Diverging microwave beam (1 km diameter)
5. Diverging microwave beam (13 km diameter)
6. Receiving antenna (on earth)
7. Conversion to electricity
8. Power transmission grid

NASA conducted a re-examination and took a 'fresh look' at the Glaser proposal in 1995-1997 [46]. The objective was to consider new technological developments which had taken place over the intervening period of about 20 years and examine if a SSP station with its associated systems could now be developed at a reasonably economical cost. The general finding was that there did appear to be a number of promising SSP concepts distinct from those of the 1970s. These concepts had been enabled by recent technological advances. The re-examination also showed that generating power in space had now became more viable than earlier. However, it was still too costly a project to be undertaken. The principal outcome of the 'fresh look' programme was the definition of a number of road maps for the possible development of SSP technologies and the initiation by NASA of a space solar power Exploratory Research and Technology (SERT) programme. This resulted in the initiation and evaluation of a wide range of selected activities to test the validity of the road maps mentioned [47].

Due to lack of funding by the US government of the SSP station as a whole, only a few smaller projects dealing with sub-systems have been executed over the past ten or fifteen years. Jaffe *et al.* [48] describe the efforts which have gone into developing designs of prototypes of sandwich modules to be used on a SSP.

Thermal and trade-off studies were conducted for each of the three functional layers: photovoltaic conversion, DC to RF conversion and power transmission via antennas, before their designs were finalized and they were tested. This was followed by assembly testing of a complete prototype. Such studies and others suggesting new concepts are continuing. For example, it has been suggested that solar concentrators be used in space stations in order to reduce the size of the PV arrays.

It is nearly fifty years since Glaser suggested the idea of building a large space solar power station. Today there are some who feel that the huge investments required will never decrease significantly. Criticisms of the concept entitled "Space solar power: an idea whose time will never come" have been published [49]. Counter opinions have also been offered suggesting that the huge government spending involved may be essential keeping the long term perspective of the energy problem in mind [50].

PHOTOVOLTAIC THERMAL COLLECTORS 8.7

A photovoltaic-thermal (PV/T) collector is a hybrid combination of a photovoltaic module and a solar thermal collector integrated together to produce electricity and collect heat simultaneously. PV cells absorb only a small part of the incident solar radiation (Section 8.2.1) to produce electricity. The remaining part is converted into heat, increasing the temperature of the cell. The increase in temperature causes a reduction in the cell efficiency. In order to reduce the temperature, a heat extraction mechanism is integrated thus producing heat output in the form of hot air or hot water. PV/T collectors thus can be designed by combining PV modules with either flat-plate or concentrating solar collectors. They are classified according to the heat removal fluid used as PV/T air collectors, PV/T water collectors, etc.

A schematic cross-section of a PV/T air collector is shown in Fig. 8.24. It consists of a PV module, which acts as the absorber of the collector. Solar radiation falls on the module after coming through a transparent cover. A parallel plate below the PV module forms a small passage through which the air to be heated flows. A sheet metal container filled with insulation is provided on the bottom and sides. Other arrangements of air flow, discussed in Section 5.2, are also possible alternatives. The transparent cover shown in Fig. 8.24 is also optional leading to an unglazed PV module.

In the case of a PV/T water collector, the top of the absorber plate of the liquid flat-plate collector (Fig. 2.1) is attached to the PV module using an appropriate adhesive. The grid of tubes fixed to the backside of the plate absorber plate carry the working fluid (water) to extract heat from the PV module. A sheet metal box with insulation on the bottom and sides holds all the components together. As with the air collector, the use of a transparent cover on top is optional.

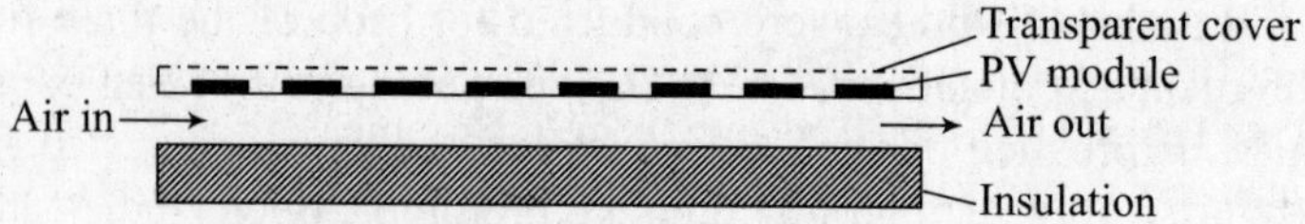

Fig. 8.24 *Cross-section of PV/T air collector*

Currently, apart from producing electricity, various types of PV/T collectors are in practical use for a number of thermal applications such as domestic water heating, space heating and process heating. Considerable R&D work on PV/T collectors has been carried out and excellent reviews are available in the literature [51-54]. In order to improve the performance of PV/T collectors, further research on PV materials to withstand stagnation and operating temperatures is needed. Other aspects worth investigating include reliable isolation of current and voltage carrying parts from the heat removal components, enhanced heat transfer between the PV module and the thermal collector and development of standards exclusively for PV/T collectors [55, 56].

Florschuetz [57] provided a thermal analysis of a PV/T flat-plate collector by extending the Hottel-Whillier-Bliss (HWB) model. He has assumed that the thermal resistance to heat flow from the PV module to the absorber plate is negligible. Consequently the temperature of the PV module is the same as the absorber plate. He has also assumed that the conversion efficiency of the PV cells (η) varies linearly with temperature (T) as follows:

$$\eta = \eta_o [1 - \beta (T - T_o)] \tag{8.10}$$

where η_o is the efficiency at a reference temperature T_o and β is a constant.

The net result is a modified form of the HWB Eq. (4.49). Subsequently, Raghuraman [58] has shown that the thermal resistance to heat flow from the PV cells to the absorber plate is not negligible because there is a layer of electrical insulation in between. He has analysed the performance of a liquid PV/T collector and an air PV/T collector in terms of a set of equations which have to be solved numerically. The results obtained agree reasonably well with the experimental data. Several investigations on the analysis of the performance of PV/T collectors have been reported since then. More recently, Hasan and Sumathy [59] have presented a review of these investigations.

PROBLEMS

8.1 The dark current density for a silicon solar cell at 40°C is 3.6×10^{-8} A/m^2 and the short circuit current density is 220 A/m^2. Calculate the voltage and current density that maximizes the power of the cell. What would be the corresponding maximum power output per unit cell area and the corresponding conversion efficiency if the global solar radiation incident on the cell is 850 W/m^2? Calculate the cell area required for an output of 36 W.

8.2 A PV system is installed near Pune (18.53°N, 73.85°E) for pumping water from a depth of 20 m. The solar cell array used consists of 20 modules, each 100 cm × 60 cm. The module 'standard test' conversion efficiency is 15 per cent. The orientation of the array is fixed and it is inclined at an angle of 18.53° with the surface azimuth angle being 0°. Assume that the values of incident solar radiation on May 15 from 0900 to 1600 h are as given in Table 4.1 and calculate the water discharge rate corresponding to each value. Assume further that this instantaneous rate is valid for half an hour on either side of the instant considered and thus calculate the amount of water pumped over the whole day from 0830 to 1630 h. Take the efficiency of the MPPT device to be 95 per cent, the efficiency of the DC-AC inverter to be 96 per cent and the efficiency of the electric motor-pump combination to be 50 per cent. Assume that the efficiency of the PV module decreases to 90 per cent of the standard test value because it is subjected to a higher temperature.

8.3 Assume that the slope of the solar cell array in Problem 8.2 is fixed at 18.53°, but the surface azimuth angle is changed in steps according to the following schedule:

IST (h)	0830–1030	1030–1430	1430–1630
γ (deg)	+45	0	–45

Calculate again the instantaneous water discharge rate on the hour and the amount of water pumped from 0830 to 1630 h. What is the percentage increase over the amount calculated in Problem 8.2?

8.4 Consider that the AC load in Fig. 8.17 is a 18 W LED street light. The PV array consists of 2 modules, each 60 cm × 40 cm, and the module 'standard test' conversion efficiency is 14.5 per cent. Such a system is installed in New Delhi and modules face south with $\beta = 34°$ and $\gamma = 0°$. Use the solar radiation data given in Table A3.1 for an average day in April and calculate the energy stored in the battery through the day. Assume:

- Battery charging efficiency = 95 per cent
- Battery discharging efficiency = 93 per cent
- Combined efficiency of MPPT and charge controller = 97 per cent
- Solar radiation is converted into electricity and stored from 0630 to 1730 h (LAT)
- The controller-cum-inverter switches the street light on at 1930 h (LAT)
- Efficiency of the DC-AC inverter = 94 per cent
- Driver efficiency of LED light = 86 per cent

Find the time when the street light will be switched off if the depth of battery discharge is 72 per cent. Assume that the efficiency of the PV module decreases to 85 per cent of the standard test value because it is subjected to a higher temperature.

8.5 A PV water pumping system is required to provide water at the rate of 11000 litres per hour. Calculate the number of PV modules required for the following conditions:

Solar flux incident on the modules	= 825 W/m^2
Size of one PV module	= 1.52 m × 0.98 m
PV module 'standard test' conversion efficiency	= 14.5 per cent
MPPT efficiency	= 94 per cent
DC-AC inverter efficiency	= 95 per cent
Efficiency of the motor-pump combination	= 45 per cent
Water depth	= 40 m

Assume that the efficiency of the PV module decreases to 90 per cent of the standard test value because it is subjected to a higher temperature.

8.6 The teaching laboratories of a college require AC power for lighting purposes for 6 hours during the day time. There are 60 tube lights to be powered by solar PV modules. Estimate the number of modules required. State your assumptions.

Given: Module 'standard test' conversion efficiency = 14.2 per cent
MPPT efficiency = 95 per cent
Size of one PV module = 1.0 m × 0.98 m
DC-AC inverter efficiency = 96 per cent

Each tube light consumes 36 W and uses an electromagnetic ballast which consumes 12 W.

Monthly average of daily solar radiation incident on the PV modules = 6.9 kWh/m^2.

8.7 A PV lighting system is installed in a building to meet the requirements of 500 W AC power for 5 hours every night. The PV array is connected through battery storage in order to meet the need. Estimate the number of modules required and the number of batteries needed for the following conditions:

Average daily global radiation available = 4.9 kWh/m^2
Battery rating 12 V; 120 A-h
Depth of battery discharge = 77 per cent
Battery charging efficiency = 94 per cent
Battery discharging efficiency = 92 per cent
Combined efficiency of MPPT and charge controller = 96 per cent
Inverter efficiency = 94 per cent
Size of each PV module = 119 cm × 53 cm
PV module 'standard test' conversion efficiency = 14.1 per cent

Assume that the efficiency of the PV module decreases to 90 per cent of the standard test value because it is subjected to a higher temperature.

How will the module and battery requirement increase if we wish to provide for an extra night following a cloudy day with no sunshine. Such an eventuality may occur on an average of once every ten days.

8.8 A solar PV LED lamp operates with a voltage of 6.4 V DC and consumes a current of 0.15 A. The lamp is required to operate for 6 hours every night. Calculate the rating and area of the PV module required which would collect enough energy through the day and operate satisfactorily through the specified period at night. Also, calculate the capacity of the battery if the battery voltage is 6 V.

Given: Minimum daily global radiation available = 3.8 kWh/m^2
PV module 'standard test' efficiency = 14.8 per cent
Battery charging efficiency = 95 per cent
Battery discharging efficiency = 93 per cent
Battery depth of discharge = 80 per cent
LED driver efficiency = 85 per cent

Assume that the efficiency of the PV module decreases to 85 per cent of the standard test value because it is subjected to a higher temperature.

8.9 Consider a grid-connected PV power plant of capacity 20 MWe with an annual plant load factor of 0.18 to be installed at a place (latitude = 29° N) near New Delhi. Estimate the module area and land area required for the plant. Use the solar radiation data of New Delhi given in Tables A3.1 and A3.2.

Given: The modules are tilted at an angle equal to the latitude and face due south.
PV module 'standard test' efficiency = 14.6 per cent
MPPT efficiency = 94 per cent
Inverter efficiency = 95 per cent

Assume that the efficiency of the PV module decreases to 85 per cent of the standard test value because it is subjected to a higher temperature.

8.10 Make suitable assumptions and calculate the PV module area required for a 5000 MW satellite solar power station operating in a geosynchronous orbit in space. Assume that: (i) the module efficiency is 16 per cent, and (ii) the overall efficiency of the devices used for converting DC power of the solar cells to microwave power and back is 70 per cent.

REFERENCES

1. *Photovoltaics Report.* 2016. Fraunhofer Institute for Solar Energy Systems ISE, Freiburg, Germany.
2. Osborne, M. 2014. The sorry state of PV manufacturing in India. (http://www.pv-tech.org/editors-blog/sorry_state_of_pv_manufacturing_in_india; accessed on July 4, 2016).
3. *Annual Report 2015-16.* Ministry of New and Renewable Energy, Government of India, New Delhi, India.
4. Green, M.A., Emery, K., Hishikawa, Y., Warta, W. and Dunlop, E.D. 2016. Solar cell efficiency tables (version 48). *Progress in Photovoltaics: Research and Applications,* 24: 905.
5. Sayyah, A., Horenstein, M.N. and Mazumdar, M.K. 2014. Energy yield loss caused by dust deposition on photovoltaic panels. *Solar Energy,* 107: 576.
6. Ghazi, S., Sayigh, A. and Ip, K. 2014. Dust effect on flat surfaces - a review paper. *Renewable and Sustainable Energy Reviews,* 33: 742.
7. Tanesab, J., Parlevliet, D., Whale, J., Urmee, T. and Pryor, T. 2015. The contribution of dust to performance degradation of PV modules in a temperate climate zone. *Solar Energy,* 120: 147.
8. *Data Sheet ISL Sapphire Series 285 W.* (http://www.indosolar.co.in; accessed on July 7, 2016)
9. Solanki, C.S. 2015. *Solar Photovoltaics: Fundamentals, Technologies and Applications,* 3rd Edition, PHI Learning Pvt. Ltd., New Delhi, India.
10. Ramanujam, J., Verma, A., Gonzalez-Diaz, B., Guerrero-Lemus, R., Canizo, C. D., Garcia-Tabares, E., Rey-Stolle, I., Granek, F., Korte, L., Tucci, M., Rath, J., Singh, U. P., Todorov, T., Gunawan, O., Rubio, S., Plaza, J. L., Dieguez, E., Hoffmann, B., Christiansen, S., and Cirlin, G. E. 2016. Inorganic photovoltaics - planar and nanostructured devices. *Progress in Material Science,* 82: 294.
11. Glunz, S.W., Preu, R. and Biro, D. 2012. Crystalline silicon solar cells - state-of-the-art and future developments. Chapter in *Photovoltaic Solar Energy Volume 1* of Comprehensive Renewable Energy (Volume editor: W.G.J.H.M. van Sark), Elsevier B.V., Amsterdam, The Netherlands.

12. www.firstsolar.com; accessed on July 25, 2016.

13. Kazmerski, L. L., 2006. Solar photovoltaics R&D at the tipping point: a 2005 technology review. *Journal of Electron Spectroscopy and Related Phenomena,* 150: 105.

14. www.solar-frontier.com; accessed on July 25, 2016

15. Saifullah, M., Gwak, J., and Yun, J.H. 2016. Comprehensive review on material requirements, present status, and future prospects for building - integrated semi-transparent photovoltaics (BISTV). *Journal of Materials Chemistry A,* 4: 8512.

16. Niu, X., Wang, M., Zhu, X., Yu, C., Li, G., Cheng, B., Liu, S., Rong, J., Chen, Z., Wu, L., Chen, C., Xu, B., Fung, K., Wronski, C. and Yang, L. 2011. Progress in research and mass production of large-scale thin film Si solar cells, *Proc. of IEEE,* 978-1-4673-0066-7/12, 001501.

17. Guha, S., Yang, J. and Yan, B. 2013. High efficiency multi-junction thin film silicon cells incorporating nanocrystalline silicon. *Solar Energy Materials and Solar Cells,* 119: 1.

18. Sai, H., Matsui, T., Koida, T., Matsubara, K., Kondo, M., Sugiyamo, S., Katayama, H., Takeuchi, Y. and Yoshida, I. 2015. Triple-junction thin-film silicon solar cell fabricated on periodically textured substrate with a stabilized efficiency of 13.6%. *Applied Physics Letters,* 106: 213902.

19. Zheng, Z., Ji. H., Yu, P. and Wang, Z. 2016. Recent progress towards quantum dot solar cells with enhanced optical absorption. *Nanoscale Research Letters,* 11: 266.

20. Guerro-Lemus, R., Vega, R., Kim, A. and Shephard, L. E. 2016. Bifacial solar photovoltaics - a technology review. *Renewable and Sustainable Energy Reviews,* 60: 1533.

21. Qin, P., Tanaka, S., Ito, S., Tetreault, N., Manabe, K., Nishino, H., Nazeeruddin, M. K. and Gratzel, M. 2014. Inorganic hole conductor-based lead halide perovskite solar cell with 12.4% conversion efficiency. *Nature Communications,* 5: 3834.

22. Park, N. 2014. Perovskite solar cells: an emerging photovoltaic technology. *Materials Today,* 18(2): 65.

23. Petrovic, M., Chellappan, V. and Ramakrishna, S. 2015. Perovskites: solar cells & engineering applications - materials and device developments. *Solar Energy,* 122: 678.

24. Liu, T., Chen, K., Hu, Q., Zhu, R. and Gong, Q. 2016. Inverted perovskite solar cells: progress and perspectives. *Advanced Energy Materials,* 1600457.

25. Habibi, M., Zabihi, F., Ahmadian-Yazdi, M. R. and Eslamian, M. 2016. Progress in emerging solution-processed thin film solar cells - part II: perovskite cells. *Renewable and Sustainable Energy Reviews,* 62: 1012.

26. Carle, J. E. and Krebs, F. C. 2013. Technological status of organic photovoltaics (OPV). *Solar Energy Materials and Solar Cells,* 119: 309.

27. Kaur, N., Singh, M., Pathak, D., Wagner, T. and Nunzi, J. M. 2014. Organic materials for photovoltaic applications: review and mechanism. *Synthetic Metals,* 190: 20.

28. Liu, X., Chen, H. and Tan, S. 2015. Overview of high-efficiency organic photovoltaic materials and devices. *Renewable and Sustainable Energy Reviews,* 52: 1527.

29. Wang, Q., Xie, Y., Soltani-Kordshuli, F. and Eslamian, M. 2016. Progress in emerging solution-processed thin film solar cells - part I: polymer solar cells. *Renewable and Sustainable Energy Reviews,* 56: 347.

30. Sugathan, V., John, E. and Sudhakar, K. 2015. Recent improvements in dye sensitized solar cells: a review. *Renewable and Sustainable Energy Reviews,* 52: 54.

31. Shalini, S., Prabhu, R.B., Prasanna, S., Mallick, T.K. and Senthilarasu, S. 2015. Review on natural dye sensitized solar cells: operation, materials and methods. *Renewable and Sustainable Energy Reviews,* 51: 1306.

32. *Solar Electricity Cost vs. Regular Electricity Cost.* 2016. (http://solarcellcentral.com/cost_page.html; accessed on July 27, 2016)

33. *Trend Analysis on Solar PV Module Prices.* 2016. (http://www.ezysolare.com/blog/knowledge-centre/trend-analysis-on-solar-pv-module-prices/; accessed on July 27, 2016)

34. *Determination of Bench Mark Capital Cost Norm for Solar Power.* 2016. (http://www.cercind.gov.in/2016/orders/SO17.pdf; accessed on July 27, 2016)

35. *Frequently Asked Questions.* 2016. Solar Energy Corporation of India. (http://www.seci.gov.in/FAQ; accessed on July 27, 2016)

36. Arora, N., Joshi, L., Tejwani, R., Chauhan, A., Venkateswaran, J., Narayanan, N. C. and Solanki, C. S. 2015. A model of localisation of solar energy for providing 'right to clean light' to rural school-going children. *Proc. 5th International Conference on Advances in Energy Research, ICAER 2015,* December 15-17, IIT Bombay, India.

37. *Energy Statistics, 23rd Issue.* 2016. Central Statistics Office, Ministry of Statistics and Programme Implementation, Government of India, New Delhi, India.

38. Yaqoot, M., Diwan, P. and Kandpal, T.C. 2015. Public financing solar lanterns versus subsidizing kerosene for domestic lighting in India: a feasibility assessment. *Renewable Energy,* 80: 564.

39. Shanks, K., Senthilarasu, S. and Mallick, T.K. 2016. Optics for concentrating photovoltaics: trends, limits and opportunities for materials and design. *Renewable and Sustainable Energy Reviews,* 60: 394.

40. Philipps, S. P., Guter, W., Welser, E., Schone, J., Steiner, M., Dimroth, F. and Bett, A.W. 2012. Present staus in the development of III-V multijunction solar cells. Chapter 1 in *Next Generation Photovoltaics.* Lopez, A.B.C., Vega, A. M. and Lopez, A. L. (eds.), Springer-Verlag, Berlin, Germany.

41. Xing, Y., Han, P., Wang, S., Liang, P., Lou, S., Zhang, Y., Hu, S., Zhu, H., Zhao, C. and Mi, Y. 2015. A review of concentrator silicon solar cells. *Renewable and Sustainable Energy Reviews,* 51: 1697.

42. Philipps, S. P., Bett, A.W., Horowitz, K. and Kurtz, S. 2016. Current status of concentrator photovoltaic (CPV) technology. *Fraunhofer ISE/NREL CPV Report 1.2,* TP-6A20-63916.

43. Glaser, P.E. 1968, Power from the sun: its future. *Science,* 162: 957.

44. Glaser, P.E. 1978. Solar power satellites: an option for power generation on earth. *Proc. of American Power Conference,* 40: 444.

45. Satellite Power System: Concept Development and Evaluation program, *Report No. DOE/ER-0023,* U.S. Department of Energy and NASA, 1978. (http://www.nss.org/settlement/ssp/library/1978DOESPS-ReferenceSystems Report.pdf; accessed on September 10, 2016)

46. Mankins, J. C. 1997. A fresh look at space solar power: new architectures, concepts and technologies. *Proc. 38th International Astronomical Federation,* paper IAF-97-R.2.03.

(http://www.nss.org/settlement/ssp/library/1977-mankins-FreshlookatSpaceSolar Power.pdf; accessed on September 10, 2016)

47. Laying the Foundation for Space Solar Power: An Assessment of NASA's Space Solar Power Investment Strategy, *Report of a Committee of the National Research Council,* 2001.

((http://www.nap.edu/catalog/10202.html; accessed on September 10, 2016)

48. Jaffe, P., Hodkin, J., Harrington, F., Person, C., Nurnberger, M., Nguyen, B., LaCava, S., Scheiman, D., Stewart, G., Han, A., Hettwer, E. and Rhoades, D. 2014. Sandwich module prototype progress for space solar power. *Acta Astronautica,* 94: 662.

49. Fetter, S. 2004. Space solar power: an idea whose time will never come. *Phys. Soc.,* 33(1): 10.

50. Globus, A. 2009. In defence of space solar power. (http://www.space. aglobus.net/papers/FetterResponse.html; accessed on September 10, 2016)

51. Zondag, H. A. 2008. Flat-plate PV-thermal collectors and systems a review. *Renewable and Sustainable Energy Reviews,* 12: 891.

52. Riffat, S. B. and Cue, E. 2011. A review on hybrid photovoltaic/ thermal collectors and systems. *International Journal of Low-Carbon Technologies,* 6: 212.

53. Michael, J. J., Iniyan, S. and Goic, R. 2015. Flat plate solar photovoltaic-thermal (PV/T) systems: a reference guide. *Renewable and Sustainable Energy Reviews,* 51: 62.

54. Buker, M. S. and Riffat, S. B. 2015. Building integrated solar thermal collectors - a review. *Renewable and Sustainable Energy Reviews,* 51: 327.

55. Fortuin, S., Hermann, M., Stryi-Hipp, G., Nitz, P. and Plater, W. 2014. Hybrid PV-thermal collector development: concepts, experiences, results and research needs. *Energy Procedia,* 48: 37.

56. Kramer, K. and Helmers, H. 2013. The interaction of standards and innovation: hybrid photovoltaic-thermal collectors. *Solar Energy,* 98: 434.

57. Florschuetz, L.W. 1979. Extension of the Hottel-Whillier model to the analysis of combined photovoltaic/thermal flat plate collectors. *Solar Energy,* 22: 361.

58. Raghuraman, P. 1981. Analytical predictions of liquid and air photovoltaic/thermal flat-plate collector performance. *Journal of Solar Energy Engineering, Trans. ASME,* 103: 291.

59. Hasan, M. A. and Sumathy, K. 2010. Photovoltaic thermal module concepts and their performance analysis: a review. *Renewable and Sustainable Energy Reviews,* 14: 1845.

Other Renewable Energy Sources

CHAPTER 9

INTRODUCTION 9.1

We now turn our attention to renewable energy sources which derive their energy indirectly from the sun. These are wind energy, energy from biomass, small hydroelectric power stations and ocean energy. The term 'ocean energy' includes four sources – tidal energy, wave energy, energy associated with temperature differences in the ocean and energy in marine currents. We describe the nature of these sources, the principles behind extracting energy from them and some of the devices used. Based on the technological developments thus far, we also estimate the potential of these sources for harnessing energy in the future. Finally, we discuss in quantitative terms whether all the renewable energy sources together (including solar thermal energy and PV conversion), can eventually meet India's future needs of electricity as estimated in Chapter 1.

WIND ENERGY 9.2

9.2.1 Introduction

❑ *Nature of Wind* Winds are caused because of two factors, (i) the absorption of solar energy on the earth's surface and in the atmosphere, and (ii) the rotation of the earth about its axis and its motion around the sun. Because of these factors, alternate heating and cooling cycles occur, differences in pressure are created, and the air is caused to move. Wind energy is thus an indirect manifestation of solar energy. The advantages of using wind energy are that its potential as a source of power is good and that its capture produces no pollution. The problems associated with utilizing it are the same as those associated with using solar energy directly. Firstly, the energy is available in dilute form, and secondly, the availability of the energy varies considerably over a day and with the seasons. For these reasons, the face areas of machines, which extract energy from the wind, have to be necessarily large and a continuous supply of mechanical or electrical power cannot be obtained from them.

❑ *History* Historically, wind energy was used for sailing ships and for generating small amounts of mechanical power through wind mills. One of the most popular and long lasting designs was that of the horizontal axis four blade Dutch wind mill. This design was used extensively in Europe from 1200 AD onwards for a few hundred years for grinding grain and for pumping water. Dutch settlers brought the design with them to the USA around 1700 AD and over the years, by about 1850 AD, the popular horizontal axis multi-blade type evolved. A few million wind mills of this type usually generating less than 1 kW of power were in use in the United States at the beginning of the last century. They were generally used to pump water or to generate electricity. With the advent of central power plants (based on hydroelectric or thermal power) and the spread of rural electrification, wind mills started to become economically unviable around 1920. Although a number of new designs were developed, wind energy ceased to be an effective means of supplying power by 1940. The new designs developed include the two- or three-blade propeller type design, and the Savonius and Darrieus types. In contrast to the earlier designs, the last two were vertical axis machines. The late 1930s also saw the construction of the largest wind machine, the Smith Putnam machine. This was a horizontal axis two-blade propeller type machine. It had a 53.3 m (175 ft) diameter rotor and generated 1250 kW of electrical power at its rated speed of 28 rpm. After a few years of intermittent operation, during which there were some major mechanical failures, the machine was finally abandoned as being economically unviable.

❑ *Recent Developments* With the advent of the oil crisis in 1973, interest in wind energy picked up considerably. A number of research and development programmes to build and test both small and large-scale wind energy conversion systems were undertaken in many countries. As a result, many commercial models ranging in capacity from a few watts to megawatts are now available. In the last two decades, wind energy has again become cost effective and the installed capacity has increased rapidly touching 432 883 MW in 2015. Most of the machines installed are of the two- or three-blade propeller type. These machines are usually located side by side in suitable wind swept locations, the clusters being referred to as wind farms. In the last few years, a significant number of machines have also been installed off-shore, the installed capacity touching 12 107 MW in 2015 [1].

Important developments have also taken place in India in the last twenty years. In the 1980s, a beginning was made with propeller type machines by erecting demonstration wind farms at five coastal locations. These had an installed capacity of 3 MW, about 0.5 MW being generated at each location using a number of 55 kW machines. Subsequently other wind farms using machines of much larger capacity have been set up (refer Plate 9.1). Typically, the sizes of

Plate 9.1 *Wind farm at Chalkewadi, Satara, Maharashtra. (Installed capacity: 28 MWe; Number of wind machines: 76).*

machines installed lie in the range of 1.2 to 3 MW and the total installed capacity in the country was 25 088 MW at the end of 2015 [2]. So far, this is all on land. Much of this capacity is in Tamil Nadu, Maharashtra, Gujarat, Rajasthan and Karnataka. The practice followed in wind farms is to connect the machines to the electricity grid and to feed power from the machines to the grid whenever available. The average cost of installation is currently about Rs 55 000 to 60 000 per kW (2015). This is considered to be reasonably cost effective and yields a payback period of 4 to 5 years at the current prices paid by the State Electricity Boards for feeding power to the grids. In addition to being connected to the grid, smaller wind machines of 0.5 to 10 kW are being used in stand-alone systems to supply localized needs. These are now manufactured in India. Wind-diesel and wind-solar PV hybrid projects have been set up at remote places where earlier electricity was being generated using diesel alone. Designs of small capacity multi-blade wind mills for pumping water have also been standardized and more than 1400 units are installed across the country.

9.2.2 Estimation of Wind Resource

Accurate mapping of the wind resource available in a country is an essential requirement for the successful use of wind energy. Since measurements of wind parameters may not be possible at all locations, the use of various types of simulation models has been resorted to for resource mapping.

As mentioned in Section 1.3.4, the annual report of MNRE for 2015-16 states that the potential for wind power in India is 49 130 and 102 788 MW at hub heights of 50 and 80 m respectively [2]. Phadke *et al.* [3] have adopted a more refined approach for estimating the potential of wind power in India. They have split the whole area of India into 5 km × 5 km cells and identified those cells where the threshold value of 200 W/m^2 for wind power density is available at a hub height of 80 m. This data has been procured from 3 Tier [4]. Further, using GIS data on topography, land use and land cover, Phadke *et al.* have separated out the cells where it would be feasible to set up wind farms. Assuming a generation capacity of 9 MW per sq km, the overall conclusion reached was that the total wind potential for India was 2006 GW with an APLF of 0.2 or more.

Hossain *et al.* [5] have followed a procedure which is similar in many respects to that adopted by Phadke *et al.* Wind speed data was obtained from measurements made by NIWE and NCEP/NCAR and a wind speed map was generated using appropriate simulation and modelling. GIS data was used for estimating the land area which could be available for wind farms. The wind energy potential was estimated for 2 MW machines with a rotor diameter of 100 m and a hub height of 100 m. It was calculated that with these machines a generation capacity of 5.7 MW could be installed per sq km and that with the land area available, the wind energy potential was 2161 GW.

Apart from the wind potential on land, a significant potential exists off-shore where the wind speeds are usually higher and steadier. Some European countries like Denmark, Germany and UK have already built wind farms off-shore. In India,

MNRE has initiated steps for estimating the off-shore wind potential. Hossain *et al.* [5] have assessed the off-shore potential to be 555 GW if wind machines are installed upto depths of 20 m. An approximate figure is also available from a study conducted with respect to some countries of the world by Liu *et al.* [6]. This research yields an off-shore potential of 1100 TWh for India, an output which is equivalent to an installed capacity of 630 GW working with an APLF of 0.20.

9.2.3 Classification and Descriptions of Wind Machines

❑ *Classification* Wind machines are generally classified in terms of the orientation of the axis of rotation of their rotors as horizontal axis machines and vertical axis machines. In a horizontal axis machine, the rotor axis is horizontal and is continuously adjusted in a horizontal plane so that it is parallel to the direction of the wind stream. On the other hand, in a vertical axis machine, the rotor axis is vertical and fixed, and is perpendicular to both the surface of the earth and the wind stream.

❑ *Rotors* Various types of rotors used in wind machines are shown in Fig. 9.1. These are (i) multi-blade type, (ii) propeller type, (iii) Savonius type, and (iv) Darrieus type. The first two are used in horizontal axis machines, and the last two in vertical axis machines. The multi-blade rotor consists of a number of curved sheet metal blades which increase in width going outwards from the centre. The number of blades usually ranges from 12 to 20. They are fixed at their inner end to a circular rim. They are also fixed near their outer edge to a second rim, which provides support. The diameter of the rotor usually ranges from 2 to 5 m.

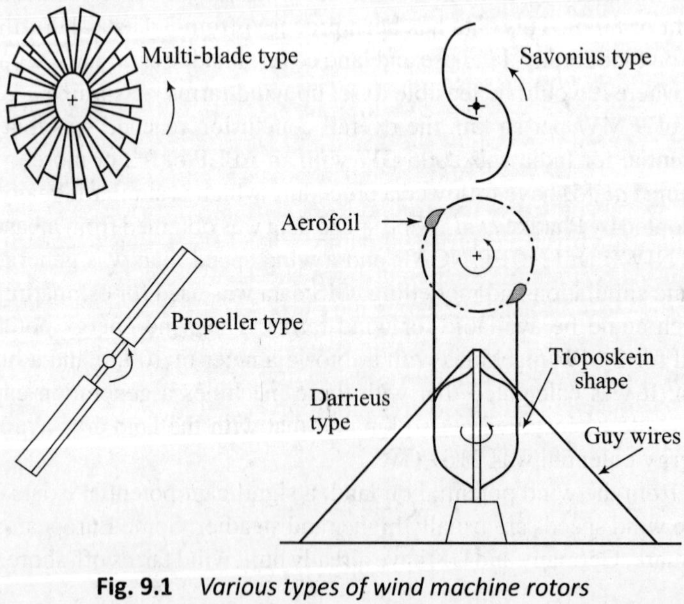

Fig. 9.1 *Various types of wind machine rotors*

In contrast to a multi-blade rotor, the propeller rotor usually has a large diameter and consists of only two or three blades made from fibreglass reinforced plastic. The blades have aerofoil sections with a high thickness-to-chord ratio and yield a high lift relative to the drag. Commercial models have rotor diameters ranging from 30 to 164 m.

The Savonius rotor consists essentially of a hollow cylinder (approximately elliptical in shape) sliced in half, the two halves being fixed to a vertical axis with a gap in between to make an S-shape. Torque is produced by the pressure difference between the two sides of the half facing the wind.

The Darrieus rotor is shaped somewhat like an egg beater. It consists of two or three blades having an aerofoil cross-section. Along the length, the blades are curved into a shape called a troposkein. The troposkein is the shape which a rope would take up if rotated about a vertical axis.

It is to be noted that both the Savonius and Darrieus type run independently of the direction of the wind because they rotate about a vertical axis. On the other hand, horizontal axis machines have to face the direction of the wind in order to generate power.

❑ *Multi-blade Type Wind Machine* A simplified sketch of a multi-blade type wind machine is shown in Fig. 9.2. The main elements are the rotor, the wind mill head (casing), the tail vane, the transmission system and the supporting structure (tower). The machine is normally used for pumping water. Thus, an additional component at the base of the tower is a reciprocating water pump.

The rotor overhangs at one end of the shaft emerging from the wind mill head. The centre of the rotor is referred to as the hub. Just behind it is the front bearing of the machine.

The transmission system consists of a power shaft, a speed reducing gear drive, a crank shaft-connecting rod mechanism and a pump rod. The gear drive reduces the rotational speed by a factor of 3 or 4. The rotational motion is then converted to a reciprocating motion by the crank shaft-connecting rod mechanism which in turn is connected by means of the pump rod to the water pump. The pump rod passes through the base of the wind mill head which sits on a ball-bearing turn table. Sometimes the motion of the rotor is not converted to a reciprocating motion. It is transmitted to the ground level as rotational motion by using bevel gears and used to drive a centrifugal pump.

The tail-vane, a simple shape cut from sheet metal, is attached to the back side of the wind mill head. It serves the purpose of aligning the rotor with the direction of the wind.

The tower is usually a steel truss construction. It serves the purpose of keeping the rotor and the wind mill head at an appropriate height from the ground where the wind speed is adequate.

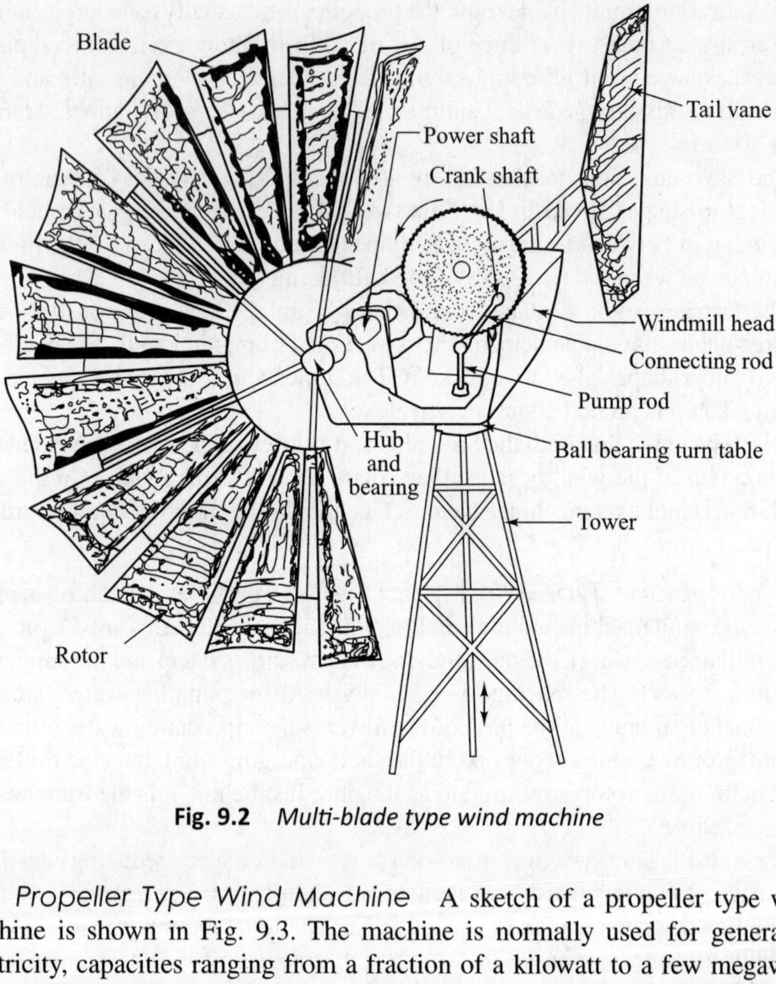

Fig. 9.2 *Multi-blade type wind machine*

❑ *Propeller Type Wind Machine* A sketch of a propeller type wind machine is shown in Fig. 9.3. The machine is normally used for generating electricity, capacities ranging from a fraction of a kilowatt to a few megawatts being available. The main elements of the machine as seen from the outside are the rotor, the nacelle and the tower.

The rotor consists of the hub and blades of the wind turbine. Generally, the blades are made from composites like fiberglass reinforced plastics. The power extracted from the wind by the blades is transmitted through the hub to a gear train and then on to a generator. The last two components, viz. the gear train and the generator are housed inside the nacelle. The nacelle also houses the rotor bearing, the braking mechanism and the roller assembly linking the nacelle to the tower. The roller assembly permits rotation of the nacelle about a vertical axis and helps to align the rotor with the direction of the wind.

As seen in Fig. 9.3, the tower is of the tubular type. This design is usually adopted for large capacity machines; In some cases, the diameter is large enough to permit ascent through an internal staircase. For small capacity machines, a steel truss design is used.

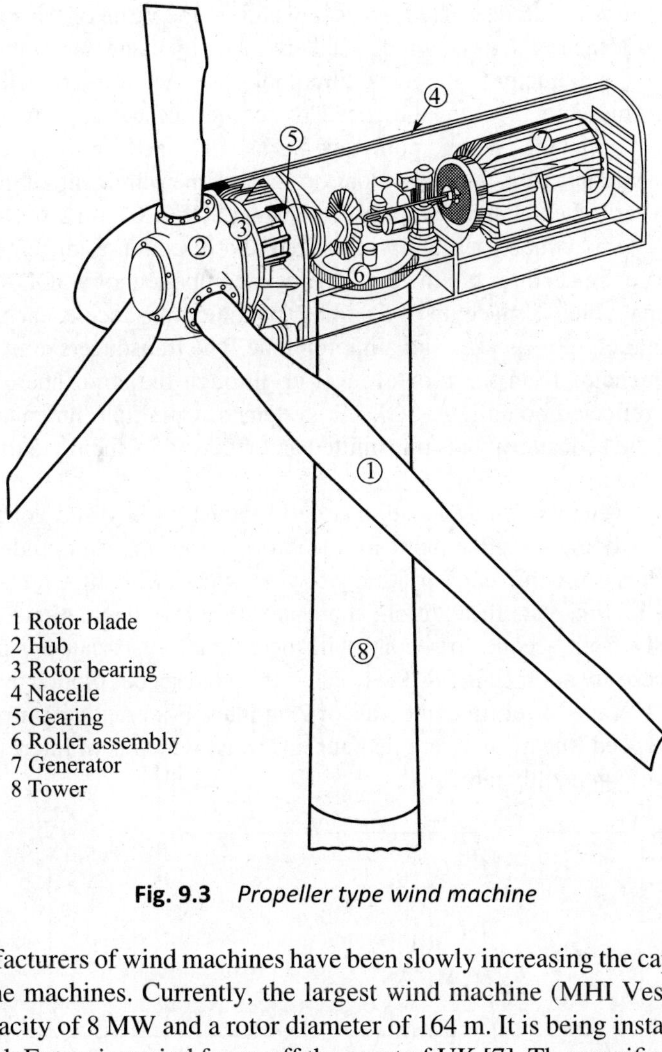

Fig. 9.3 *Propeller type wind machine*

Manufacturers of wind machines have been slowly increasing the capacity and size of the machines. Currently, the largest wind machine (MHI Vestas V164) has a capacity of 8 MW and a rotor diameter of 164 m. It is being installed in the Buro Bank Extension wind farms off the coast of UK [7]. The specifications of a large wind machine installed recently in India are as follows [8]:

Rated capacity	: 2.1 MW	Rotor diameter	: 97 m
Cut-in wind speed*	: 3.5 m/s	Hub height	: 90 m
Rated wind speed*	: 11 m/s	Power control	: Active blade pitching
Cut-out wind speed*	: 20 m/s	Generator type	: Asynchronous
Number of blades	: 3		

9.2.4 Wind Data

Wind velocity is a vector quantity, specified by its magnitude and direction. From the point of view of extracting energy from the wind, the horizontal component of

*These terms are defined later in Section 9.2.4.

the magnitude is of interest*. Hence, by convention, the value of this component is tabulated. Measurements of wind speed are made by using anemometers. The rotating cup anemometer is the most commonly used wind instrument. Various methods are used to measure the speed of rotation of the cups to which the wind speed is related. The anemometers are usually fixed on tall mast towers. Satellite communication permits monitoring and measurement of the speed. In recent years, there has been a shift to using SODAR (sonic detection and ranging) systems which measure wind speed and direction at different heights and eliminate the need for tall towers. The main component of a SODAR system is its antenna which is made up of a number of sonic transducers, each of which is a combination of a speaker and a microphone. The transducers emit pulses of sound frequencies from the ground level up through the atmosphere and then detect the reflected sounds. A SODAR system has multiple antennas and an analysis of the acoustic signals transmitted and received by them yields the wind speed and direction.

The wind velocity at any location usually varies rapidly and continuously, the variation being irregular both in terms of period and amplitude. Typical measurements showing the variation of wind speed with time at a location are shown in Fig. 9.4. In addition, there are turbulent fluctuations occurring continuously from second to second with occasional gusts causing peaks and valleys. These are not seen in Fig. 9.4 because the wind speeds plotted in the figure are values averaged over time intervals of 2 minutes. For practical purposes, it is the hourly speed (the mean over one hour intervals) which is of interest and this is the quantity generally tabulated.

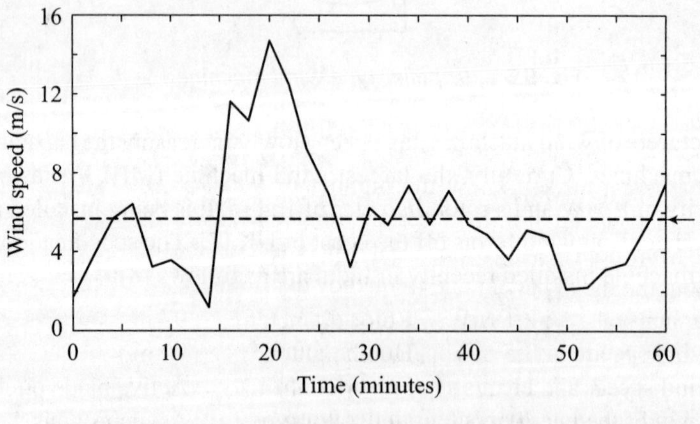

Fig. 9.4 *Typical variation of wind speed with time. Values plotted are averages over 2 minute intervals. Mean hourly speed for data presented = 5.6 m/s.*

❑ *Presentation of Data* In the past, measurements of wind speed in India were made for many years at various stations by the Indian Meteorological Department. These have been analyzed and data showing the diurnal variation

*The direction of the wind is not of much interest and will not be mentioned.

of the mean hourly wind speed (in kmph) for each month and for the year are available for 37 stations across the country [9]. A study of the tables shows that in many locations, the diurnal variation of the wind speed passes through a maximum in the late afternoon or early evening hours (Fig. 9.5). This feature could be advantageously exploited if in future large MW-sized hybrid power stations involving both solar (PV and thermal) and wind energy are built. The fact that the maximum for solar radiation occurs around noon while that for-wind speed occurs five or six hours later, would result in a smaller size for the solar thermal storage system needed for 24×7 operation (see Section 8.6.6). Currently, under the wind resource assessment programme of MNRE, NIWE in association with state nodal agencies has installed 806 dedicated wind monitoring stations [2]. The measured data have been published by NIWE in the form of handbooks, titled as Wind Energy: Resource Survey of India, which can be procured from them.

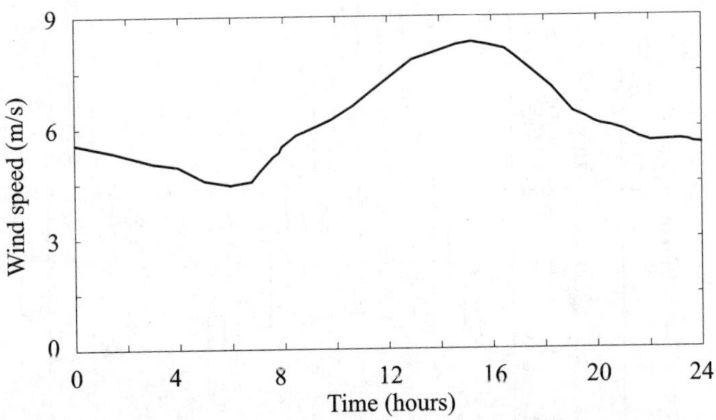

Fig. 9.5 *Typical diurnal variation of mean hourly wind speed at Thoothukudi in July (Adapted from Mani [10])*

It is of interest also to know the number of hours in a month or in a year when the wind speed lies in a certain range. This information can also be computed from hourly wind speed data. It is usually presented in the form of tables giving the mean percentage frequency distributions of hourly wind speeds in 2 kmph intervals. Values for two locations, Kandla in Gujarat and Indore in Madhya Pradesh, are given in Appendix 5. Kandla and Indore have strong winds throughout the year and are suitable places for locating wind machines from the point of view of the energy content in the wind.

The number of hours per month or per year during which the hourly wind speed lies in a certain range can readily be calculated by multiplying the frequency data by the total number of hours in the month or in a year. Thus, for example, from Table A5.1 we see that at Kandla, the wind speed is zero for (0.012 × 744 =) 8.9 h in January and for (0.032 × 8760 =) 280.3 h in the whole year. At Indore (Table A5.2), the wind speed lies in the range of 18 to 20 kmph for (0.087 × 744 =) 64.7 h in January and for (0.081 × 8760 =) 709.6 h in the whole year.

Based on the data given in Table A5.1, the annual frequency distribution of hourly wind speed for Kandla is plotted in Fig. 9.6. It is called the *speed-frequency distribution.* The distribution passes through a maximum for a particular range of speeds. For Kandla, the maximum occurs for hourly wind speeds in the range of 18 to 20 kmph. These occur for 710 h in a year. It is also seen that there are many calm periods (zero wind speed) at Kandla adding up to 280 h in the year. The point A marked on the y-axis indicates the total extent of the calm periods. The shape of the speed-frequency distribution at most locations is similar to the one drawn in Fig. 9.6. Speed-frequency distributions can also be drawn on a monthly basis to study the characteristics of wind speed available in a particular month.

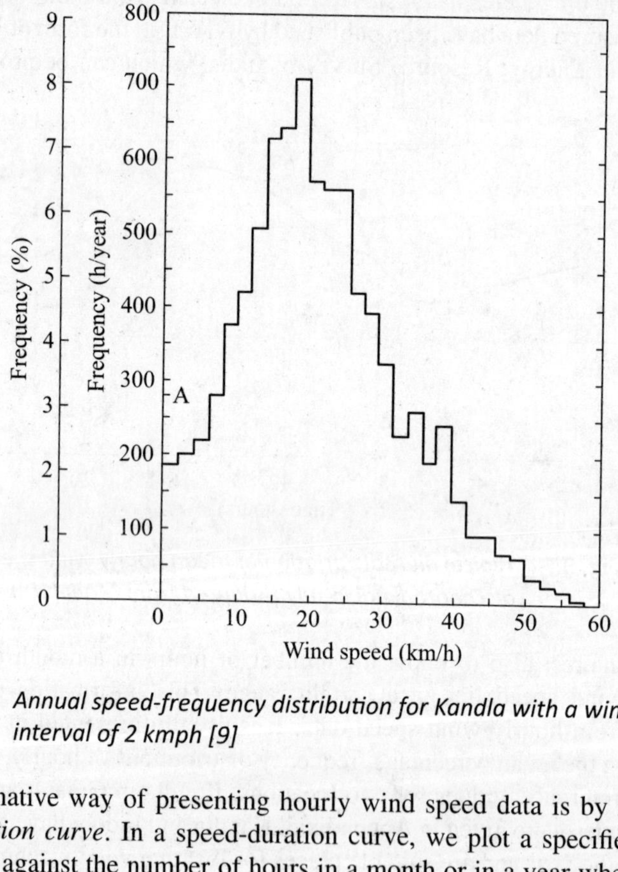

Fig. 9.6 *Annual speed-frequency distribution for Kandla with a wind speed interval of 2 kmph [9]*

An alternative way of presenting hourly wind speed data is by means of a *speed-duration curve*. In a speed-duration curve, we plot a specified value of wind speed against the number of hours in a month or in a year when the wind speed equals or exceeds the specified value. The usual convention is to plot the duration on the *x*-axis and the wind speed on the *y*-axis. The annual speed-duration curve* for Kandla is plotted in Fig. 9.7. It is seen that the curve begins

*Although the word 'curve' is used, one obtains a step-wise variation because of the use of the discrete interval of 2 kmph in wind speed. The height of the steps would reduce and a smooth curve would be obtained in the limiting case as the interval tends to zero. A curve could also be obtained by joining the mid-points of the horizontal parts of the steps.

with the point having the coordinates (8760, 0). The curve coincides with the x-axis for a time period corresponding to the calm period of zero wind speed. Thereafter, the curve rises to the left with decreasing slope, passes through a point of inflection and finally merges with the y-axis. Curves having a similar shape are obtained at most locations. Speed-duration curves can also be drawn on a monthly basis.

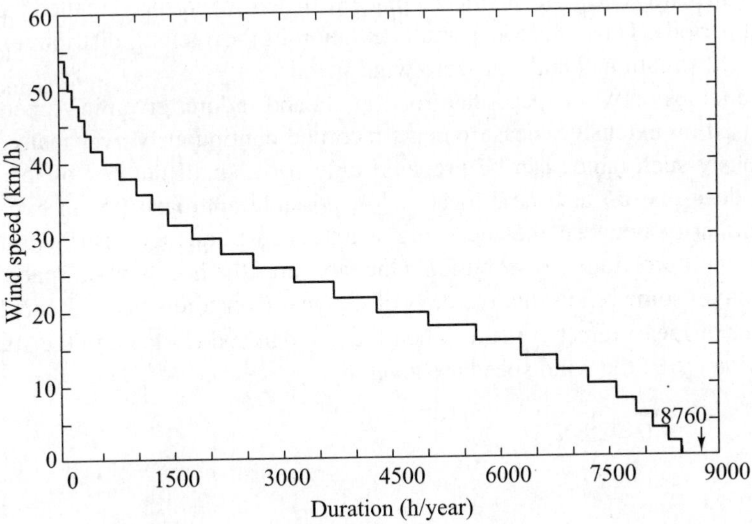

Fig. 9.7 *Annual speed-duration curve for Kandla*

The annual speed-frequency distribution or annual speed-duration curve is useful for assessing the overall suitability of a place from the point of view of locating wind machines. However, winds are seasonal and the speeds vary significantly over the year. Thus, monthly speed-frequency distributions or monthly speed-duration curves are needed for detailed analysis and design of wind machines and associated systems.

❑ *Statistical Models* Various probability distributions have been compared with annual wind speed-frequency distributions (like Fig. 9.6) drawn on the basis of measured data and it has been found that the Weibull distribution fits measured distributions with a reasonable degree of accuracy. The Weibull distribution is given by the equation [11]

$$f(V) = \frac{k}{c}\left(\frac{V}{c}\right)^{k-1} \exp[-(V/c)^k] \tag{9.1}$$

where $V =$ wind speed, and

$f(V) =$ probability density function.

k and c are parameters called the *Weibull shape factor* and *scale factor* respectively.

The cumulative distribution function $F(V)$ is obtained by integrating the probability density function and is given by

$$F(V) = \int_0^V f(u)\,du = 1 - \exp[-(V/c)^k] \qquad (9.2)$$

It is seen that $F(0) = 0$ and $F(\infty) = 1$. In reality, $F(0)$ will not be zero since there will always be some hours at a place with zero wind speed (called the calm or lull period). Thus, there is a small deviation of the Weibull distribution from the actual situation at and near-zero wind speed.

The tables on wind speed data for Kandla and Indore, given in Appendix 5, are based on extensive measurements recorded continuously over many years. Obviously such tables can be prepared only for a small number of locations. What does one do at a new location? A possible approach lies in exploiting the similarity between the measured wind speed-frequency distributions and the Weibull probability distribution. One measures the hourly wind speed at the location on some representative days of the year over a few years. Let N be the number of measurements made. Then the average speed ($\overline{V}$) and the standard deviation (σ) of the wind speed are given by

$$\overline{V} = \frac{1}{N}\sum_{i=1}^{N} V_i \qquad (9..3)$$

and
$$\sigma^2 = \frac{1}{(N-1)}\sum_{i=1}^{N}(V_i - \overline{V})^2 \qquad (9.4)$$

For the Weibull distribution, it can be shown that the parameters k and c are related to $\overline{V}$ and σ^2 by the following expressions [11]:

$$\overline{V} = c\,\Gamma(1 + 1/k) \qquad (9.5)$$

$$\sigma^2 = \overline{V}^2\left[\frac{\Gamma(1+2/k)}{\Gamma^2(1+1/k)} - 1\right] \qquad (9.6)$$

where Γ is the gamma function.

Thus k and c can be calculated by simultaneously solving Eqs (9.5) and (9.6) and the annual wind speed-frequency distribution obtained for the location under consideration by using Eq. (9.1).

Example 9.1

It is proposed to set-up a wind farm at a promising location on the Deccan plateau for which long-term data on wind speed is not available. In order to assess the wind potential and select suitable wind machines, a large number of discrete measurements of hourly wind speed are made over some representative days for a couple of years. The measurements yield the following values of the average wind speed and the standard deviation:

$$\bar{V} = 5.8 \text{ m/s}$$

$$\sigma = 3.5 \text{ m/s}$$

Plot the annual wind speed-frequency distribution under the assumption that it is a Weibull distribution.

Solution

We have from Eqs (9.5) and (9.6),

$$5.8 = c\,\Gamma(1 + 1/k)$$

$$(3.5)^2 = (5.8)^2 \left[\frac{\Gamma(1+2/k)}{\Gamma^2(1+1/k)} - 1\right]$$

Solving these using the gamma function table [12], we get the Weibull parameters k and c as

$$k = 1.7 \quad \text{and} \quad c = 6.5 \text{ m/s}$$

Substituting the values of k and c in Eq. (9.1), we get

$$f(V) = \left(\frac{1.7}{6.5}\right)\left(\frac{V}{6.5}\right)^{0.7} \exp[-(V/6.5)^{1.7}]$$

For V = 0.5 m/s

$$f(V) = \left(\frac{1.7}{6.5}\right)\left(\frac{0.5}{6.5}\right)^{0.7} \exp[-(0.5/6.5)^{1.7}]$$

$$= 0.2615 \times 0.1661 \times 0.9873 = 0.0429 \text{ s/m}$$

Thus for the interval 0 to 1 m/s, frequency = $f(V)\,\Delta V = 0.043 \times 1 = 0.043$, i.e. 4.3 percent.

Proceeding in a similar manner for other intervals, the following values of frequency are obtained.

Speed interval (m/s)	0–1	1–2	2–3	3–4	4–5	5–6	6–7	7–8
Frequency (%)	4.29	8.63	11.00	11.96	11.84	10.96	9.62	8.07
	8–9	9–10	10–11	11–12	12–13	13–14	14–15	15–16
	6.51	5.07	3.82	2.79	1.98	1.37	0.92	0.60

Figure 9.8 shows the annual speed-frequency distribution obtained. The distribution has the shape and characteristics displayed by the measured distribution shown in Fig. 9.6. The only deviation is that the Weibull distribution does not indicate the calm period (zero wind speed).

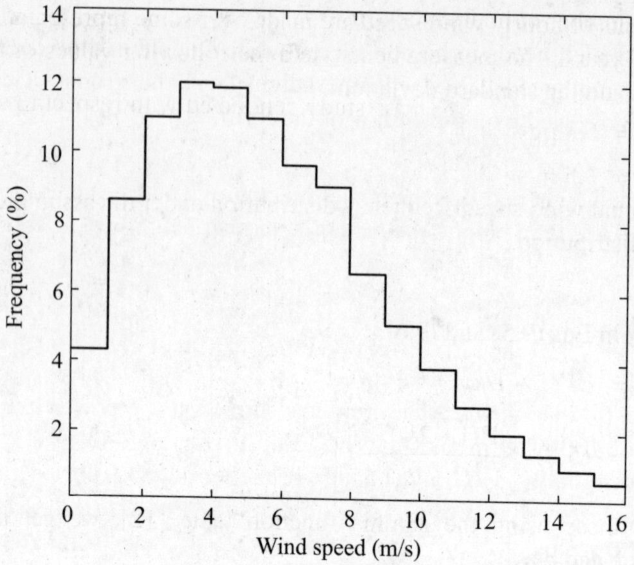

Fig. 9.8 *Example 9.1*

❑ *Energy in the Wind* Consider the wind flowing with a speed V. The power density, (i.e. the power per unit area normal to the wind) is the kinetic energy flowing per unit area and is given by

$$\frac{P}{A} = \frac{1}{2}(\rho V)V^2 = \frac{1}{2}\rho V^3 \tag{9.7}$$

where ρ = air density and V = wind speed.

The energy content of the wind per unit area for a specified period is obtained by integrating the right hand side of Eq. (9.7) with respect to time.

$$\frac{E}{A} = \int_0^{\tau} \frac{P}{A} dt = \frac{1}{2}\rho \int_0^{\tau} V^3 dt \tag{9.8}$$

Since the data is available in terms of the hourly wind speed, Eq. (9.8) is written as a summation. Thus the energy content per square metre for a month (E_m/A) is given by

$$\frac{E_m}{A} = \frac{1}{2}\rho \sum_{i=1}^{N_m} V_i^3 \Delta t \tag{9.9}$$

where N_m = number of hours in the month, ρ = density corresponding to the average air temperature for the period, V_i = hourly wind speed, and Δt = time interval of 1 hour. Similarly, the energy content per square metre for a year (E_a/A) is given by

$$\frac{E_a}{A} = \frac{1}{2}\rho \sum_{i=1}^{N_a} V_i^3 \Delta t \tag{9.10}$$

where N_a= number of hours in a year.

A factor which needs to be considered while calculating the energy content is that the wind speed varies with the height from the ground. Generally the variation is governed by a power law of the form

$$V \propto z^{\alpha}$$

where z is the height and α is the *power law index*. Thus if values of wind speed are known at a certain height, the values at another height can be found from the relation

$$\frac{V_2}{V_1} = \left(\frac{z_2}{z_1}\right)^{\alpha} \tag{9.11}$$

if the value of α is known. Measurements show that the power law is valid up to heights of 100–150 m and that the values of α range from 0.1 to 0.4. The value at a location depends on the nature of the terrain and other parameters like elevation, time of day, season, etc. With a smooth terrain (flat plain land with no obstructions), the value of α is around 0.1. With a moderately rough terrain (grass, crops), it varies from 0.1 to 0.2. With a rough terrain (rural, woods), it is around 0.2, while with a very rough terrain (urban), it varies from 0.25 to 0.40. Values of α for Indian locations have been given by Mani and Mooley [9] in the form of tables and maps. However, this data is difficult to use because it gives the variation of α with the time of the day as well as with the seasons. A review of correlations for calculating the power law index is given by Manwell *et al.* [11].

Example 9.2

Calculate the monthly energy content of the wind per square metre for the following situation:

Location : Indore
Month : May
Height above ground : 10.9 m

Take $\rho = 1.20$ kg/m^3.

Solution

A reasonably good estimate of the energy content can be obtained by using the data in Table A5.2 even though the interval range is 2 kmph. We take values of V_i in Eq. (9.9) to be the average values in each interval, viz. 1, 3, 5, . . . kmph and assume that the wind speed exists at this average value for the number of hours specified in Table A5.2 for that interval. Substituting into Eq. (9.9), we have,

$$\begin{aligned}\frac{E_m}{A} &= \frac{1}{2} \times 1.20 \times \frac{744}{100} [1.6 \times 1^3 + 0.9 \times 3^3 + 1.3 \times 5^3 + 2.0 \times 7^3 \\ &\quad + 1.3 \times 9^3 + 3.9 \times 11^3 + 4.0 \times 13^3 + 6.3 \times 15^3 + 6.2 \times 17^3 \\ &\quad + 6.5 \times 19^3 + 10.3 \times 21^3 + 7.4 \times 23^3 + 8.0 \times 25^3 + 4.3 \times 27^3\end{aligned}$$

$$+ 5.1 \times 29^3 + 7.5 \times 31^3 + 4.5 \times 33^3 + 5.7 \times 35^3 + 4.1 \times 37^3$$
$$+ 1.7 \times 39^3 + 2.2 \times 41^3 + 0.8 \times 43^3 + 0.7 \times 45^3 + 0.1 \times 47^3]$$
$$= 8298622 \left[\frac{\text{kg}}{\text{m}^3} \times \text{h} \times \frac{\text{km}^3}{\text{h}^3} \right]$$

In order to convert the value of (E_m/A) into kWh/m², we divide by $(3.6^3 \times 1000)$. Thus

$$\frac{E_m}{A} = 177.9 \text{ kWh/m}^2$$

It should be noted that a correction factor for height was not needed in this example.

The energy content in the wind can be visualised by plotting the power density-duration curve. Since $(P/A) \propto V^3$, this curve is obtained from the speed-duration curve by plotting V^3 (multiplied by an appropriate constant) instead of V on the y-axis. A typical annual power density-duration curve is shown in Fig. 9.9. It is easy to visualise that the total area under the curve represents the energy content in the wind for a year.

The energy available for the rotor of a given wind machine is less than the total area under the curve in Fig. 9.9. In order to understand the reasons for this, we first define three wind speeds for a wind machine. These are the cut-in speed, the design (or rated) speed, and the cut-out (or furling) speed. The *cut-in speed* for a machine is the wind speed below which the machine does not rotate and no power is produced. The *design speed* is the wind speed at which a machine produces its rated output. Usually the output is held constant at the rated value for wind speeds greater than the design speed. This is achieved by some kind of governing mechanism. The *cut-out speed* is the wind speed at which it is advisable to shut down the wind machine in order to avoid mechanical damage.

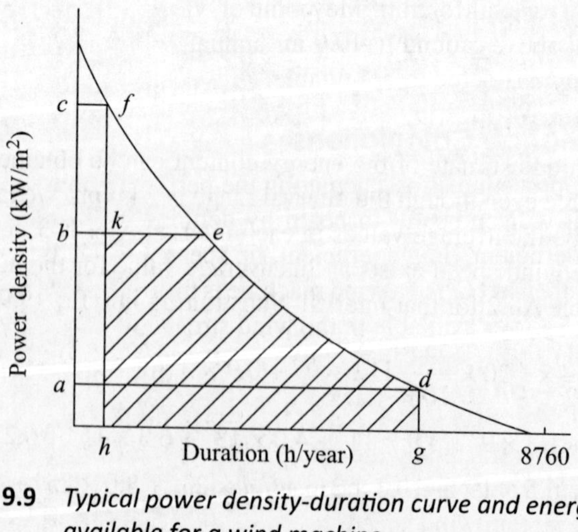

Fig. 9.9 *Typical power density-duration curve and energy available for a wind machine*

Referring again to Fig. 9.9, let the points *a, b* and *c* represent the power density values corresponding to the cut-in, design and cut-out speeds respectively. We draw horizontal lines from these points to intersect the power density-duration curve at points *d, e* and *f,* and drop perpendiculars *dg* and *fh* on the *x*-axis. From Fig. 9.9, it is clear that the energy associated with the area under the curve to the right of *dg* is lost because of wind speeds less than the cut-in speed. Similarly the energy associated with the area to the left of *fh* is lost because of wind speeds greater than the cut-out speed. Also, since the output of the machine is held constant at the rated value for wind speeds in excess of the design speed, the energy associated with the area above the line *be* is not used. Thus, the actual energy available is given by the shaded area *kedgh.*

Example 9.3

Use the data given in Example 9.2 and calculate the actual energy available for a wind machine for which the cut-is speed in 14 kmph, the design speed is 36 kmph and the cut-out speed is 90 kmph.

Solution

$$\frac{E_m}{A} = \frac{1}{2} \times 1.20 \times \frac{744}{100} [6.3 \times 15^3 + 6.2 \times 17^3 + 6.5 \times 19^3 + 10.3 \times 21^3$$

$$+ 7.4 \times 23^3 + 8.0 \times 25^3 + 4.3 \times 27^3 + 5.1 \times 29^3 + 7.5 \times 31^3$$

$$+ 4.5 \times 33^3 + 5.7 \times 35^3 + (4.1 + 1.7 + 2.2 + 0.8 + 0.7 + 0.1) \times 36^3]$$

$$= 7\,606\,803 \left[\frac{\text{kg}}{\text{m}^3} \times \text{h} \times \frac{\text{km}^3}{\text{h}^3}\right] = 163.0 \text{ kWh/m}^2$$

It has been mentioned earlier that Kandla and Indore are both suitable places for locating wind machines from the point of view of the energy content in the wind. In general, locations having an annual energy content in excess of 1000 kWh/m^2 are considered to be suitable.

9.2.5 Performance Calculations

We now describe procedures for calculating the performance of wind machines. In this context, it will be useful to begin by defining a few terms, viz. power coefficient, lift coefficient, drag coefficient, tip speed ratio and solidity.

The *power coefficient* (C_p) of a wind machine is the ratio of the power extracted by the rotor to the power available in the wind stream.

$$C_p = P_e \Big/ \frac{1}{2} \rho A V_\infty^3 \qquad (9.12)$$

where P_e = power extracted by the rotor

A = swept frontal area of the machine, and

V_∞ = free stream wind speed

The *lift coefficient* (C_L) of a blade of a wind rotor is the ratio of the lift force on the blade to the force of the free stream wind. For a stationary blade,

$$C_L = F_L \Big/ \frac{1}{2} \rho A_b V_\infty^2 \tag{9.13}$$

where A_b is the projected area of the blade facing the wind and F_L is the lift force. F_L is perpendicular to the direction of the incoming airflow and arises due to the unequal pressure on the two surfaces of the blade.

The *drag coefficient* (C_D) of a blade of a wind rotor is the ratio of the drag force on the blade to the force of the free stream wind. For a stationary blade,

$$C_D = F_D \Big/ \frac{1}{2} \rho A_b V_\infty^2 \tag{9.14}$$

where F_D is the drag force. It is parallel to the direction of the oncoming air flow and arises due to the viscous friction forces at the surface of blade as well as the unequal pressure on the blade surfaces.

The *tip speed ratio* (λ) is the ratio of the speed of the blade tip to the free stream wind speed.

$$\lambda = \omega R / V_\infty \tag{9.15}$$

where ω = angular velocity of the rotor, and
R = tip radius

This definition holds for a horizontal axis machine. For a vertical axis machine, the peripheral speed at the middle of the blade length is used in the numerator instead of the speed of the blade tip.

The *solidity* (γ) of a wind machine is the ratio of the blade area to the swept frontal area (face area) of the machine. This definition holds for a horizontal axis machine and the following expression is obtained under the assumption that the blade length is equal to the tip radius,

$$\gamma = Nc / \pi R \tag{9.16}$$

where N = number of blades,
c = mean chord of the blades.

For a vertical axis machine, the expression (Nc/R) is used for the Darrieus type and ($Nc/2R$) for other types where R is the mean radius.

The variation of the power coefficient with tip speed ratio and other parameters has been the subject of considerable theoretical and experimental investigation. The earliest derivation was due to Lanchester [13] and Betz [14], who extended propeller theory [15] to derive an upper limit for the value of the power coefficient of a wind rotor. Lanchester and Betz assumed ideal conditions of a one-dimensional, steady, incompressible, inviscid flow. They also made a number of other simplifying assumptions, including the assumption that the number of blades in the rotor was infinite and showed that the upper limit for the power coefficient was (16/27) = 0.593. Glauert [16] and other investigators included the effects of wake rotation, which had been neglected by earlier theories. They obtained numerical values of the maximum possible power coefficient as a function of the tip speed ratio. These results are presented in Fig. 9.10. They show that the higher the tip speed ratio, the greater the maximum value of C_p and that the power coefficient approaches the Lanchester - Betz limit as $\lambda \to \infty$.

The results for the maximum C_p can also be obtained from the following empirical equation:

$$C_p = \left(\frac{16}{27}\right) \exp(-0.3538\ \lambda^{-1.2946}) \tag{9.17}$$

Equation (9.17) is a useful result because it yields an upper bound on the value of C_p regardless of the type of wind machine.

Equation (9.17) has been modified for propeller and multi-blade rotors by taking the effect of drag into account. It has been shown that the power coefficient is then given by [17]

$$C_p = \frac{16}{27} [\exp(-0.3538\ \lambda^{-1.2946}) - \varepsilon\,\lambda] \tag{9.18}$$

where $\varepsilon = C_D/C_L$ is the ratio of the drag coefficient and the lift coefficient. Equation (9.18) is valid for values of $\lambda > 1$. Typically, the value of ε is in the range from 0.008 to 0.02 for propeller rotors having blades with aerofoil cross-sections. For multi-blade rotors having curved plates as blades, the value of ε is in the range of 0.05 to 0.1 [17].

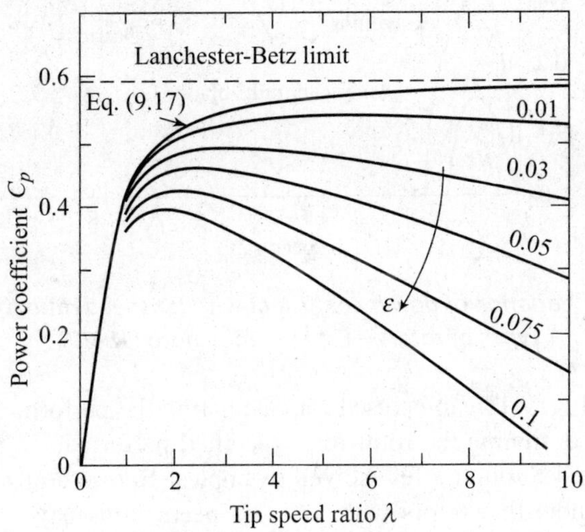

Fig. 9.10 *Theoretical variation of power coefficient with tip speed ratio – Eqs (9.17) and (9.18) [17]*

The variation of C_p with λ for various values of ε is also shown in Fig. 9.10. It is seen that for a given value of ε, the value of C_p increases with λ, passes through a maximum and then decreases. Also as $\varepsilon \to 0$, the value of C_p tends to the upper bound value given by Eq. (9.17).

It may be noted that the actual rotor of a propeller wind machine has a finite number of blades. Also, there are losses from the tips of the blades. Equation (9.18) gets further modified when these effects are considered.

Typical experimental results obtained with various types of wind machine rotors – propeller, Savonius, Darrieus and multi-blade – are presented in Fig. 9.11. It is seen that:

1. In all the cases, the power coefficient passes through a maximum at a particular value of tip speed ratio.
2. Excepting the propeller type, the rise and fall of the power coefficient around the maximum value is quite rapid.
3. The highest values of C_p are obtained with the propeller-type rotor.
4. The multi-blade type and Savonius type are characterised by low values of λ (around one), while the propeller type and Darrieus type are characterised by high values of λ ranging from 4 to 8.

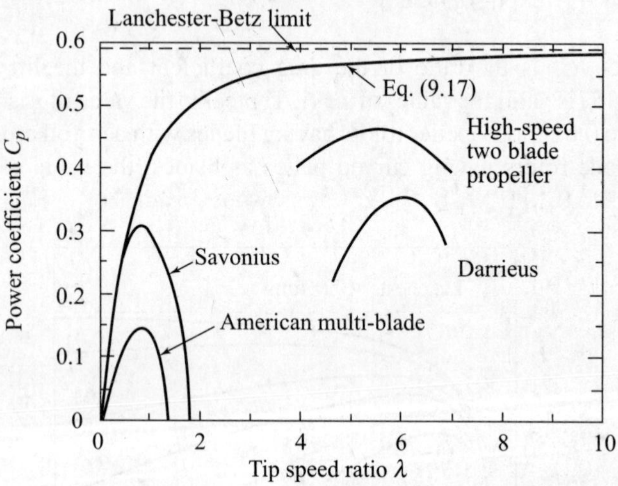

Fig. 9.11 *Variation of power coefficient with tip speed ratio for various types of rotors – Experimental data [18-20]*

The data in Fig. 9.11 can be used for calculating the performance of a given machine or for designing the rotor for a specified performance. The procedure will be illustrated through a few solved examples. Before doing so, it may be worthwhile to note that rotors with low tip speed ratios have high values of solidity, while rotors with high tip speed ratios have low values of solidity. The values can be approximately obtained from Fig. 9.12. It should also be noted that the power generated by a wind machine is proportional to the cube of the wind speed. Since wind speeds vary significantly within localities, the power and energy output of a wind machine is critically dependent on the precise location and height of the machine. Careful siting of a machine is therefore a very important factor in ensuring good performance.

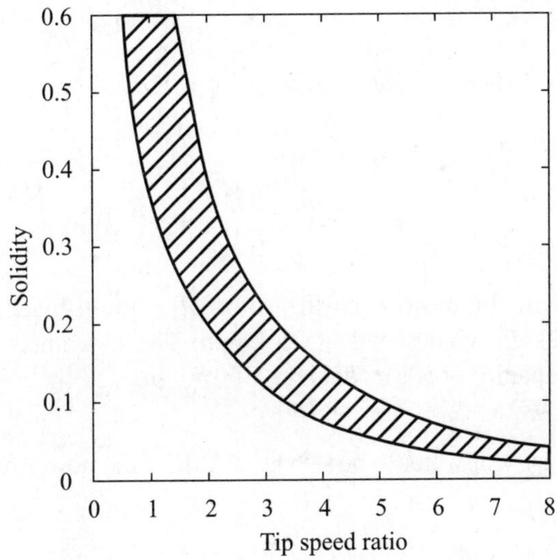

Fig. 9.12 *Variation of solidity with tip speed ratio*

Example 9.4

A propeller wind machine has a rotor diameter of 60 m. It is operating at a location having a wind speed of 35 kmph and rotating at 20 rpm. Calculate theoretically, the power which the machine can extract from the wind if (a) only wake rotation is considered, (b) both wake rotation and the effects of drag are considered. For part (b) assume that the value of $\varepsilon = 0.012$.

Solution

$$\text{Angular velocity } \omega = 20 \text{ rpm} = \frac{2\pi}{60} \times 20 \text{ rad/s} = 2.094 \text{ rad/s}$$

$$\text{Tip speed ratio } \lambda = \frac{2.094 \times 30}{35 \times 1000/3600} = 6.463$$

(a) From Eq. (9.17),

$$C_p = \left(\frac{16}{27}\right) \exp(-0.3538 \times 6.463^{-1.2946}) = 0.5742$$

Taking the density of air to be 1.2 kg/m^3, the power extracted (P_e)

$$= 0.5742 \times \frac{1}{2} \times 1.2 \times \pi \times 30^2 \times \left(\frac{35 \times 1000}{3600}\right)^3 \text{ W} = 0.895 \text{ MW}$$

(b) From Eq. (9.18),

$$C_p = \frac{16}{27}[\exp(-0.3538 \times 6.463^{-1.2946}) - 0.012 \times 6.463] = 0.5282$$

$$P_e = 0.823 \text{ MW}$$

Example 9.5

Calculate the main dimensions of the rotor of a multi-blade wind machine operating at a design wind speed of 25 kmph. The machine operates a water pump having a capacity of 5.1 m^3/h and a lift of 9 m.

Solution

Assuming the density of water to be 996 kg/m^3, the ideal power required to pump the water

$$= \frac{5.1}{3600} \times 996 \times 9.81 \times 9 = 124.6 \text{ W}$$

We take the efficiency of the reciprocating water pump to be 0.6 and the transmission efficiency from the rotor to the pump to be 0.9. Thus, the power required at the rotor

$$= \frac{124.6}{0.9 \times 0.6} = 230.7 \text{ W}$$

We will use the data plotted in Fig. 9.11 and assume that the rotor is delivering the maximum value of C_p at the design wind speed. From Fig. 9.11,

$$C_{p,\,max} = 0.15$$

The corresponding value of $\lambda = 0.80$. Using the definition of C_p, we have

$$0.15 \times \left[\frac{1}{2} \times 1.2 \times \pi R^2 \times \left(\frac{25 \times 1000}{3600}\right)^3\right] = 230.7$$

where R = rotor radius and the density of air is taken as 1.2 kg/m^3. Solving, we get,

$$R = 1.561 \text{ m}$$

As stated earlier, the number of blades in a multi-blade type usually varies from 12 to 20, and the solidity is high. We assume that the number of blades is 18 and that the solidity is 0.5. Making a small allowance for the size of the hub at the centre of the rotor, we assume that the blade length = 1.4 m. Then the mean chord length is calculated from the definition of solidity. Thus,

$$18 \times 1.4 \times c = 0.5 \times \pi \times 1.561^2$$

$$c = 0.152 \text{ m}$$

The angular velocity of the rotor can be calculated from the definition of tip speed ratio. We get

$$\omega = \frac{\lambda V_\infty}{R} = 0.80 \times 25 \times \frac{1000}{3600} \times \frac{1}{1.561} = 3.56 \text{ rad/s} = 34.0 \text{ rpm}$$

Example 9.6

Find the electric power output of a three-bladed propeller-type wind machine operating at a constant rotor speed of 20 rpm. The cut-in speed, design speed and the cut-out speed of the machine are 14, 36 and 90 kmph respectively. The mechanical efficiency of the wind machine is 93 per cent and the generator efficiency is 97 per cent. The machine is installed at a height of 60 m from the ground and the diameter of its rotor is 62 m. Plot the variation of the electrical power output with wind speed. Given: $\varepsilon = 0.011$, $\rho = 1.2$ kg/m^3.

Solution

Angular velocity of the rotor = 20 rpm = 2.094 rad/s
For the cut-in speed, tip speed ratio

$$\lambda = \omega R/V_\infty = \frac{2.094 \times 31}{14/3.6} = 16.7$$

We will calculate the value of C_p from the theoretical expression. Thus from Eq. (9.18),

$$C_p = \frac{16}{27}[\exp(-0.3538 \times 16.7^{-1.2946}) - 0.011 \times 16.7] = 0.478$$

Thus $$P_e = 0.478 \times \frac{1}{2} \times 1.2 \times \pi \times (31)^2 \times \left(\frac{14}{3.6}\right)^3 = 50\,925 \text{ W}$$

Electrical power output $P_{elec} = 50\,925 \times 0.93 \times 0.97$ W = 45.9 kW

Repeating the calculations for different values of V_∞, we get the following values of P_{elec}.

V_∞ (kmph)	18	24	30	36
λ	13.0	9.7	7.8	6.5
C_p	0.50	0.518	0.527	0.532
P_{elec} (kW)	102.1	250.8	498.4	869.3

The results obtained are plotted in Fig. 9.13. It is seen that there is no output up to the cut-in speed, at which speed there is a step jump in the output. Thereafter the output increases at a rate proportional to the cube of the wind velocity until the design speed is attained. At wind speeds greater than 36 kmph, the electrical output is constant until a value of 90 kmph, when the cut-out speed is attained. Then the output drops to zero. It should be noted that in actual practice a smooth curve without discontinuities in slope at the points *A*, *B* and *C* will be obtained.

In this example:

1. It is given that the machine operates at a constant angular velocity. This feature determines the operating point on the $C_p - \lambda$ curve as the wind velocity varies. Sometimes the operating point is determined by some other characteristic. For example, since the wind machine's power is transmitted to some load, sometimes the characteristics of the load may determine the operating point on the $C_p - \lambda$ curve.
2. The theoretical expression, Eq. (9.18), has been used for calculating the values of the power coefficient. It would be more accurate to use an experimentally obtained $C_p - \lambda$ curve, if it is available for the specific rotor and aero-foil cross-section used.

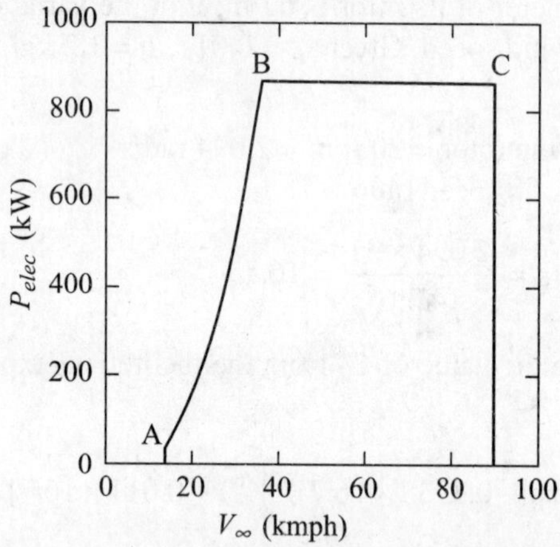

Fig. 9.13 *Example 9.6 –Variation of electrical power output with wind speed for a propeller type wind machine*

Example 9.7

Find the annual energy output of the wind machine described in Example 9.6 if it is located in Kandla. Assume that the value of the power law index is constant and equal to 0.3. Hence calculate the annual energy conversion efficiency and the annual plant load factor.

Solution

We will use the data given in Table A5.1. Since this data is based on the measurements made at a height of 33.3 m, we will adjust it to a height of 60 m by multiplying the wind speeds by the factor $(60/33.3)^{0.3}$. Thus a wind speed of 12 kmph at 33.3 m corresponds to 14.32 kmph at 60 m and a wind speed of 30 kmph at 33.3 m corresponds to 35.80 kmph at 60 m. The values of 14.32 and 35.80 will be taken to be the cut-in and design speeds since they are very close to the values specified for the wind machine.

In order to determine the annual energy output in any interval, say 12–14 kmph, we will take the average wind speed, i.e. 13 kmph. The corresponding wind speed at 60 m height is

$$13 \times (60/33.3)^{0.3} = 15.5 \text{ kmph}$$

Following the procedure of Example 9.6, we calculate P_{elec} available from the machine.

$$P_{elec} = 63.8 \text{ kW}$$

The annual energy output in the interval 12-14 kmph

$$= P_{elec} \times \text{Number of hours/year}$$

$$= 63.8 \times 5.8 \times 8760/100 \text{ kWh}$$

The annual energy output of the machine is obtained by summing over all speed intervals. Thus the annual energy output of the machine.

$$= 87.6\,[63.8 \times 5.8 + 100.4 \times 7.2 + 149.1 \times 7.3 + 211 \times 8.1 + 288 \times 6.5$$
$$+ 377.5 \times 6.4 + 488.5 \times 6.4 + 619.7 \times 4.8 + 770.4 \times 4.5 +$$
$$869.3(3.7 + 2.6 + 3.0 + 2.2 + 2.8 + 1.6 + 1.1 + 1.1 + 0.8 + 0.7$$
$$+ 0.4 + 0.3 + 0.2 + 0.1)]$$

$$= 2\,786\,244 \text{ kWh}$$

Following the procedure of Example 9.2, the annual energy content of the wind at Kandla at a height of 60 m can be calculated. We get

$$E_a/A = 3453.3 \text{ kWh/m}^2$$

The diameter of the rotor is 62 m. Thus the annual energy content of the wind incident on the machine.

$$= 3453.3 \times \frac{\pi}{4} \times (62)^2 = 10\,425\,770 \text{ kWh}$$

Annual energy conversion efficiency of the machine

$$= \frac{2\,786\,244}{10\,425\,770} \times 100 = 26.7 \text{ per cent}$$

The annual plant load factor is defined as the ratio of the annual energy output of a machine to the energy output which would be obtained if the design wind speed was available through out the year.

Hence the annual plant load factor $= \dfrac{2\,786\,244}{869.3 \times 8760} \times 100 = 36.6$ per cent.

9.2.6 Concluding Remarks

Wind energy is a renewable source and it is worth reiterating that the growth in installed capacity of wind energy conversion systems in the world in the last twenty years has been spectacular. This is true for India also. The treatment of the subject in this book has been necessarily brief and of an introductory nature. The reader is referred to the books by Manwell *et al.* [11], Spera [21], Mathew [22], Nelson [23] and Jain [24] for more details.

We conclude this section with a discussion of some general issues of importance.

1. Wind machines may be installed as single units or in clusters in wind farms. They may be a part of large electrical networks, isolated diesel-wind hybrid systems, isolated wind-solar PV hybrid systems or stand-alone power generating units for specific loads. In all cases, issues pertaining to installation at a proper site, operation and control, maintenance, grid connection, safety, etc. assume importance and need careful attention.
2. As machines become larger in capacity, an issue of importance is whether to use fixed speed wind machines or variable speed machines. Fixed speed machines operate at a fixed speed beyond the design wind speed and deliver a constant power. The grid connection, therefore, becomes simpler. On the other hand, variable speed machines can generate more power. However, these machines require complex and costly power electronics. In spite of these constraints, several commercial manufacturers are opting for variable speed machines, especially in off-shore installations.
3. Many environmental issues have also arisen in recent years [25–27]. Wind machine noise, risks to birds, electromagnetic interference, land-use impact, etc. are some issues of concern in both on-shore and off-shore wind farms. Impact on the land around wind machines has also been an issue of concern in on-shore wind farms in India. This is particularly true on forest land in hilly areas where construction of access roads has resulted in serious long-term damage to vegetation and the loss of green cover.

Off-shore wind installations are being built abroad at an increasing pace. In a few years, they will be set up in India as well. It will be useful to learn from the experience of other countries before these installations are set up along the Indian coast line. Major environmental concerns are the effect on marine life and the risk of collisions of seabirds with propellers of wind machines. These issues can be site specific. They will need to be studied and addressed in advance by taking appropriate mitigating steps.

ENERGY FROM BIOMASS 9.3

Plant matter created by the process of photosynthesis is called *biomass*. Photosynthesis is a naturally occurring process which derives its energy requirement from solar radiation. In its simplest form, the reaction of this process can be represented as follows

$$H_2O + CO_2 \xrightarrow{\text{Energy}} CH_2O + O_2$$

It is seen that in the process, water and carbon dioxide are converted into organic material.

The term 'biomass' includes all plant life – trees, agricultural plants, bush, grass and algae, and their residues after processing. The residues include crop residues (such as straw, stalks, leaves, roots, etc.) and agro-processing residues (such as oilseed shells, groundnut shells, husk, bagasse, molasses, coconut shells, saw dust, woodchips, etc.). They may be obtained from forests, woods, agricultural lands, arid lands and even waste lands in a planned or unplanned manner. The term 'biomass' is also generally understood to include human waste, and organic fractions of sewage sludge, industrial effluents and household wastes.

There are a variety of ways of obtaining energy from biomass. These may be broadly classified into three groups, viz. (i) thermochemical route, (ii) biochemical route, and (iii) oil extraction. *Thermochemical conversion* includes processes like combustion, gasification and pyrolysis. Combustion refers to the conversion of biomass to heat and power by directly burning it, as occurs in boilers. Gasification is the process of converting solid biomass with a limited quantity of air into producer gas, while pyrolysis is the thermal decomposition of biomass in the absence of oxygen. The products of pyrolysis are charcoal, a condensable liquid and gaseous products, the relative proportions of these products depending on the chemical composition of biomass used and on the operating conditions.

Biochemical conversion includes anaerobic digestion to produce biogas and fermentation to obtain alcohol fuels. The third approach is *oil extraction*. Edible and non-edible oils can be extracted from a variety of grains and seeds. They can be directly used as liquid fuels or transesterified to produce biodiesel which is a good substitute for conventional diesel oil.

The various ways for obtaining energy from biomass are listed in Fig. 9.14. They will be discussed in the following sections.

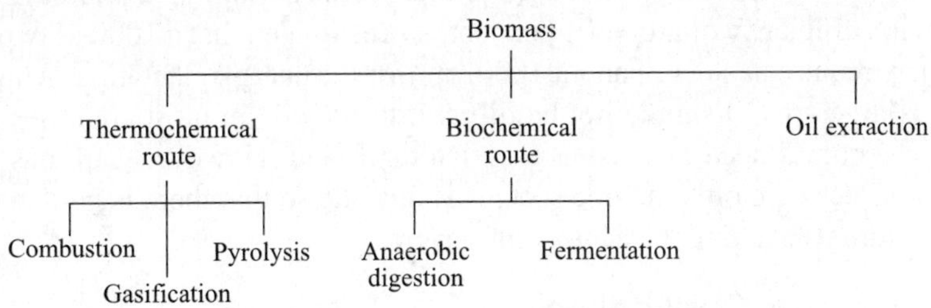

Fig. 9.14 *Biomass conversion routes*

Biomass is a sustainable source of energy. There is no net release of carbon dioxide in growing biomass and using it subsequently for energy conversion process. Besides, the use of biomass results in very low sulphur emission to the environment compared to fossil fuels. The total potential of biomass power in India is estimated to be about 25 000 MW [2]. Till the end of 2015, 4678 MW grid-connected power was installed and it is reported that there was about 780 MW capacity in the form of stand-alone units not connected to the grid [2].

9.3.1 Combustion

One way of using a biomass resource is to directly burn it, the fuel most commonly used being wood. Wood has always been one of man's primary sources of energy. Even today, in India a substantial part of the cooking energy requirements of millions of village households is met by burning wood, agricultural waste and dung cakes. The cooking is done on stoves (chulhas) or on open fires. Chulhas are generally made from mud and clay, and numerous designs have evolved over the years. Most of the traditional designs make poor use of the heat released by combustion and are characterised by low values of efficiency, usually less than 10 per cent. For this reason, considerable effort has been spent in recent years in developing improved designs which cause the combustion of the fuel to be more complete and also transfer the heat of combustion more efficiently to the cooking pots. Efficiencies in the range of 15 to 20 per cent are usually obtained in such chulhas. Over 35 million improved chulhas have been installed in the country under the National Programme of Improved Chulhas and 40 models approved by MNRE [2]. The programme is now handled by the State Governments under the Unnat Chulha Abhiyan programme.

An interesting approach for the large-scale planned use of wood is the 'energy plantation' approach. In this scheme, selected species of trees are planted and harvested over regular intervals of time in a phased manner so that wood is continuously available for cooking or allied purposes. Some of the trees grown in India for this purpose are eucalyptus, babool and casuarina.

It has been suggested that electrical power be produced by the energy plantation approach, the wood grown in this manner being used as a fuel for the boilers of a conventional power plant. The technology of biomass-based electric power plants is well established in USA and Europe and there are over 500 such plants generating power in the range of 5 to 50 MW. These plants use wood, wood waste and various types of agricultural waste. Assuming a photosynthetic conversion efficiency of around 1 per cent, it is estimated that a 1000 MW power plant may require an area of about 1000 km^2 for the energy plantation. Although this is a large area, it should not be difficult to provide in most countries since the land required need not displace agricultural land. However, care has to be taken in selecting more than one species at any site so that there is no danger of monoculture weakening the ecological system.

9.3.2 Biomass Gasification

Biomass gasification is a process of partial combustion in which solid biomass usually in the form of pieces of wood or agricultural residue, is converted into a combustible gas mixture. Partial combustion occurs because the air supply is intentionally kept less than the amount required for the full combustion of biomass. As a result, a gaseous mixture of carbon monoxide, carbon dioxide, hydrogen and nitrogen called producer gas is obtained.

A schematic diagram of a gasifier is shown in Fig. 9.15. The particular type shown is the down draft type. Typically the gasifier is a vertical cylindrical vessel of varying cross section. The biomass is fed in at the top at regular intervals of time and is converted through a series of processes into producer gas and ash as it moves down slowly through various zones of the gasifier.

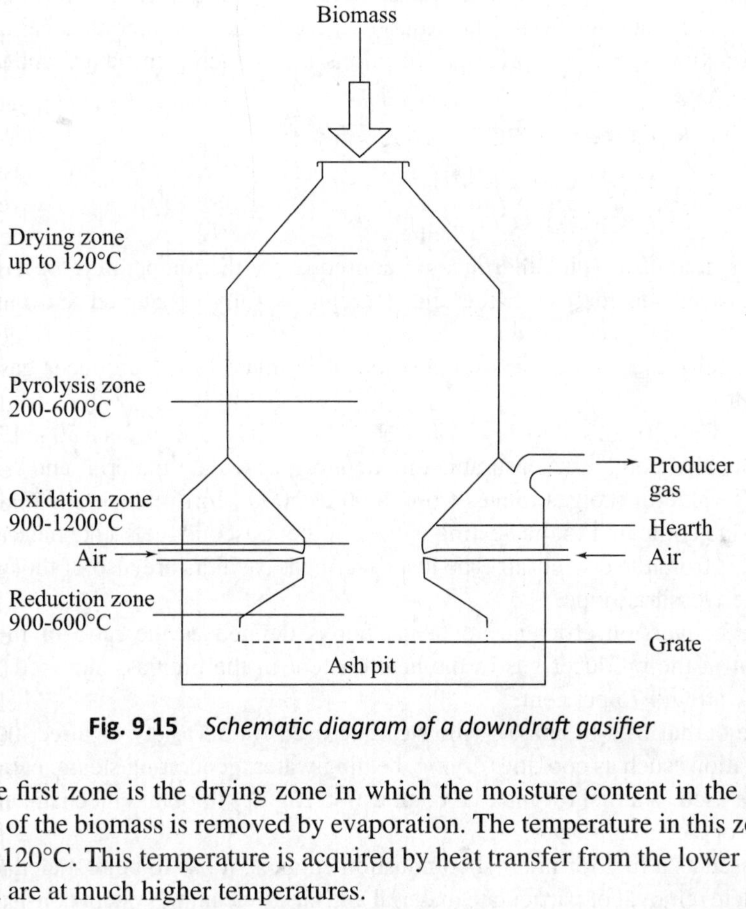

Fig. 9.15 *Schematic diagram of a downdraft gasifier*

The first zone is the drying zone in which the moisture content in the upper layers of the biomass is removed by evaporation. The temperature in this zone is about 120°C. This temperature is acquired by heat transfer from the lower zones which are at much higher temperatures.

The dried biomass moves down to the second zone called the pyrolysis zone which is at temperatures ranging from 200° to 600°C from top to bottom. Throughout this zone, the biomass loses its volatiles. In addition, in the lower part of this zone, when the temperature reaches 400°C, a self-sustaining exothermic reaction takes place in which the structure of the biomass breaks down. As a result, water vapour, methanol, acetic acid and significant amounts of hydrocarbon tars are evolved. The remaining solid is called char (carbon). All the products, gaseous and solid, pass on to the zone below.

The third zone is called the oxidation (or combustion) zone. A predetermined quantity of air is drawn into this zone through nozzles and temperatures ranging from 900° to 1200°C are attained. Combustion reactions similar to those occurring in a normal stove or furnace take place in this zone and a portion of the char and the pyrolysed gases coming from the second zone are burnt. The principal reactions taking place are exothermic and oxidising in nature, and the resultant products are carbon dioxide and water vapour.

These products of combustion pass on to the fourth and last zone called the reduction zone, along with the unburnt pyrolysis gases and char. This zone is at temperatures ranging from 900° to 600°C, the highest temperature being near the oxidation zone. The principal reactions are reducing in nature and are as follows:

$$CO_2 + C \rightarrow 2CO$$

$$C + H_2O \rightarrow CO + H_2$$

$$C + 2H_2O \rightarrow CO_2 + 2H_2$$

These reactions are endothermic and consequently the temperature of the zone progressively decreases. At the end, the char is fully consumed and the final products are producer gas and ash.

Typically the volumetric composition of biomass based producer gas is as follows:

$$CO\ 20\text{–}22\%,\ H_2\ 15\text{–}18\%,\ CH_4\ 2\text{–}4\%,\ CO_2\ 9\text{–}11\%,\ N_2\ 50\text{–}54\%.$$

The gas also contains measurable amounts of particulate material and tar. The heating value of the gas ranges from 4000 to 5000 kJ/m^3, which is a relatively low value compared to the heating value of other gaseous fuels, like natural gas.

Apart from the downdraft type just described, gasifiers are also of the updraft and the crossdraft type.

The conversion efficiency of a gasifier is defined as the ratio of the heat content in the producer gas to the heat content in the biomass supplied and is usually around 75 per cent.

The output of a biomass gasifier can be used for a variety of direct thermal applications such as cooking, drying, heating water, generating steam, etc. It can also be used as a fuel for internal combustion engines to obtain mechanical shaft power or electrical power.

If used as a fuel for internal combustion engines, it has to be first cleaned for complete removal of particulate material and tar. A cleaning system consisting of a cyclone, a scrubber and a filter is used for the purpose. If the engine is a spark ignition engine, it can operate with producer gas alone. The gas is sucked from the gasifier and cleaner unit by the engine suction along with a proportionate amount of air. It is then compressed, ignited and exhausted as in a petrol engine. If the engine is a compression ignition engine, it operates in the 'dual-fuel' mode. Here too the engine sucks in a mixture of producer gas and air during the suction stroke. This mixture is compressed and a small amount of diesel sprayed in. Combustion initiates with the diesel droplets and then spreads to the mixture of the gaseous fuel and air. The phrase 'dual-fuel' implies that both diesel and producer gas are simultaneously used.

India is one of the leading countries in the world in the field of biomass gasification. Biomass gasifier systems are available in a wide range of capacities

Plate 9.2 *Gasifier installed in Bengaluru for generating 120 kWe. (Courtesy: Ankur Scientific Energy Technologies Pvt. Ltd., Vadodara, Gujarat).*

and standard facilities for testing and evaluating gasifiers have been set up. A typical gasifier is shown in Plate 9.2. Systems having a wide range of outputs are available both for thermal applications and electric power generation.

Example 9.8

A biomass gasifier is used to run a compression-ignition engine. The engine operates in the dual-fuel mode with 80% diesel replacement. The gasifier-engine system produces 350 kW of power. Calculate the biomass feeding rate to the gasifier if the efficiency of the engine is 35 per cent and the calorific value of biomass is 16 800 kJ/kg. Given that the efficiency of the gasifier is 0.75.

Solution

A block diagram of the system is shown in Fig. 9.16.

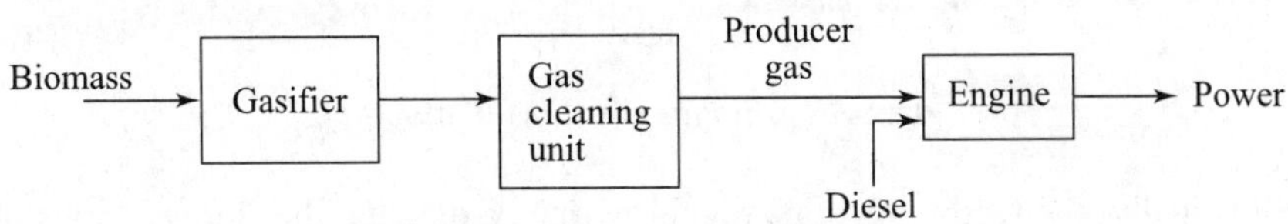

Fig. 9.16 *Example 9.8*

Power input to the engine = 350/0.35 = 1000 kW

The system replaces 80 per cent of diesel. Hence power supplied by gasifier

$$= 1000 \times 0.8 = 800 \text{ kW}$$

Input to gasifier and gas cleaning unit = 800/0.75 = 1066.7 kW

$$\text{Biomass feeding rate} = \frac{1066.7}{16\,800} \text{ kW/(kJ/kg)} = 0.0635 \text{ kg/s} = 228.6 \text{ kg/h}$$

9.3.3 Biogas

An alternative way of obtaining gaseous fuel is through the anaerobic fermentation of wet livestock (or human) wastes to produce biogas [28], a mixture of methane (45 to 70 per cent) and carbon dioxide. The production of biogas is of particular significance in India because of its large cattle population. It will, therefore, be worthwhile to describe a typical plant in detail. A sketch is shown in Fig. 9.17. The plant consists essentially of two parts: the digestor, which is a well containing the animal waste in the form of a slurry, and the dome which floats on the slurry and serves as the gas holder.

The digestor is normally below ground level and two pipe lines lead to its bottom – one for feeding the animal waste slurry, and the other for the spent slurry called sludge to come out after it has undergone fermentation.

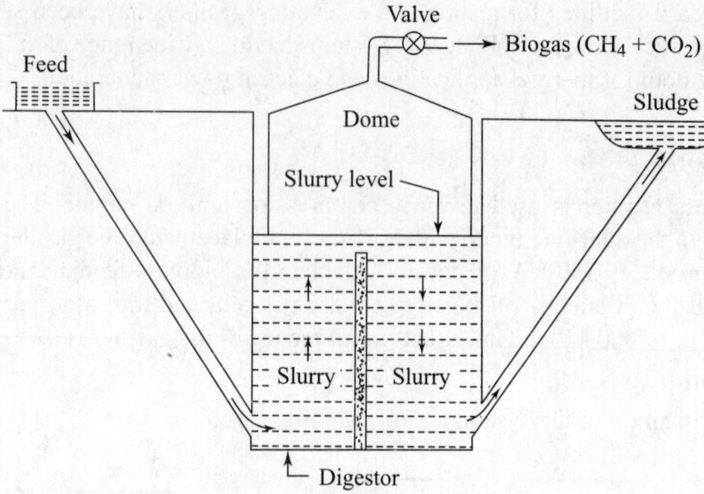

Fig. 9.17 *Biogas plant (KVIC model)*

Typically, the feeding is done regularly once a day and the sludge comes out with the build up of gas pressure in the dome. It is worth noting that the sludge retains the nitrogen, phosphorus and potassium originally present in the animal waste and is an excellent fertilizer. The diameter of the digestor ranges from 1.2 to 6 m, while its depth ranges from 3 to 6 m. A vertical partition wall divides it into two equal parts and serves to direct the flow of the slurry.

The gas generation process takes place slowly and in two stages. In the first stage, the complex, organic substances contained in the waste are acted upon by a certain kind of bacteria called acid formers and are broken up into small-chain simple acids. In the second stage, these acids are acted upon by another kind of bacteria which produce methane and carbon dioxide.

The calorific value of biogas ranges from 16000 to 25000 kJ/m^3. It is an excellent fuel for cooking and lighting. When blended with diesel, it is a very good alternate fuel for compression-ignition engines and can yield diesel savings of 70 to 80 per cent. Modified designs of compression-ignition engines which run on biogas alone, have also been developed.

The following are some approximate rules used for sizing biogas plants or for estimating their performance:

1. 1 kg of dry cattle dung produces approximately 1 m^3 of biogas.
2. 1 kg of fresh cattle dung (gobar) contains 8 per cent dry biodegradable mass.
3. 1 kg of fresh cattle dung has a volume of about 0.9 litres.
4. 1 kg of fresh cattle dung requires an equal volume of water for preparing slurry.
5. Typical retention time of slurry in a biogas plant is 40 days.

Example 9.9

An engine-generator system running on biogas is installed to produce 2 kW of electric power. Estimate the volume of the digestor of the biogas plant required, if cowdung is used as the feed material.

Given: Calorific value of biogas : 20 000 kJ/m^3
Efficiency of generator : 90.1 per cent
Engine efficiency : 30 per cent

Solution

A block diagram of the system is shown in Fig. 9.18.

$$\text{Power input to the engine} = \frac{2}{0.901 \times 0.30} = 7.4 \text{ kW}$$

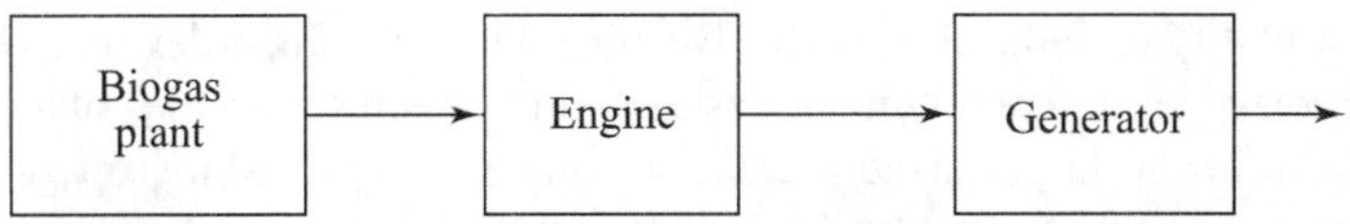

Fig. 9.18 *Example 9.9*

Quantity of biogas required = 7.4/20 000 = 0.00037 m^3/s

For continuous operation, biogas required = 0.00037 × 3600 × 24
= 31.96 m^3/day

Dry cowdung required = 31.96 kg/day

Mass of fresh cowdung required = 31.96/0.08 = 399.6 kg/day

Volume of fresh cowdung required = 399.6 × 0.9 = 359.6 litres/day

An equal volume of water is added to get slurry.

Volume of slurry = 359.6 × 2 = 719.2 litres/day

This much slurry should be stored for 40 days. Therefore, volume of the digestor

= 719.2 × 40 = 28 768 litres, i.e. 29 m^3.

The potential of biogas as a heat energy resource is large. India has about 191 million cattle and about 109 million buffaloes yielding a huge quantity of dung every year [29]. Even if a small part of this dung is processed in biogas plants, it could generate a large amount of gas. Besides, the fertilizer value of the dung will remain unaltered. In addition to using dung, biogas plants based on poultry droppings, industrial and commercial wastes and effluents have also been developed.

Biogas plants are available in different sizes and have capacities ranging from about 1 to 150 m^3 per day. At the moment, there are approximately 4.8 million biogas plants installed in India. Most of them have small capacities [2]. Various designs of biogas plants have been developed. Under the National Biogas and Manure Management programme of MNRE, four models have been approved [2]. They are KVIC (Khadi and Village Industries Commission) model, Deenbandhu model, Pre-fabricated model and Bag-type model. The use of the last two models is limited to remote locations where it is difficult to transport construction materials. The design shown earlier in Fig. 9.17 is called the KVIC model. The Deenbandhu model is the most popular design. It has a hemispherical fixed-dome type gas holder (Fig. 9.19), unlike the floating-dome of the KVIC design. The dome is made from pre-fabricated ferrocement or reinforced concrete and attached to the digester, which has a curved bottom. The slurry is fed from a mixing tank through an inlet pipe connected to the digester. After fermentation, the biogas collects in the space under the dome. It is taken out for use through a pipe connected to the top of the dome, while the sludge, which is a byproduct, comes out through an opening in the side of the digester. About 90 per cent of the biogas plants in India are of the Deenbandhu type. In order to assure better utilization of animal and human wastes and to obtain economies of scale, it is desirable to install larger sized plants for community use. Many operating parameters such as temperature and alkalinity of the slurry, sludge liquidity and build up of scum on the surface of the slurry require to be controlled in order to obtain good performance. Such control is obviously easier to achieve in community sized plants.

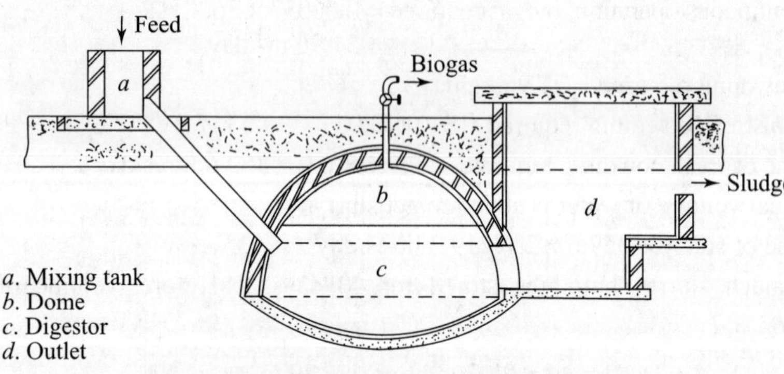

Fig. 9.19 *Biogas plant (Deenbandhu model)*

9.3.4 Bio-ethanol and Bio-diesel

❑ *Bio-ethanol* The conversion of certain types of biomass by fermentation into ethanol has assumed significance because of its potential importance in the transport sector. A variety of edible biomass materials (sugars and starches) containing high levels of glucoses, or precursors to glucose, can be used as feedstock once they are broken down to fermentable sugars. Crops such as sugar cane, sugar beet byproducts (like molasses) and sweet sorghum are high in glucose (a 6-carbon sugar) and are easily fermented. Fermentation involves microorganisms which use the glucose for food and produce ethanol and other byproducts. A specific yeast called Baker's yeast is commonly used [30]. After the fermentation, the ethanol is separated from the other byproducts and purified by distillation.

Alcohols can also be produced from a number of starch crops such as barley, wheat, corn, potato, etc. Starch molecules are long chains of glucose molecules. Thus they can also be fermented into ethanol after first breaking the chains into simple glucose molecules. This requires a process called hydrolysis in which the starch is mixed with water and forms a slurry which is heated to rupture the chains. Enzymes are added during the heating cycle. Figure 9.20 shows a block diagram for production of ethanol from edible biomass. This alcohol is sometimes referred to as a first generation biofuel because it is made from edible foods.

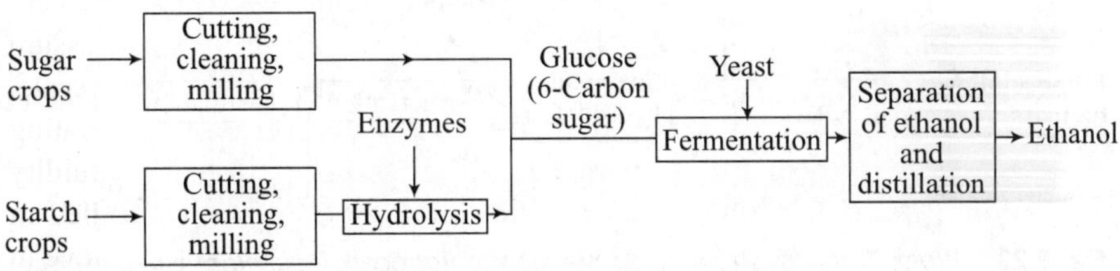

Fig. 9.20 *Block diagram showing conversion of edible biomass to ethanol*

When blended with petrol, ethanol is a good alternate fuel for automotive engines. This fact has received considerable attention as a means of overcoming the oil crisis. However, if one examines the requirement of land for growing the agricultural products concerned, it is obvious that the method can be of substantial benefit only to a country having a large surplus of land. Brazil has thus adopted this method on a large scale and produced about 17 million tons of ethanol in 2007 mostly from sugarcane for use as an alternate fuel. The United States also produced about 21 million tons of ethanol from corn in 2007. However, the position in India is quite different since the availability of agricultural land is limited. Consequently, the amount of ethanol which India can hope to produce from edible biomass for use in transportation will always be small relative to its needs. This is the case in many other countries as well.

An alternative is to produce ethanol from cellulosic biomass which is non-edible. The alcohol thus produced is referred to as a second generation biofuel. This route holds great potential due to the abundance of cellulosic biomass and the fact that its use does not interfere with the human food chain. Cellulosic biomass comprises wood, crop residues, forest residues, etc. A large fraction of municipal solid waste (paper, cardboard etc.) is also cellulosic biomass. Several technical processes have been developed for conversion of cellulosic biomass first into a fermentable sugar and then into ethanol. This is an area in which considerable research and development work is in progress to develop large scale cost-effective technologies.

Cellulosic materials are often referred to lignocellulosic materials and are essentially made up of three components – cellulose (30 to 35 per cent), hemicellulose (25 to 30 per cent) and lignin (10 to 20 per cent). Cellulose molecules are long chains of glucose molecules like starch molecules. However their structure is different from that of starch molecules. Hemicellulose is also composed of long chains of sugar molecules, but has both 6-carbon (glucose) and 5-carbon sugar molecules. Lignin contains no sugars, but encloses the cellulose and hemi-cellulose molecules and generally helps to give rigidity to the biomass.

A block diagram showing the steps involved in converting lignocellulosic biomass into ethanol is given Fig. 9.21. They are as follows:

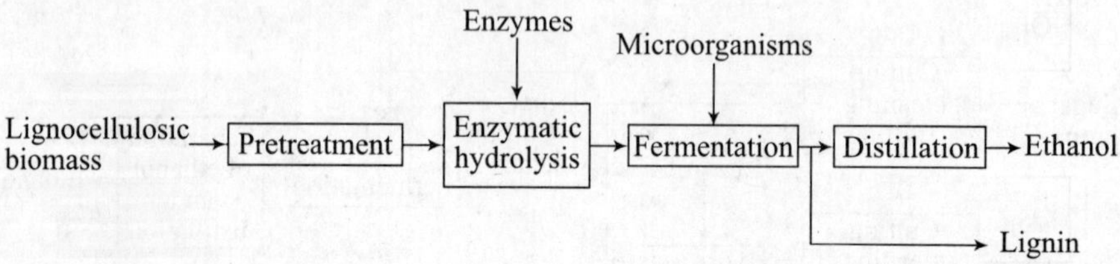

Fig. 9.21 *Block diagram showing conversion of lignocellulosic biomass to ethanol*

1. Pretreatment: Physical (like chipping and milling) and thermochemical processes are used. These reduce the particle size of the biomass to about 1 mm or less and help to break away the lignin from the cellulose and hemicellulose.
2. Hydrolysis: During this reaction, the cellulose and hemicellulose are converted into smaller 6- and 5- carbon sugar molecules. Specific enzymes have been developed for the purpose.
3. Fermentation: Fermentation requires the presence of microorganisms to convert the sugar molecules into ethanol. Specific microorganisms have been developed for the purpose. In order to make the technology more efficient and cost effective, the hydrolysis and fermentation steps are done in the same reactor.
4. Distillation: The end products after the fermentation are a mixture of ethanol and water along with other byproducts. The byproducts are separated and the ethanol-water mixture subjected to fractional distillation to obtain ethanol of the desired purity. More details are given by Achimas and Euverink [31] and Maurya *et al.* [32].

The processes developed thus far by a number of investigators have not succeeded in producing ethanol from lignocellulosic biomass in an economical manner on a large scale. However it is clear that the potential for doing so exists.

❑ *Bio-diesel* Bio-diesel is a methyl or ethyl ester of a fatty acid made from used or unused vegetable oils (edible or non-edible) and from animal fat. It can be used as a fuel in diesel engines in blended form or in pure form. The use of bio-diesel in engines results in substantial reduction in emission of unburned hydrocarbons, carbon monoxide and particulate matters. Figure 9.22 shows a block diagram for the production of bio-diesel. Edible vegetable oils, such as palm, soyabean, sunflower, peanut, olive oil, etc. and non-edible oils, such as jatropha, karanja, neem, etc. are converted to esters. The chemical process is called as transesterification. In this process, triglycerides in the oil are reacted with methanol or ethanol to produce esters and glycerol. Animal fats can also be transesterified to produce bio-diesel. The transesterification process is described in detail by Bankovic-Ilic *et al.* [33].

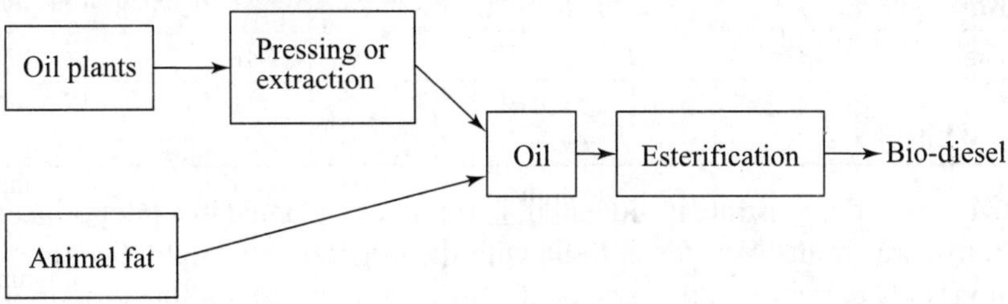

Fig. 9.22 *Block diagram showing production of bio-diesel*

The properties of a few bio-diesel fuels made from edible and non-edible oils are shown in Table 9.1. The table also shows the properties of conventional diesel and a 20 per cent bio-diesel blend. It can be seen that the properties of the bio-diesel fuels compare well with the properties of diesel, while the properties of the blend are very close to the properties of diesel. Therefore, bio-diesel fuels and blends are rated as good alternatives to diesel. Studies have shown that engines running with bio-diesel fuels emit less particulate matter, hydrocarbon and carbon monoxide. However, there is a marginal increase in NO_x emission. This can be taken care of by optimizing the engine parts or using De-NO_x catalyst.

In India, edible oils are in short supply. On the other hand, it is endowed with a variety of non-edible vegetable oil-producing trees and shrubs, which can be used for bio-diesel production. Jatropha is a widely occurring variety in India. It grows practically all over the country under a variety of agro-climatic conditions. For this reason, it was identified in 2003 as being the most suitable inedible oil seed under the national Bio-diesel Mission of the Government of India. The process of utilizing jatropha consists of the following steps:

1. Cultivating the jatropha plantation
2. Harvesting the nuts
3. Grinding the oil seeds
4. Extracting the plant oil
5. Transesterification
6. Purifying to make bio-diesel

Table 9.1 Properties of bio-diesel fuels [34-36]

Bio-diesel	Kinematic viscosity at 40°C (mm^2/s)	Cetane number	Lower heating value (MJ/kg)	Cloud point (°C)	Power point (°C)	Flash point (°C)	Density (kg/litre)
Edible oils:							
Palm	5.7	62	33.5	13		164	0.880
Peanut	4.9	54	33.6	5		176	0.883
Soyabean	4.5	45	33.5	1	–7	178	0.885
Sunflower	4.6	49	33.5	1		183	0.860
Non-edible oils:							
Jatropha	5.2		39.2	13		175	0.865
Karanja	4.7	55	37.1	7	4	183	0.876
Diesel	3.06	50	43.8		–16	76	0.855
20% Bio-diesel blend	3.2	51	43.2		–16	128	0.859

In order to demonstrate its potential, jatropha was planted in a pilot project on 1700 hectares in many states in India with the target of eventually planting it in millions of hectares of barren, arid land. However, only 0.5 million hectares were planted by 2010. It was observed that although the jatropha plant was hardy, the seed yield was significantly lower than the expected yield when the land was not fertile and the rainfall inadequate. Consequently, the cost of production of bio-diesel was too high to be economical.

The use of jatropha as a fuel was also demonstrated by using it for supplying electricity to villages. It was also tested as a fuel in railway trains and transport buses. However, the whole programme to produce and encourage the use to jatropha on a large scale nationwide has essentially slowed down because of low yield of the seed and the consequent high cost of the oil. This is true in India as well as in other countries like China and Tanzania.

Recent developments indicate that some form of jatropha hybrids have been developed in USA [37]. In India, tree-borne oilseeds like pongania (karanja), neem, kusum and mahua have been identified as possible alternatives [38]. All of these grow extensively in India and in some cases, are native to India. The reader is referred to Borugadda and Goud [39] and Kumar and Sharma [40] for more details.

SMALL HYDROPOWER 9.4

Harnessing electricity from the force of falling water is a well-established way of obtaining power and large hydroelectric power stations provide a significant part of India's electricity output. In this section, we focus briefly on small hydropower plants (SHP) which have an important niche to fill for providing local power in remote areas and for utilizing the potential of small rivers, streams and canals. Another important reason why the development of SHPs needs to be pursued is that the construction of any large hydropower station today raises a lot of environmental concerns.

9.4.1 Classification, Potential and Installed capacity

There is no uniform definition of a SHP on the global scale. Different countries have different norms [41]. As far as India is concerned, plants with a capacity of less than 25 MW are considered as SHPs. According to MNRE, they are further classified as follows [2]:

Classification	Capacity
Micro	Upto 100 kW
Mini	100 to 2000 kW
Small	2 to 25 MW

Micro small hydropowcr plants are primarily built for electrification of a small cluster of villages. Mini plants are either for local electrification or are connected to the grid. Small plants are connected to the grid. Based on systematic studies of river water and canal flows, the total potential of SHPs in the country is estimated to be 19 749 MW from 6474 sites [2]. Fifty per cent of this capacity is located in Himachal Pradesh, Uttarakhand, Jammu and Kashmir and Arunachal Pradesh. By the end of 2015, the installed capacity was 4177 MW of grid-connected power. In addition, off-grid power equivalent to 17.21 MW was being generated through microhydel units [2]. SHP is one of the good options for providing electricity to people in remote and hilly areas where providing a grid connection is difficult and also uneconomical.

9.4.2 Layout Schemes

Various types of schemes are used for setting up small hydro projects [42], one of the most common being the run-of-the-river (ROR) scheme. A ROR scheme is one in which the natural flow and existing elevation drop of a river are used. In this scheme, there is either no water storage or only a limited amount is stored behind a low weir. A typical ROR project comprises a diversion structure on the river, an intake channel for the diverted water, a desilting tank followed by another channel, and then a forebay tank. All these are at the higher elevation. From here, the water flows down through the penstock pipe to the power generating station. After giving up its energy, the flowing water flows through the tail race channel back into the river. Standardized turbine generator units are used, the Francis turbine being the most common. The use of standard items simplify the planning and development and help to reduce the cost of the plant. Run-of-the-river SHPs have a lower environmental impact than SHPs of the same size having a dam and water storage.

Apart from ROR schemes, SHPs which utilise the fall available in many large canals have also been built. The layout of canal-based projects is similar in many aspects to that of a ROR scheme.

9.4.3 Turbine Selection

The important parameters which characterize a turbine are the power generated (P), the head of water available (H), the rotational speed of the machine (N), the specific speed of the turbine (N_s), diameter of the runner in the turbine (D), the volume rate of water (Q) and the efficiency of the turbine (η).

The specific speed of a turbine is an important parameter and we will derive an expression for it. Consider a water turbine rotating at a speed of N rpm under a head H (in metres) and generating a power P (in HP). It can be shown that the quantity $[N\,(P)^{1/2}/H^{5/4}]$ for the turbine remains the same if we consider geometrically and dynamically similar models of the turbine. Now, consider a 'similar' model operating under a head of 1 m and producing 1 HP, and assume that it is rotating with a speed of N_s rpm.

Then, we have

$$N_s\,(1)^{1/2}/(1)^{5/4} = N\,(P)^{1/2}\,/\,H^{5/4}$$

or

$$N_s = N\,(P)^{1/2}\,/\,H^{5/4} \tag{9.19}$$

N_s, which is the speed of the model, is called the *specific speed* of the turbine under consideration.

The specific speed is an important parameter because its value helps in identifying the type of turbine to be used. Each type has its own characteristic limits of N_s. For example, the specific speed of a Francis turbine lies in the range of 70 to 400 rpm, while that of a Kaplan turbine lies in the range of 400 to 800 rpm. More information is available on this aspect from Indian Standard (IS 12837) [43]. It should be noted that the above range of values apply if the units used are metres for head and HP for power.

Example 9.10

A hydraulic turbine with an output of 8 MW (10 723 HP) is to be erected at a ROR site where the available head is 40 m. Specify the type of turbine to be installed. What will be the water flow rate through this turbine if $\eta = 0.9$?

Solution

We will assume that an 8-pole generator with a synchronous speed of 375 rpm is connected to the turbine.

Specific speed of the turbine (N_s)

$$= \frac{375 \times (10\,723)^{1/2}}{(40)^{5/4}} = 386\,\text{rpm}$$

We select a Francis turbine because the value of N_s is between 70 and 400.

Taking the density of water to be 998 kg/m^3, the volume flow rate of water

$$= \frac{\text{Power output of the turbine}}{\text{Efficiency of the turbine} \times \text{Density of water} \times g \times \text{Head}}$$

$$= \frac{8 \times 10^6}{0.90 \times 998 \times 9.81 \times 40}$$

$$= 22.7 \text{ m}^3/\text{s}$$

9.4.4 Environmental Concerns

Most people believe that SHPs are a safe substitute for large-sized hydroelectric power plants. The environmental issues associated with the construction and operation of a SHP are in fact small compared with those associated with a large plant. However, the issues are not trivial and need to be addressed with care during the planning and operation of a small plant. For example, the disruption caused by building roads and putting up power transmission lines may be less in the case of SHP. However, on a per megawatt basis, the impact is likely to be the same. The environmental impacts of SHPs may be classified as being ecological and physical in nature. The ecological impact is concerned with the effect on aquatic flora and fauna, while the physical impact is the deleterious effect on water quality in the stream and in the ground below, reduction in recharge and the reduction in sediment capacity of the river stream.

In a ROR scheme, there is a significant distance between the point where water is diverted from the river and the point where the tail water re-enters the river. A large volume of water bypasses this length of the river when the SHP is in operation. This can have a severe ecological impact on this stretch because the reduction in flow affects the plant life in the water as well as on the banks and the fish (if they exist) in the river. It has been suggested that SHPs be designed in such a way that a minimum water flow called the ecological flow (e-flow) should be released at all times.

Finally, it needs to be mentioned that there is a limit to the number of ROR schemes and other type of hydroelectric schemes which can be or should be installed in one river basin. The totality of the effect of having a large number of schemes in series along a river needs to be considered. This is well described in reference [41] in the discussion on the hydroelectric projects in the river basins of the Alaknanda and Bhagirathi rivers, which flow through the Himalayas and eventually merge to form the Ganga at Devprayag. Apart from 29 large hydel plants, 40 SHPs with a capacity of 442.3 MW are planned. Some plants are operational, some under construction, some in the clearance stage and some under development. Although environmental clearances are not needed for SHPs in India today, the combined effect of having a large number of projects in one river system was studied by an inter-ministerial group (IMG) in 2013. Many of the plants (both operational and under construction) had been designed with an e-flow regulation of only 10 per cent of the lean flow, while some had provided for no e-flow regulation. The IMG felt that this could lead to serious consequences when all the plants were operational. They felt that it would be necessary to have increased e-flow norms ranging from 20 to 25 percent for most of the year. These values would more closely mimic the natural flow in the Alaknanda and Bhagirathi rivers, and would ensure that the health of the rivers is maintained. At the same time, the IMG recognized the economic importance of the projects. It estimated that there would be a power loss of 11 to 23 per cent in the SHPs on the Alaknanda and 8 to 20 percent on the Bhagirathi projects because of the increased e-flow recommended.

OCEAN ENERGY 9.5

Ocean energy refers to energy harnessed from the oceans and the seas of the world. The different forms in which the energy is available are tidal energy, wave energy, the energy associated with temperature differences in the ocean and the energy in marine currents. At the end of 2015, the installed capacity of the global ocean energy was about 530 MW [44]. Most of this capacity is due to tidal power plants. A few wave energy conversion devices and one commercial plant based on the temperature differences in the ocean has been installed. We discuss each of these energy sources one by one in the following sections.

9.5.1 Tidal Energy

Tides are generated primarily by the gravitational attraction between the earth and the moon acting upon the oceans of the earth. They arise twice a day. In mid-ocean, the tidal range is only a metre or less, but in some coastal estuaries, it is much greater. This is due to the amplification of the tidal wave as it moves up the narrowing channel of the estuary.

Basically, in a tidal power station, a barrage with sluice gates is built across an estuary. As the water level rises twice a day, it is allowed to flow in through the sluice gates. At high tide, the gates are closed so that the water is impounded in a basin behind the barrage. Thus as the tide ebbs and the water level in the open sea falls, a head of water is created between the water in the basin and the water outside. This head is used to obtain work by making the trapped water flow through water turbines. The water turbines are connected to electrical generators so that electrical power is obtained for about 4 to 5 hours in every 12 hours cycle when the tide is ebbing. This method of operation is referred to as the 'ebb generation' mode. Apart from this mode, it is possible to generate power when the tide is flowing towards the shore. This is called the 'flood generation' mode. It is also possible to have a two-way (or bi-directional) scheme in which power is generated both during ebbing and flooding, and for 9 to 10 hours in every 12 hour cycle. Of the three modes, 'ebb generation' is preferred because of its simplicity even though it generates less power [45].

Jeffreys [46] has estimated that the rate of tidal energy dissipation for the world is about 3×10^6 MW. However, only a small fraction of this can be exploited since, for practical purposes, one needs a certain minimum difference of level between the high and low tides. In addition, the geography of a location has to be suitable from the point of view of the civil construction involved. Waters and Aggidis [45] have listed potential locations world-wide and estimated that 117 539 MW of electricity can be generated from these sites if tidal power stations are built. The major sites with capacities exceeding 5000 MW are Mezen, Tugur and Penzhinsk (Russian Federation), Turnagain Arm (USA), Severn (UK), Gulf of Khambhat (India), Cobequid (Canada), and San Jose and Golfo Nuevo (Argentina). In an overall sense, it should be noted that tidal power is a small resource compared to the energy requirements of the world and can at best supply the needs of a few regions.

The first commercial tidal power station in the world was constructed in France in 1965 across the mouth of the La Rance estuary. It has a capacity of 240 MW. The average tidal range at La Rance is 8.4 m and the 720 m long dam built across the estuary encloses an area of 22 km^2. It provides a yearly output of 480 GWh [45]. Another large tidal plant is located at the Shiswa Lake in South Korea. It is in operation since 2011 and has a capacity of 254 MW. Apart from these two, a few small units have been built in Canada, China, Russia, the Netherlands, Scotland and France [44, 45], the total installed capacity till 2014 being 518 MW [45, 47]. Feasibility studies have also been carried out at suitable locations all over the world. Large size commercial power plants have not been constructed mainly because building the long barrages required for tidal power stations is capital intensive, resulting in relatively high unit costs per installed MW. The building of barrages also affects the existing eco-system in the estuary and raises many environmental issues. In order to reduce environmental effects, tidal lagoons and tidal reefs are being considered in place of barrages [45]. However, no such commercial projects have as yet been constructed.

In India, the two main potential regions for exploiting tidal energy are the Gulf of Khambhat and the Gulf of Kutch in Gujarat. The variation of the range at three sites in these regions is given in Table 9.2.

Table 9.2 Variation of the tidal range at three locations in India

Site	Spring tide range (m)	Neap tide range (m)	Average (m)
Bhavnagar (Gulf of Khambhat)	10.29	3.60	7.00
Navalakhi (Gulf of Kutch)	7.17	3.74	5.45
Kandla (Gulf of Kutch)	6.32	3.68	5.00

Calculations show that it should be possible to have an installed capacity of 900 MW generating 1.6 TWh/year in the Gulf of Kutch and an installed capacity of 7000 MW generating 15 TWh/year in the Gulf of Khambhat. A number of preliminary studies have been conducted in and around the Gulf of Khambhat and the feasibility of having tidal power generation there has been examined. In addition, a feasibility study for a small capacity 3 MW station in the Sunderbans area in West Bengal has been completed. The average tidal range in the Sunderbans is only about 3 m which is much less than the values in the Gulfs of Kutch and Khambhat. Given this and the fact that the proposed plant has a capacity of only 3 MW, it is not likely to be economically viable. However, the construction and running of a small-sized power station will help in acquiring good operational experience.

9.5.2 Wave Energy

Wave energy arises because of the interaction of the wind with the surface of the oceans. Hence it is one of the indirect ways of utilizing solar energy. The amount of energy transferred from the wind to the waves, and hence the size of the resulting waves, depends on the wind speed, the length of time for which the wind blows and the distance over which it blows. Thus, the power available varies with the amplitude and frequency of the waves and can be estimated from the formula

$$P = 0.55\, H^2 t_p \qquad (9.20)$$

where P is the power in kW per metre length of wave front, H is the significant wave height in metres, and t_p is the time period in seconds. The significant wave height is defined as the average height of the highest one-third waves.

Estimates of the global potential of wave energy vary. The resource assessments are based on field data such as buoy data and radar measurements, usually followed by numerical modelling. A list of modelling approaches and tools used is given by Uihlein and Magagna [48]. According to the IEA report of 2006, the global wave power potential is estimated to be in the broad range of 8000 – 80 000 TWh/year [49]. Recently, Taveira-Pinto *et al.* [50] report the global resource to be 32 000 TWh/year. The resource is unevenly distributed over the globe. The highest levels of wave activity, with annual average power levels between 20 and 70 kW for every metre of wave front, are found between the latitudes of ~30° and ~60° in both hemispheres. High resources are located along the coast of Western Europe, off the coasts of Canada and USA and the southern coasts of Australia and South America [51]. No estimates are available for India. Wave energy is difficult to collect because of wide fluctuations in amplitude and frequency of the waves at any location. Considerable ingenuity is therefore required in building cost-effective devices to extract the energy.

Devices for Wave Energy Conversion

Wave energy technology is the most advanced amongst technologies for exploiting the energy available in the oceans and a number of devices for converting wave energy into electricity have been developed. Depending on their location with respect to the shoreline, these devices are classified into three groups, viz. shoreline devices, near shore devices and offshore devices. Shoreline devices are located on the shoreline itself. As a result, they are easier to install and maintain and do not require deep-water moorings and long underwater electrical cables. An example of this device is the oscillating water column (OWC), described later in this section. Nearshore devices are set-up in shallow water, typically 10 to 25 m deep. The OWC has also been developed for such applications. Usually, such plants are incorporated into a breakwater, meant for harbour or coastal protection. Offshore devices are located in water depths exceeding 40 m and exploit the high power levels occurring in deep water. The Pelamis is an example of this device and is described later in this section. Another classification is also used for wave energy conversion devices. This is based on the working principle of the devices. Accordingly, the devices are classified into three groups, viz. oscillating water column, oscillating bodies and overtopping. Examples in each group are OWC, Pelamis and Wave Dragon respectively [52].

❑ *Oscillating Water Column System* The most common wave energy conversion device is the oscillating water column system (Fig. 9.23). It consists basically of a chamber in the sea exposed to the wave action through an opening on the side. The air inside the chamber is alternately compressed or expanded because of wave action. Consequently, pulsating, bi-directional air movement occurs through an opening at the top of the chamber. The bi-directional flow can be converted through a system of four valves to a uni-directional flow. As shown in Fig. 9.23, this is achieved by having two valves open and two closed at any given instant of time. The uni-directional flow drives an air turbine, which in turn is coupled to an electrical generator. A number of OWC devices have been installed and tested all over the world [44, 53]. The first commercial plant with a capacity of 296 kW was installed in the Basque County of Spain in 2011 [44].

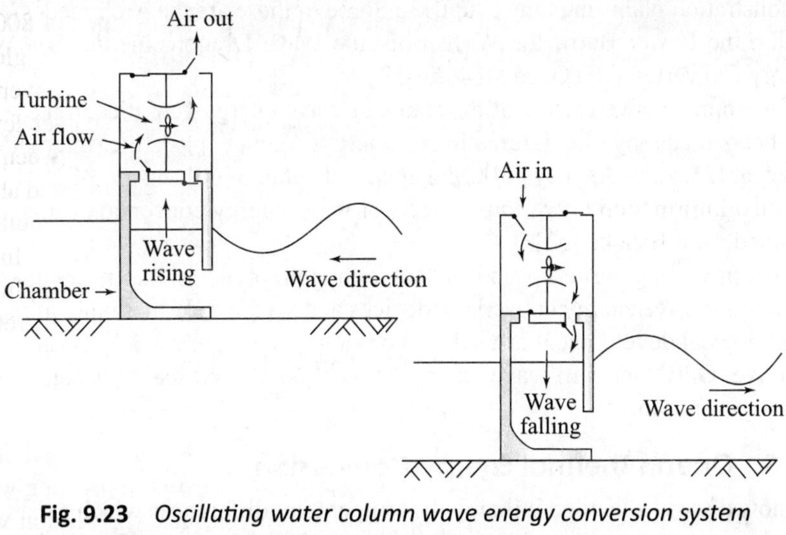

Fig. 9.23 *Oscillating water column wave energy conversion system*

In India, a good beginning was made for utilizing wave energy by designing, building and operating an OWC system at Thiruvananthapuram in 1991 [54].

The average wave power potential at the site is about 13 kW/m. The station is located at the tip of an existing breakwater wall at a water depth of 10 m. The chamber is approximately 10 m × 10 m and 15 m in height. In order to reduce the mechanical complexity associated with having four valves, which are required to open and close with a period of a few seconds, a turbine which rotates in the same direction independent of the direction of air flow has been used. Such a turbine uses a symmetrical aerofoil concept. At the Thiruvananthapuram station, the turbine is 2 m in diameter and is coupled to a matching induction generator. The system was designed to deliver an average power of 75 kW for eight months from April to November and 25 kW for four months from December to March. Subsequently, the plant was modified and operated at a lower power level of 18 kW.

❑ *Pelamis System* The Pelamis system is a semi-submerged device. It consists of a number of cylindrical sections, which are connected by hinged joints. These joints move with respect to each other in the waves. This motion of the joints is resisted by hydraulic rams, which pump high-pressure oil through hydraulic motors via smoothing accumulators. The hydraulic motors drive electrical generators to produce electricity. Several devices can be connected together and linked to the shore through a single seabed cable. The machine is held in position by a mooring system.

The world's first commercial wave energy power plant using Pelamis systems was built at Agucadoura, Portugal. It produced 2.25 MW of power using three systems, each producing 750 kW [55]. The plant has been subsequently closed down due to technical reasons.

❑ *Other Systems* In addition to the OWC system and the Pelamis system, other systems have also been developed and are being tested in pilot or demonstration plants in many countries. Some of these are the Archimedes Wave Swing, the Power Buoy, the Wave Bob, the Wave Dragon, Inertial Sea Wave Energy Converter, CETO, etc. [44, 56-59].

A comprehensive review of the status of wave energy activities in the world has been made by the International Energy Agency [49], the International Renewable Energy Agency [47], Magagna and Uihlein [60] and the REN 21 [44]. Detailed information on various aspects of wave energy conversion can also be obtained from Brooke [53].

The preceding write-up shows that although progress has been made in developing wave energy conversion devices and systems, further innovation and technological development is needed in order to make their use reliable and cost-effective. Only then will wave energy be used on a large scale for commercial power generation.

9.5.3 Ocean Thermal Energy Conversion

We now discuss a method utilizing the temperature differences which exist naturally between the upper and lower layers of water in the ocean. Tropical oceans collect and store very large amounts of solar energy. Utilization of this energy with its associated temperature difference, and its conversion into work forms the basis of ocean thermal energy conversion (OTEC) systems. According to the estimates of Taveira-Pinto *et al.*, the global resource is about 61 300 TWh/year [50].

Historically, it is worth noting that the concept of using the ocean temperature difference to produce power was successfully demonstrated first by Claude in 1930. Claude designed and constructed a small open cycle system off the coast of Cuba. Although the system built by him worked, no net power was obtained because the power required for pumping cold water from the ocean depth was larger than the power developed by the turbine. More recently, two pilot plants have been built. The first plant, called a mini-OTEC plant, was built on board a US Navy barge located off the coast of Hawaii. The system operated for sometime in 1979 producing a small amount of net power. The second plant was installed in the Republic of Nauru in the Central Pacific Ocean in 1981. During testing, it developed a maximum gross power output of 120 kW and a net output of 31.5 kW. Both plants worked on the principle of a closed Rankine cycle. In addition to the above two closed cycle plants, an open cycle OTEC plant for producing power and desalinated water was operated from 1993 to 1998 in Hawaii. In India, an open cycle plant producing 100 000 litres of desalinated water per day has been operating at Kavaratti in Lakshadweep since 2005.

Principle of Working and Efficiency

OTEC systems utilise the temperature difference between water at the surface and the lower levels of the ocean. This difference is around 20°C and occurs over a few hundred metres. A typical temperature profile is shown in Fig. 9.24. It is seen that the temperature at the surface is fairly constant for the first few metres because of mixing. Subsequently it decreases and asymptotically approaches the value at the lower level.

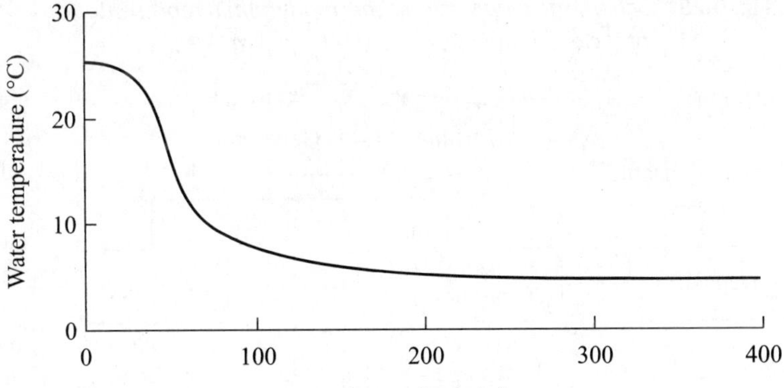

Fig. 9.24 *Typical variation of temperature with ocean depth*

The systems proposed work either on a closed Rankine cycle or an open Claude cycle. The closed cycle system uses a low boiling point working fluid like ammonia, propane, R22, etc. Warm surface sea water is pumped through an evaporator where the working fluid in the form of a high pressure liquid absorbs heat and is converted into vapour. The vapour drives a turbine-generator producing electricity. Then the vapour is condensed in a condenser using cold water brought up from the ocean depths through a cold water pipe. The condensed liquid is pressurised before being fed into the evaporator again, thereby closing the cycle. A schematic diagram and a thermodynamic *T-s* representation is shown in Fig. 9.25.

The open cycle system uses sea water as the working fluid. A schematic diagram is shown in Fig. 9.26. The warm surface sea water is degasified and then evaporated in a flash evaporator under vacuum, thereby producing low pressure steam. The steam is expanded in a turbine, which is coupled to a generator producing electricity. The cold water from the deep sea is used to condense the steam in a condenser. Thus desalinated water is obtained as a useful by-product.

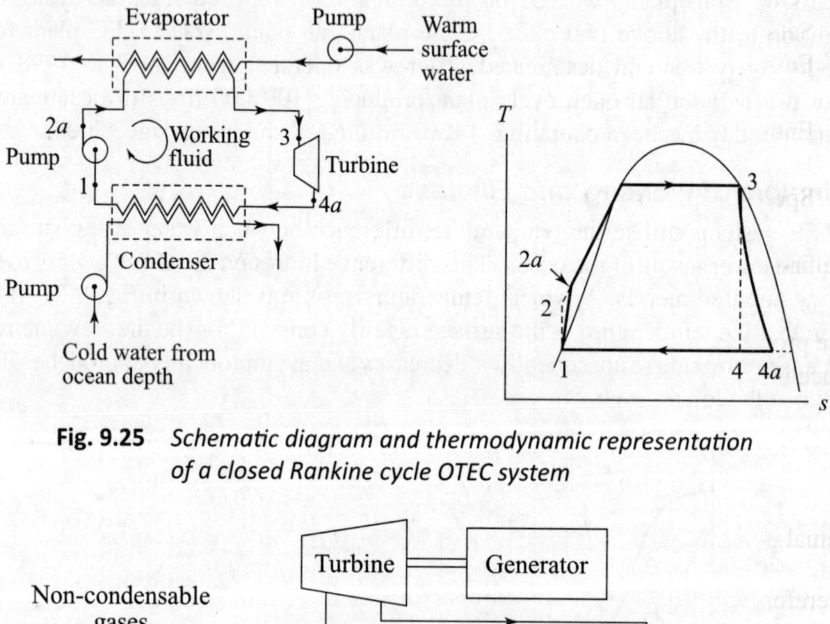

Fig. 9.25 *Schematic diagram and thermodynamic representation of a closed Rankine cycle OTEC system*

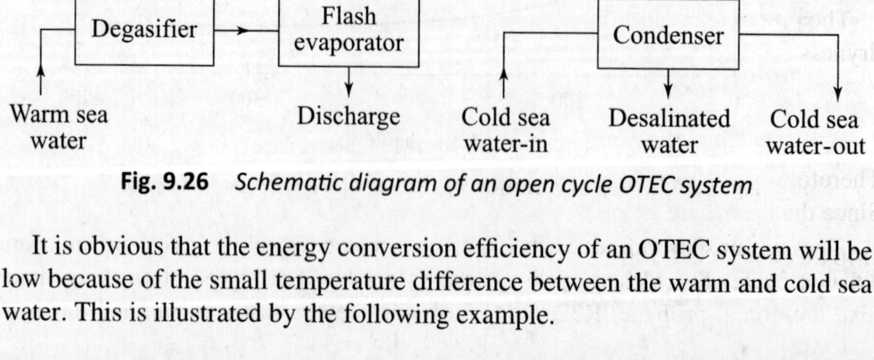

Fig. 9.26 *Schematic diagram of an open cycle OTEC system*

It is obvious that the energy conversion efficiency of an OTEC system will be low because of the small temperature difference between the warm and cold sea water. This is illustrated by the following example.

Example 9.11

Calculate the efficiency of a closed cycle OTEC system using ammonia as the working fluid and installed at a location where the warm and cold sea water streams are at temperatures of 29° and 8°C respectively. Make an allowance of about 5°C for the temperature difference required in the evaporator and condenser for transferring heat, and assume that the ammonia is evaporating at 24°C and condensing at 13°C. Take the isentropic efficiencies of the turbine and pump to be 90 and 80 per cent respectively.

Solution

The thermodynamic cycle is shown on a *T-s* diagram in Fig. 9.25. The ideal cycle with isentropic expansion and compression is 1–2–3–4, while the actual cycle because of non-isentropic expansion and compression is 1–2*a*–3–4*a*. The relevant properties of ammonia are as follows:

Temperature (°C)	13	24
Saturation pressure (bar)	6.813	9.725
Enthalpy – saturated liquid (kJ/kg)	241.5	293.7
–saturated vapour (kJ/kg)	1456.2	1464.6
Entropy –saturated liquid (kJ/kg-K)	0.929	1.107
–saturated vapour (kJ/kg-K)	5.175	5.048
Specific volume of saturated liquid (m^3/kg)	1.612×10^{-3}	1.655×10^{-3}

We first determine the enthalpies at the points 1, 2*a*, 3 and 4*a*

h_1 = enthalpy of the saturated liquid at 13°C = 241.5 kJ/kg

The process 1 to 2 represents isentropic compression of a liquid.
Since the liquid is essentially incompressible, ideal pump work

$$= \left(\frac{1.612 + 1.655}{2}\right) \times 10^{-3} \times (9.725 - 6.813) \times 10^2 = 0.48 \text{ kJ/kg}$$

$$\text{Actual pump work} = \frac{0.48}{0.8} = 0.60 \text{ kJ/kg}$$

Therefore $\quad h_{2a} = 241.5 + 0.60 = 242.1$ kJ/kg

h_3 = enthalpy of saturated vapour at 24°C = 1464.6 kJ/kg

The process 3 to 4 represents isentropic expansion of a vapour. We obtain the dryness fraction at 4 by equating the values of entropy at 3 and 4. Thus,

$$5.048 = 0.929 + x_4(5.175 - 0.929)$$

$$x_4 = 0.970$$

Therefore, $\quad h_4 = 241.5 + 0.970(1456.2 - 241.5) = 1419.8$ kJ/kg

Since the isentropic efficiency of the turbine is 0.9,

$$(1464.6 - h_{4a}) = 0.9(1464.6 - 1419.8)$$

$$h_{4a} = 1424.3 \text{ kJ/kg}$$

$$\text{Net work} = (h_3 - h_{4a}) - (h_{2a} - h_1)$$
$$= (1464.6 - 1424.3) - (242.1 - 241.5) = 39.7 \text{ kJ/kg}$$
$$\text{Heat input} = h_3 - h_{2a} = 1464.6 - 242.1 = 1222.5 \text{ kJ/kg}$$
$$\text{Efficiency} = \frac{39.7}{1222.5} \times 100 = 3.25\%$$

In spite of the low energy conversion efficiency, the concept of an OTEC system may eventually turn out to be economically viable because both the collection and storage of solar energy is being done free by nature.

System Designs

Major system designs for large size OTEC plants were carried out in the seventies by five organisations in USA: TRW, Lockheed, University of Massachusetts (UMASS), Carnegie Mellon University (CMU) and Sea Solar Power, Inc. (SSPI). All the systems were based on a closed Rankine cycle and were proposed to be located offshore on large floating platforms or inside floating hulls. The TRW and Lockheed designs were based on available state-of-the-art know. On the other hand, the other studies were based on novel heat exchanger configurations for the boiler and condenser. The investigators felt that these configurations could be developed and that this developmental work was essential if the OTEC system was to become economically competitive.

McGowan [61] has made a comparison of the thermal parameters of the design. This is given in Table 9.3. It will be seen that three designs were for net power outputs of 100 MW, one was for 160 MW and one for 400 MW. The sites for which the plants were designed were different, and the temperature difference available varied from 17.8° to 22.2°C. Nevertheless, the net cycle efficiency was about the same in all the cases and ranged from 2.1 to 2.4 per cent.

Table 9.3 OTEC systems – Comparison of thermal cycle parameters of various investigators [61]

Parameter	TRW	Lockheed	UMASS	CMU	SSPI
Net power output (MW)	100	160	400	100	100
Gross power output (MW)	125	240	480	140	125
Number of modules	4	4	16	4	4
Working fluid	Ammonia	Ammonia	Propane	Ammonia	R-12/41
Specific working fluid flow rate (kg/s/MW net)	32.6	38.4	116	34.5	234
Site ΔT (°C)	21.8	18.5	17.8	22.2	20.0
Working fluid ΔT (°C)	11.3	12.0	9.4	11.1	10.0
Turbine efficiency (per cent)	90	90	90	90	90

(*Contd.*)

Table 9.3 (*Contd.*)

Net cycle efficiency (per cent)	2.4	2.1	2.4	2.3	2.3
Cold water flow (m^3/s)	370	1800	1470	430	330
Cold water ΔT (°C)	2.7	1.0	2.6	2.0	3.3
Hot water flow (m^3/s)	440	1370	4520	515	620
Hot water ΔT(°C)	2.3	1.35	0.9	2.0	1.7

Ammonia was chosen as the working fluid by three investigators, while R-12/41 and propane were chosen by the other two. From a heat transfer and thermodynamic standpoint, ammonia is a superior fluid. It yields a much smaller value of the specific fluid flow rate, which in turn leads to smaller component sizes. However, other considerations like safety and materials compatibility are also important and have probably prompted the use of the other two fluids.

Because of the low cycle efficiency, the heat to be transferred in the boiler and condenser is large. In addition, the temperature difference between the sea water and the working fluid in these heat exchangers has to be restricted to very small values. For these reasons, very high flow rates are required for the sea water both in the boiler and the condenser. This results in high pumping power requirements and is reflected in the gross power outputs indicated in Table 9.3, which are 20 to 50 per cent higher than the net power outputs.

A second important consequence is that both the evaporator and the condenser are much larger in size than similar components in conventional practice. It will also be noted that the materials suggested for these heat exchangers are titanium or an alloy of copper and nickel. This is necessitated because of the corrosive nature of sea water. As noted earlier, TRW and Lockheed opted for available know-how. For this reason, they selected shell-and-tube configurations for the heat exchangers. On the other hand, UMASS selected a plate-fin compact heat exchanger design which would need to be developed, while CMU proposed a shell-and-tube design with certain augmentative features like externally fluted tubes.

The problem of conveying large volumes of cooling water from the lower depths is also of considerable importance. Three different materials, fibreglass reinforced plastic, prestressed concrete and aluminium were suggested for the cold water pipe.

None of the system designs described in the preceding paragraphs were executed primarily because of the large costs involved. However interest in OTEC systems continues and a number of feasibility studies, simulations and sub-system concepts have been reported.

Results on Pilot Plants

As mentioned earlier, two pilot plants operating on the closed cycle have been built so far. The results obtained with these systems will now be briefly discussed [62, 63].

The mini-OTEC plant in Hawaii employed a Rankine cycle with ammonia as the working fluid.. The plant was designed to produce 50 kW of gross power with a design sea water temperature difference of 21°C and a net generated power of about 15 kW. A description of the main components is given in Table 9.4.

In the power cycle, the ammonia was vapourized at 903 kPa with an inlet sea water temperature of 27°C. The evaporator heat load was approximately 2.7 MW and the ammonia vapour flow rate 2.2 kg/s. After expanding through the turbine, the vapour was condensed at a pressure of 630 kPa, the condenser heat load being about 2.6 MW.

Table 9.4 Description of components used in Mini-OTEC [62]

Component	Description
Evaporator and condenser	Plate-type heat exchangers having an area of 408 m^2 each. Titanium plates, 0.6 mm thick, with a plate separation of 3.9 mm.
Turbine	28 200 rpm single stage, radial inflow turbo- expander
Generator	3600 rpm, 3-phase synchronous/rotating field type with a shaft mounted brushless exciter. Output voltages 120/208 V at 60 Hz. Generated power 50 kW at a power factor of 0.8 and a full load efficiency of 87.6%
Cold water pipe	High density polyethylene, 0.56 m I.D., 0.61 m O.D., 661 m long

During a typical run, 53 kW of gross power was generated. Out of this, 35 kW was parasitic power resulting in a net power output of 18 kW. 70 per cent of the parasitic power load was used in pumping the warm and cold water streams.

A possible method of rating an OTEC plant is the ratio of the gross-to-net power. Mini-OTEC had a gross-to-net ratio of about 3, which was much larger than the values of 1.2 to 1.5 typical of a full scale OTEC plant (Table 9.3). The main reason for this difference was the relatively low efficiency values of many of the components used. These were off-the-shelf items not specifically designed for the job at hand. For example, the turbine-generator set had an overall efficiency of only 52 per cent. With an efficiency of 82 per cent, which is feasible, an increase of 30 kW in the net power would have been obtained. Similarly, an additional 10 kW could have been obtained with improved pumping efficiencies.

Unlike the Hawaii plant, the plant at Nauru was an on-shore plant. This type of plant is advantageous from the point of view of construction costs. However, it requires a longer cold water pipeline resulting in a need for more pumping power. At Nauru, a length of 945 m was required to reach a depth of 580 m. The plant was designed to produce 100 kW of gross power with a sea water temperature difference of 21.7°C. R22 was the working fluid operating in a Rankine cycle. The specifications of the main components are given in Table 9.5.

In the cycle, the R22 was vaporized at a pressure of 1040 kPa and a temperature of 24.8°C. The vapour flow rate was 20.5 kg/s. After expansion, the vapour was condensed at a pressure of 745 kPa and a temperature of 13.1 °C.

Table 9.5 Specifications of components used in Nauru pilot plant [63]

Component	Specification
Evaporator	Horizontal tube type
Condenser	Vertical tube type
Turbine	Axial flow, 3000 rpm
Generator	Air cooled horizontal rotating field type, 3000 rpm. Output voltage 415 V at 50 Hz
Cold water pipe	Polyethylene, 0.7 m diameter, 945 m long

During the test runs, a maximum gross power output of 120 kW and a net power output of 31.5 kW was obtained. These results were better than the design expectations.

The operation of the mini-OTEC and Nauru plants demonstrated the technical feasibility of OTEC power generation using a closed Rankine cycle. The operational results also provided confidence in the performance levels expected from future full-scale power plants.

Based on the previous experience, a 100 kW closed cycle OTEC plant has been installed in Hawaii and connected to the electric grid in 2015 [44]. It uses deep ocean water at 5°C and surface sea water at 25°C to run a closed cycle ammonia-based plant. The cold water is taken from a depth of 670 m. This is the only operating commercial plant in the world.

We now turn our attention to a description of the two open cycle plants mentioned earlier. An OTEC plant was erected at the Natural Energy Laboratory in Hawaii in 1993. It was a land-based experimental facility designed for an output of 210 kW with 26°C warm surface water and 6°C deep water temperature [64]. About 10 per cent of the steam produced in the flash evaporator (Fig. 9.25) was diverted to a surface condenser for the production of desalinated water. The plant operated for six years till 1998. It achieved a gross power output of 255 kW, a corresponding net power of 103 kW and produced 0.4 litres per second of desalinated water.

In India, efforts were initially made to build a closed cycle OTEC plant with a 1 MW output off the coast of Tamil Nadu. However, power generation does not seem to be viable at the moment. Consequently, an open cycle system for producing desalinated water only has been commissioned. The plant located at Kavaratti (Lakshadweep) has been giving an output of 100 000 litres per day. The plant is housed in a structure on the shore, the temperatures of warm and cold water at the site being 28°C and 13°C respectively [65].

Recent Advances

Several research institutions and industrial companies are pursuing the development of OTEC technology. Some of the salient developments can be summarised as follows.

- Efforts are being made to identify more efficient working fluids based on the experience of other types of plants, especially geothermal energy plants.

- In order to increase the overall efficiency of an OTEC plant, various designs of multi-stage heat exchangers are being evaluated.
- Development of durable materials and solutions to bio-fouling and corrosion problems are being pursued.
- The concept of multiple-product OTEC systems is being developed. In addition to generating electricity and producing desalinated water, the possibility of re-circulating or dedicating cooling water for air-conditioning or circulating residual cooling water for aqua-culture and agriculture is being explored.
- Other issues receiving attention are the environmental impact due to the discharge of large volumes of water, the possible release of toxic chemicals (such as ammonia and chlorine), and the impingement of organisms on intake screens [66].

Needless to say, cost remains the major issue from the point of view of large scale commercialisation.

9.5.4 Energy in Marine Currents

Marine currents are formed due to a number of the reasons, viz. (i) temperature and salinity differences between different parts of the sea, (ii) coriolis effects caused by the rotation of the earth, and (iii) the rise and fall of tides. They are also referred to as tidal streams. Although the kinetic energy in marine currents is generally diffuse, it gets concentrated at many locations off-shore where the sea flows are channelled and accelerated because of the presence of straits and islands. These locations are suitable from the point of view of extracting energy from marine currents and converting it into electricity. At such locations, the flows which drive the currents are reasonably predictable. Consequently the quantity of energy resource available as a function of time can also be predicted by making suitable measurements.

As an energy resource, marine currents are analogous to wind energy. Thus the kinetic energy of marine currents can be extracted by using a horizontal axis turbine in much the same way that a wind turbine extracts energy from the wind. Utilizing marine currents requires only that the turbine be held in the flow stream by some type of anchoring system. It should be noted that there is no need to impound water with the help of dams as is the case with the tidal energy. At present, some marine energy pilot plans are in operation and a few are being installed [44, 67, 68].

The energy resources available from suitable locations all over the world are estimated to be about 800 TWh/year [49]. This has been assessed by making direct measurements and by using various modelling techniques [48, 67]. The suitable locations identified are along the coast of Britain and Ireland, between the Channel islands and France, in the Straits of Messina between the mainland of Italy and Sicily, around the Greek islands and along the coasts of Canada, Japan, Russia, Australia, China and USA [51, 59]. No estimate is available for India.

Equation (9.12) given earlier for calculating the power extracted by the rotor of a wind machine is also valid for marine tidal current machines. For these machines, the value of the power coefficient lies in the range 0.35 – 0.51.

Example 9.12

Calculate the power generated by a horizontal axis marine current propeller turbine having a rotor diameter of 8 m if the speed of the water current available is 3 m/s. Take $C_p = 0.41$.

The wind speed available at a nearby location on shore is 11.2 m/s. Calculate and compare the size of a wind turbine rotor which would generate the same power as the marine current turbine. Assume that the value of C_p is still 0.41.

Solution

Substituting the values of $V_\infty = 3$ m/s, rotor diameter = 8 m, power coefficient (C_p) = 0.41 and density of water = 998 kg/m^3 in Eq. (9.12), we get

$$P_e = 0.41 \times \left[\frac{1}{2} \times 998 \times \frac{\pi}{4} \times 8^2 \times 3^3\right] \times 10^3 = 277.7 \text{ kW}$$

Taking the density of air 1.2 kg/m^3, the diameter of the rotor of a wind machine generating the same power is given by

$$0.41 \times \left[\frac{1}{2} \times 1.2 \times \frac{\pi}{4} \times D^2 \times 11.2^3\right] = 277.7 \times 10^3$$

$$D = 32.0 \text{ m}$$

Note that the diameter of the wind machine rotor is four times that of the marine current turbine's rotor. This is because the density of the air is much less than that of water and in spite of the fact that the wind speed at the location is more than three times the marine current speed.

CAN RENEWABLE ENERGY SOURCES MEET INDIA'S FUTURE NEEDS OF ELECTRICITY? 9.6

9.6.1 Looking Ahead

Thus far, we have discussed solar energy (photovoltaic conversion and solar thermal energy) in detail from Chapters 2 to 8 and other renewable energy sources (wind energy, small hydroelectric power, biomass, etc.) briefly in Sections 9.2 to 9.5. India's future needs of electricity have also been estimated in Chapter 1 (Section 1.3.6) to be 6800 TWh assuming that the mean annual per capita need reaches a reasonable value of 4000 kWh. This need would arise about 50 years from now around 2060 or 2070 and could stabilize after that.

Recognizing that fossil fuels will eventually be consumed (Fig. 1.16), the question which we will now pose is the following: "Can renewable energy sources eventually meet India's future needs of electricity?" Quite obviously, this question and the answer to it would be of interest because of its long term policy implications.

9.6.2 Estimating the Potential

The technological developments which have occurred so far indicate that the major contributors are likely to be photovoltaic conversion, solar thermal power plants, hydroelectric power and wind energy. We will, therefore, essentially focus our attention on these renewable sources and estimate their potential for generating electricity [69]. We consider the resources one-by-one.

❑ *Solar Power (PV Conversion and Thermal)* The growth in solar PV power in India in the last five years has been spectacular and the installed capacity has crossed 5000 MW. In comparison, very few solar thermal power plants have been set up because they arc more costly. However, the position is likely to change in the future as the advantage of being able to store thermal energy more easily is realized and then cost reduction would be expected. What is the grid-connected capacity that can be installed in the long run via these two routes? Keeping in mind that only barren, uncultivated land can be used for this purpose, it is clear that in India availability of land is going to be the important constraint. We will focus our attention on the land area required for setting up PV plants.

In Example 8.5, we obtained a value of 1.46 ha/MW. Based on the data available from many existing MW-sized PV plants, it is known that about 2 ha/MW has been used. In order to be conservative, we will use this value. This number has also been suggested by the Solar Energy Corporation of India [70]. Further, we will assume that this value is also valid for solar thermal power plants. Data on a few existing solar thermal power plants show that this is a reasonable assumption.

The total amount of barren and uncultivated land in India is about 200 000 sq km. Most of this land is in Rajasthan, Gujarat and parts of the Deccan Plateau. Acquiring land for any industrial purpose is difficult and it is estimated that while it may be possible to acquire 5 or 10 per cent of the land, it will be difficult to acquire more than 20 per cent for generating solar power. We will use 20 per cent as a reasonable upper bound.

The other issue to be considered is the value of the annual plant load factor (APLF). The value of APLF is low for a solar plant because sunshine is only available for nine or ten hours and the average value during this period is only about half the peak value of radiation attained around noon. Typical values of APLF for plants performing well lie in the range 0.17 – 0.19. We will use a value of 0.18.

Thus, the upper limit on the solar power generation capacity

$$= \text{Area of land which can be utilized / Area required per MW}$$

$$= [0.20 \times 200\,000 \times 10^2 / 2] \text{ MW}$$

$$= 2000 \text{ GW}$$

$$\text{Annual production of electricity} = 2000 \times 8760 \times 0.18 \times 10^3 \text{ TWh}$$
$$= 3153.6 \text{ TWh}$$

❑ *Decentralized Solar PV Systems* Apart from MW-scale power plants connected to the grid, PV conversion is also used to generate power in decentralized stand-alone kW-scale systems which are located mostly on roof tops or in small open spaces and supply the needs of a house or a few houses. Typically such systems would have a capacity of about 3 kWe, require 30 m^2 of area and generate 4.5 MWh/year [71]. It is reasonable to expect that they would be installed in large numbers in India. India is likely to have about 425 million households in 2070 when its population would have stabilized. Half of the households would be in urban areas where it would be difficult to install roof-top systems. Even in the other half in rural areas, only those who are economically well-off and have a suitable roof would have some interest in installing PV systems. Keeping all these factors in mind, an optimistic estimate is that with some financial incentives, it may be eventually possible to induce at most about 10 per cent of the rural households, i.e. 20 million households, to have solar roof-top systems. The electricity generated by 20 million roof-top systems each producing 4.5 MWh/year = $20 \times 10^6 \times 4.5$ MWh = 90 TWh

❑ *Hydroelectric Power* In Section 1.3.2, it was indicated that the reserves of hydroelectric power for large capacity plants have been estimated to be 148 700 MW. For small plants, a potential of 19 749 MW has been identified (Section 9.4.1). Presently, the installed capacity of large plants is about 25 per cent of the potential. It has become increasingly clear that exploitation of all the potential is improbable because of the large amount of land submerged behind the dams, the consequent displacement of people and the destruction of flora and fauna. However, in the future it should be possible to utilize at least 40 per cent of the reserves and eventually perhaps upto 80 per cent. It will be difficult to exceed this value because of environment constraints. The average value of the APLF is at present 0.37.

Utilizing these values, the upper limit of annual production of hydroelectric power (large and small units)

$$= 0.80\,(148\,700 + 19\,749) \times 8760 \times 0.37 \times 10^{-6} \text{ TWh}$$
$$= 436.8 \text{ TWh}$$

❑ *Wind Energy* Next to photovoltaic conversion, the most promising developments in the field of renewable energy have been in wind energy. As stated in Section 9.2.2, the potential on land has been estimated by to be 2006 GW [3]. We will use this estimate. Combining it with a conservative value of 0.20 for APLF, the annual production of the electricity works out to be 2108.7 TWh if 60 per cent of the potential is exploited.

The potential of wind energy off-shore could be as much as 1100 TWh [6]. Assuming again that a maximum of 60 per cent of the resource can be eventually utilized, we obtain an annual production figure 660 TWh.

❑ *Biomass Power* It has been stated earlier in Section 9.3 that India's total potential for generating electricity from biomass is only about 25 000 MW. Since the availability of biomass is seasonal, an APLF of 0.3 is assumed. We assume further that the potential will be fully harnessed. This yields an annual electricity production = $(25\,000 \times 8760 \times 0.3 \times 10^{-6})$ TWh = 65.7 TWh.

❑ *Wave Energy, Ocean Thermal Energy Conversion and Energy in Marine Currents* Development of systems for extracting electricity from wave energy has been described in Section 9.5.2. Although some commercial activity has taken place, the chances that a significant amount of electricity on the TWh scale will be generated in India are negligible.

With regard to OTEC, there is only one 100 kW plant in operation in the world (Section 9.5.3) and it does not seem likely that OTEC systems will contribute any significant amount of electrical power in India.

Utilizing the energy available in marine currents has not received much attention in India (Section 9.5.4). Developments thus far do not give any indication that this resource will deliver significant amounts of electrical energy.

❑ *Tidal Energy* As far as tidal energy is concerned, a potential of 16.6 TWh has been identified at two locations in India (Section 9.5.1). We will assume that tidal power stations will eventually be constructed at these two sites and their potential realized.

9.6.3 Total Potential of All Renewable Sources to Supply Electricity

The total potential of all renewable sources to supply electrical energy in India in the future is obtained by summing up the estimated production of the individual sources. A compilation is presented in Table 9.6. The data presented show that in the long run, solar power (PV and thermal) and wind energy followed by hydroelectric power are likely to be the major contributors to the renewable energy scene. The calculations also show that with the assumed upper limits, the total production could be 6531.4 TWh. It is to be noted that the total is an estimate and that it is sensitive to the upper limits assumed. Consider for example, the contributions of the two principal contributors, viz. solar power (No. 1a) and wind energy (No. 3). If it is assumed that 25 per cent of barren and uncultivated land could be acquired for installing solar power plants, the annual electricity production would be 3942 TWh. Also, if wind energy could be eventually exploited to 70 per cent of its potential (on land and off-shore), the estimated annual electricity production would increase to $(2108.7 + 660) \times (70/60) = 3230.2$ TWh. The net effect of these two increases would be to increase the total annual production to 7781.3 TWh. Similarly, if the percentages were assumed to be 15 and 50 per cent for solar power and wind energy respectively, we would obtain a total annual production of 5281.7 TWh. These production figures indicate the sensitivity of the total to the percentage values assumed.

Nevertheless, all the three annual production figures, 5281.7, 6531.4 and 7781.3 TWh, are close to the annual requirement of 6800 TWh. Thus, it is clear that renewable energy sources will play an important role in supplying India's future needs and may eventually come close to fulfilling that role to a large extent.

Table 9.6 Total potential of all renewable sources for generating electrical energy in the future in India. Adapted from Sukhatme [69]

No	Renewable source	Annual electricity production (TWh)	Estimates based on the following upper limts
1a	Solar power (PV + thermal)	3153.6	Use of 20% of barren and uncultivated land
1b	Decentralized roof-top PV	90	20 million rural households
2	Hydroelectric power (large and small units)	436.8	Utilization of 80% of potential
3	Wind energy - on land - off shore	 2108.7 660	Utilization of 60% potential
4	Biomass	65.7	Utilization of full potential
5	Ocean energy: (Tidal energy, wave energy, OTEC and marine currents)	16.6	Utilization of full potential estimated for tidal energy. Remaining three sources have negligible potential
	Total	6531.4	

9.6.4 Technical Challenges in Using Renewable Energy Sources

The fact that renewable energy sources may be able to meet or come close to meeting the future annual demand of electrical energy is clearly not enough. The demand has to be met not only on an overall basis, but continuously round-the-clock. This is an exacting requirement. We have seen earlier that both solar and wind power are intermittent suppliers. On a sunny, cloudless day, solar energy is available for only about 8 or 9 hours with a peak around noon (see Fig. 3.8). Consequently, electricity is generated only for those hours unless the sun's energy is stored. Similarly, at any location, adequate wind energy is generally available for only 5 to 6 hours in a day and electricity is produced only if the wind speed is above the cut-in speed. In contrast, there is a demand throughout the 24 hours of a day. Typically, the demand curve has two gentle peaks, one around noon and the other in the late evening, and passes through a minimum value in the early hours of the morning around 3 am. Thus many technical challenges will have to be overcome if in the future, renewable sources alone have to meet our electricity needs fully. A major challenge will be to coordinate the supply of thousands of relatively small distributed solar and wind power sources using appropriate software and to take that supply with the help of a robust and extensive transmission grid from where it is produced to where it will be consumed. A similar challenge on the demand side will be to develop computer-controlled load controlling capabilities. Grids which perform such tasks are called 'smart' grids. They would need to be designed and developed.

A further challenge will be to develop energy storage systems both for short-term and long-term storage. Short-term storage systems help to even out the daily mismatch between the hourly variation of the power supplied by solar and wind energy systems and the variation of the demand. On the other hand, long-term storage systems are needed to store energy on days when supply exceeds demand and to provide energy to match demand during days when there are lulls in wind speed or periods of cloudy weather.

As of now, it appears that short-term energy storage systems could be pumped hydroelectric power storage systems, electric battery storage systems, thermal energy storage tanks or compressed air storage systems. Long-term energy storage systems would probably be chemical storage. Hydrogen or methane gas is produced when excess electricity is available. It is stored at a suitable temperature and pressure and burnt to generate energy when it required.

In the long run even if the technical challenges can be resolved, the real issue will be the cost of the energy produced. Is the economics affordable? It is appropriate therefore that the last chapter of this book should concern itself with making economic analyses of renewable energy systems.

PROBLEMS

9.1 Fit a Weibull distribution to the wind speed data of Kandla (Table A5.1) and compare it with the measured data in Fig. 9.6.

[Procedure:

1. First note that Eq. (9.2) can be written as
   $\ln[-\ln\{1 - F(V)\}] = k \ln V - k \ln c$
2. From the measured data in Table A5.1 obtain values of the cumulative frequency $F(V)$ for intervals 0-2 kmph, 0-4 kmph, etc.
3. Plot $\ln[-\ln\{1 - F(V)\}]$ against $\ln V$ and fit a straight line to the points obtained.
4. The slope and intercept of the line yield the values of k and $-k \ln c$.]

9.2 The average wind speed and standard deviation from a discrete set of measurements at a place are found to be 11.6 m/s and 3.5 m/s respectively. Plot the wind speed-frequency distribution assuming that it is a Weibull distribution. Compare the results with those presented in Fig. 9.8.

9.3 Using the data given in Table A5.1, plot the speed-frequency distribution and speed-duration curve for the months of January and July for Kandla. Compare the variations obtained with the annual values plotted in Figs 9.6 and 9.7.

9.4 Calculate the annual energy content of the wind at Kandla at a height of 23 m. Take $a = 0.3$.

9.5 Calculate the energy available for a wind machine at Kandla at a height of 23 m if the cut-in speed is 14 kmph, the design speed is 36 kmph, and the cut-out speed is 90 kmph.

9.6 Assume that the annual speed-duration curve of a location can be expressed in the simple linear form

$$V = 40[1 - (t/8760)]$$

where V is in kmph measured at a height of 10 m and t is in hours. Calculate the annual energy content in the wind at the location at a height of 20 m. Assume that the value of the power law index is 0.25.

9.7. Evaluate the potential of Mumbai as a location for generating electrical power from wind energy. Given the following annual percentage frequency distribution of hourly wind speed:

V (kmph)	0	0-2	2-4	4-6	6-8	8-10
Frequency (%)	11.2	6.7	9.3	10.8	10.2	11.1
	10-12	12-14	14-16	16-18	18-20	20-22
	8.6	8.4	8.6	5.2	4.4	2.1
	22-24	24-26	26-28	28-30	30-32	
	1.5	1.1	0.4	0.3	0.1	

Height of sensor from ground = 15.3 m.

9.8 The following wind data is available for a place for the month of August.

V (kmph)	0	0-2	2-4	4-6	6-8	8-10
Frequency (%)	7.5	2.5	2.1	3.5	6.8	4.5
	10-12	12-14	14-16	16-18	18-20	20-22
	10.6	12.0	13.4	9.0	8.0	7.5
	22-24	24-26	26-28	28-30	30-32	32-34
	4.5	3.7	1.6	1.4	0.8	0.6

Height of sensor from ground = 10 m

Calculate the monthly energy available to the rotor of a wind machine per square metre of its face area, if the hub height of the rotor is 30 m above the ground.

Given: Cut-in, design and cut-out speeds respectively are 9, 29 and 58 kmph.

Power law index = 0.15

Density of air = 1.18 kg/m^3

9.9 A multi-blade wind mill lifts 1.03 m^3/h of water through a head of 28 m when the wind speed is 3.3 m/s. Calculate the power coefficient if the rotor diameter is 4.5 m, given that the transmission efficiency = 0.95 and pump efficiency = 0.7.

9.10 Find the annual energy output of the wind machine of Example 9.6 if it is located in Indore. Assume $a = 0.25$.

9.11 A biomass gasifier is used to run a compression-ignition engine. The engine operates in the dual fuel mode with 85% diesel replacement. The biomass feed rate for the gasifier is 250 kg/h. Calculate the power produced by the engine.

9.12 An engine-generator system running on biogas is installed to produce electric power. Calculate the power produced if cow-dung is used as the feed material and the volume of the digester is 40 m^3. Use the approximate rules for estimating performance given in Section 9.3.3.

Given: Calorific value of biogas = 21 000 kJ/m^3

Efficiency of the generator = 90 per cent

Efficiency of the engine = 33 per cent

9.13 A Francis turbine operates at a ROR site where the water head is 30 m. The water flow rate through the turbine is 18 770 kg/s and the speed of the generator connected to the turbine is 300 rpm. Calculate the power output and the specific speed of the turbine.

Given: Efficiency of the turbine = 0.91

9.14 It is proposed to install an OTEC plant of capacity 100 MW (net power) in a suitable location off the coast of India. Use the data of Table 9.3 and make an approximate calculation to estimate the magnitude of the input thermal power required.

REFERENCES

1. Global Wind Report. 2015. Global Wind Energy Council, Brussels, Belgium.
2. *Annual Report 2015-16.* 2016. Ministry of New and Renewable Energy, Government of India, New Delhi, India.
3. Phadke, A., Bharvirkar, R. and Khangura, J. 2012. Reassessing wind potential estimates for India: economic and policy implications, *Report No. LBNL- 5077E Revision* 1, Lawrence Berkeley National Laboratory; http://ies.lbl.gov/ India_Wind_potential.
4. http://www.3tier.com/en/package_detail/wind_speed_gis_data_layer/
5. Hossain, J., Sharma, D., Mishra, N. K., Ansari, M. Z. and Kishore, V.V.N. 2016. Re-assessment of wind energy potential with new technology in India. *Wind Engineering*, 40: 379.
6. Liu X., McElroy, M.B. and Kiviluoma, J. 2009. Global potential for wind-generated electricity, Proceedings National Academy of Sciences (PNAS) (http://www.pnas.org/content/106/27/10933.full; accessed on June 9, 2016)
7. DONG Energy installs world's largest wind turbine. 2016. (http://www.cleantechnica.com/2016/09/12dong_energy_installs_worlds_largest_wind_turbine; accessed on October 17, 2016).
8. *S97 2.1 MW platform.* Suzlon Energy Limited, Pune, India. (http://www.suzlon.com/pdf/media_nit/S97_Sl 1 l_productBrochure.pdf; accessed on October 18,2016)
9. Mani, A. and Mooley, D.A. 1983. *Wind Energy Data for India.* Allied Publishers, New Delhi, India.
10. Mani, A. 1990. *Wind Energy Resource Survey for India –I*, Allied Publishers, New Delhi, India.
11. Manwell, J.E., McGowan, J.G. and Rogers, A.L. 2002. *Wind Energy Explained: Theory, Design and Application.* John Wiley, England, UK.
12. Kreyszig, E. 2000. *Advanced Engineering Mathematics.* 8th Edition, John Wiley & Sons (Asia) Pvt. Ltd., Delhi, India.
13. Lanchester, F.W. 1915. Contribution to the theory of propulsion and the screw propeller. *Trans. of the Institution of Naval Architects*, LVII, March 25: 98
14. Betz, A. 1920. Das maximum der theoretisch moeglichen Ausnuetzung des windes durch windmotoren. *Zeitschrift fuer das gesam to turbinenwesen,* Heft 26, September 26.
15. Froude, W. 1878. On the elementary relation between pitch, slip and propulsive efficiency. *Trans. Institute of Naval Architects*, 19: 47.

16. Glauert, H. 1935. *Airplane propellers. Aerodynamic Theory* (ed. W.F. Durand), Div. L, Chapter XI, Springer Verlag, Berlin, Germany.

17. Jansen, W.A.M. and Smulders, P.T. 1986. *Rotor design for horizontal axis windmills.* CWD 77-1, Consultancy Services Wind Energy Developing Countries, The Netherlands.

18. Johnson, G.L. 1985. *Wind Energy Systems,* Prentice Hall, New Jersey, USA.

19. Kloeffler, R.G. and Sitz, E.L. 1946. *Electrical energy from winds. Experiment Station Bulletin 52*, Kansas State College of Engineering, Manhattan, Kansas, USA.

20. Savonius, S.J. 1931. The S-rotor and its applications. *Mechanical Engineering,* 53(5): 333.

21. Spera, D.A. 1994. *Wind Turbine Technology: Fundamental Concepts of Wind Turbine Engineering*, ASME Press, New York, USA.

22. Mathew, S. 2006. *Wind Energy :Fundamentals, Resource Analysis and Economics,* Springer Verlage, Berlin, Germany.

23. Nelson, V. 2009. *Wind Energy: Renewable Energy and the Environment*, CRC Press, New York, USA.

24. Jain, P. 2011. *Wind Energy Engineering*, McGraw Hill, New York, USA.

25. Bailey, H., Brookes, K. and Thompson, P.M. 2014. Assessing environmental impacts of offshore wind farms: Lessons learned and recommendations for the future. *Aquatic Biosystems*, 10:8.

26. Wang, S., Wang, S. and Smith, P. 2015. Quantifying impacts of onshore wind farms on ecosystem at local and global scales. *Renewable and Sustainable Energy Reviews*, 52: 1424.

27. Kaldellis, J.K., Apostolou, D., Kapsali, M. and Kondili, E. 2016. Environmental and social footprint of offshore wind energy: Comparison with onshore counterpart. *Renewable Energy*, 92: 543.

28. NIR Board, 2004. *Handbook on Biogas and its Applications*. National Institute of Industrial Research, New Delhi, India.

29. *Salient Features of 19th Livestock Census*. Press Information Bureau, Government of India.

(http://www.pib.nic.in/newsite/PrintRelease.aspx?relid=l09280; accessed on November 4, 2016)

30. Badger, P.C. 2002. *Ethanol from cellulose: A general review*, in *Trends in new crops and uses,* J. Janick and A. Whipkey (eds.), ASHS Press, Alexandria, VA, p. 17-21

31. Achinas, S. and Euverink, G.J.W. 2016. Consolidated briefing of biochemical ethanol production from lignocellulosic biomass. *Electronic J of Biotechnology*, 23: 44.

32. Maurya, D.P., Singla, A. and Negi, S. 2015. An overview of key pretreatment processes for biological conversion of lignocellulosic biomass to bioethanol. *3 Biotech*, 5:597.

33. Bankovic-Ilic, I.B., Stamenkovic, O.S. and Veljkovic, V.B. 2012. Biodiesel production from non-edible plant oils. *Renewable and Sustainable Energy Reviews*, 16: 3621.

34. Barwal, B.K. and Sharma, M.P. 2005. Prospects of biodiesel production from vegetable oils in India. *Renewable and Sustainable Energy Reviews*, 9: 363.

35. Ramesh, D., Samapathrajan, A. and Venkatachalam, P. 2007. Production of biodiesel from jatropha curcas oil by using pilot diesel plant (http://www.jatropha.de/journal/Pilot_Plant_for_Biodiesel-leafletl.pdf; accessed on May 10,2007).

36. Singh, R.K., Kumar, A. K. and Sethi, S. 2006. Preparation of Karanja oil methyl ester. *Offshore World*, April-May 2006. (http://dspace.nitrkl.ac.in/dspace; accessed on May 10, 2007).

37. Lane, J. 2014. Jatropha biofuel around the world: a 13-country tour of development activity. (http://www.renewableenergyworld.com/articles/2014/09/jatropha-biofuel-around-the-world-a-13-country-tour-of-development-activity.html; accessed on November 22, 2016)

38. *India – Biofuels Annual 2016*, GAIN Report no. IN 6088, Global Agricultural Information Network, USDA Foreign Agricultural Service, New Delhi, India.

39. Borugadda, V.B. and Goud, V.V. 2012. Biodiesel production from renewable feedstocks: Status and opportunities. *Renewable and Sustainable Energy Reviews*, 16: 4763.

40. Kumar, M. and Sharma, M.P. 2016. Selection of potential oils for biodiesel production. *Renewable and Sustainable Energy Reviews*, 56: 1129.

41. *Small Hydropower*, Chapter 3, State of Renewable Energy in India: A citizen's report. 2014. Centre for Science and Environment, New Delhi, India.

42. Kumar, A. 2016. Personal Communication, Indian Institute of Technology, Roorkee, Uttarakhand, India.

43. *IS 12837*- 1989. Hydraulic turbines for medium and large power houses – guidelines for selection. Bureau of India Standards, New Delhi, India.

44. *Renewable 2016: Global Status Report*, 2016. Renewable Energy Policy Network for the 21st Century, Paris, France.

45. Waters, S. and Aggidis, G. 2016. Tidal range technologies and state of the art in review. *Renewable and Sustainable Energy Reviews*, 59: 514.

46. Jeffreys, H. 1973. Tidal friction, *Nature*, 246: 346.

47. *Tidal Energy Technology Brief*, 2014. International Renewable Energy Agency, Bonn, Germany.

48. Uihlein, A. and Magagna, D. 2016. Wave and tidal current energy: A review of the current state of research beyond technology. *Renewable and Sustainable Energy Reviews*, 58: 1070.

49. IEA Report. 2006. *Review and analysis of ocean energy systems development and supporting policies,* AEA Energy and Environment on behalf of Sustainable Energy, Ireland.

50. Taveira-Pinto, F., Tglesias, G., Rosa-Santos, P. and Deng, Z.D. 2015. Preface to special topic: marine renewable energy. *Journal of Renewable and Sustainable Energy*, 7: 061601.

51. EU Report. 2006. *Ocean energy conversion in Europe*, Centre for Renewable Energy Sources, Greece.

52. Falcao, A.F. de O. 2010. Wave energy utilization: A review of the technologies, *Renewable and Sustainable Energy Reviews*, 14: 899.

53. Brooke, J. 2003. *Wave Energy Conversion*, Elsevier, London.

54. Ravindran, M. and Koola, P.M. 1991. Energy from sea waves - The Indian wave energy programme, *Current Science*, 60: 676.

55. Pelamis, World's First Commercial Wave Energy Project, Agucadoura, Portugal, (http:// Power-technology.com/projects/pelamis; accessed on May 25, 2007)

56. *CETO device*, Carnegie Wave Energy limited, Australia, (http://www.carnegiewave.com; accessed on April 22, 2016)

57. Cagninei, A., Raffero, M., Bracco, G., Giorcelli, E., Mattiazzo, G. and Poggi, D. 2015. Productivity analysis of the full scale inertial sea wave energy converter prototye: A test case in Pantelleria island. *Journal of Renewable and Sustainable Energy*, 7: 061703.

58. *Technological status of wave energy*, World Energy Council. (http://www.worldenergy.org; accessed on May 25, 2007)

59. Pontes, M.T. and Flacao, A. 2001. Ocean energies: resources and utilization, *World Energy Council, 18th Congress*, Buenos Aires.

60. Magagna, D. and Uihlein, A. 2015. *2014 JRC ocean energy status report*, Joint Research Centre, European Commission, Petten, The Netherlands.

61. McGowan, J.G. 1976. Ocean thermal energy conversion, a significant solar resource, *Solar Energy*, 18: 81.

62. Owens, W.L. and Trimble, L.C. 1981. Mini-OTEC operational results, *Journal of Solar Energy Engineering, Trans*. AS ME., 103:233.

63. Mitsui, T. Ito, F., Seya, Y. and Nakamoto, Y. 1983. Outline of the 100 kW OTEC pilot plant in the Republic of Nauru. *IEEE Trans on Power Apparatus and Systems*, PAS-102: 3167.

64. Vega, L.A. 1999. Ocean thermal energy conversion (http://www.otecnews.org/articlcs/vega; accessed on May 26, 2007).

65. Low Temperature Thermal Desalination (LTTD) Plant at Kavaratti (http ://www.niot.res.in; accessed on May 26, 2007)

66. Pele, R. and Fujita, R.M. 2002. Renewable energy from the ocean, *Marine Policy*, 26: 471.

67. Bajaj, A.S. 2011. Generating electricity from the oceans. *Renewable and Sustainable Energy Reviews*, 15: 3399.

68. Benelghali, S., Benbouzid, M. and Charpentier, J.F. 2007. Marine tidal current electric power generation technology: State of the art and current status. *IIEEE IEMDC'07*, May 2007, Antalya, Turkey, 2, pp. 1407-1412.

69. Sukhatme, S.P. 2012. Can India's future needs of electricity be met by renewable energy sources? A revised assessment. *Current Science*, 103(10): 1153.

70. *Frequently Asked Questions*, Solar Energy Corporation of India Limited, New Delhi, India.
(http://www.seci.gov.in/content/innerpage/fre.php; accessed on October 28, 2016)

71. Chokshi, A.H. 2012. Nuclear riddles: TINA and NIMBY. *Current Science*, 102: 1096.

Economic Analysis of Renewable Energy Systems

CHAPTER 10

INTRODUCTION 10.1

The value of an energy system must ultimately be judged on the basis of its economy. In this chapter, we define some economic quantities and describe some of the methods used for making evaluations. A simple way to assess the economic viability of a renewable energy system is to calculate the payback period. This is described in Section 10.3. We then derive expressions for determining the present worth of money and calculating the repayment of a loan in equal instalments (Sections 10.4 and 10.5). The economic analysis of a stand-alone renewable energy system is presented in Section 10.6 and expressions are derived for the annual cost, the cumulative cost and the cumulative saving. Following this, we obtain expressions for the same parameters for an add-on system (Section 10.7) and introduce the concept of levelized cost of electricity (Section 10.8). Next we describe the clean development mechanism in the Kyoto Protocol which provides a means for making more saving in renewable energy systems installed in developing countries (Section 10.9). The chapter concludes (Section 10.10) with a discussion on the limitations of the economic analyses described and the need for doing a sensitivity analysis on the results presented. The description of methods for making economic evaluations given in this chapter are brief and elementary in nature. For more details, the reader is referred to books by Grant *et al.* [1], Degarmo *et al.* [2] and Mathew [3].

INITIAL AND ANNUAL COSTS 10.2

The costing of an energy system is usually considered in two parts – an initial cost and an annual operating cost. As is evident by now, a renewable energy system is characterized by a relatively high initial cost and a relatively low annual cost as compared to a conventional fossil fuel based system. On the other hand, an energy system using a conventional energy source is characterized by a low initial cost and a high annual cost. Thus an economic analysis of an energy system has to consider both these aspects.

❑ *Initial Cost* The initial cost of a solar system is the cost of buying the equipment and installing it. Generally, it is site dependent. With regard to the initial cost of a stand-alone renewable energy system it is generally not designed to meet the complete energy demand of the application for which it is installed. For example, providing a solar hot water system to supply 100 per cent of the hot water requirement of an establishment at all times would result in a system which is oversized for most of the time. The resulting initial cost would make the investment uneconomical. The renewable energy system is thus often used in conjunction with an auxiliary sub-system using conventional energy. The auxiliary source helps to meet unusual high demand situations. It also takes care of situations which arise when the renewable energy is not available in adequate measure because of adverse weather conditions.

In order to promote the use of renewable energy, the burden of the initial cost as well as running expenses is reduced by offering a number of incentives such as capital subsidy, accelerated depreciation and preferential tariff. The nature of these incentives keeps changing from time to time. They vary with the renewable source and are offered by the Central as well as State Governments. For example, in the case of wind power, incentives have been offered in the form of 80 per cent accelerated depreciation as well as a 10 year tax holiday on income from generation of electricity. In addition, there is a concessional customs duty on import of specified components and exemption from excise duty for manufacture of wind electric generators. Some states also offer preferential feed-in tariffs. However now there is no capital subsidy.

For solar PV roof-top systems, central financial assistance of 15 per cent of the total cost is available [4]. In addition, there are fiscal incentives in the form of concessional import duty, excise duty exemption, accelerated depreciation and a tax holiday on income earned as feed-in tariff.

In contrast to wind energy and solar PV roof-top systems, solar water heaters which were among the earliest to enter the renewable energy scene, are now considered to be established technologies with reasonable returns. Earlier the central government encouraged the installation of these systems with a financial subsidy and an excise duty exemption. These incentives have been withdrawn in 2014 and 2015.

❑ *Annual Cost* The net annual cost of an energy system is the sum of a number of factors, some of which are costs and some benefits. The yearly costs generally include (1) the maintenance cost, manpower cost, local taxes, etc., (2) cost of fuel/electricity, (3) repayment of loan taken (if applicable). For a renewable energy system, item (2) refers to the consumption of fuel/electricity by the auxiliary sub-system. Benefits are available in the form of certain types of tax deductions. These serve as incentives for installing the system by helping to reduce the annual cost. They are of various kinds. For example, they may be related to the interest payable on the loan taken or to the annual depreciation permitted on the system. Thus, the net annual cost of the system may be stated as follows:

Net annual cost of an energy system (AC)

= [Yearly costs] - [Yearly benefits]

= [Maintenance charges, local taxes, manpower, etc.

+ Cost of fuel/electricity + Repayment of loan]

– [Tax deductions as available] (10.1)

Equation (10.1) is a general equation applicable to most situations. However, it is possible that some more terms in the form of costs or benefits may have to be added to the right hand side of the equation for some specific cases. Also, in some cases, some of the terms may not be applicable.

PAYBACK PERIOD 10.3

We will first discuss the concept of a payback period. Calculating the payback period is a simple way of assessing the economic viability of a renewable energy system. The payback period (n_p) is the number of years needed for the rupee savings on account of fuel/electricity or the income derived from power fed to the grid to equal the cost of the system. It is a useful and easy quantity to calculate and decide whether one should go ahead for a low investment system. It is also a useful first step even for higher investment systems.

Example 10.1

A 5 kW PV system installed on an office building roof costs Rs 400 000. It helps to save 7200 units of electricity annually. If the cost of electricity is Rs 8.00 per unit, calculate the payback period.

Solution

Payback period = 400 000/(7200 × 8.00) = 6.9 years

Example 10.2

Consider the 100-litre solar water heating system of Example 2.1. Calculate the payback period if the cost of the system is Rs 20 000.

Solution

Reduction in monthly electricity bill = Rs 553

Payback period = 20 000/ (553 × 12) = 3.0 years

PRESENT WORTH CALCULATION 10.4

Individuals and companies prefer money today to money in the future. To account for this, cash flows are reduced to compute the equivalent present worth. This practice is referred to as discounting. The present worth of money required some time in the future is obtained by calculating how much would have to be invested at a rate called the *discount rate* today in order that the required money would be available when needed in the future.

The present worth (PW) of a rupee needed n years from today is given by $1/(1 + d)^n$ where d = discount rate. Assume that payments A_1, A_2 …, A_n, are due after 1, 2,…n years. The present worth of these payments is $A_1/(1 + d)$, $A_2/(1 + d)^2$,…, $A_n/(1 + d)^n$ respectively.

The sum (S) of these present worths is given by

$$S = \sum_{j=1}^{n} A_j / (1+d)^j \qquad \text{...(10.2)}$$

Example 10.3

Repayments of Rs 20 000 and Rs 25 000 are due at the end of the first year and second year of operation on the loan taken for installing an energy system. Calculate the present worth of the total amount due if the discount rate is 9 per cent.

Solution

Present worth of the total amount to be paid

$$= (20\,000/1.09) + (25\,000/1.09^2)$$

$$= 18\,349 + 21\,042$$

$$= \text{Rs } 39\,391$$

Frequently, the payments A_1, A_2, etc. are related and increase at a constant rate i. Thus $A_2 = A_1(1 + i)$, $A_3 = A_1(1 + i)^2$, etc. Equation (10.2) then becomes

$$S = A_1 \sum_{j=1}^{n} (1+i)^{j-1} / (1+d)^j = \frac{A_1}{(d-i)}\left[1 - \left(\frac{1+i}{1+d}\right)^n\right] \qquad (10.3)$$

REPAYMENT OF LOAN IN EQUAL ANNUAL INSTALMENTS 10.5

We now derive an expression for the annual instalment due in repayment of a loan. Assume that the initial investment required for a given situation is C and that a fraction f_l of this amount is taken as loan. Assume further, that the lending institution charges an interest rate d_l and that the agreement made is that the loan is to be paid back in equal annual instalment of A over a period of n_l years. By taking the value of each of the annual repayments at the time of making the initial investment, it follows that

$$f_l C = \sum_{j=1}^{n_l} A/(1+d_l)^j = \frac{A}{d_l}\left[1 - \{1/(1+d_l)^{n_l}\}\right]$$

Therefore,

$$A = \frac{d_l f_l C}{[1 - \{1/(1+d_l)^{n_l}\}]} \tag{10.4}$$

Every annual repayment consists of two components – a part repayment on the loan and a payment of interest on the outstanding loan. The first component increases from year to year, while the second component decreases. Since tax deductions are generally permissible on the interest component, it is useful to derive an expression for the interest component. It can be shown that after j years, the interest component

$$= \left[1 - \frac{(1+d_l)^{j-1} - 1}{(1+d_l)^{n_l} - 1}\right] d_l f_l C \tag{10.5}$$

ECONOMIC ANALYSIS OF A STAND-ALONE RENEWABLE ENERGY SYSTEM 10.6

Statement of the Problem

Consider a stand-alone renewable energy installation with an auxiliary sub-system.

1. Assume that the cost of the installation as a whole is C of which a fraction f_l is taken as a loan. As in Section 10.5, assume that the interest rate on the loan is d_l and that it is to be paid back in equal annual instalments over a period of n_l years.
2. Let the annual energy load to be met be E and assume that renewable energy supplies a fraction F of this load, the rest $(1 - F)E$ being supplied by the auxiliary sub-system. Assume that the cost of this conventional energy is c_f per unit of energy and that it increases at the rate of i_f every year.
3. The installation will require some annual expenditure by way of maintenance, manpower and local taxes. We assume that the cost associated with these items of expenditure is M in the first year of operation and that it increases at the rate of i_m every year. The rates i_m and i_f are related to inflation.

4. Finally we assume that tax deductions are allowed both on the interest component of the annual loan repayment instalment as well as on depreciation of the system. The depreciation is assumed to be at a uniform rate r_d per year. The income tax rate is r_t.

Derive expressions for the net annual cost (AC), the cumulative cost (CC) and the cumulative saving (CS).

Solution

❑ *Net Annual Cost* Expressions for the quantities occurring on the right hand side of Eq. (10.1) are first noted down.

Costs:

In any year j,

Cost of fuel/electricity for auxiliary $= c_f(1 + i_f)^{j-1}(1 - F)E$

Maintenance charges + Manpower + Local taxes $= (1 + i_m)^{j-1}M$

From Eq. (10.4),

Annual repayment on loan

$$= \frac{d_l f_l C}{[1 - \{1/(1 + d_l)^{n_l}\}]} \quad \text{if } j \le n_l$$

$$= 0 \quad \text{if } j > n_l$$

Benefits:

From Eq. (10.5),

Tax deduction on the interest component of loan repayment

$$= \left[1 - \frac{(1 + d_l)^{j-1} - 1}{(1 + d_l)^{n_l} - 1}\right] r_t d_l f_l C \quad \text{if } j \le n_l$$

$$= 0 \quad \text{if } j > n_l$$

Tax deduction on depreciation $= r_t r_d C$ if $j \le (1/r_d)$
$= 0$ if $j > (1/r_d)$

Thus net annual cost in the year j, $(\text{AC})_j$

= Costs – Benefits

$$= c_f(1 + i_f)^{j-1}(1 - F)E + (1 + i_m)^{j-1}M + \frac{d_l f_l C}{[1 - \{1/(1 + d_l)^{n_l}\}]}$$

$$- \left[1 - \frac{(1 + d_l)^{j-1} - 1}{(1 + d_l)^{n_l} - 1}\right] r_t d_l f_l C - r_d r_t C \tag{10.6}$$

Equation (10.6) is valid for $j \le n_l$ and $j \le (1/r_d)$.

If $j > n_l$, the third and fourth terms on the right hand side of Eq. (10.6) are zero, while if $j > (1/r_d)$, the fifth term is zero

The present worth of the net annual cost in the year j is obtained by dividing the right hand side by $(1 + d)^j$ where d is the discount rate.

❑ *Cumulative Cost* The present worth of the net annual cost over a period of n years is called the *cumulative cost* (CC). It is obtained by summing the present worth of the net annual cost over each year and adding the initial down payment on the system. Thus

$$\text{CC} = \text{Initial down payment} + \sum_{j=1}^{n} \text{PW of (AC)}_j$$

$$= (1-f_l)C + c_f\,(1-F)E\sum_{j=1}^{n}\frac{(1+i_f)^{j-1}}{(1+d)^j} + M\sum_{j=1}^{n}\frac{(1+i_m)^{j-1}}{(1+d)^j}$$

$$+\frac{d_l f_l C}{\left[1-\left\{\dfrac{1}{(1+d_l)^{n_l}}\right\}\right]}\sum_{j=1}^{n_l}\frac{1}{(1+d)^j}$$

$$-\,r_t d_l f_l C\sum_{j=1}^{n_l}\frac{1}{(1+d)^j}\left[1-\frac{(1+d_l)^{j-1}-1}{(1+d_l)^{n_l}-1}\right] - r_t r_d C\sum_{j=1}^{1/r_d}\frac{1}{(1+d)^j}$$

$$= (1-f_l)C + \frac{c_f(1-F)E}{(d-i_f)}\left[1-\left(\frac{1+i_f}{1+d}\right)^n\right]$$

$$+\frac{M}{(d-i_m)}\left[1-\left(\frac{1+i_m}{1+d}\right)^n\right]$$

$$+\frac{d_l f_l C}{[1-\{1/(1+d_l)^{n_l}\}]}\frac{1}{d}\left[1-\frac{1}{(1+d)^{n_l}}\right]$$

$$-\,r_t d_l f_l C\left[\frac{(1+d_l)^{n_l}}{(1+d)^{n_l}}\frac{1}{d}\left\{\frac{(1+d)^{n_l}-1}{(1+d_l)^{n_l}-1}\right\}\right.$$

$$\left.-\frac{1}{\{(1+d_l)^{n_l}-1\}}\frac{1}{(d-d_l)}\left\{1-\left(\frac{1+d_l}{1+d}\right)^{n_l}\right\}\right]$$

$$-\frac{r_t r_d C}{d}\left[1-\frac{1}{(1+d)^{(1/r_d)}}\right] \tag{10.7}$$

Equation (10.7) is valid for $n \geq n_l$, $n \geq (1/r_d)$, $d \neq i_f$, $d \neq i_m$ and $d \neq d_l$. If $n < n_l$, the summations for load repayment (fourth term) and for tax deduction on interest component of loan repayment (fifth term) will be from $j = 1$ to $j = n$. The fourth term then becomes

$$\frac{d_l f_l C}{[1-\{1/(1+d_l)^{n_l}\}]}\frac{1}{d}\left[1-\frac{1}{(1+d)^n}\right]$$

and the fifth term becomes

$$r_t d_l f_l C\left[\frac{(1+d_l)^{n_l}}{(1+d)^n}\frac{1}{d}\left\{\frac{(1+d)^n-1}{(1+d_l)^{n_l}-1}\right\}-\frac{1}{\{(1+d_l)^{n_l}-1\}}\frac{1}{(d-d_l)}\left\{1-\left(\frac{1+d_l}{1+d}\right)^n\right\}\right]$$

Similarly, if $n < (1/r_d)$, the summation for tax deduction on depreciation will be from $j = 1$ to $j = n$, and sixth term on the right hand side in Eq. (10.7) becomes

$$[r_t r_d\, C/d]\,[1-\{1/(1+d)^n\}]$$

Finally, it should be noted that situations corresponding to $d = i_f$, $d = i_m$ or $d = d_l$ are special cases, for which summations can be readily done and the second third and fifth terms in Eq. (10.7) suitably modified (Problem 10.2).

❑ *Cumulative Saving* The principal advantage which occurs because of the use of a stand-alone renewable energy system is the saving in the use of conventional fuel. This saving is not reflected in the cumulative cost. We will take account of this saving by defining a term called the cumulative saving (CS).

Cumulative saving (CS) = (Cumulative saving due to non-use of conventional energy system) – (Cumulative cost, Eq. 10.7)

$$= c_f E \sum_{j=1}^{n}\frac{(1+i_f)^{j-1}}{(1+d)^j} - (\text{CC})$$

$$= \frac{c_f E}{(d-i_f)}\left[1-\left(\frac{1+i_f}{1+d}\right)^n\right] - (\text{CC}) \qquad (10.8)$$

Cumulative saving is also referred to as *net present value* (NPV).

Example 10.4

A solar system having an array of flat-plate collectors is installed in a factory for providing hot water at 70°C. The system costs Rs 250 000 and is set up with an initial down payment of 20 per cent of the investment, the balance 80 per cent being taken as a soft loan to be repaid in equal annual instalments over a 5 year period at an interest rate of 5 per cent.

The solar system supplies 90 per cent of the annual energy required for heating water in the factory, the balance of 10 per cent being supplied by an auxiliary sub-system. The conventional fuel used in the auxiliary sub-system costs Rs 4700 during the first year of operation. This cost increases at the rate of 5 per cent every year.

The annual expenditure required by way of maintenance, manpower, etc. is Rs 9000 in the first year, and this expense increases at the rate of 5 per cent every year.

Tax deductions are permissible only on depreciation, which is allowed at the rate of 25 per cent each year, and the company income tax rate is 30 per cent.

Assuming that the market discount rate is 10 per cent, calculate the net annual cost (AC) and the cumulative saving (CS) from year to year over a period of 8 years.

Solution

In the present case, $n_l = 5$ years, $(1/r_d) = 4$ years and $n = 8$ years. Also $d = 0.10$, $d_l = 0.05$, $i_f = 0.05$ and $i_m = 0.05$ and $r_t = 0.30$

Cost of the solar system (C) = Rs 250 000

Loan ($f_l\, C$) = Rs (0.8 × 250 000) = Rs 200 000

Initial down payment = Rs 50 000

$c_f E$ = Rs 47 000

Year 1:

Cost of conventional fuel used in auxiliary sub-system = Rs 4700

Cost of maintenance, manpower, etc. = Rs 9000

$$\text{Annual repayment of loan} = \frac{0.05 \times 200\,000}{[1 - \{1/(1+0.05)^5\}]} = \text{Rs } 46\,915$$

Tax saving on depreciation = 250 000 × 0.3/4 = Rs 18 750

Saving due to non-use of conventional energy system = Rs 47 000

Net annual cost = (4700 + 9000 + 46 915) – (18 750 + 47 000)

= Rs (–) 5855

Present worth of net annual cost = Rs (–) 5323

Cumulative saving at the end of the 1st year = – 50 000 + 5323 = Rs (–) 44 677

The calculations for 8 years are given in Table 10.1. It is seen that the cumulative saving (net present value) at the end of 8 years is Rs 41 285. This is a good saving.

The variation of CS with the number of years is plotted in Fig. 10.1. It is seen that the value of CS is negative at the start and increases slowly for the first four years because the benefits are higher than the costs by amounts ranging from Rs 5323 to Rs 7585. From the fourth to the fifth year, the CS decreases by about Rs 3550 because the tax deduction on depreciation is not available after four years. However, once the loan is paid back at the end of five years, the CS increases steadily. It becomes positive between the sixth and seventh year and attains the value of Rs 41 285 after 8 years.

Table 10.1 Calculations for Example 10.4

Column	(1)	(2)	(3)	(4)	(5)	(6)	(7)	(8)	(9)	(10)	(11)
Year	Initial payment	Cost of conventional fuel used in auxiliary	Maintenance cost	Repayment on loan	Tax saving on depreciation	Saving due to non-use of conventional energy system	Costs (2) + (3) + (4)	Benefits (5) + (6)	Net annual cost (7) – (8)	PW of (9)	Cumulative saving
0	50 000										–50 000
1		4700	9000	46 195	18 750	47 000	59 895	65 750	–5855	–5323	–44 677
2		4935	9450	46 195	18 750	49 350	60 580	68 100	–7520	–6215	–38 462
3		5182	9923	46 195	18 750	51 820	61 300	70 570	–9270	–6965	–31 497
4		5441	10 419	46 195	18 750	54 410	62 055	73 160	–11 105	–7585	–23 912
5		5713	10 940	46 195		57 130	62 848	57 130	5718	3550	–27 462
6		5999	11 487			59 990	17 486	59 990	–42 504	–23 992	–3470
7		6298	12 061			62 980	18 359	62 980	–44 621	–22 898	19 428
8		6613	12 664			66 130	19 277	66 130	–46 853	–21 857	41 285

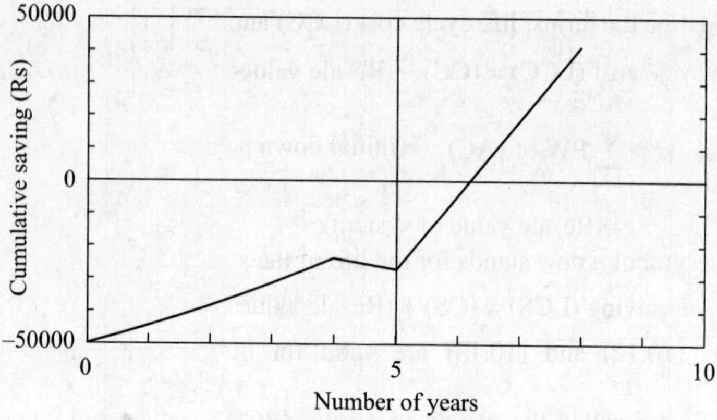

Fig. 10.1 *Variation of CS with number of years – Example 10.4*

ECONOMIC ANALYSIS OF AN ADD-ON RENEWABLE ENERGY SYSTEM 10.7

We now take up for consideration add-on systems which produce electricity by using a renewable energy source and are connected to the grid. Consider a wind energy or PV system connected to the electricity grid. It feeds power into the grid whenever the wind speed or solar radiation is adequate and the owner is paid for the amount of energy fed. For such a system, the costs are associated with the repayment of the loan and maintenance, manpower, local taxes, etc. The benefits are the tax deductions available and the income derived from the power fed to the grid. Thus, Eq. (10.1) for the net annual cost would be modified as follows:

Net annual cost (AC) = [Yearly costs] – [Yearly benefits]
= [Maintenance charges, manpower, local taxes, etc. + repayment on loan] – [Tax deductions as available + Income from power fed to the grid] (10.9)

Once again, it is necessary to caution the reader that although Eq. (10.9) is fairly general in its applicability, situations may arise in specific cases requiring the addition or deletion of some terms on the right hand side of the equation.

The cumulative cost (CC) over n years is again defined by the expression

$$\text{CC} = \sum_{j=1}^{n} \text{PW (AC)}_j + \text{(Initial down payment on system)} \tag{10.10}$$

Since there is no term involving saving due to non-use of a conventional energy system to be incorporated (as was the case for a stand-alone system),

CS (or NPV) for an add-on system = – CC for the system (10.11)

We now define the terms, life cycle cost (LCC) and life cycle saving (LCS).

$$\text{Life cycle cost (LCC)} = \text{(CC)} - \text{(Resale value of system)}$$

$$= \sum_{j=1}^{n} \text{PW of } (\text{AC})_j + \text{(Initial down payment on system)}$$

$$- \text{(Resale value of system)} \quad (10.12)$$

where the symbol n now stands for the life of the system in years.

$$\text{Life cycle saving (LCS)} = \text{(CS)} + \text{(Resale value of system)} \quad (10.13)$$

Equations (10.12) and (10.13) are valid for both stand-alone and add-on systems.

An indicator called the *benefit-cost ratio* (BCR) is often used to judge the economic viability of a system. The benefit-cost ratio is defined as the ratio of the sum of the present worth of all annual benefits to the sum of the present worth of all annual costs and the initial investment. A viable project requires that the BCR be greater than one.

Mention was made earlier that the owner of the add-on system is compensated for the electrical energy fed into the grid. The rate of compensation is called the feed-in tariff (FIT) and its value is fixed by the state in which the wind machine or PV system is located. FIT is a cost-based price which varies with the renewable source. It is offered as a long term contract and helps to accelerate investment in renewable energy sources.

Example 10.5

A wind energy conversion system of capacity 1 MW is installed at a cost of Rs 45 000 000 and connected to the electricity grid. It works at an annual capacity factor of 20 per cent Assume:

1. The Electricity Board purchases the electrical energy at the feed-in-tariff of Rs 5.00 per kWh in the first year. This price decreases by 2 per cent of the initial price every year.
2. 85 per cent of the initial cost of Rs 45 000 000 is financed through a loan borrowed at the rate of 5 per cent to be repaid in equal instalments in 5 years.
3. The rate of depreciation is 80 per cent in the first year and the remaining 20 per cent in the second year. The corporate tax rate is 31 per cent.
4. No tax saving is allowed on the interest component of the loan repayment.
5. The annual operation and maintenance costs in the first year are 2 per cent of the total cost of the system. This expense increases at the rate of 6 per cent every year.

Calculate the cumulative saving year-by-year over a period of 10 years assuming a discount rate of 10 per cent. Also calculate the benefit-cost ratio at the end of 10 years.

Solution

In this case, Eqs (10.9) – (10.11) are applicable. We will solve the problem proceeding from one year to the next.

Year 1:

Electricity generated every year = 1000 × 8760 × 0.2 kWh

Income from sale of electricity = 1000 × 8760 × 0.2 × 5.00 = Rs 8 760 000

Depreciation = 0.80 × 45 000 000 = Rs 36 000 000

Tax saving = 36 000 000 × 0.31 = Rs 11 160 000

Annual loan payment, from Eq. (10.4)

$$= \frac{0.05 \times 0.85 \times 45\,000\,000}{[1 - \{1/(1+0.05)^5\}]}$$

= Rs 8 834 786

Maintenance and operation cost = 0.02 × 45 000 000 = Rs 900 000

Thus benefits = 8 760 000 + 11 160 000 = Rs 19 920 000

Costs = 8 834 786 + 900 000 = Rs 9 734 786

Net annual cost = (9 734 786 – 19 920 000) = Rs (–) 10 185 214

Present worth of net annual cost = (–) 10 185 214 / 1.10 = Rs (–) 9 259 285

Cumulative cost at the end of the 1st year, Eq (10.10)

= Rs (–) 9 259 285 + Initial down payment

= (–) 9 259 285 + 6 750 000 = Rs (–) 2 509 285

Cumulative saving at the end of the 1st year, Eq (10.11)

= Rs 2 509 285

The calculations for all years are given in Table 10.2. It is seen from the last column that the investment in the wind energy system has been very profitable and that the cumulative saving (i.e. net present value) is throughout positive right from the end of the first year. It increases sharply in the first two years because of the high rate of depreciation permitted. Thereafter it decreases somewhat for the next three years because the loan is being repaid. However it continues to be positive. Subsequently the cumulative saving keeps on increasing steadily for the next five years.

Table 10.2 Calculations for Example 10.5

Year	Down payment	Income from sale of electricity	Tax savings on depreciation	Loan	Operation and maintenance	Benefits	Costs	Net annual cost	PW of benefits	PW of costs	PW of net annual cost	Cumulative saving
0	6 750 000											–6 750 000
1		8 760 000	11 160 000	8 834 786	900 000	19 920 000	9 734 786	–10 185 214	18 109 091	8 849 805	–9 259 285	2 509 285
2		8 584 800	2 790 000	8 834 786	954 000	11 374 800	9 788 786	–1 586 014	9 400 661	8 089 906	–1 310 755	3 820 040
3		8 409 600		8 834 786	1 011 240	8 409 600	9 846 026	1 436 426	6 318 257	7 397 465	1 079 208	2 740 832
4		8 234 400		8 834 786	1 071 914	8 234 400	9 906 700	1 672 300	5 624 206	6 766 409	1 142 203	1 598 629
5		8 059 200		8 834 786	1 136 229	8 059 200	9 971 015	1 911 815	5 004 129	6 191 216	1 187 087	411 542
6		7 884 000			1 204 403	7 884 000	1 204 403	–6 679 597	4 450 312	679 854	–3 770 458	4 182 000
7		7 708 800			1 276 667	7 708 800	1 276 667	–6 432 133	3 955 834	655 132	–3 300 701	7 482 701
8		7 533 600			1 353 267	7 533 600	1 353 267	–6 180 333	3 514 480	631 309	–2 883 171	10 365 872
9		7 358 400			1 434 463	7 358 400	1 434 463	–5 923 937	3 120 680	608 352	–2 512 327	12 878 199
10		7 183 200			1 520 531	7 183 200	1 520 531	–5 662 669	2 769 435	586 231	–2 183 204	15 061 403

The benefit-cost ratio at the end of 10 years

$$= \frac{\text{PW of benefits}}{\text{PW of costs} + \text{Initial down payment}} = \frac{\text{Rs } 62\,267\,085}{\text{Rs } 40\,455\,679 + \text{Rs } 6\,750\,000}$$

$$= 1.319$$

The value of BCR is greater than unity and will increase as the number of years of operation exceeds ten.

The above situation (Example 10.5) in which CS is positive right from the end of the first year is unusual and occurs because of the high rate of depreciation (80 per cent) allowed in the first year. Usually for a given renewable energy system, CS is negative at the beginning. Thereafter, its value increases, crosses over from being negative to a positive value after a certain number of years. It then goes on increasing with time.

While planning for a new system, it is often desired that LCS should be zero. The discount rate for which this will happen, other parameters being constant, is called the *internal rate of return* (IRR).

LEVELIZED COST OF ELECTRICITY 10.8

A term often used for comparing the economic viability of different energy systems (renewable or non-renewable) is the *levelized cost of electricity* (LCOE). LCOE is the sum of all costs incurred over the life time of an energy system (minus any benefits) divided by the electrical energy produced by the system over its life time.

The costs and benefits are all discounted. A feed-in tariff is not included in the benefits. Thus LCOE may be considered as the minimum cost at which the electricity must be sold in order to break even over the life time of the system.

Example 10.6

Consider again the data given for the wind energy conversion system of Example 10.5. Delete the feed-in tariff component and assume that the life time of the system is 20 years with a resale value of Rs 4 000 000. Calculate the life cycle cost (LCC) and the levelized cost of electricity (LCOE). Assume that (i) the wind machine operates with an average APLF of 0.16 over its life time, and (ii) the annual operation and maintenance costs are three times the values used in Example 10.5.

Solution

Year 1:

Tax saving on depreciation = 0.31 × 0.80 × 45 000 000 = Rs 11 160 000

Loan repayment using Eq. (10.4) = Rs 8 834 786

Operation and maintenance cost = 3 × 900 000 = Rs 2 700 000

Net annual cost = (8 834 786 + 2 700 000) – 11 160 000 = Rs 374 786

PW of net annual cost = 374 786/1.10 = Rs 340 715

Cumulative cost = (6 750 000 + 340 715) = Rs 7 090 715

The remaining calculations are shown in Table 10.3.

PW of resale price = 4 000 000/1.10^{20} = Rs 594 575

LCC = (63 110 933 – 594 575) = Rs 62 516 358

LCOE = 62 516 358/(1000 × 8760 × 20 × 0.16)= Rs 2.23 / kWh

Table 10.3 Calculations for Example 10.6

Year	Down payment	Tax saving on depreciation	Loan repayment	Operation and maintenance costs	Net annual cost	PW of net annual costs	Cumulative cost
0	6 750 000						
1		11 160 000	8 834 786	2 700 000	374 786	340 715	7 090 715
2		2 790 000	8 834 786	2 862 000	8 906 786	7 360 980	14 451 695
3			8 834 786	3 033 720	11 868 506	8 916 984	23 368 679
4			8 834 786	3 215 743	12 050 529	8 230 673	31 599 352
5			8 834 786	3 408 688	12 243 474	7 602 234	39 201 586
6				3 613 209	3 613 209	2 039 562	41 241 148
7				3 830 002	3 830 002	1 965 397	43 206 545
8				4 059 802	4 059 802	1 893 928	45 100 473
9				4 303 390	4 303 390	1 825 057	46 925 530
10				4 561 593	4 561 593	1 758 692	48 684 222
11				4 835 289	4 835 289	1 694 739	50 378 961
12				5 125 406	5 125 406	1 633 112	52 012 073
13				5 432 930	5 432 930	1 573 726	53 585 799
14				5 758 906	5 758 906	1 516 500	55 102 299
15				6 104 441	6 104 441	1 461 355	56 563 654
16				6 470 707	6 470 707	1 408 214	57 971 686
17				6 858 950	6 858 950	1 357 007	59 328 875
18				7 270 487	7 270 487	1 307 661	60 636 536
19				7 706 716	7 706 716	1 260 110	61 896 646
20				8 169 119	8 169 119	1 214 287	63 110 933

CLEAN DEVELOPMENT MECHANISM 10.9

The emission of green house gases (GHG), namely carbon dioxide, methane, nitrous oxide, hydrofluorocarbons, sulphur hexafluoride and perfluorocarbons due to rapid industrialization and urbanization has resulted in an increase in the temperature of the earth's surface. This phenomenon is commonly known as global warming. In order to tackle this problem, emission reduction targets have been fixed under the Kyoto Protocol of the United Nations Framework Convention on Climate Change. According to this protocol, the developed nations were required to bring down the GHG emissions by the end of 2012 to a value which is 5.2 per cent less than that of the year 1990. In order to achieve this reduction target, the Kyoto Protocol provides a few mechanisms. The clean development mechanism (CDM) is one of them [5]. While the developed countries can initiate emission reduction domestically, they can also participate in projects that reduce GHG emission in developing countries under CDM and take credit for such emission reductions into their account for complying with the Kyoto Protocol. Thus CDM is an innovative international financial co-operation mechanism, which aims at mitigating the harmful effects of climate change and promoting sustainable development. Under this mechanism, carbon has become a tradeable commodity. Consequently, projects undertaken to improve energy efficiency or installation of renewable energy systems qualify for carbon trading. The annual revenue generated would be a part of the financial benefits of the system and can be included in the calculation of annual savings. Thus the expression for calculating savings of a renewable energy system can be modified by adding a term on savings under CDM.

If the projects are of a small size, CDM provides for bundling of similar projects so that the common expenditures for preparation of relevant documents and getting necessary verifications and certifications can be shared. In order to facilitate various activities under CDM, a designated national authority has been created in India under the name of CDM India [6].

The annual saving under CDM depends on the nature of project and therefore varies widely. As an example, it may be noted that under a bundled solar cooking project in India, a private company obtained the benefit of trading 500 t of CO_2 at the rate of 11.5 Euro/t from the German government [7].

Although CDM has been widely supported, it has also been the subject of criticism. Critics have said that many of the renewable energy projects supported under the mechanism did not satisfy the requirement of 'additionality'. A project is considered to be additional, if it has been executed because of the funds received under CDM. It has been pointed out that most of the funded projects would have been built anyway without the income received under CDM. Indeed some of them were already under construction or had been completed when their proposals for funding under CDM were submitted. For such projects, the use of CDM allows the industrialized country to emit more than the reduced target set for it under the Kyoto Protocol. At the same time, the execution of the project in the developing country causes no more change on the ground there than would have occurred in any case.

CONCLUDING REMARKS 10.10

Two methods – one based on the payback period approach and the other on the calculation of cumulative saving – have been described in this chapter. Of these, the second obviously gives a more realistic cost-benefit picture and is to be preferred. This is because it takes account of many relevant factors like fuel costs, income from power fed to the grid, loan repayments, maintenance costs, tax deductions, inflation rates, etc. On the other hand, the commonly used simple definition of payback period takes a limited view and is concerned only with the fuel savings or income derived from power fed to the grid vis-à-vis the initial cost of the system.

Finally, it is important to keep in mind the limitations of the economic analyses described in this chapter. An economic analysis on viability is a look into the future. It is a prediction based on current costs and rates. It is also based on future estimates (intelligent guesses) of certain parameters like the discount rate and the inflation rate. For this reason, it is desirable to perform a sensitivity analysis and calculate CS for different values of the parameter. For example, if the calculation is performed for a discount rate of 10 per cent, it may be desirable to repeat the calculation for values of 9 and 11 per cent to test the sensitivity of the final result to variations in the value of d and to perhaps use the more conservative value of CS for decision making.

PROBLEMS

10.1 Consider a renewable energy system whose installation cost is C. Assume that the savings on account of the cost of fuel / electricity saved in the first year is B and that this saving increases at the rate of i_f every year. If the discount rate is d, show that the payback period is given by the expression

$$n_p = \frac{\ln\left[1 - \frac{C(d - i_f)}{B}\right]}{\ln\left[\frac{1 + i_f}{1 + d}\right]}$$

10.2 Consider the special situations in which $d = i_f$, $d = i_m$ or $d = d_l$ and determine modified expressions for the second, third and fifth terms in Eq. (10.7).

10.3 Calculate the LCS for the system of Example 10.4 if the life time of the system is taken to be 20 years and the resale value is Rs 60 000.

10.4 Examine the calculations presented in Table 10.1 for Example 10.4 and determine the payback period for the solar hot water system with no discounting.

10.5 An industrial solar energy system is installed at a cost of Rs 400 000 for pre-heating boiler feed water. Assume:

- The cost of fuel saved in the first year is Rs 56 000. This cost increases at the rate of 10 per cent in subsequent years.
- The entire initial cost of Rs 400 000 is financed through a loan borrowed at the rate of 12.5 per cent.
- In order to encourage the use of solar energy, the Government permits depreciation at the high rate of 75 per cent in the first year and the remaining 25 per cent in the second year. The corresponding tax rate is 60 per cent.
- The annual repayment of the loan is not in equal instalments, but is adjusted every year to such a value that the annual savings are zero. This is done until the loan is fully paid.
- No tax saving is allowed on the interest component of the loan repayment.
- The annual expense associated with maintenance and other recurring items is Rs 11 000 in the first year. This expense increases at the rate of 6 per cent every year.

Calculate the cumulative saving over a period of 8 years assuming a discount rate of 10 per cent.

10.6 A flat-plate collector hot water system having an area of 30 m^2 and costing Rs 250 000 is set up with an initial payment of 20 per cent of the investment. The balance is taken as a loan to be repaid annually in equal instalments over a 10 year period at an interest rate of 12 per cent. The fuel cost saved in the first year as a consequence of installing the solar system is Rs 50 000. This cost inflates at the constant rate of 5 per cent per year. The cost of maintaining the system is Rs 12 500 in the first year and this increases at the constant rate of 5 per cent per year. The corporate tax rate is 31 per cent and tax deductions are permissible on the interest component of the annual repayment of the loan as well as on depreciation which is allowed at the rate of 33 1/3 per cent per year. The cost of an equivalent conventional system is Rs 20 000. Calculate the CS over a period of 15 years if the market discount rate is (i) 10 per cent (ii) 15 per cent, and (iii) 30 per cent.

10.7 A solar energy system costs Rs 494 000 and supplies 70 per cent of the annual energy consumption of 1000 GJ. It is bought with a 20 per cent down payment, the balance being borrowed at the rate of 16 per cent. The loan is to be returned in equal instalments over a period of 8 years. Fuel costs are Rs 110 per GJ in the first year and are expected to go up at the rate of 6 per cent every year. Assume that (i) maintenance charges are Rs 20 000 in the first year and increase at the rate of 5 per cent per year (ii) depreciation is allowed at the rate of 20 per cent every year and that the tax rate is 50 per cent, and (iii) tax deductions are permissible both on the depreciation and on the interest component of the annual repayment of the loan. The resale value of the equipment at the end of its life cycle of 15 years is Rs 100 000. Calculate the LCS if the discount rate is 11 per cent.

10.8 The cost of a solar flat-plate collector system is given by equation

$$C = \text{Rs}\ (8000\,A_c + 9500)$$

where A_c is the collector area in m^2. Such a system with an area of 20 m^2 is installed in a certain location and supplies 67 per cent of the annual energy use of 100 000 MJ. The other details of the system are as follows:

Initial down payment	: 20 per cent
Interest rate on bank loan	: 15.5 per cent
Period for repayment of loan in equal annual instalments	: 5 years
Fuel cost in the 1st year	: Rs 600 per GJ
Rate of increase in fuel cost every year	: 5 per cent
Maintenance charges	: Negligible
Depreciation rate	: 20 per cent
Tax rate	: 55 per cent

Calculate the cumulative saving over a period of 15 years if tax deductions are permissible on the depreciation and on the interest component of the annual loan repayment. Assume that the market discount rate is 11 per cent.

10.9 Repeat the calculation of CS for the system specified in Problem 10.8 for a number of values of collector area ranging from 10 to 50 m^2. Given that the solar fraction, i.e. the fraction of energy supplied by the solar system out of the annual energy requirement, varies with the collector area in the following manner:

Collector area A_c (m^2)	10	20	30	40	50
Solar fraction	0.43	0.67	0.80	0.86	0.90

Plot the variation of solar fraction and CS with the collector area, and comment on the nature of the variations obtained.

10.10 A wind energy conversion system of capacity 2.1 MW is installed at a cost of Rs 11.5 crore for supplying power to an electricity grid. It works at an annual capacity factor of 20 per cent. Assume:

(i) The Electricity Board purchases the electrical energy at the rate of Rs 4.00 per kWh and this price increases by 5 per cent every year.

(ii) 80 per cent of the initial cost of the system is financed through a loan borrowed at the rate of 12 per cent to be repaid in equal instalments in 5 years.

(iii) The rate of depreciation is 80 per cent in the first year and the remaining 20 per cent in the second year. The corporate tax rate is 33 per cent.

(iv) Tax saving is allowed on the interest component of the loan repayment.

(v) The annual operation and maintenance costs in the first year are 2 per cent of the total cost of the system. This cost increases at the rate of 5 percent every year.

Calculate the cumulative cost over a period of 8 years and 15 years assuming a discount rate of 11 per cent.

10.11 A solar PV lighting system costing Rs 2 500 000 is used to supply power of 10 kW to a remote village instead of a diesel-generator set costing Rs 90 000. Assume:

(i) The cost of fuel saved at the end of the first year is Rs 175 000. This cost increases at the rate of 6 per cent in each subsequent year.

(ii) 80 per cent of the initial cost is financed through a soft loan borrowed at the rate of 5 per cent to be repaid in equal instalments in 5 years.
(iii) The rate of depreciation is 80 per cent in the first year and the remaining 20 per cent in the second. The corporate tax rate is 31 per cent.
(iv) Tax saving is allowed on the interest component of the loan repayment.
(v) The annual operation and maintenance costs in the first year are Rs 25 000. This expense increases at the rate of 6 per cent every year.

Calculate the cumulative saving over a period of (a) 10 years and (b) 15 years assuming a discount rate of 11 per cent. Comment on the results obtained.

10.12 A solar PV system of capacity 10 kW is installed at a cost of Rs 2 500 000 and connected to the electricity grid. It supplies power for an average of 6 hours every day. Assume:
(i) The Electricity Board purchases the electrical energy at the rate of Rs 12.00 per kWh and that this price increases by 5 per cent every year.
(ii) 80 per cent of the initial cost is financed through a loan borrowed at the rate of 12 per cent to be repaid in equal instalments in 5 years.
(iii) The rate of depreciation is 10 per cent per year and the corporate tax rate is 31 per cent.
(iv) The annual operation and maintenance costs in the first year are Rs 20 000. These expenses increase at the rate of 4 per cent every year.

Calculate the cumulative cost over a period of 14 years assuming a discount rate of 10 per cent. Also calculate the benefit cost ratio for the same period.

10.13 In the solved Example 10.5, the cumulative saving on a wind energy conversion system connected to the grid was calculated over a period of 10 years. The discount rate was assumed to be 10 per cent. Test the sensitivity of the answer to the value of the discount rate by repeating the calculation for a discount rate of 9 per cent and a discount rate of 11 per cent.

REFERENCES

1. Grant, E.L, Ireson, W.G. and Leavenworth, R.S. 1982. *Principles of Engineering Economy*, 7th Ed. John Wiley, New York, USA.
2. Degarmo, E.P., Sullivan, W.G., Bontadelli, J.A. and Wicks, E.M. 1997. *Engineering Economy*, Prentice Hall International, Inc., New Jersey, USA.
3. Mathew, S. 2006. *Wind Energy: Fundamentals, Resource Analysis and Economics*. Springer, Berlin, Germany
4. *Frequently Asked Questions*. 2016. Solar Energy Corporation of India. (http://www.seci.gov.in/FAQ; accessed on July 27, 2016)
5. *Clean Development Mechanism*, United Nations Framework Convention on Climate Change. (http://cdm.unfccc.in)
6. *CDM India*, Designated National Authority, Ministry of Environment and Forests, Government of India. (http://cdmindia.nic.in)
7. Gadhia, D. 2007. Personal communication. Gadhia Solar Energy Systems Pvt. Ltd., Valsad, Gujarat, India.

Appendices

APPENDIX 1

Answers

Chapter 3

1. (a) 5779 K
   (b)

λ (μm)	0.2	0.3	0.4	0.5018	0.6
e_λ (W/m^2-μm)	99.3	829.4	1569.6	1794.6 (max)	1668.6
	0.7	0.8	0.9	1.0	2.0
	1414.8	1150.8	920.1	732.1	102.3

2. 3.84×10^{26} W, 4.50×10^{-10}, 1.73×10^{17} W
3. 38.94°
4.

LAT (h)	0550	0600	0800	1000	1200
θ (deg)	90	87.7	58.8	29.9	5.8

5. 24.01°
6. May14, July 30
7. $\psi = (\pi - \beta - 2\phi + 2\delta)/3$
8.

	Srinagar	Kolkata	Vishakhapatnam	Thiruvananthapuram
June 21	14.28	13.39	13.06	12.49 h
Dec 21	9.72	10.61	10.94	11.51 h

9. 11.59 h, 10.29 h
10. Sunrise 78.2°; sunset –62°
11. 1.372 m, 55.4° west of north
12. 1.50 m. Makes an angle of 215.4° on the wall with the line pointing east.
13. 1.85 m
14. 0.416
15. Monthly average value = 40 034 kJ/m^2-day
    Value on July 17 = 40 056 kJ/m^2-day
16. $H_o = I_{sc}\,[1 + 0.033\cos(360\,n/365)]\,[\omega_s \sin\delta \sin(\phi - \beta) + \cos\delta \cos(\phi - \beta) \sin\omega_s]$
17. 18 323 kJ/m^2–day, 3937 kJ/m^2–day
18. $a = 0.34$; $b = 0.44$

19.

ω (deg)	$\bar{I}_g$ (kJ/m^2–h) Predicted	$\bar{I}_g$ (kJ/m^2–h) Measured	$\bar{I}_d$ (kJ/m^2–h) Predicted	$\bar{I}_d$ (kJ/m^2–h) Measured
67.5	1187	1058	439	619
52.5	1894	1760	637	842
37.5	2551	2556	799	965
22.5	3058	3110	913	1008
7.5	3335	3427	973	1012
–7.5	3335	3463	973	983
–22.5	3058	3236	913	932
–37.5	2551	2765	799	850
–52.5	1894	2081	637	716
–67.5	1187	1296	439	551

20.

LAT (h)	0700	0800	0900	1000	1100	1200
I_g (W/m^2)	150	387	613	795	911	951
I_d (W/m^2)	65	102	126	140	149	152

21. At 1200 h (LAT):

β (deg)	0	21.1	29.4	60	90
r_b	1	1.136	1.148 (max)	0.988	0.563

At 1700 h (LAT):

β (deg)	0	21.1	50.5	60	90
r_b	1	1.370	1.572 (max)	1.550	1.213

22. 777.5 W/m^2, 33.4%

23. 532.8 W/m^2, 676.8 W/m^2

Chapter 4

3.

θ (deg)	30	45	60
$(\tau\alpha)_b$	0.725	0.711	0.649

4. Eq. (4.36): 3.23 W/m^2–K, Eq. (4.37): 3.43 W/m^2-K, exact: 3.50 W/m^2–K

6.

ε_p	0	0.2	0.4	0.6	0.8	1.0
U_t (W/m^2–K)	2.78	3.97	4.89	5.62	6.20	6.69

7. 43.5%, 83.0°C

8. (a) 54.7, 46.0, 37.2% (b) 57.1, 42.7, 28.2% (c) 207.5°C, 138.7°C

9. $q_u = F_o A_p[S - U_L (T_{fo} - T_a)]$, $q_u = F_m A_p[S - U_L \{(T_{fi} + T_{fo})/2 - T_a)\}]$, where $1/F_o = 1/F_R - A_p U_L / (\dot{m}\, C_p)$ and $1/F_m = 1/F_R - A_p U_L/(2\, \dot{m}\, C_p)$

11. (i) $\gamma = 0°$, $\beta_{opt} = 11.7°$ (ii) $\gamma = 180°$, $\beta_{opt} = 7.8°$ (iii) $\gamma = 0°$, $\beta_{opt} = 27.5°$

12. (i) $\gamma = 0°$, $\beta_{opt} = 23.8°$(ii) $\gamma = 0°$, $\beta_{opt} = 24.8°$

14.

Material	Copper	Galvanised iron	Aluminium
F'	0.880	0.856	0.890

16. (a) 4.05 W/m^2–K, 0.873 (b) 0.883
17. 66.4°C
18. 0.951
19. 51.6°C, 120.3°C
20. $a_0 = 0.743$, $a_1 = 4.473$ W/m^2-K and $a_2 = 0.067$ W/m^2-K^2
21. 42.2°C

Chapter 5

1. 52.2%, 55.8°C, 37 N/m^2
2. 54.1%, 56.0°C, 74 N/m^2
3. 57.8%, 56.6°C, 81 N/m^2
4.

U_b	Neglected	Included
Useful heat gain rate (W)	999.2	1014
Efficiency (%)	45.5	46.2

5. 70.15°C, 51.3%, 31.85 N/m^2
6. 34.2%, 63.4°C, 15.8 N/m^2
8. 444.2 W/m^2
9. 516.8 W/m^2
10. 52.1%, 78.1%, 67.0%
11. 40.6%

Chapter 6

2. June 21: $r_b = 0.9329, 0.6235, 0.3840$
   Dec 21: $r_b = 1.1405, 0.4694, 0.7622$
3.

LAT (h)	0800	0900	1000	1100	1200
β (deg)	18.29	8.88	3.96	1.53	0.80

Values of β are symmetric about solar noon.
$\gamma = 180°$ throughout.

4.

LAT (h)	0600	0700	0800	0900	1000	1100	1200
γ (deg)	180	180	180	180	0	0	0
β (deg)	71.47	27.26	9.49	2.09	1.45	3.13	3.63
θ (deg)	75.10	68.98	56.81	43.10	28.89	14.48	0

Sunrise at 0540 h (LAT).

5. 52.7°, 1.652
6. 2.83 W/m^2-K
7. 6.15 W/m^2-K
8. 59.8°C, 13.24 W/m^2-K
9. 58.1°C, 13.43 W/m^2-K. Change in $U_l = 1.4\%$
10. 1368 W, 166.2°C, 35.8%
11. 144.3°C, 4060 W, 51.1%
12. $U_l = 2.59$ W/m^2-K, $T_{fo} = 187.8°C$, $\eta_i = 45.1\%$
    $\eta_{ib} = 60.4\%$, $\Delta p = 0.71$ cm of water
13. 44.2, 40.2, 35.7%
14. (i) $\eta_i = 25.6\%$, $\eta_{ib} = 34.2\%$ (ii) $\eta_i = 28.9\%$, $\eta_{ib} = 38.7\%$
16. $q_u = 803$ kW; $n_{ib} = 44.0\%$
17. Daily efficiency = 41.5%
19. 1.70, 17.0 cm, 18.6 cm, 0.382 m^2
20. 7.6 h
21. 321, 2351

22. 7.57 m. Receiver is 7.57 m in diameter and 15.14 m in height.
Concentration ratio = 195

Chapter 7

1. 47.6, 45.9, 44.6, 43.6, 43.0°C; 0532 h

3.

Time (LAT)	1000	1100	1200	1300	1400	1500
ρ (kg/m^3)	848.00	846.45	839.75	830.85	818.43	805.55
C_p (kJ/kg-K)	2.234	2.238	2.255	2.279	2.310	2.342
T_l (°C)	283.1	296.5	314.3	337.4	360.9	375.8
	1600	1700	1800	1900	2000	2100
	798.10	797.00	797.65	820.25	826.13	826.24
	2.363	2.366	2.364	2.306	2.291	2.291
	378.0	376.7	334.1	291.7	291.6	291.5
	2200	2300	2400	0100	0200	0300
	826.35	826.46	826.57	826.68	826.79	826.90
	2.291	2.291	2.290	2.290	2.290	2.290
	291.4	291.3	291.2	291.1	291.0	290.9
	0400	0500	0600	0700	0800	0900
	827.01	827.12	827.23	827.34	827.45	827.56
	2.289	2.289	2.289	2.289	2.288	2.288
	290.8	290.7	290.6	290.5	290.4	290.3

5. 77.0, 71.5, 64.1°C

6.

x (m)	0	0.25	0.5	0.75	1.0
T_s (°C)	70.0	61.0	40.1	28.8	25.6
T_f (°C)	70.0	64.1	44.0	30.4	26.0

$F = 0.43$, $\Delta p = 1.25$ cm of water

7. (a) 4.21 W/K (b) 5.88 W/K **8.** 106.7 MJ, ratio = 0.849

Chapter 8

1. 0.5265 V; 209.3 A/m^2, 110.2 W; 0.327 m^2

2.

IST (h)	0900	1000	1100	1200	1300	1400	1500	1600
$\dot{m}$ (kg/s)	2.019	2.683	3.211	3.444	3.420	3.068	2.477	1.816

Water pumped from 0830 to 1630 h = 80 018 litres

3.

IST (h)	0900	1000	1100	1200	1300	1400	1500	1600
$\dot{m}$ (kg/s)	2.432	3.032	3.211	3.444	3.420	3.068	2.772	2.164

Water pumped from 0830 to 1630 h = 85 093 litres.
Increase = 6.3%

4. 356.3 Wh; Street light will be switched off at 0613 h.

5. 19

6. 20
Appropriate battery bank needed since the solar radiation changes throughout the day.

7. 12; 5

8. 2 Wp; 160 cm^2; 1.52 A-h

9. 11.97 ha; 30.47 ha

10. 32.7×10^6 m^2

Chapter 9

1. $k = 1.696$, $c = 7.15$ m/s

Interval (m/s)	0–1	1–2	2–3	3–4	4–5	5–6	6–7	7–8
Frequency (%)	3.7	7.5	9.6	10.7	10.9	10.4	9.5	8.3
	8–9	9–10	10–11	11–12	12–13	13–14	14–15	15–16
	7.0	5.7	4.5	3.5	2.7	2.0	1.4	1.0

2. $k = 3.69$, $c = 12.856$ m/s

Interval (m/s)	0–1	1–2	2–3	3–4	4–5	5–6	6–7	7–8
Frequency (%)	0.0	0.0	0.3	0.9	1.7	2.8	4.2	5.9
	8–9	9–10	10–11	11–12	12–13	13–14	14–15	15–16
	7.6	9.2	10.4	11.0	10.8	9.9	8.3	6.5
	16–17	17–18	18–19	19–20				
	4.6	2.9	1.7	0.8				

4. 1457 kWh/m^2

5. 1254 kWh/m^2

6. 3031 kWh/m^2

7. Annual energy content = 226 kWh/m^2. Poor potential.

8. 66.75 kWh/m^2

9. 0.343

10. 126 863 kWh

11. 360.3 kW

12. 3.21 kW

13. 5030 kW; 351 rpm

14. 5682 MW

Chapter 10

3. Rs 246 571

4. 5.3 years

5. Rs 130 847

6. (a) Rs 226 034 (b) Rs 158 242
(c) Rs 70 111

7. Rs 333 418

8. Rs 292 316

9.

Collector area (m^2)	10	20	30	40	50
CS (Rs)	197 454	292 316	324 974	318 047	299 810

10. Rs 333 086, Rs (–) 42 422 517

11. (a) Rs (–) 249 475 (b) Rs 139 978

12. Rs (–) 207 308, 1.0744

13. Rs 16 115 358; Rs 14 087 900

APPENDIX 2

List of Symbols

A swept frontal area of wind machine, m^2; annual instalment of loan repayment, Rs

A_a effective aperture area, m^2

A_b blocked area m^2; projected blade area of wind machine, m^2

A_c collector gross area, m^2 ; total collector area in an array, m^2 ; area of solar cell, m^2

A_{con} surface area of concentrator, m^2

A_g ground area, m^2

A_j annual payments due on a system, Rs

A_m area of mirror, m^2

A_p area of absorber, m^2

A_s shaded area, m^2

AC net annual cost, Rs

APLF annual plant load factor

AS annual savings, Rs

b width of absorber surface of CPC, m; radial gap in annulus, m

BCR benefit-cost ratio

c velocity of light, m/s; Weibull scale factor; mean chord of blades, m,

c_f cost of energy per unit, Rs/J

C area concentration ratio; initial cost, Rs

C_1 constant in Eq. (6.37)

C_2 constant in Eq. (6.42)

C_3 constant in Eq. (6.46)

C_c circumference of collector gross area, m; charge capacity of thermal storage device, J

C_d discharge capacity of thermal storage device, J

C_D drag coefficient

C_L lift coefficient

$C_{m,2D}$ maximum concentration ratio of a two-dimensional (line-focus) concentrator

$C_{m,3D}$ maximum concentration ratio of a three-dimensional (point-focus) concentrator

C_p specific heat, kJ/kg-K; power coefficient for a wind machine

C_{pf} specific heat of fluid, kJ/kg-K

C_{ps} specific heat of solid, kJ/kg-K

CC cumulative cost, Rs

CS cumulative savings, Rs

d equivalent spherical diameter, m; average diameter of the bed material, m; discount rate

d_e equivalent diameter, m

d_l interest rate for lending

D diameter of packed bed, m; diameter of the runner in turbine, m

D_{ci} inner diameter of glass cover, m

D_{co} outer diameter of glass cover, m

D_i inner diameter of tube, m

D_o outer diameter of tube, m

D_p diameter of spherical absorber, m

D_λ percentage of extraterrestrial solar radiation up to a wavelength λ

e charge of an electron, J/V

e_b emissive power of a black surface, W/m^2

e_λ monochromatic extraterrestrial solar radiation flux, W/m^2-μm

$e_{b\lambda}$ monochromatic spectral emissive power of a black surface, W/m^2-μm

E equation of time correction, minutes; energy/annual energy load, J, eV

E_a/A energy content in wind per unit area over a year, kWh/m^2

E_g band gap energy, eV

E_L elevation, km

E_m/A energy content in wind per unit area over a month, kWh/m^2

f friction factor; probability density function, s/m

f_l fraction of initial cost taken as loan

F fraction of energy stored; fraction of annual energy load supplied by solar system; cumulative distribution function

F_D drag force, N

F_L lift force, N

F_R collector heat-removal factor

F' collector efficiency factor

FF fill factor

FIT feed-in tariff, Rs/kWh

g acceleration due to gravity, m/s^2

G mass velocity, kg/s-m^2

h heat transfer coefficient, W/m^2-K; Planck's constant, erg-s; enthalpy, kJ/ kg

h_{c1-c2} heat transfer coefficient between first and second cover, W/m^2-K

h_e effective heat transfer coefficient defined in Eqs (5.10), (5.22), (5.40), W/m^2-K

h_f heat transfer coefficient on inside surface of tube, W/m^2-K

h_{fb} heat transfer coefficient between bottom plate and air stream, W/m^2-K

h_{fc} heat transfer coefficient between cover and air stream, W/m^2-K

h_{ff} heat transfer coefficient between fin surface and air stream, W/m^2-K

h_{fp} heat transfer coefficient between absorber plate and air stream, W/m^2-K

h_{p-c} heat transfer coefficient between absorber plate and glass cover, W/m^2-K

h_{p-c1} heat transfer coefficient between absorber plate and first cover, W/m^2-K

h_r radiative heat transfer coefficient, W/m^2-K

h_v volumetric heat transfer coefficient, W/m^3-K

h_w wind heat transfer coefficient, W/m^2-K

H height of chimney tower, m; height of CPC, m; height of LFR receiver, m; height of central tower, m; length over which tape is twisted through 180°, m; water head, m; wave height, m

H_b daily beam radiation on a horizontal surface, kJ/m^2-day

H_d daily diffuse radiation on a horizontal surface, kJ/m^2-day

H_g daily global radiation on a horizontal surface, kJ/m^2-day

H_o daily extraterrestrial radiation on a horizontal surface, kJ/m^2-day

H_T daily radiation on a tilted surface, kJ/m^2-day

i_f rate at which energy cost increases annually

i_m rate at which operation and maintenance cost increases annually

I current through a load, A

I_b instantaneous/hourly beam radiation on a horizontal surface, W/m^2 or kJ/m^2-h

I_{bn} instantaneous/hourly beam radiation on a surface normal to the direction of the rays or direct normal irradiance (DNI), W/m^2 or kJ/m^2-h

I_d instantaneous/hourly diffuse radiation on a horizontal surface, W/m^2 or kJ/m^2-h

I_g instantaneous/hourly global radiation on a horizontal surface or global horizontal irradiance (GHI), W/m^2 or kJ/m^2-h

I_j p-n junction current, A

I_m solar cell current for maximum power, A

I_o instantaneous/hourly extraterrestrial radiation on a horizontal surface, W/m^2 or kJ/m^2-h; reverse saturation or dark current, A

I_{sc} solar constant, W/m^2; short circuit current, A

I'_{sc} extraterrestrial flux on any day of the year, Eq. (3.1), W/m^2

I_{sh} shunt resistance current, A

I_T instantaneous/hourly flux incident on top cover of collector, on aperture plane of concentrator, or on solar cell, W/m^2 or kJ/m^2-h

IRR internal rate of return

j j-factor = $(h_w/\rho C_p V_\infty)$ $\text{Pr}^{2/3}$; any arbitrary year

k thermal conductivity, W/m-K; Boltzmann constant, J/K; Weibull shape factor

k_a thermal conductivity of adhesive, W/m-K

k_{eff} effective thermal conductivity, W/m-K

k_f thermal conductivity of fin material/fluid, W/m-K

k_i thermal conductivity of insulation, W/m-K

k_p thermal conductivity of absorber plate material, W/m-K

k_T clearness index = (I_g/I_o)

K extinction coefficient, m^{-1}

K_T clearness index = (H_g/H_o)

$K_{\tau\alpha}$ incidence angle modifier

L spacing between absorber plate and cover, or between absorber plate and bottom plate, or between covers, m; length of concentrator, m; length of packed bed, m

L^* characteristic dimension of flat-plate collector cover, m

LAT local apparent time, h

L_1 length of absorber plate, m

L_2 width of absorber plate, m

L_3 height of flat-plate collector casing, m

L_b length of blocked area, m

L_f fin height, m

L_i size of image, m

L_s length of shaded area, m

LCC life cycle cost, Rs

LCS life cycle saving, Rs

LCOE Levelized cost of electricity, Rs/kWh

m mass, kg; air mass; average number of reflections in a CPC; ideality factor of PV cell

$\dot{m}$ mass flow rate, kg/s

$\dot{m}_{load}$ mass flow rate to load, kg/s

$(mC)_c$ heat capacity of cover, kJ/K

$(mC)_e$ effective heat capacity of collector, kJ/K

$(mC)_p$ heat capacity of absorber plate, tubes, bottom and side insulation, and water in tubes, kJ/K

M number of transparent covers; annual operation and maintenance cost of system, Rs

MACE mean annual per capita needs of electricity, kWh/year

n day of the year; number of years; number of reflector-rows

n_1, n_2 refractive indices

n_l period over which loan is paid back, years

n_p payback period, years

n_t life time of a system, years

N number of tubes; number of mirrors; number of blades, rotational speed, rpm

N_a number of hours in a year, h

N_m number of hours in a month, h

N_s specific speed of water turbine, rpm

NPV net present value, Rs

Nu average Nusselt number (= hL/k) where L is the characteristic dimension

p pitch, m

Δp pressure drop, cm of water

P power from a PV cell, W; wave power per unit length, W/m; power generated by a water turbine, W

P' production rate, Mt

P/A power density, W/m^2

P_e power extracted by wind machine, W

P_{elec} electrical power output, W

P_m maximum power output of a PV cell, W

Pr Prandtl number (= $C_p\mu/k$)

PW present worth

q_{abs} rate of energy absorbed, W

q_b rate of heat loss from bottom, W

q_l rate of heat loss, W

q_{load} rate of withdrawal or extraction of energy to load, W

q_r rate of heat loss by radiation, W

q_s rate of heat loss from sides, W

q_t rate of heat loss from top, W

q_u rate of useful heat gain, W

Q volume flow rate, m^3/s
Q_u useful heat gain, kJ
r growth rate in production, per cent
r_b tilt factor for instantaneous/hourly beam radiation
$(r_b)_{av}$ average tilt factor for N mirrors
r_d tilt factor for instantaneous/hourly diffuse radiation; depreciation rate
r_r tilt factor for instantaneous/hourly reflected radiation
r_t income tax rate
R proved reserves of coal, Mt; rotor radius, m
Ra Rayleigh number = $(g\beta \Delta T L^3/\nu^2)$ (Pr) where L is the characteristic dimension
Ra* modified Rayleigh number defined by Eq. (6.35)
Re Reynolds number (= VL/ν) where L is the characteristic dimension
R_b tilt factor for daily beam radiation
R_d tilt factor for daily diffuse radiation
R_L load resistance, ohm
R_r tilt factor for daily reflected radiation
R_s series resistance, ohm
R_{sh} shunt resistance, ohm
S incident solar flux absorbed in the absorber plate or tube, W/m^2; sunshine hours per day, h
S_{max} day length, h
t time, h
t_p time period for charging or discharging, h; time period of a wave, s
Δt finite time interval, h
T temperature, K
ΔT temperature difference, K
T_a ambient temperature, K
$\bar{T}_a$ average daily ambient temperature, K
T_{av} average of T_{pm} and T_{bm}, K
T_{bm} mean bottom plate temperature, K
T_c temperature of cover (subscripts 1, 2, etc. added when there is more than one cover), K
T_f local fluid temperature, K
T_{fi} temperature of fluid at inlet, K
T_{fo} temperature of fluid at outlet, K
$\bar{T}_f$ arithmetic mean of inlet and outlet fluid temperature, K
T_i temperature of make-up liquid, K; initial temperature of solid in bed, K
ΔT_i step change in inlet temperature of transfer fluid, K
T_l liquid temperature in a well mixed tank, K
T_{li} liquid temperature in a well mixed tank at $t = 0$, K
T_{l1}, T_{l2} liquid temperatures in a tank with two well mixed sections, K
T_o reference temperature, K
T_p local absorber surface temperature, K
T_{pm} mean absorber surface temperature, K
T_{pi} initial value of T_{pm}, K
T_{po} absorber plate temperature above tube, K
T_s local temperature of solid, K
T_{sky} temperature of sky, K
TSC theoretical storage capacity of thermal storage device, J
TCE tonnes coal equivalent
TOE tonnes oil equivalent
U_b bottom loss coefficient, W/m^2-K
U_l overall loss coefficient, W/m^2-K
U'_l loss coefficient defined by Eq. (5.21), W/m^2-K
U''_l equivalent overall loss coefficient defined by Eq. (5.20), W/m^2-K
U_s side loss coefficient, W/m^2-K
U_t top loss coefficient, W/m^2-K
$(UA)_e$ product of overall heat transfer coefficient and area of heat exchanger coil, W/K
$(UA)_t$ product of overall heat transfer coefficient and area of tank, W/K
$(UA)_{t1}$ product of overall heat transfer coefficient and area of tank section 1, W/K

$(UA)_{t2}$ product of overall heat transfer coefficient and area of tank section 2, W/K

V velocity, m/s; volume, m^3; voltage, V

$\overline{V}$ average wind speed, m/s

V_∞ free stream wind speed, m/s

V_m solar cell voltage for maximum power, V

V_{oc} open circuit voltage, V

w span of mirrors, m; width of reflector-row, m

w_b width of blocked area, m

w_s width of shaded area, m

W pitch of tubes, m; aperture, m; fin spacing, m

X tape twist ratio (= H/D_i); dimensionless distance

z height from ground, m

Greek Symbols

α absorptivity of absorber surface for solar radiation; thermal diffusivity, m^2/s; power law index in Eq. (9.11)

α_a solar altitude angle

α_v solar elevation angle measured with reference to horizontal

α'_v solar elevation angle measured with reference to equatorial plane

α_λ monochromatic absorptivity

β slope or tilt; coefficient of volume expansion, K^{-1}

β_i tilt angle

β_{opt} optimum tilt

γ surface azimuth angle; solidity of a wind machine

γ_i surface azimuth angle of ith surface

γ_{int} intercept factor

γ_s solar azimuth angle

Γ gamma function

δ declination; phase lag

δ_a average thickness of bonding adhesive, m

δ_b thickness of bottom insulation, m

δ_c thickness of transparent cover, m

δ_f thickness of fin, m

δ_p thickness of absorber plate, m

δ_s thickness of side insulation, m

ε void fraction; ratio of drag coefficient and lift coefficient

ε_b emissivity of bottom plate surface for long wavelength radiation

ε_c emissivity of cover for long wavelength radiation

ε_p emissivity of absorber surface for long wavelength radiation

ε_λ monochromatic emissivity

η solar cell efficiency, daily collection efficiency; turbine efficiency

η_i instantaneous collection efficiency

η_{ib} instantaneous collection efficiency based on beam radiation

η_{max} maximum efficiency of a solar chimney

η_o optical efficiency; efficiency at reference temperature T_o

θ, θ_1 angle of incidence

θ_2 angle of refraction

θ_a half acceptance angle

θ_e angular error associated with reflection

θ_s angle subtended by the sun at the earth

θ_z zenith angle

λ wavelength, μm; latent heat, kJ/kg; tip speed ratio

μ viscosity, N-s/m^2

μ_f viscosity of fluid, N-s/m^2

ν kinematic viscosity, m^2/s

ρ reflectivity; density, kg/m^3

ρ_I, ρ_{II} reflectivities of two components of polarization

ρ_b reflectivity for beam radiation

ρ_d reflectivity for diffuse radiation

ρ_e effective reflectivity of a CPC

ρ_f density of fluid, kg/m^3

ρ_s density of solid, kg/m^3

σ Stefan–Boltzmann constant, W/m^2-K^4; standard deviation of wind speed, m/s

Ω_i angle in the E-W plane between the vertical and the projection of the reflected ray of ith reflector-row on the E-W plane

τ	transmissivity of cover or covers; time period, h; dimensionless time
τ_a	transmissivity based on absorption
τ_b	beam optical depth
τ_d	diffuse optical depth
τ_{ab}	transmissivity based on absorption for beam radiation
τ_{ad}	transmissivity based on absorption for diffuse radiation
τ_r	transmissivity based on reflection and refraction
τ_{rI}, τ_{rII}	transmissivity based on reflection and refraction of two components of polarisation
τ_{rb}	transmissivity based on reflection and refraction for beam radiation
τ_{rd}	transmissivity based on reflection and refraction for diffuse radiation
$(\tau\alpha)_b$	transmissivity–absorptivity product for beam radiation
$(\tau\alpha)_d$	transmissivity–absorptivity product for diffuse radiation
$(\tau\alpha)_{av}$	average value of transmissivity–absorptivity product for beam and diffuse radiation
ϕ	latitude; plate effectiveness
ϕ_f	fin effectiveness
ϕ_r	rim angle
ψ	inclination of reflector; fraction of ground area covered by mirrors; angle in the E-W plane between the vertical and the projection of the sun's ray in the E-W plane
ω	hour angle; frequency, s^{-1}; angular velocity, rad/s
ω_s	hour angle corresponding to sunrise or sunset on a horizontal surface
ω_{st}	hour angle corresponding to sunrise or sunset on a tilted surface
ω_t	hour angle corresponding to a specified time
N.B.	A bar over a symbol is used to indicate a monthly average, unless otherwise mentioned.

Abbreviations

BOS	Balance of system
CBM	Coal-bed methane
CDM	Clean development mechanism
CERC	Central Electricity Regulatory Commission
CIGS	Copper indium gallium diselenide
CLFR	Compact linear Fresnel reflector
CPV	Concentrator photovoltaics
CSP	Concentrated solar power
DoD	Depth of discharge
DSSC	Dye sensitized solar cell
ETC	Evacuated tube collector
EWT	Emitter wrap-through
FBTR	Fast breeder test reactor
FPC	Flat-plate collector
FTO	Fluorine-doped tin oxide
GHG	Green house gas
HDI	Human development index
HWB	Hottel-Whillier-Bliss
IMD	Indian Meteorological Department
IMG	Inter-Ministerial Group
IPCC	Inter-Governmental Panel on Climate Change
ITER	International thermonuclear energy reactor
KVIC	Khadi and Village Industries Commission
LFR	Linear Fresnel reflector
LNG	Liquefied natural gas
MNRE	Ministry of New and Renewable Energy
MPPT	Maximum power point tracking
MWCNT	Multiwall carbon nanotubes
MWT	Metal wrap-through

NIWE	National Institute of Wind Energy
OPV	Organic photovoltaic cell
OSC	Organic solar cell
OTEC	Ocean thermal energy conversion
OWC	Oscillating water column
PERC	Passivated emitter and rear cells
PERL	Passivated emitter and rear locally diffused
PESC	Passivated emitter solar cells
PHWR	Pressurized heavy water reactor
PTC	Parabolic trough collector
PV	Photovoltaic
PVD	Physical vapour deposition
PV/T	Photovoltaic-thermal
PWR	Pressurized water reactor
QDT	Quasi-dynamic test
ROR	Run-of-the river
SEGS	Solar electric generating system
SHP	Small hydroelectric power
SODAR	Sonic detection and ranging
SSP	Satellite solar power
UTC	Unglazed transpired collector

APPENDIX 3

Radiation Data*

Table A3.1 Monthly average hourly global radiation ($\bar{I}_g$), daily global radiation ($\bar{H}_g$) and sunshine hours ($\bar{S}$).

Location: New Delhi (28°35' N, 77°12' E)

	$\bar{I}_g$														$\bar{H}_g$ (kWh/m^2-day)	$\bar{S}$ (h)
	6	7	8	9	10	11	12	13	14	15	16	17	18	19		
J	0.000	0.004	0.096	0.271	0.433	0.556	0.618	0.615	0.552	0.430	0.268	0.099	0.005	0.000	3.987	7.6
F	0.000	0.020	0.163	0.364	0.540	0.670	0.727	0.728	0.659	0.534	0.358	0.165	0.021	0.000	5.001	9.0
M	0.001	0.066	0.260	0.475	0.655	0.781	0.845	0.844	0.769	0.635	0.458	0.249	0.064	0.001	6.138	8.2
A	0.010	0.130	0.339	0.554	0.727	0.848	0.910	0.903	0.833	0.699	0.523	0.318	0.122	0.010	6.935	8.6
M	0.030	0.177	0.383	0.581	0.746	0.865	0.925	0.920	0.844	0.719	0.547	0.352	0.158	0.027	7.287	8.0
J	0.036	0.164	0.339	0.510	0.657	0.757	0.809	0.813	0.750	0.639	0.483	0.318	0.159	0.036	6.544	5.9
J	0.026	0.130	0.275	0.425	0.549	0.640	0.672	0.671	0.606	0.518	0.412	0.273	0.141	0.032	5.334	5.8
A	0.012	0.101	0.243	0.392	0.512	0.607	0.640	0.659	0.595	0.510	0.392	0.255	0.107	0.013	5.053	5.6
S	0.003	0.079	0.248	0.431	0.587	0.698	0.757	0.747	0.689	0.571	0.417	0.244	0.074	0.003	5.602	7.0
O	0.000	0.036	0.202	0.406	0.581	0.701	0.764	0.758	0.692	0.569	0.391	0.191	0.033	0.000	5.355	8.8
N	0.000	0.009	0.128	0.319	0.495	0.620	0.685	0.680	0.613	0.490	0.315	0.125	0.010	0.000	4.523	9.2
D	0.000	0.003	0.085	0.259	0.425	0.543	0.605	0.605	0.538	0.418	0.255	0.086	0.003	0.000	3.843	8.0

N.B. Time indicated is LAT. Value of $\bar{I}_g$ given for a particular time corresponds to the radiation (in kWh/m^2-h) incident on a horizontal surface during the one hour preceding the time. Multiply by 3600 to obtain an hourly value (in kJ/m^2-h)

*Adapted from A. Mani, Handbook of Solar Radiation Data for India, Allied Publishers, New Delhi (1981).

Table A3.2 Monthly average hourly diffuse radiation ($\bar{I}_d$) and daily diffuse radiation ($\bar{H}_d$)

Location: New Delhi (28°35'N, 77°12'E)

	$\bar{I}_d$														$\bar{H}_d$
	6	7	8	9	10	11	12	13	14	15	16	17	18	19	(kWh/m^2-day)
J	0.000	0.003	0.051	0.103	0.140	0.160	0.174	0.172	0.162	0.137	0.100	0.051	0.003	0.000	1.240
F	0.000	0.013	0.076	0.126	0.159	0.183	0.191	0.190	0.180	0.160	0.125	0.075	0.013	0.000	1.474
M	0.001	0.043	0.113	0.159	0.189	0.210	0.222	0.226	0.215	0.197	0.163	0.113	0.041	0.001	1.861
A	0.009	0.083	0.153	0.202	0.236	0.262	0.274	0.279	0.274	0.250	0.212	0.156	0.079	0.008	2.473
M	0.027	0.114	0.190	0.244	0.279	0.303	0.315	0.318	0.308	0.286	0.247	0.189	0.107	0.022	2.922
J	0.033	0.124	0.215	0.286	0.342	0.382	0.399	0.399	0.377	0.340	0.279	0.203	0.114	0.028	3.540
J	0.023	0.096	0.178	0.257	0.316	0.358	0.372	0.365	0.333	0.291	0.241	0.174	0.101	0.023	3.135
A	0.009	0.074	0.153	0.224	0.277	0.317	0.334	0.331	0.302	0.265	0.210	0.149	0.075	0.010	2.727
S	0.002	0.053	0.123	0.178	0.222	0.256	0.269	0.263	0.248	0.217	0.167	0.118	0.048	0.002	2.152
O	0.000	0.026	0.087	0.128	0.154	0.169	0.179	0.177	0.169	0.154	0.126	0.083	0.021	0.001	1.465
N	0.000	0.007	0.057	0.099	0.122	0.137	0.145	0.146	0.142	0.124	0.099	0.055	0.006	0.000	1.141
D	0.000	0.002	0.045	0.092	0.123	0.144	0.155	0.156	0.147	0.125	0.092	0.043	0.002	0.000	1.117

N.B. Time indicated is LAT. Value of $\bar{I}_d$ given for a particular time corresponds to the radiation (in kWh/m^2-h) incident on a horizontal surface during the one hour preceding the time. Multiply by 3600 to obtain an hourly value (in kJ/m^2-h)

Table A3.3 Monthly average hourly global radiation ($\bar{I}_g$), daily global radiation ($\bar{H}_g$) and sunshine hours ($\bar{S}$)

Location: Mumbai (19°07'N, 72°51'E)

	$\bar{I}_g$														$\bar{H}_g$ (kWh/m^2-day)	$\bar{S}$ (h)
	6	7	8	9	10	11	12	13	14	15	16	17	18	19		
J	0.000	0.018	0.154	0.358	0.528	0.660	0.735	0.746	0.685	0.560	0.383	0.175	0.022	0.000	5.031	9.3
F	0.000	0.032	0.198	0.407	0.599	0.737	0.823	0.835	0.769	0.644	0.456	0.230	0.042	0.000	5.755	9.7
M	0.001	0.062	0.252	0.465	0.654	0.801	0.888	0.901	0.832	0.704	0.513	0.281	0.072	0.001	6.448	9.5
A	0.004	0.095	0.288	0:492	0.695	0.845	0.937	0.955	0.893	0.759	0.563	0.333	0.113	0.005	6.991	9.6
M	0.012	0.124	0.294	0.489	0.710	0.864	0.952	0.962	0.899	0.768	0.578	0.360	0.150	0.017	7.262	9.3
J	0.011	0.093	0.217	0.364	0.508	0.605	0.666	0.678	0.638	0.548	0.412	0.253	0.108	0.015	5.177	5.6
J	0.007	0.069	0.168	0.286	0.399	0.498	0.541	0.555	0.515	0.433	0.325	0.199	0.082	0.009	4.061	2.4
A	0.003	0.057	0.162	0.290	0.407	0.478	0.554	0.540	0.493	0.410	0.293	0.176	0.066	0.005	3.976	2.4
S	0.001	0.051	0.184	0.329	0.466	0.576	0.650	0.693	0.669	0.571	0.401	0.222	0.068	0.002	4.878	5.1
O	0.000	0.038	0.195	0.382	0.550	0.663	0.749	0.773	0.733	0.616	0.429	0.222	0.050	0.001	5.443	8.2
N	0.000	0.022	0.166	0.367	0.547	0.669	0.731	0.741	0.671	0.545	0.371.	0.174	0.024	0.000	5.074	8.9
D	0.000	0.013	0.142	0.342	0.517	0.647	0.717	0.719	0.652	0.524	0.353	0.155	0.017	0.000	4.794	8.9

N.B. Time indicated is LAT. Value of $\bar{I}_g$ given for a particular time corresponds to the radiation (in kWh/m^2-h) incident on a horizontal surface during the one hour preceding the time. Multiply by 3600 to obtain an hourly value (in kJ/m^2-h)

Table A3.4 Monthly average hourly diffuse radiation ($\bar{I}_d$) and daily diffuse radiation ($\bar{H}_d$)

Location: Mumbai(19°07'N,72°51'E)

	$\bar{I}_d$														$\bar{H}_d$
	6	7	8	9	10	11	12	13	14	15	16	17	18	19	(kWh/m^2-day)
J	0.000	0.010	0.064	0.104	0.133	0.155	0.164	0.161	0.150	0.133	0.107	0.070	0.012	0.000	1.279
F	0.000	0.020	0.083	0.127	0.161	0.178	0.184	0.177	0.161	0.143	0.121	0.083	0.023	0.000	1.467
M	0.001	0.041	0.109	0.162	0.192	0.209	0.207	0.202	0.189	0.172	0.146	0.104	0.040	0.000	1.776
A	0.002	0.067	0.145	0.203	0.237	0.254	0.252	0.242	0.222	0.201	0.174	0.130	0.062	0.002	2.187
M	0.009	0.087	0.172	0.234	0.268	0.280	0.281	0273	0.259	0.236	0.199	0.153	0.086	0.011	2.508
J	0.010	0.076	0.164	0.243	0.304	0.339	0.367	0.359	0.336	0.301	0.244	0.166	0.079	0.011	2.969
J	0.008	0059	0.147	0,234	0.309	0.373	0.396	0.404	0.368	0.316	0.244	0.160	0.067	0.008	3.065
A	0.001	0.048	0.145	0.249	0.333	0.377	0.432	0.414	0.380	0.327	0.233	0.143	0.060	0.002	3.156
S	0.001	0.045	0.144	0.227	0.288	0.345	0.366	0.359	0.339	0.296	0.224	0.139	0.049	0.001	2.860
O	0.000	0.028	0.115	0.167	0.207	0.232	0.244	0.249	0.220	0.190	0.147	0.096	0.031	0.000	1.982
N	0.000	0.012	0.067	0.101	0.124	0.142	0.148	0.144	0.142	0.128	0.103	0.065	0.012	0.000	1.193
D	0.000	0.009	0.061	0.092	0.113	0.129	0.137	0.139	0.133	0.119	0.098	0.061	0.010	0.000	1.106

N.B. Time indicated is LAT. Value of $\bar{I}_d$ given for a particular time corresponds to the radiation (in kWh/m^2-h) incident on a horizontal surface during the one hour preceding the time. Multiply by 3600 to obtain an hourly value (in kJ/m^2-h)

APPENDIX 4

Properties

Table A 4.1 Properties of water

T °C	ρ kg/m^3	C_p kJ/kg-K	$\mu \times 10^6$ N-s/m^2	k W/m-K	Pr	$\beta \times 10^4$ 1/K	$\nu \times 10^6$ m^2/s	$\sigma \times 10^4$ N/m
0	999.9	4.212	1787.8	0.551	13.67	–0.63	1.789	756
10	999.7	4.191	1305.3	0.575	9.52	+0.70	1.306	742
20	998.2	4.183	1004.2	0.599	7.02	1.82	1.006	727
30	995.7	4.174	801.2	0.618	5.42	3.21	0.805	712
40	992.2	4.174	653.1	0.634	4.31	3.87	0.659	696
50	988.1	4.174	549.2	0.648	3.54	4.49	0.556	679
60	983.2	4.179	469.8	0.659	2.98	5.11	0.478	662
70	977.8	4.187	406.0	0.668	2.55	5.70	0.415	644
80	971.8	4.195	355.0	0.675	2.21	6.32	0.365	628
90	965.3	4.208	314.8	0.680	1.95	6.95	0.326	608
100	958.4	4.220	282.4	0.683	1.75	7.52	0.295	589
110	951.0	4.233	258.9	0.685	1.60	8.08	0.272	569
120	943.1	4.250	237.3	0.686	1.47	8.64	0.252	549
130	934.8	4.267	217.7	0.686	1.36	9.19	0.233	528
140	926.1	4.287	201.0	0.685	1.26	9.72	0.217	507
150	917.0	4.313	186.3	0.684	1.17	10.3	0.203	485
160	907.4	4.346	173.6	0.683	1.10	10.7	0.191	463
170	897.3	4.380	162.8	0.679	1.05	11.3	0.181	441
180	886.9	4.417	153.0	0.675	1.00	11.9	0.171	420
190	876.0	4.459	144.2	0.670	0.96	12.6	0.165	398
200	863.0	4.505	136.3	0.663	0.93	13.3	0.158	376

Table A 4.2 Properties of dry air at atmospheric pressure

T °C	ρ kg/m^3	C_p kJ/kg-K	$\mu \times 10^6$ N-s/m^2	k W/m-K	Pr	$\nu \times 10^6$ m^2/s
0	1.293	1.005	17.2	0.0244	0.707	13.28
10	1.247	1.005	17.7	0.0251	0.705	14.16
20	1.205	1.005	18.1	0.0259	0.703	15.06
30	1.165	1.005	18.6	0.0267	0.701	16.00
40	1.128	1.005	19.1	0.0276	0.699	16.96
50	1.093	1.005	19.6	0.0283	0.698	17.95
60	1.060	1.005	20.1	0.0290	0.696	18.97
70	1.029	1.009	20.6	0.0297	0.694	20.02
80	1.000	1.009	21.1	0.0305	0.692	21.09
90	0.972	1.009	21.5	0.0313	0.690	22.10
100	0.946	1.009	21.9	0.0321	0.688	23.13
120	0.898	1.009	22.9	0.0334	0.686	25.45
140	0.854	1.013	23.7	0.0349	0.684	27.80
160	0.815	1.017	24.5	0.0364	0.682	30.09
180	0.779	1.022	25.3	0.0378	0.681	32.49
200	0.746	1.026	26.0	0.0393	0.680	34.85
250	0.674	1.038	27.4	0.0427	0.677	40.61
300	0.615	1.047	29.7	0.0461	0.674	48.33
350	0.566	1.059	31.4	0.0491	0.676	55.46
400	0.524	1.068	33.0	0.0521	0.678	63.09
500	0.456	1.093	36.2	0.0575	0.687	79.38
600	0.404	1.114	39.1	0.0622	0.699	96.89
700	0.362	1.135	41.8	0.0671	0.706	115.4
800	0.329	1.156	44.3	0.0718	0.713	134.8
900	0.301	1.172	46.7	0.0763	0.717	155.1
1000	0.277	1.185	49.0	0.0807	0.719	177.1

APPENDIX 5

Wind Speed Data**

Table A5.1 Percentage frequency distribution of hourly wind speed

Location: Kandla Port

Interval (kmph)	J	F	M	A	M	J	J	A	S	0	N	D	Annual
00	1.2	2.8	3.1	3.6	2.2	4.2	1.7	2.0	3.9	7.1	4.6	2.2	3.2
00-02	2.0	2.2	2.6	2.1	0.7	1.8	0.9	0.6	1.2	4.3	4.4	2.4	2.1
02-04	2.4	2.3	2.9	2.3	0.6	1.5	1.0	0.9	2.3	3.9	4.8	2.7	2.3
04-06	3.1	2.5	3.0	2.1	1.1	1.9	1.6	1.1	2.2	4.2	4.5	3.0	2.5
06-08	3.2	3.5	3.7	2.5	1.4	1.7	1.3	1.6	3.1	5.8	6.5	4.4	3.2
08-10	5.2	4.4-	5.0	4.0	1.4	2.2	2.1	2.5	4.5	7.3	7.6	6.5	4.3
10-12	6.5	5.6	5.3	4.4	1.7	2.2	2.2	2.6	4.7	7.8	8.1	7.5	4.8
12-14	8.2	6.8	6.4	5.0	2.4	2.6	2.6	3.9	5.6	8.7	9.7	8.4	5.8
14-16	9.2	9.5	8.1	6.4	3.2	3.6	3.4	5.5	7.6	10.3	10.2	10.3	7.2
16-18	10.9	9.1	8.2	6.6	3.4	4.0	3.9	4.7	7.6	9.1	9.0	11.2	7.3
18-20	12.2	11.0	9.3	7.1	4.7	4.6	4.4	6.5	8.9	8.7	9.3	11.5	8.1
20-22	9.6	8.8	7.8	6.6	4.6	3.7	4.4	4.8	6.7	5.8	6.0	9.0	6.5
22-24	8.5	9.1	8.2	7.3	5.2	4.0	4.4	5.4	5.9	4.9	6.1	7.8	6.4
24-26	6.5	6.8	7.5	7.5	6.9	5.6	6.2	7.8	8.3	3.8	4.2	5.4	6.4
26-28	4.5	5.0	5.3	5.5	5.4	5.3	5.5	7.4	5.7	2.1	2.5	3.2	4.8
28-30	2.8	4.1	3.8	6.1	6.4	5.8	6.4	7.1	5.0	1.8	1.3	2.5	4.5
30-32	1.8	2.4	2.8	4.9	5.8	5.2	7.1	6.9	4.2	1.1	0.6	1.2	3.7
32-34	0.8	1.1	1.8	3.2	5.1	4.7	5.6	4.8	2.5	0.7	0.3	0.3	2.6
34-36	0.7	1.2	1.5	3.5	6.4	5.2	7.3	5.2	2.3	0.8	0.3	0.4	3.0
36-38	0.2	0.5	1.0	2.4	5.3	4.6	5.6	3.9	2.1	0.4		0.1	2.2
38-40	0.2	0.6	1.1	2.7	6.0	5.9	7.6	5.4	2.8	0.2			2.8
40-42		0.3	0.6	1.5	3.9	3.6	4.4	2.9	1.0	0.3			1.6
42-44	0.3	0.2	0.4	1.1	3.2	2.9	2.6	1.7	0.5	0.1			1.1
44-46		0.2	0.3	0.9	3.0	3.0	2.8	1.9	0.7	0.2			1.1
46-48			0.1	0.4	2.6	2.6	1.7	1.0	0.5	0.1			0.8
48-50			0.2	0.2	2.7	1.8	1.4	0.8		0.5			0.7
50-52					1.9	1.4	0.7	0.4					0.4
52-54					1.0	1.5	0.5	0.3					0.3
54-56					1.1	0.9	0.5	0.2					0.2
56-58					0.4	0.6	0.1	0.1					0.1
58-60					0.3	0.6	0.1	0.1					
60-62						0.3							
62-64						0.2							

Sensor 33.3 m above ground

**A. Mani and D.A. Mooley, Wind Energy Data for India, Allied Publisher, New Delhi (1983)

Table A5.2 Percentage frequency distribution of hourly wind speed

Location: Indore Airport

Interval (kmph)	J	F	M	A	M	J	J	A	S	O	N	D	Annual
00	10.0	16.4	9.9	9.3	3.2	2.3	2.3	3.3	7.1	14.9	6.9	8.0	7.7
00-02	3.8	1.7.	1.9	1.4	1.6	1.1	0.6	0.3	2.4	3.0	5.9	1.9	2.1
02-04	5.3	1.9	3.0	0.9	0.9	0.6	0.5	0.4	3.1	3.4	4.1	2.2	2.1
04-06	5.1	4.1	2.7	1.5	1.3	0.7	1.1	0.4	3.3	4.2	4.5	4.4	2.7
06-08	4.3	3.6	2.1	2.2	2.0	1.4	1.2	0.8	3.1	3.9	4.7	5.8	2.9
08-10	2.6.	0.5	1.1	0.6	1.3	0.4	0.9	0.4	0.6	1.2	1.7	1.2	1.0
10-12	6.8	5.4	3.8	4.4	3.9	2.9	3.4	1.7	5.2	5.2	6.2	8.9	4.8
12-14	6.9	6.4	5.4	3.2	4.0	3.9	3.5	2.2	4.5	5.9	7.9	8.5	5.1
14-16	9.2	8.9	9.0	5.1	6.3	7.0	7.7	4.2	6.5	8.2	10.4	10.8	7.7
16-18	9.2	10.0	8.8	5.2	6.2	7.1	8.3	5.4	7.5	9.6	13.6	9.9	8.3
18-20	8.7	9.2	9.0	5.3	6.5	6.2	9.2	6.0	7.2	8.3	12.9	9.1	8.1
20-22	10.8	11.4	10.1	10.2	10.3	10.4	11.3	10.5	10.1	9.7	9.7	10.5	10.4
22-24	6.0	5.7	7.7	6.8	7.4	7.7	9.5	7.4	7.3	7.3	4.5	7.5	7.2
24-26	4.9	4.9	8.2	8.2	8.0	9.6	10.8	10.1	9.5	6.8	2.9	5.6	7.6
26-28	1.6	3.4	3.2	7.1	4.3	6.4	3.6	4.6	4.4	3.1	1.5	3.2	4.0
28-30	1.9	1.9	4.9	5.8	5.1	5.2	7.4	7.9	4.3	2.4	1.1	1.5	4.3
30-32	1.6	3.0	3.8	8.5	7.5	8.6	7.0	10.5	7.0	1.8	0.7	0.7	5.2
32-34	0.5	0.6	2.0	4.5	4.5	3.8	3.7	6.1	2.5	0.7	0.4	0.2	2.6
34-36	0.4	0.6	1.5	4.0	5.7	3.8	3.0	5.6	2.5	0.4	0.3		2.4
36-38	0.4	0.3	1.2	2.5	4.1	3.7	2.6	5.1	1.2		0.1		1.9
38-40		0.1	0.5	1.1	1.7	2.0	1.1	2.4	0.6				0.8
40-42			0.2	1.4	2.2	3.1	0.7	2.2	0.1				0.5
42-44				0.3	0.8	0.6	0.2	0.8					
44-46				0.2	0.7	0.4		0.7					
46-48					0.1			0.3					

Sensor 10.9 m above ground

Subject Index

Author Index

A

B

C

D

L

M

Q

R

S

T

U

V

W

X

Y

Z